Nonlinear Optical Communication Networks

WILEY SERIES IN MICROWAVE AND OPTICAL ENGINEERING

KAI CHANG, Editor
Texas A&M University

A complete list of the titles in this series appears at the end of this volume.

Nonlinear Optical Communication Networks

EUGENIO IANNONE

FRANCESCO MATERA

ANTONIO MECOZZI

MARINA SETTEMBRE

A WILEY-INTERSCIENCE PUBLICATION

JOHN WILEY & SONS, INC.

Contents

Preface

Optical technologies in telecommunication networks are rapidly advancing, and the optical fiber is no longer the only optical network element. A large number of optical devices and systems that accomplish diverse network functions are now mature for application. Electrical regenerators are starting to be replaced by optical amplifiers in both long distance and local networks, and in five to seven years the first optical switching systems in transport networks will be probably introduced.

Along with the massive introduction of optical technologies in telecommunication networks have been emerging wide ranging problems that affect not only transmission but also routing strategies, standards, operation, and management. Transmission problems are particularly important, since the high bit-rate signals that travel over long-distance links go through a large variety of optical devices used for transmission control and other network functions such as routing or switching. In all these circumstances fiber transmission cannot be described as a linear phenomenon, and the interplay between fiber-related effects and the behavior of the other optical components makes the system design quite complex.

Within this framework, the purpose of this work can be summarized as follows:

- To present the basic design principles of network transmission architectures where optical technologies are deployed and fiber nonlinearities cannot be neglected.
- To assess the performance of optical transmission systems in such working regimes.

The book has been organized so as to offer the reader an intuitive basis for the phenomena determining system performance, an explanation of the main design aspects of the most promising nonlinear optical transmission systems, and a methodological approach for both the design and the analysis of a generic system even if not explicitly considered in this book. However, the book is not intended as a design handbook, something that would be difficult to prepare in such a rapidly evolving field; rather it can be used as a means to compare the different choices that are offered to solve a particular problem.

To accomplish this, after a description of the various phenomena, the book supplements its presentation of the main tools for system analysis and performance evaluation by a rigorous physical analysis. Whenever a result or a formula is given without direct derivation, a full reference is provided to the original paper(s) where all the details can be found. The more cumbersome mathematical derivations that are fundamental to understanding the key issues in the book are placed in the Appendixes.

This book was mainly conceived for engineers and scientists working in the telecommunications field; therefore the reader is assumed to be familiar with the basic issues of optical communications and with the usual optical devices adopted in communication systems (semiconductor laser, photodiode, etc.). On the other hand, no previous knowledge of nonlinear optics in optical fibers is assumed, since this is not a common background of the telecommunication expert.

The book is divided into three parts. Chapters 2 to 4 review optical fiber propagation, optical amplifiers, and optical transmission systems working in a linear fiber propagation regime. The discussion in Chapter 2 deals at length with nonlinear fiber propagation.

Chapters 5 to 8 focus on optical transmission systems operating in the nonlinear fiber propagation regime. A wide variety of systems is analyzed (with and without in-line amplifiers, conventional and soliton systems, TDM and WDM systems), highlighting each system's best placement in the communication network. Results obtained from theoretical models, simulation analysis, and experiments complement each other to give an idea of both state-of-the-art system performance and possible improvements.

Lastly, Chapter 9 is entirely devoted to signal propagation in optically switched networks. Local area networks, regional networks, and geographical networks are considered so as to give a complete idea of transmission potentials and problems that may be encountered.

We sincerely hope the reader will find this book useful, and we encourage suggestions of every kind that can be taken into account for future editions. We also wish to acknowledge all our colleagues whose suggestions have been invaluable.

Nonlinear Optical Communication Networks

CHAPTER ONE

Introduction

With the spread of telephony in the first part of this century came the development of a worldwide telecommunication network, intended mainly to support the telephone system. Subsequently other telecommunication services were added and either integrated in the telephone network or implemented by dedicated networks.

In the last 20 years data transmission has assumed particular importance [1]. Increasing computation speed and market penetration of electronic computers have pushed an already increasing demand for data communication networks. This demand cannot be completely satisfied by the telephone network and a number of dedicated data communication systems have been built in Europe, the United States, and Japan. Networks designed for data communications have different characteristics from telephone networks; for example, they are generally based on packet switching, whereas the telephone network is based on circuit switching [2].

Besides the continual rise in data transmission demand, another decisive change in the telecommunication market has occurred with the introduction of image-based telecommunication services. In particular, cable television (CATV) has experimented a rapid growth and special distribution networks have been introduced to provide CATV.

The most recent development in telecommunication services, which has been generating a rapid growth of the required transmission capacity, is the introduction of multimedia services. Multimedia services promise to integrate moving images, static images, text, and sound in an interactive environment, and they have a large market, mainly due to the introduction of INTERNET, a worldwide data telecommunication network that is able to support such wide-ranging services.

In this environment further growth of the telecommunication market requires a more ordered and flexible development of the telecommunication infrastructure. This can be attained by merging the different dedicated net-

1

works into a single network that is able to support all services. This network is called an integrated services digital network (ISDN) [3].

In the 1980s a large amount of research and standardization activity was devoted to developing technologies and standards for the ISDN; today ISDN is maintained by all the major telecommuncation companies but has not replaced dedicated networks.

The telecommunication traffic increased rapidly in the mid-1990s. High demand has continued for some time; it is predicted, for example, that the number of basic-rate ISDN connections will grow by a factor ten from 1995 to 2010 [4].

The large amount of traffic itself has increased the spread of integrated networks and put serious pressure on these networks concerning transmission technologies, switching systems, and network management.

In creating this worldwide integrated network, optical technologies have had a major role. A huge increase of the transmission capacity was obtained by introducing optical transmission systems, and a similar revolution is foreseen with the introduction of optical switching. The diffusion of optical technologies is expected to improve network flexibility and expandability; these characteristics enable integrated networks to respond effectively to fluctuations in user demand with a minimum need of new economic investment.

This Chapter provides an introduction to the role of optical technologies in the integrated telecommunication network. It also elucidates the connection between the issues analyzed in this book with other aspects of optical transmission network analysis.

1.1   TELECOMMUNICATION NETWORK AREAS

The whole telecommunication network covers a very large geographic area: Potentially it is all the world. To obtain such a large coverage, the network is organized in different areas having different functions. All areas cooperate to the final task of the network: to guarantee perfect communication among a large number of end users.

The principle structure of the network is shown in Fig. 1.1. Some of the main network areas shown in the figure are explained below.

Local Area
Networks in the local area mainly connect electronic calculators placed within small distances such as within a single building or among a few adjacent buildings. Local area networks (LANs) are not generally part of any public network but are owned by private organizations. However, they can be connected to a public network. The connection can be realized in the distribution area, in the access area, or even directly in the transport area depending on the traffic that the LAN exchanges with the public network. Due to the different

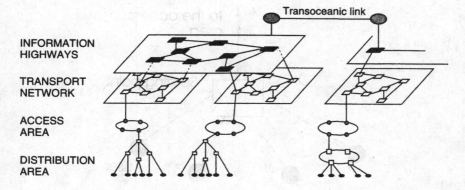

FIGURE 1.1 Simplified diagram of a telecommunication network divided into its main areas.

position a LAN can assume within the global network, the local area is not shown in Fig. 1.1.

The required capacity of a LAN is directly related to the calculation speed of the connected computers; when increasing computer speed, the capacity requirements for the LANs also increase.

A wide variety of possible topologies has been considered for LANs, such as the bus, the ring, or the star.

Distribution Area

End users are connected to a public telecommunication network by what is called a distribution area, or distribution network. The function of this network area is to send signals coming from the end users to a local exchange and to distribute signals from the local exchange to the end users. When the distribution network uses copper pairs or coaxial cables (e.g., as in CATV), the topology is often a star, as shown in Fig. 1.2, or a cluster of stars. In optical distribution networks these topologies can be combined (see Section 1.2.2). A typical distance covered by a distribution network is few kilometers.

Access Area

The access area (or access network) multiplexes the signals coming from the end users to obtain higher-speed signals. The traffic directed inside the area covered by the access network is routed by the access nodes, demultiplexed, and sent to an opportune distribution network. Signals directed outside this area are passed to the transport network. In the other direction the access network receives, by the transport network, high-speed signals coming from other access areas, demultiplexes them, and feeds the distribution network.

Typical areas covered by an access network are a city or a small region where large cities are far away. The signal path along the access network usually extends some tens or hundreds of kilometers, so the usual physical

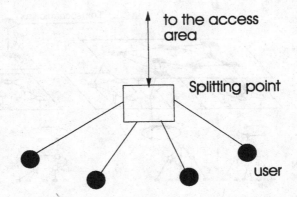

FIGURE 1.2 Distribution network with a star topology.

topology for the access network proposed in various contexts is the ring. A diagram of a ring access network is shown in Fig. 1.3. Of course the ring is not the only possible topology, and a number of alternative topologies have been proposed.

Transport Area

The transport area network (or transport network or core network) has the task to route large capacity fluxes without breaking them into the component streams. It receives signals from access areas, multiplexes them up to the desired capacity, routes them to their destination and, when the destination is reached, sends the high-speed signals to the access area network.

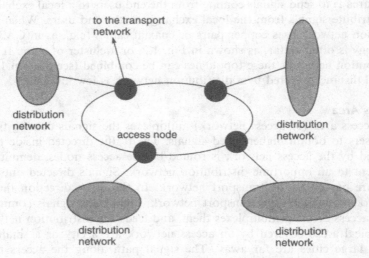

FIGURE 1.3 Access network with a ring topology.

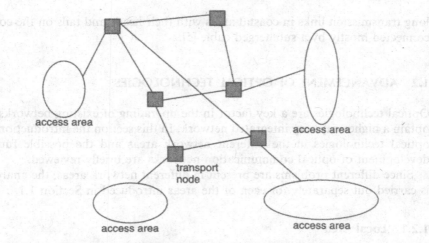

FIGURE 1.4 Transport network with a mesh topology.

Generally the transport network covers the area of a nation, so a path through the transport network can be as long as 1000 to 2000 km, with network nodes spaced from 50 to few hundred kilometers. A typical topology of the transport network is the mesh shown in Fig. 1.4.

Information Highways
Information highways connect different transport networks. In general, the information highways network covers a very large geographic area (e.g., all of Europe or the United States). Very-high-capacity signals are conveyed along information highways for distances of a few thousands kilometers in a mesh network structure.

Submarine Links
Submarine links connect different transport or highways networks. Transmission cables, amplifiers, and signal regenerators designed to operate undersea have much more rigid requirements then their terrestrial counterparts and are very costly. Submarine systems are thus generally point-to-point systems designed to minimize the number of submerged equipment and to optimize system performances.

Presently the longest links in the worldwide communication network are the transoceanic links connecting Europe with the United States and the United States with Japan. Links crossing the Atlantic Ocean are about 6000 km long, while links crossing the Pacific Ocean are about 9000 km long.

Besides transoceanic links, submarine links with lengths of 100 to 400 km are widely used in the Mediterranean and in the Asian Pacific areas. They hold promise for reliable communications in other parts of the world in the form of long transmission links in coastal areas with their heads and tails on the coast connected mostly by a submerged cable [5].

1.2 ADVANCEMENT OF OPTICAL TECHNOLOGIES

Optical technologies are a key factor in the upgrading of existing networks to obtain a higher-capacity integrated network. In this section the introduction of optical technologies in the different network areas and the possible future development of optical communication networks are briefly reviewed.

Since different problems are present in different network areas, the analysis is carried out separately for each of the areas introduced in Section 1.1.

1.2.1 Local Area

Until recently LANs processed data signals from electronic computers. Advancements in the transmission capacity required by LANs were pushed by advancements in computer performances: This has caused the interest in optical transmission. At first the introduction of optical technology in the design of LANs was limited to substituting electrical data links with optical data links. High-speed LANs based on this principle are now on the market [6].

However, the tremendous increase of computational speed of personal computers and workstations is driving the trend toward very-high-capacity LANs. Today the need to add information specifically addressed to human operators, for example, voice and video, is emerging. Multimedia services that merge data, voice, and video are expected in the near future.

To respond to this need, newly designed optical LANs have emerged, allowing a more efficient exploitation of optical technologies [7].

Further improvement in LAN capacity will be made by optical processing of data signals. Then the LAN nodes will directly process high-speed data (40–100 Gbit/s) transmitted along optical fibers. Two enabling technologies are needed before such networks can be put into practice: devices for simple optical processing of digital data (e.g., packet header recognition and packet routing) [8] and devices for optical time division multiplexing (OTDM) (e.g., multiplexers and demultiplexers) [9].

1.2.2 Distribution Area

Before 1990 the distribution of signals completely depended on electrical technology. For the distribution of telephone signals the adopted transmission media were generally paired copper wires, whereas in CATV networks coaxial cables were used.

Starting in the early 1990s, the major telecommunication companies began to introduce optical transmission. The application of optical transmission in distribution areas was pushed by the increased capacity that was needed to guarantee new services to the end user, for example, CATV, video on demand, and connections with geographical data networks. Each solution is conditioned by the particular requirements that the transmission devices in the distribution network have to fulfill. The main requirement is to guarantee optimal use of the network infrastructure, since its implementation meant a major investment for the network operator. Another important requirement, in the case of fiber cables, is to optimize the capacity of the optical fiber by multiplexing in a single fiber as many signals as possible. Such a requirement, which is not present in a copper distribution network, is an important issue for the fiber network design.

There are two main classes of possible architectures for fiber distribution networks:

- *Fiber to the curb* (FTTC) [10]. In this network an optical link is used to convey the signal from the local exchange to a point of the distribution network near the location of a group of end users (e.g., a certain building). Here an optical receiver makes the optical to electrical conversion, the electrical signal is demultiplexed, and the end users are connected by copper pairs or coaxial cables. In this architecture the optical fiber, the optical transmitter, and the optical receiver are used simultaneously by different users. When the distribution is by coaxial cables, the network architecture is also called a hybrid fiber coaxial (HFC) architecture.

- *Fiber to the home* (FTTH) [11]. In this architecture the link between the local exchange and the end user is entirely handled by optical fibers, and the end user's equipment incorporates the optical receiver.

Because of the high cost of the optical infrastructure FTTC seems to be more suitable for the first stage in the introduction of optical technologies in a distribution area, this also makes sense because passage from FTTC to FTTH is quite simple and can be done as the traffic demand requires it.

Different topologies are possible for the optical distribution network: star, multiple star, tree, or ring. However, to select the most suitable architecture a large number of factors have to be taken into account.

From an economic point of view the selected topology must allow the most efficient configuration of the available network infrastructure that keeps pace with user demand. The network must also be flexible and open to technological change as upgrades become available [12].

1.2.3   Access and Transport Area

The problems with access and transport networks are quite similar despite the geographical differences of these networks.

In both the transport and access networks the communication links are almost entirely based on optical technology due to the presence of high-performance and shared systems. Optical fibers connect electrical network nodes where switching and routing take place. Regeneration along these optical links is generally attained by electronic regenerators providing not only signal amplification but also signal reshaping and retiming.

The first important step in extending optical technology in these network areas is to replace electronic regenerators with optical amplifiers. The introduction of optical amplifiers is desirable for two good reasons: (1) Optical amplifiers are much cheaper than electronic repeaters. When the repeaters have to be changed, such as to upgrade link capacity, it is more economical to install optical amplifiers. (2) Optical amplifiers are *transparent* devices: The only conditions for the amplifier working is that the input optical power be compatible with the desired gain, and the signal bandwidth be less than the amplifier bandwidth. Thus the signal bit rate can be changed without changing the amplifiers, and in many cases even WDM can be introduced to upgrade the link, maintaining the installed amplifiers. In effect optical amplifiers enhance the network flexibility.

The price to be paid for the introduction of optical amplifiers is the lack of signal reshaping and retiming along the link. The presence of very long optically amplified links poses new transmission problems that have to be analyzed.

A second step in exploiting optical technology is upgrading the network link capacity by adopting optical frequency division multiplexing (WDM). The first WDM point-to-point systems have been marketed in the last two years, and their proliferation is expected to be rapid. By the WDM a single-link capacity of the order of 40 Gbit/s can be obtained using state-of-the-art technologies.

A third fundamental step in the development of optical technologies will be the introduction of optical switching and routing. In the general structure of transport networks, as standardized by the ITU-T [13], switching and routing of high-speed signals are performed in the so-called path layer of the network [14]. The main network elements of the path layer are the cross-connect and the add/drop multiplexer; the cross-connect is basically a switch that is able to route a signal path entering from a certain input to the desired output; the add/drop multiplexer is a device that processes multiplexed signals: It is able to extract from the high-capacity data stream a low-capacity component and to replace it with a locally generated channel. A functional diagram of the cross-connect switch is shown in Fig. 1.5 and that of an add/drop multiplexer is shown in Fig. 1.6.

A network node throughput of the order of 100 Gbit/s is not far away. As service demand increases rapidly, operators must respond with the same rapidity. This demand would be hard to satisfy if routing and switching depended completely on electrical technologies: That would require a great increase in the dimension of the electronic switching fabric, thus decreasing the cross-connect and add/drop reliability and increasing their cost.

A robust transport network can be achieved by introducing all-optical switching and routing [15]. These operations must be performed by optical cross-connects (OXC) and optical add/drop multiplexers replacing the functions of their electrical counterparts.

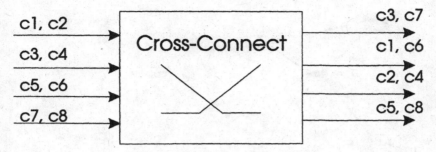

FIGURE 1.5 Functional scheme of a cross-connect (XC); $c_1, c_2, \ldots, c_8$ represent the channels transmitted along the network; two channels are multiplexed for each transmission line.

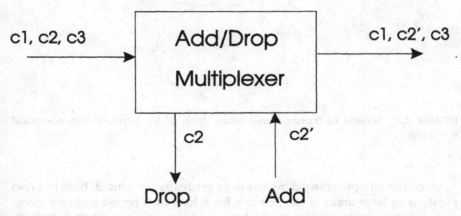

FIGURE 1.6 Functional scheme of an add/drop multiplexer (ADM); c_1, c_2, c_2', c_3 are channels present in the access network, c_2 is dropped by the ADM, while c_2' is added to the flux in the access area.

To enable demultiplexing and routing at lower hierarchical levels, the OXC can be interfaced with an electrical digital cross-connect (DXC). However, for the high-speed optical signal that has to be directly delivered to the access network, this can be done, for example, by an OADM. An illustration of transport and access network areas with all-optical switching is provided in Fig. 1.7. In such a network the OXCs and OADMs can perform not only routing and switching but also monitoring of the network and failure restoration. Indeed, if these functions are directly realized by optical elements, different transmission techniques will be integrated over the same optical platform, enhancing network feasibility and flexibility. For example the plesiochronous hierarchy, PDH [16], the synchronous hierarchy, SDH [17], and the asynchronous transfer mode, ATM [18] could coexist in the same network.

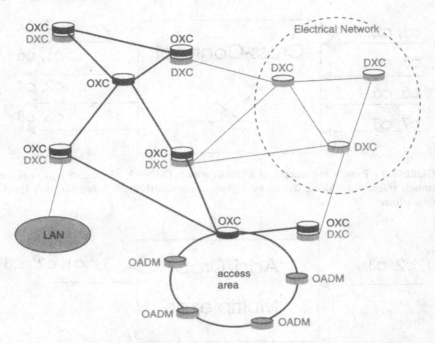

FIGURE 1.7 Scheme of transport and access areas of the network with all-optical switching.

Of course all-optical switching has to be gradually introduced, both to avoid paralysis of large areas of the network for a long time period and to concentrate the investment in areas in which demand increases require the new technology.

A possible evolutionary path is shown in Fig. 1.8. Starting from a network in which optical links connect DXCs (Fig. 1.8a), the optical switching is at first introduced as optical islands in areas of large traffic (Fig. 1.8b). As traffic increases, the optical switching proceeds until only electrical islands survive in the network (Fig. 1.8c).

1.2.4 Information Highways

Because of the required high transmission capacity, this network area can be realized only by adopting optical transmission technologies. In the foreseeable future an overall capacity of 20 to 100 Gbit/s will be required over distances of a few thousands of kilometers. Under these conditions transmission issues will be critical. Therefore the network structure of the highways should be as simple as possible.

In the beginning the highways area might be composed of point-to-point high-capacity links connecting different transport networks. The electrical path layer of the transport network might provide initially an interface between the highways and the transport network. As the traffic increases, the highways area could assume the characteristics of a network area, with a layered structure similar to that of the transport network.

A possible network topology of the highways is a mesh network with a limited number of optical nodes performing almost static routing of very-high-capacity signals. The overall capacity of a single optical link between adjacent nodes of the highway area could be from 40 to over 100 Gbit/s.

Since electronic devices (as signal regenerators and switches) able to process such high-capacity signals are far too complex and expensive, the highways will probably need to depend on optical technologies. Signal regeneration along the transmission lines could be obtained by optical amplifiers and routing along the network by high-capacity OXCs or OADMs.

Both WDM and all-optical TDM can be considered as possible multiplexing technologies.

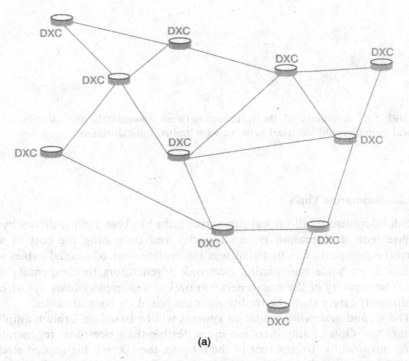

(a)

FIGURE 1.8 Evolution of the transport network towards the introduction of all-optical switching: (a) electrical transport network.

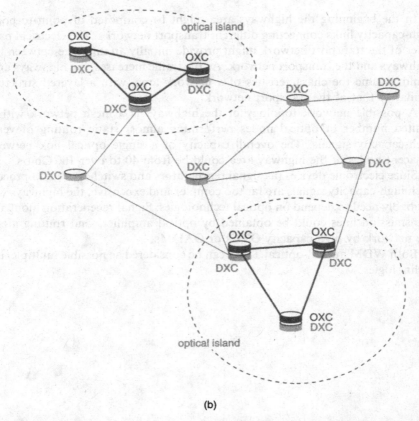

(b)

FIGURE 1.8 Evolution of the transport network towards the introduction of all-optical switching: (b) electrical network with limited optical islands.

1.2.5 Submarine Links

The development of all-optical submarine links has been mainly driven by the requirement of increasing system capacity and decreasing the cost of submerged equipment. The first step was the replacement of coaxial cables with optical fibers while maintaining electronic regenerators to compensate fiber loss. The capacity of the first-generation optical undersea systems was already significantly larger than the traditional ones based on coaxial cables.

The second-generation undersea systems will be based on Erbium amplifier technology. Optical amplifiers are more flexible than electronic regenerators. Being intrinsically transparent to the bit rate, they permit high-speed electronics to be used only at the transmitter and the receiver located at coastal stations.

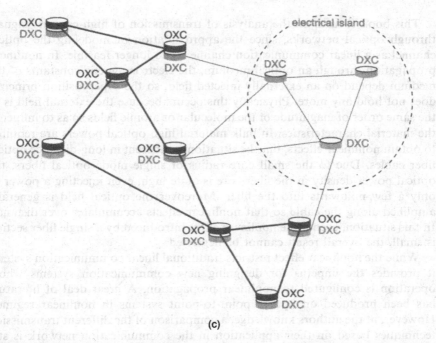

(c)

FIGURE 1.8 Evolution of the transport network towards the introduction of all-optical switching: (c) optical network with limited electrical areas.

The target of the design of an undersea system is the transmission of a 2.5-to 20-Gbit/s signal over a distance that may be as long as 9000 km. One can easily understand that the effort made to achieve such extreme performance has led to significant progress in the technology of optical communications.

Recent progress in optical booster and preamplifier technology will make possible the design and installation of submarine links spanning 100 to 400 km with no in-line amplifiers or with remote pumped Erbium or Raman amplifiers. The absence of active undersea elements will permit a much simpler operation of the transmission system.

1.3 ORGANIZATION OF THE BOOK

An optical network is a complex system; a large number of different functions must cooperate for a network to function smoothly: transmission, routing and switching, control and management, and so on. As a result the network must be analyzed from several new points: economic validity, management, routing and switching strategies, transmission techniques, and performances.

This book focuses on the analysis of transmission of high-capacity signals through optical networks, since the approximation of modeling the optical channel as a linear communication channel is no longer feasible. In nonlinear propagation through an optical medium, the electromagnetic constants of the medium depend on an externally injected field, so the superposition principle does not hold any more. Physically this occurs because the external field is of the same order of magnitude of the molecular or atomic fields so as to influence the material characteristics. In bulk material high optical powers are required to obtain nonlinear effects, but the situation is different in long-distance optical fiber cables. Due to the small core radius of single mode optical fibers, the optical power density in the fiber core is quite high, even injecting a power of only a few milliwatts into the fiber. Moreover the optical field is generally amplified along the cable so that nonlinear effects accumulates over distance. In this situation, even if the nonlinear effect introduced by a single fiber section is small, the overall result cannot be neglected.

While the nonlinear effect restricts traditional linear communication systems it provides the impetus for designing new communication systems whose operation is configured to nonlinear propagation. A great deal of literature has been produced on optical point-to-point systems in nonlinear regimes. However, at the authors knowledge, a comparison of the different transmission techniques based on their application in the communication network is still lacking. On the other hand, in the view of the telecommunication engineer, it is not so important to state what system achieves the highest capacity. Most important is to understand what transmission techniques are suitable for the considered application and how they compare. We have tried to build the book starting from this last point of view.

Optically switched networks have a certain characteristic that causes transmission through an optical network to be different from transmission in a point-to-point system: their transparency. Whatever technology is adopted to realize an optical network, optical digital processing will be not available for a long time. Presently routing has to be done by broadcasting all signals and selecting the desired signal at the end node (passive networks) or by opportune analog systems (e.g., the OXCs) providing no signal regeneration (switched networks).

Switched networks are very different from networks with electrical nodes, where regeneration occurs inside the nodes so that the transmission of a signal through the network is simply the concatenation of several independent transmissions through point-to-point links connecting the network nodes. In an optical network the optical node appears, from a transmission point of view, as an in-line device, perhaps with gain, introducing signal distortion, noise, and possibly causing crosstalk among the channels that feed it.

In both passive and switched networks nonlinear propagation is significantly influenced by the broadcast strategy and by the structure of the in-line nodes, respectively.

1.3.1   Overview of the Chapters

The nine chapters composing this book are grouped into four parts:

- The Introduction (Chapter 1);
- The analysis of nonlinear systems (Chapter 2 dealing with fiber propagation, Chapter 3 dealing with optical amplifiers, and Chapter 4 dealing with transmission systems performance evaluation methods);
- The structure and performances of different point-to-point optical transmission techniques (Chapter 5 presenting a theoretical analysis of soliton transmission systems, Chapter 6 presenting a comparison between transmission techniques that do not adopt in-line optical amplification, Chapter 7 presenting a comparison between transmission techniques adopting time division multiplexing and in-line optical amplification, and Chapter 8 describing transmission techniques adopting WDM and in-line optical amplification);
- Transmission through all-optical network, including in-line devices for switching and routing of the optical signal (Chapter 9).

In the following there are some brief details on the structure of the chapters.

In Chapter 2, which deals with optical fibers, are reviewed the characteristics of signals propagation in a single mode optical fiber in a linear regime, with a particular attention to dispersive effects and random mode coupling. Dispersive effects have a main role in determining the performances of optical systems, while random mode coupling causes both fluctuations of the field polarization state at the fiber output and polarization dispersion; polarization state fluctuations are important when analyzing coherent receivers, while polarization dispersion is an important performance impairment of very-high-speed optical transmission systems.

Optical amplifiers are analyzed in Chapter 3; the simple theory of the two-level optical amplifier is presented in the first part of the chapter to introduce the main phenomena regarding optical amplifiers while, in the second part of the chapter, semiconductor amplifiers and erbium doped fiber amplifiers are described and analyzed by means of simple models. A more detailed physical analysis of such devices is not within the scope of this book, and a full reference list is provided for the interested reader.

In Chapter 4 the block diagram of a generic optical transmission system is discussed. Different optical multiplexing technologies are compared on the basis of their technological maturity. In the second part of the chapter the main parameters for the performance evaluation of optical systems are introduced: The eye penalty, the amplitude signal-to-noise (ASN) ratio, the Q factor, and the system error probability are defined and compared. Some of the techniques used to evaluate the Q factor and the error probability are also introduced along with some simple applications.

Chapter 5 present the analytical theory of soliton systems. Soliton systems are an interesting case of optical transmission systems operating in a nonlinear propagation regime for which a complete and satisfactory analytical theory can be carried out. All the important phenomena that can be observed during the propagation of optical solitons in a fiber link containing in-line optical amplifiers are analytically described assuming an amplifier spacing sufficiently small to adopt the average soliton approximation (a typical value of the amplifier spacing in such systems is 30 km). Also included are the methods used to stabilize soliton propagation, and their effectiveness is discussed.

In Chapter 6 high-capacity systems without in-line optical amplifiers are considered which are important in transport and in access areas as far as electronic switching and routing are adopted. Such systems are also widely used as short submerged links in certain geographical areas. A theoretical analysis is very difficult for such systems, which are characterized by the propagation of high-power optical signals over a few hundreds of kilometers of fiber; thus simulation and experimental results are extensively used to analyze and compare the different techniques.

Very-long-distance systems using in-line amplifiers are dealt with in Chapters 7 and 8. Chapter 7 describes and analyzes time division multiplexed systems, while an analysis of frequency division multiplexed systems is provided in Chapter 8. A simulation of the system behavior is presented in these chapters in order to compare the different transmission techniques. Even soliton systems are discussed in this chapter, since simulation is needed if all the effects arising during solitons propagation are to be considered together. Moreover, when the signal propagation is strongly nonlinear, the distinction between soliton and nonsoliton systems tends to vanish. Important and recent experimental results are also reviewed in these chapters.

Finally in Chapter 9 transmission in all-optical networks is considered. We make no effort to review the main technologies enabling the introduction of optical signal processing, since we deal only with transmission issues. The introduction of optical switching and routing gives rise to new transmission problems that are not present in point-to-point links. In order to describe and analyze these problem, we introduce a abstract transmission model that can be used to describe the signal path in the different network areas. We can particularize this model to represent local area networks, access area networks, and transport area networks in order to analyze specific problems arising in these network areas.

REFERENCES

1. A large literature exists on telephone networks and their use for data transmission. In the following are some references from which a more extended bibliographic research could start: For detailed description of telephone networks comprising data transmission, see

R. L. Freeman, *Telecommunication System Engineering, Analog and Digital Network Design*, Wiley, New York, 1980.
For description of telephone networks and a detailed analysis of switching nodes, see
GRINSEC (Groupe des Ingénieurs du Secteur Commutation du CNET), *Electronic Switching*, North-Holland, Amsterdam, 1983.

2. It is beyond the scope of this book to analyze the different switching techniques adopted by communications networks. A review of circuit switching can be found in the GRINSEC book cited in ref. [1]. For a survey on packet switching technologies, see, for example,
R. D. Rosner, *Packet Switching*, Lifetime Learning Publications, Belmont, MA, 1982.

3. For a general description of the structure of ISDN, see, for example,
P. Bocker, *ISDN: Integrated Services Digital Network*, Springer, Berlin, 1990.
J. M. Griffits, *ISDN Explained*, Wiley, New York, 1990.
For the standard criteria, at least for CCITT (now ITU-T), refer to CCITT study group XVIII, Recommendations I.412, I.413, I.430, I.431, I.432, G.960, G.961, G96Y, G96X, 1992.

4. A study of traffic patterns was carried out in the European research project of the telecommunications market RACE R2091 URSA; for some important project results, see, for example,
M. Hopkins, G. Louth, H. Biley, R. Yellon, A. Ajibulu, and M. Niva, A multi-faceted approach to forecasting broadband demanded traffic, *IEEE-Communications Magazine*, 33: 36–42, 1995.

5. A good example of optical links designed using terrestrial transmitters and receivers is included in the Italian transport network. See, for example,
L. Lattanzi, P. Rosa, S. Cascelli, and F. Balena, Up to 200 km repeaterless submarine optical fiber links at 1.55 μm in the Italian long distance network, *Proceedings of ECOC'89*, pp. 454–457, Gothenburg, Sweden, September 10–14, 1989.

6. The most assessed type of high-speed LAN adopting optical fibers is the FDDI. This type of LAN is described, for example, in
W. Stallings, *Local and Metropolitan Area Networks*, Prentice Hall, Englewood Cliffs, NJ, 1997, ch. 8.

7. A large literature has been produced about topologies and protocols for very high-capacity LANs adopting optical transmission and electronic processing. Here we report some examples from which the interested reader can start to obtain a more complete reference list:
B. Glance, T. L. Koch, O. Scaramucci, K. C. Reichmann, L. D. Tzeng, U. Koren, and C. A. Burrus, Densely spaced FDM optical coherent system with near quantum limited sensitivity and computer controlled random access channel selection, *Electronics Letters*, 25: 883–885, 1989.
N. Mehravari, Performance and protocol improvement for very high speed optical fiber local area networks using a passive star topology, *IEEE-Journal of Lightwave Technology* 8: 520–530, 1990.
Y. Birk, Fiber optic bus-oriented single-hop interconnections among multi-tranceivers stations, *IEEE-Journal of Lightwave Technology* 9: 1657–1663, 1991.
G. De Marchis, E. Iannone, F. Renzi, and M. Todaro, A new protocol for high

performance fiber optical star networks based on FDM technique, *Journal of Optical Communications* 14: 5–12, 1993.

8. Different devices have been proposed for such functions, based both on fiber and semiconductor technology. For some examples, see

   M. N. Islam, C. E. Soccolich, and A. B. Miller, Low-energy ultrafast fiber soliton logic gates, *Optics Letters* 15: 909–911, 1990.

   M. Settembre and F. Matera, All optical implementation for high-capacity TDM networks, *Fiber and Integrated Optics* 12: 173–186, 1993.

   D. Nesset, N. C. Tatham, L. D. Westbrook, and D. Cotter, Ultrafast all-optical AND gate at the same wavelength using four wave mixing in a semiconductor laser amplifier, *Proceedings of ECOC'94*, September 25–29, Florence, pp. 529–532, 1994.

   A. D'Ottavi, E. Iannone, and S. Scotti, Address recognition in all-optical packet switching by FWM in semiconductor amplifiers, *Microwave and Optical Technology Letters* 10: 228–230, 1995.

9. For examples of all-optical time division multiplexing and switching, see

   D. M. Spirit, D. E. Wickens, T. Widdowson, G. R. Walker, D. L. Williams, and L. C. Blank, 137 km, 4 × 5 Gbit/s optical time division multiplexed unrepeated system with distributed erbium fiber preamplifier, *Electronics Letters* 28: 1218–1219, 1992.

   Y. Shimazu, M. Tsukada, Ultrafast photonic ATM switch with optical output buffers, *IEEE-Journal of Lightwave Technology* LT-10: 265–271, 1992.

   For an impressive demonstration of the potential capacity of OTDM systems, see, for example,

   S. Kawanishi, H. Takara, O. Kamatani, and T. Morioka, 100 Gbit/s, 500 km, optical transmission experiment, *Electronic Letters* 31: 737–741, 1995.

10. For a general introduction to FTTC strategies is presented, see, for example,

    C. J. Brunet, Hybridizing the local loop, *IEEE-Spectrum* 31: 28–33, 1994.

    R. C. Menendez, D. L. Waring, and D. S. Wilson, High-bit-rate copper and fiber-to-the-curb video upgrade strategies, *Proceedings of the IEEE International Conference on Communications (ICC)*, May 23–26, 1993.

    For a specific example of a plan to realize FTTC, see C. Carroll, Development of integrated cable/telephony in the United Kingdom, *IEEE-Communications Magazine* 33: 48–60, 1995.

    For a comparison among different approaches, see

    W. Pugh and G. Boyer, Broadband access: Comparing alternatives, *IEEE-Communications Magazine* 33: 34–47, 1995.

11. For a general introduction of FTTC, see, for example,

    P. Kaiser and P. W. Shumate, Fiber to the home, *Optics News* 15: 14–20, 1989.

    P. W. Shumate, Optical fibers reach into the home, *IEEE-Spectrum* 26: 43–47, 1989.

    For an example of a concrete project, see

    T. Miki and R. Komiya, Japanese subscriber loop network and fiber optic loop development, *IEEE-Communications Magazine* 29: 60–67, 1991.

    For an almost complete review of the main issues arising when FTTH is considered, see the different papers presented at the workshop "Fiber to the home," Brussels, June 26, 1990.

12. For an economic evaluation of the investment needed to adopt optical technologies in the distribution area, see, for example,

    A. Del Pistoia, A. Gambaro, U. Mazzei, and G. Roso, A market driven approach for the deployment of optical loop, *Proceedings of ISSLS'91*, Amsterdam, 22–26 April, 1991, pp. 240–246.

13. ITU-T Recommendation, G. 803, "Architectures of transport networks based on the synchronous digital hierarchy (SDH)," 03/93, 1993.
    For the ITU-T standard layered architecture for the transport network, see also K. Sato, S. Okamoto, and H. Hadama, Network performance and integrity enhancement with optical path layer technologies, *IEEE-Journal on Selected Areas in Communications* 12: 159–170, 1994.

14. The concept of layered structure is a fundamental concept in the communications engineering. For an example of the numerous books and papers in which the concept of a layered structure with a client-server association among layer elements is rigorously described, see H. Zimmermann, OSI reference model—The ISO model of architecture for open systems interconnections, *IEEE-Transactions on Communications* 28: 425–432, 1980.

15. The concept of optical path layer has been introduced and discussed in several papers. For example, see
    K. Sato, S. Okamoto, and H. Hadama, Network performance and integrity enhancement with optical path layer technologies, *IEEE-Journal on Selected Areas in Communications*, 12: 159–170, 1994.
    K. Sato, S. Okamoto, and H. Hadama, Optical path layer technologies to enhance B-ISDN performance, *Proceedings ICC'93*, pp. 1300–1307, 1993.
    G. R. Hill et al., A transport network layer based on optical network elements, *IEEE-Journal of Lightwave Technology* 11: 667–679, 1993.
    R. Sabella, E. Iannone, and E. Pagano, Optical transport networks employing all-optical wavelength conversion: Limits and features, *IEEE-Journal on Selected Areas in Communications*, Special issue on Optical WDM Networks 15: 968–978, 1996.

16. The principles of PDH are described in many books. The standardization of PDH is studied by the ITU-T group SG15, and the main ITU-T (former CCITT) recommendation on PDH is the recommendation G.954. The longitudinal compatibility in PDH system is analyzed in the recommendation G.955.

17. The SDH transmission mode is described in numerous books and papers, see, for example,
    IEEE-Lightwave Telecommunication Systems, Special issue on SDH, 2: 1991.
    IEEE-Communication Magazine, Special issue on Global Deployment of SDH-Compliant Networks, 28: 1990.
    A complete standardization of SDH transmission mode has been produced by the CCITT (now ITU-T). See CCITT Recommendation, Blue Book, G 707, G 708, G 709, 1989.

18. For a general description of the ATM transfer mode, see, for example,
    M. de Prycker, *Asynchromous Transfer Mode.*, Prentice Hall, London, 1995.
    The ATM transfer mode has been proposed for the implementation of ISDN in several papers. See, for example,
    K. Sato, S. Ohta, and I. Tokizawa, Broadband ATM network architecture based on virtual path, *IEEE-Transaction on Communications* 38: 1212–1222, 1990.
    The standardization of ISDN based on the ATM transfer mode is provided by the CCITT I-Series Recommendations (B-ISDN), 1992.

CHAPTER TWO

Optical Fiber Propagation

Since the late 1970s when the first low-loss optical fibers were designed and produced, the telecommunications environment has been rapidly changing [1]. Such fibers had a core dimension of the order of one hundred microns; they worked in multimode regimes, having high dispersion, and this critically affected their transmission performance.

A great improvement was obtained with the introduction of the single-mode fibers, thanks to the advances in fibers technology which permit fibers to be processed that have a core radius of the order of several microns. Since such fibers had a refractive index profile with a step shape, they were called *step-index fibers*. At the beginning of the 1980s several megameters of step-index fibers were installed, incorporating high-capacity fiber links in the telecommunication network. Such fibers have very low attenuation around the wavelength of 1.3 µm (called the *second-transmission window*; typical attenuation 0.3–0.4 dB/km) and 1.5 µm (called the *third-transmission window*; typical attenuation 0.2–0.25 dB/km) [2]. The dispersion of step-index fibers is zero around 1.3 µm, while in the other window at 1.5 µm the dispersion is not negligible. This problem has been recently solved with the introduction of the dispersion shifted (DS) fibers that have a zero chromatic dispersion at 1.5 µm. This result is reached by a complex shape of the refractive index profile [3].

This chapter discusses some fundamental characteristics of optical fibers: Only single-mode fibers are considered, since only this type of fiber is now adopted in telecommunication systems.

The chapter is divided into two main sections. Section 2.1 describes linear propagation of signals in optical fibers, including fiber loss, polarization mode dispersion, chromatic dispersion, and third-order chromatic dispersion.

Section 2.2 considers nonlinear effects in optical fibers; after a review of the main nonlinear effects (Brillouin scattering, Raman scattering, and the Kerr effect), the discussion focuses on the Kerr effect which is the most important nonlinear effect in optical communication systems. Some characteristics of signal propagation in linear and in nonlinear propagation regimes are presented and the concept of soliton is introduced.

2.1   LINEAR PROPAGATION IN OPTICAL FIBERS

2.1.1   Propagation Modes and Single-Mode Fibers

An optical fiber is a silica glass thread constituted by a *core* surrounded by a *cladding*. The working principle of the optical fiber is based on the total reflection of the light. In particular, the light is directed into the core and it is reflected by the cladding wall allowing it to propagate into the core. To enable this, the refractive index of the cladding is smaller than that of the core.

A complete study of the propagation of the electromagnetic field in optical fibers is beyond the scope of this book [4]. In this section we give a simple description of the light propagation based on geometrical optics in order to give the reader a useful understanding of the working principles of an optical fiber. Such a description constitutes a good approximation in the case of the multimode fiber, where the light wavelength is much smaller than the core dimension. On the other hand, it does not hold in the case of single-mode fibers, where the light wavelength is of the order of the core radius.

Let us consider the scheme of a step-index fiber shown in Fig. 2.1. In such a fiber the refractive index assumes a constant value r_1 in the core and a constant value $r_2 < r_1$ in the cladding. The behavior of the light rays in a step-index fiber is illustrated in Fig. 2.2. When a light ray incises on the fiber core with an angle θ_i, it is refracted inside the core with an angle θ_r according the relationship

$$r_0 \sin(\theta_i) = r_1 \sin(\theta_r) \tag{2.1}$$

where r_0 is the refractive index of the medium surrounding the fiber. The refracted ray hits the core-cladding interface with an angle θ_c. If the angle satisfies the relationship $\sin(\theta_c) > r_2/r_1$, the ray is refracted in the cladding (unguided ray); conversely, if $\sin(\theta_c) < r_2/r_1$, the ray is reflected inside of the

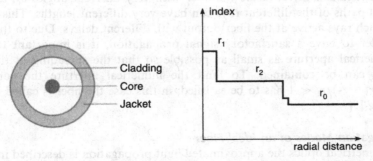

FIGURE 2.1   Transverse section and index profile of a step-index optical fiber.

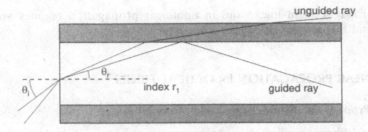

FIGURE 2.2 Guided and unguided rays in the geometrical optics approximations.

core. If the condition $\sin(\theta_c) < r_2/r_1$ is verified in the first reflection, it is also verified every time the ray hits the core-cladding interface; therefore the light remains confined inside the core and can propagate for a long distance. This is the principle of the optical fiber. The angle θ_{cr} that verifies the condition

$$\sin(\theta_{cr}) = r_2/r_1 \qquad (2.2)$$

is called the *critical angle*. The *input angle* θ_{ir} of a ray hitting the core-cladding interface at the critical angle is given by

$$\sin \theta_{ic} = \frac{r_1 \cos(\theta_{cr})}{r_0} = \frac{\sqrt{r_1^2 - r_2^2}}{r_0} \qquad (2.3)$$

The quantity $r_0 \sin \theta_{ic}$ is called the *numerical aperture* (NA) of the fiber: It represents the capacity of the fiber to accept light. In the fibers designed for telecommunications $r_1 \approx r_2$; thus the numerical aperture can be written as

$$NA = r_1 \sqrt{2 \frac{r_1 - r_2}{r_1}} = r_1 \sqrt{2\Delta} \qquad (2.4)$$

When much power must be coupled into the fiber, the numerical aperture has to be high. However, this choice is detrimental from a point of view of signal propagation. The reason is that when the numerical aperture is large, the guided rays can travel inside the fiber with very different angles θ_c, and the optical paths of the different rays can have very different lengths. This means that such rays arrive at the fiber output with different delays. Due to this fact, in order to have a satisfactory signal propagation, it is important to have a numerical aperture as small as possible so that the spreading of the rays delays can be contained. To limit the numerical aperture the condition $\Delta = (r_1 - r_2)/r_1 \ll 1$ has to be satisfied; in this case the fiber is called *weakly guiding* [5].

Propagation Modes of an Ideal Fiber
In geometrical optics the approximated light propagation is described in terms of rays, but in the more accurate description based on Maxwell equations the

light propagation can be described in terms of propagation modes [4]. The electrical and the magnetic field of a mode in an optical fiber can be written as

$$\mathbf{E}_m(x, y, z, t) = \xi_{Em}(x, y) \, e^{i(\beta_m z - \omega t)}$$
$$\mathbf{H}_m(x, y, z, t) = \xi_{Hm}(x, y) \, e^{i(\beta_m z - \omega t)}$$

$$(2.5)$$

In expression (2.5) the index m represent the mode and ω the field's angular frequency. In the expression of modes, the dependence on the transverse coordinates x and y is separated from the dependence on the coordinate z, which represents the axis along which propagation occurs. Moreover the modes depend only on z by means of the exponential term where z is multiplied by the propagation constant β_m.

Since each mode travels with a different group velocity, a multimode pulse launched into the fiber broadens during propagation: Such an effect is called *modal dispersion*. Fiber dispersion can be minimized to propagating only one mode, and the introduction of *single-mode* fibers can make a significant improvement in the performance of optical communication systems.

Transverse electromagnetic waves cannot propagate rigorously in optical fibers; the hybrid propagation modes are generally described in polar coordinates and labeled as EH and HE modes. In the case of fibers designed for telecommunications, the index profile depends only on the distance ρ from the fiber axis, and the weakly guiding condition ($\Delta \ll 1$) holds. Under these assumptions, the longitudinal components of the fields have a negligible amplitude with respect to the transverse ones, so the approximation of transverse electromagnetic waves can be applied. Moreover the components of the electric and magnetic fields are locally related by the same equation for plane waves in a bulk medium, allowing a great simplification of the mathematical formalism.

There always exists in such fibers a fundamental mode (mode HE_{11}) that does not present a frequency cutoff. In general, the index profile of the fiber has a rotational symmetry. In this case the fundamental mode HE_{11} has a double degeneracy corresponding to two orthogonal states of polarization.

Due to the existence of a mode without cutoff, a weakly guiding fiber behaves as a single mode waveguide if the optical frequency is smaller than the lowest cutoff frequency, ν_{cutoff}. In the case of step-index fibers

$$\nu_{cutoff} = \frac{2.405c}{2\pi\rho_c \sqrt{r_1^2 - r_2^2}}$$

$$(2.6)$$

where ρ_c is the core radius and r_1 and r_2 the refractive indexes of the core and of the cladding, respectively; c represents the speed of light in vacuum.

To minimize fiber loss, the more interesting optical bandwidths for fiber transmission are those around 1.3 µm (the so-called second transmission window) and around 1.55 µm (the so-called third transmission window). To operate at these wavelengths in single-mode regime commercial fibers have a cutoff of about 1 μm.

An alternative approach to determining the fiber modes that yield to HE and EH modes uses the cartesian components of the field instead of the polar ones and focuses on the weakly guiding property of the fiber in order to simplify the resulting equations. This procedure leads to the definition of two sets of linearly polarized modes (LP_{mx} and LP_{my}, with $m = 0, 1, ...$) [6] that are polarized along the x and y axes. The LP_{mx} and LP_{my} modes are degenerate, sharing the same propagation constant. Obviously the alternative descriptions of the field are not uncorrelated: It can be shown [7] that every LP mode can be obtained as a superposition of the HE and EH modes.

The description of the fiber mode structure using the LP modes holds only in the approximation of a weakly guiding fiber. Under this scenario, an LP mode launched into the fiber will propagate without any change in the spatial structure at any fiber cross section, retaining the original linear polarization state. In contrast, if the exact description of HE and EH modes is used, the same input field is considered as the superposition of two different modes, which travel with very close but different velocities; thus the linear polarization input state will not be retained all along the fiber. In the case of conventional fibers, the LP modes picture leads to errors by far smaller than those due to other approximations, for example, perfect fiber circular symmetry, perfectly straight fiber axis, and absence of micro defects in the fiber material. In particular, the fiber defects generate coupling between LP modes by far stronger than that arising from the weakly guiding approximation.

In terms of the LP approximation, two degenerate, linearly polarized, fundamental modes exist; they are commonly called LP_{0x} and LP_{0y} and can be written as

$$\mathbf{E}_{0x} = \xi(\rho) \, e^{i(\beta z - \omega t)} \mathbf{x}$$
$$\mathbf{E}_{0y} = \xi(\rho) \, e^{i(\beta z - \omega t)} \mathbf{y} \tag{2.7}$$

where ρ is the transverse coordinate, and $\mathbf{x}$, $\mathbf{y}$ are the unit vectors of the transverse axes. The exact expression of the modes is quite complex and depends on the index's profile. However, independently of the index profile, $\xi(\rho)$ can be represented with a good approximation by a gaussian shape of standard deviation ρ_0.

The parameter ρ_0 depends on the fiber structure. For example, in the case of the step-index fiber $\rho_0 = \rho_c / \ln V^2$, where $V = (\omega/c)\rho_c\sqrt{r_1^2 - r_2^2}$ is called the normalized frequency.

A measure of the effective width of the fiber fundamental mode, alternative to ρ_0, is the so-called guided mode radius ρ_m; it is defined as

$$\rho_m^2 = \frac{\displaystyle\int_0^\infty \rho|\mathbf{E}(\rho)|^4 \, d\rho}{\displaystyle\int_0^\infty \rho[d|\mathbf{E}(\rho)|^2/d\rho]^2 \, d\rho} \tag{2.8}$$

Since the definition (2.8) has been standardized by the ITU-T (the international standardization agency in the field of telecommunications) [8] and it is generally used instead of ρ_0, in this book we will use ρ_m. Typical values of the effective mode radius in different types of fibers used for telecommunications are reported in Table 2.1.

2.1.2 Propagation in Single-Mode Uniform Fibers

So far we have described the modes of ideal fibers. A real fiber always shows some imperfections, but the modes of an ideal fiber are useful to study the evolution of the electromagnetic field. In particular, if geometric defects are small enough, we can suppose that the transverse mode shape remains unaltered. Conversely, the propagation term, containing the dependence on z, is significantly affected by fiber imperfections.

The simplest example is that of a fiber showing uniform imperfection along the longitudinal direction (e.g., an elliptic core, stress, or bending) and uniform loss. In this case, due to the fiber imperfection, the degeneration between the linearly polarized modes LP_{0x} and LP_{0y} is removed: the fiber shows the phenomenon of *birefringence*. Equation (2.7) still holds, but with two different propagation constants β_x and β_y and with a loss coefficient α that is generally independent of field polarization. Therefore the expression of the electrical field of the LP_0 modes can be written as

$$\mathbf{E}_{0x} = \xi(\rho)\, e^{-\alpha/2z} e^{i(\beta_x z - \omega t)} \mathbf{x}$$
$$\mathbf{E}_{0y} = \xi(\rho)\, e^{-\alpha/2z} e^{i(\beta_y z - \omega t)} \mathbf{y}$$

$$(2.9)$$

Due to birefringence, the polarization of a monochromatic field shows periodic evolution during propagation with a period equal to $L_b = 2\pi/\Delta\beta$, called *beat length*, where the quantity $\Delta\beta = \beta_x - \beta_y$ measures fiber birefringence. In other words, the field shows the same state of polarization (SOP) at z and at $z + L_b$.

A signal propagating through a single-mode fiber can be generally expressed as a Fourier decomposition, obtaining

$$\mathbf{E}(\rho, z, t) = \sum_{k=x,y} \int_{-\infty}^{+\infty} c_k(\omega)\xi(\rho)\, e^{-\alpha/2z} e^{i(\omega t - \beta_k z)}\, \mathbf{u}_k\, d\omega \qquad (2.10)$$

where $\mathbf{u}_k = \mathbf{x}, \mathbf{y}$ represent the axes unit vectors. The coefficients $c_k(\omega)$ take into account the input condition; it can be found by vector multiplication ($\times$) of the electric field at the fiber input for the magnetic field representing the mode and by an inverse Fourier transform of the result. The obtained expression is

$$c_k(\omega) = \frac{\psi}{4\pi P_o} \int_{-\infty}^{+\infty} e^{-i\omega t} \int_{ST} [\mathbf{E}(x, y, 0, t) \times \mathbf{u}_k] \cdot \mathbf{z}\xi^*(x, y)\, dx\, dy\, dt \qquad (2.11)$$

TABLE 2.1 Typical parameters of different optical fibers.

		Step index ($\lambda = 1.55\ \mu m$)	Step index ($\lambda = 1.3\ \mu m$)	Dispersion shifted ($\lambda = 1.55\ \mu m$)	Dispersion compensating ($\lambda = 1.55\ \mu m$)
Attenuation (dB/km)	α	0.25	0.4	0.25	0.5
Core refraction index (dimensionless)	r_0	1.5	1.5	1.5	1.5
Effective mode radius (μm)	ρ_m	4	4.7	3.8	—
Dispersion parameter (ps/nm/km)	D	15.6	$-3.3/3.3$	$-2.3/2.3$	-62.5
Dispersion coefficient (ps^2/km)	β_2	-20	$-3/3$	$-3/3$	80
Third-order dispersion coefficient (ps^3/km)	β_3	0.05	0.063	0.063	0.02
Birefringence (m^{-1})	$\Delta\beta$	0.1	0.1	0.1	—
Polarization mode dispersion (ps/$\sqrt{km}$)	$\Delta\tau$	0.1	0.1	0.1/1	—
Random mode coupling characteristic length (m)	L_h	100	100	100	50
Real part of the third-order susceptibility (W^{-1})	$\chi^{(3)}_{xxxx}$	1.56×10^{-19}	1.82×10^{-20}	1.9×10^{-19}	—
Nonlinear coefficient (km^{-1} W^{-1})	γ	2	2	2.7	3
Brillouin gain coefficient (m^{-1} W^{-1})	g_B	6.02×10^{-12}	4.6×10^{-12}	6.02×10^{-12}	—
Brillouin gain bandwidth (MHz)	$\Delta\nu_B$	16	25	16	—
Raman gain coefficient (m^{-1} W^{-1})	g_R	7×10^{-15}	8.3×10^{-15}	7×10^{-15}	—

where ST represents the transverse fiber section, P_o the optical power, and ψ a constant relating the optical power density and the square of the electrical field. The value of ψ depends on the measurement unit system adopted; for example, in the MKS system $\psi = \sqrt{\varepsilon/\mu}$, where ε and μ are the dielectric and magnetic constant of the fiber. Sometimes a system where the electric field is measured in $W^{1/2}m^{-2}$ allows the notation to be simplified; in this system $\psi = 1$. Generally we will use the MKS system; we will demonstrate where the system in which $\psi = 1$ is used.

Generally, the function $\xi(x, y)$ and the loss coefficient α can be assumed to be constant in the signal bandwidth. As a consequence equation (2.10) becomes

$$\mathbf{E}(\rho, z, t) = \xi(\rho) e^{-\alpha/2z} \sum_{k=x,y} \mathbf{u}_k \int_{-\infty}^{+\infty} c_k(\omega)\, e^{i(\omega t - \beta_k z)}\, d\omega$$

$$= \xi(\rho)[A_x(z, t)\mathbf{x} + A_y(z, t)\mathbf{y}] \qquad (2.12)$$

where the functions $A_k(t, 0)$ represent the components of the input signals.

Equation (2.12) allows the evolution of the electrical field along the fiber to be analyzed by considering only the function $\mathbf{A} = A_x\mathbf{x} + A_y\mathbf{y}$, and not the transverse shape of the mode. The functions $A_k(z, t)$ can be easily expressed in the Fourier domain starting from the initial conditions by means of the relationship

$$\mathbf{A}(z, \omega) = e^{-\alpha/2z} \begin{bmatrix} e^{-i\beta_x(\omega)z} & 0 \\ 0 & e^{-i\beta_y(\omega)z} \end{bmatrix} \begin{bmatrix} A_x(0, \omega) \\ A_y(0, \omega) \end{bmatrix} [\mathbf{x}\ \mathbf{y}]$$

$$= e^{-\alpha/2z}[\beta]\mathbf{A}(0, \omega) \qquad (2.13)$$

In equation (2.13) the dependence of the propagation constants on ω has been explicitly indicated. Two main phenomena are taken into account in equation (2.13): loss and dispersion.

Attenuation in optical fibers is due to absorption and scattering. Absorption is a very complex phenomenon. It is governed by the laws of energy exchange at the atomic level: photons, traveling along the fiber, can transfer their energy to the glass, exciting some energy levels. In the middle-infrared region (above 1.8 µm), the energy is transferred mainly by vibrational transitions. In the ultraviolet region, the absorption is mainly due to electronic and molecular transitions.

Because of intrinsic absorption processes, silica fibers are normally used in wavelength regions that range between 800 and 1600 nm: Outside these regions the absorption is too high to allow propagation over long distances.

Further, absorption arises not only from the pure material forming the fiber (intrinsic absorption) but also from the impurities such as Fe^{3+} or OH^- ions (extrinsic absorption) resulting from the production process. Extrinsic absorption was responsible for the high attenuation (>20 dB/km) in early fibers.

Scattering loss, which is another factor influencing the total loss, is also caused by numerous phenomena. Here again, it is possible to distinguish between intrinsic scattering (Rayleigh scattering) and extrinsic scattering due to imperfections in the fiber or in the protective jacket.

Rayleigh scattering is caused by variations in the refractive index of the transmission medium over distances shorter than the wavelength. These variations are induced by the production process when the localized fluctuations in the density of the molten material become fixed on cooling. The resulting loss is proportional to ω^4, that is λ^{-4}.

Scattering is almost the only type of loss in modern fibers. Figure 2.3 shows a plot of the fiber attenuation versus wavelength for a commercially available fiber: In the operating region the attenuation is very close to the Rayleigh lower limit. The possible presence of residual OH^- absorption peaks limits the signal transmission to three transmission windows, roughly $0.85\,\mu m$, $1.3\,\mu m$, and $1.5\,\mu m$. Typical values for the attentuation in different kinds of fibers are reported in Table 2.1.

There are several contributing factors to dispersion. Expanding $\beta_x(\omega)$ and $\beta_y(\omega)$ in Taylor series, we have

$$\beta_k(\omega) \approx \beta_k(\omega_0) + \beta_k'(\omega_0)(\omega - \omega_0) + \frac{\beta_k''(\omega_0)}{2}(\omega - \omega_0)^2 + \frac{\beta_k'''(\omega_0)}{6}(\omega - \omega_0)^3 \quad (2.14)$$

where the apexes means derivative with respect to ω and $k = x, y$.

The term $\beta_k(\omega_0)$ causes a phase shift, and the difference $[\beta_x(\omega_0) - \beta_y(\omega_0)]z$ is responsible for the polarization evolution of the field. The term $\beta_k'(\omega_0)z$ induces

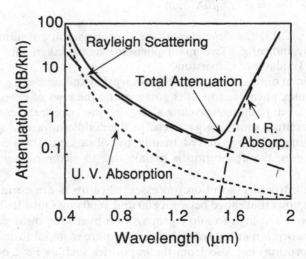

FIGURE 2.3 Main components causing attenuation in a single-mode optical fiber. In the plot the attenuation is shown versus the wavelength.

a group delay $T_k = z/v_{gk}$, where $v_{gk} = 1/\beta_k'(\omega_0)$ is the group velocity. A pulse injected into the fiber along a SOP that does not coincide with the SOP of one mode broadens, due to the fact that the two modes constituting the field have different group velocities. The pulse broadening is measured by the quantity $[\beta_x'(\omega_0) - \beta_y'(\omega_0)]z$, which is called the differential group delay, while the term

$$\Delta\beta'(\omega_0) = \beta_x'(\omega_0) - \beta_y'(\omega_0) \qquad (2.15)$$

is called the polarization mode dispersion (PMD) of the uniform fiber. In Section 2.1.3 we will show that the simple model of a uniform fiber introduced in this section is not suitable for a correct description of the PMD in a real fiber. This is due to the presence of other important effects, as random coupling between the polarization modes, that are not described by equations (2.14).

The term $\beta_k''(\omega_0)$ causes the group velocity of a monochromatic wave to depend on the wave frequency. This means that when a pulse is transmitted along the fiber, its frequency components have different group velocities, so the pulse broadens during propagation. This spreading of the group velocity is known as chromatic dispersion or group velocity dispersion (GVD). In common fibers, the LP_0 modes have the same GVD, generally labeled as β_2 and measured in ps^2/km.

The broadening ΔT of a pulse at the output of a fiber of length z can be approximately measured by the difference in the arrival times of the extreme frequency components. If the pulse bandwidth is $\Delta\omega$, ΔT is given by

$$\Delta T = \frac{dT}{d\omega}\Delta\omega = z\frac{d}{d\omega}\left(\frac{1}{v_g}\right)\Delta\omega = z\beta_2\Delta\omega \qquad (2.16)$$

In optical communications the bandwidth is frequently measured as $\Delta\lambda$ instead of $\Delta\omega$, where $\Delta\omega = -2\pi c\,\Delta\lambda/\lambda^2$. Adopting $\Delta\lambda$, equation (2.16) can be rewritten as

$$\Delta T = \frac{dT}{d\lambda}\Delta\lambda = z\frac{d}{d\lambda}\left(\frac{1}{v_g}\right)\Delta\lambda = -\frac{2\pi c}{\lambda^2}z\beta_2\Delta\lambda = zD\Delta\lambda \qquad (2.17)$$

where the parameter $D = -2\pi c\beta_2/\lambda^2$ is the *dispersion parameter* (measured in ps/nm/km), and this is the usual way to measure the chromatic dispersion.

Chromatic dispersion essentially consists of two contributions, one due to the material of the fiber and the other to the waveguide structure. Material dispersion is due to the fact that the refraction index of the fiber material depends on frequency, while waveguide dispersion depends on guided propagation and is present even if material dispersion vanishes. In fact the transverse mode shape depends critically on the ratio ρ_c/λ so that, when the operation wavelength changes, the power ratio between the light traveling in the core and in the cladding regions is changed as well. Since the phase velocity in the two regions is different, the group velocity of the mode is wavelength dependent.

The overall dispersion coefficient D can be written as the sum of D_M (material contribution), D_W (waveguide contribution) and of a mixed contribution D_{MW} that is generally quite smaller than D_M and D_W [9]. The value of the waveguide contribution can be negative or positive; on this ground, the guiding structure of the fiber can be designed so to shift the wavelength at which D is zero away from the value of about 1.3 μm typical of step-index fibers. Dispersion-shifted fibers, having a zero dispersion wavelength around 1.55 μm, are based on this principle.

The behavior of the fiber dispersion versus the wavelength is reported in Fig. 2.4 for several different commercial fibers.

Finally the term $\beta_k'''(\omega_0)$ is known as third-order dispersion; since in standard fibers generally $\beta_x'''(\omega_0) = \beta_y'''(\omega_0)$, third-order dispersion is labeled simply as β_3. This term is important around the wavelength at which β_2 is equal to zero. As for second-order dispersion, even third-order dispersion can be written starting from the expression of the optical bandwidth in terms of $\Delta\lambda$ obtaining the third-order dispersion coefficient $D_3 = -2\pi c\beta_3/\lambda^2$ which is measured in ps/nm^2 km.

Typical values for the various parameters related to fiber dispersion are provided in Table 2.1.

2.1.3 Propagation in Single-Mode Fibers with Random Mode Coupling

The uniform fiber model, while accurately describes fiber attenuation and chromatic dispersion, does not reproduce the behavior of PMD in real fibers.

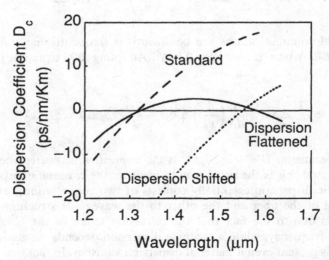

FIGURE 2.4 Fiber dispersion versus wavelength for several kinds of commercial fibers.

The random fluctuations in the fiber structure induce a random power exchange between the modes: Such an effect is generally called *random mode coupling* [10].

Due to random mode coupling, even if the optical power is coupled to the fiber along a single mode, after a short fiber span (typically a few meters) it is randomly distributed between the LP_0 modes. Therefore even signals coupled along a single mode suffer PMD differently than in the case of the uniform fiber.

Fiber propagation in a real fiber can be described by a vector equation similar to equation (2.13) holding for the uniform fiber. In writing this equation, it is useful to separate deterministic effects (e.g., attenuation and chromatic dispersion) from the random mode coupling, thus obtaining [11]

$$A(z,\omega) = [M_1][M_2]A(0,\omega)$$
$$= e^{-\alpha/2z}e^{-i(\beta_2\omega^2 z/2 + \beta_3\omega^3 z/6)}[M_1][I]A(0,\omega) \qquad (2.18)$$

where $[M_1]$ is a z-dependent random matrix taking into account birefringence and random mode coupling, $[I]$ is the unit matrix and $[M_2] = e^{-\alpha/2z}e^{-i(\beta_2\omega^2 z/2 + \beta_3\omega^3 z/6)}[I]$ takes into account attenuation, chromatic dispersion and third order dispersion.

Equation (2.18) can be rewritten as

$$A(z,\omega) = [M_2]\tilde{A}(z,\omega) \qquad (2.19)$$

where $\tilde{A}(z,\omega) = [M_1]A(0,\omega)$ represents the longitudinal field in the absence of attenuation, chromatic dispersion, and third-order dispersion. To describe the impact of random mode coupling and PMD, the analysis can be restricted to $\tilde{A}(z,\omega)$, which is determined by the input field and by the matrix $[M_1]$. This matrix, generally called the Jones matrix, has to be unitary, since $|\tilde{A}(z,\omega)| = |A(0,\omega)|$ in each fiber section. This property implies the following relations: $m_{11} = m_{22}^* = m_1$, $m_{12} = m_{21}^* = m_2$, and $|m_1|^2 + |m_2|^2 = 1$ where m_{ij} are the components of $[M_1]$.

To determine the components of the Jones matrix, the space evolution equations for $\tilde{A}(z,\omega)$ have to be solved. These equations can be obtained starting from the wave equation. In particular, applying the rotating wave and the slowly varying envelope approximations and writing the field as in equation (2.12) the wave equation can be separated in a part determining the transverse mode profile and in a couple of equations governing the evolution of the components of $A(z,\omega)$. The equation for $\tilde{A}(z,\omega)$ is obtained by neglecting in the equations for $A(z,\omega)$ the terms representing attenuation and GVD and phenomenologically introducing the random mode coupling. This results in

$$\frac{d\tilde{A}_x}{dz} = [-i\beta_x(\omega_0) + k_{11}]\tilde{A}_x + k_{12}\tilde{A}_y$$
$$\qquad (2.20)$$
$$\frac{d\tilde{A}_y}{dz} = k_{21}\tilde{A}_x + [-i\beta_y(\omega_0) + k_{22}]\tilde{A}_y$$

where the coupling coefficients k_{hj} are complex random numbers that satisfy the relation $|k_{hj}| \leq 1$. In a conventional fiber, random mode coupling is weak, so $|k_{hj}| \ll \beta_x(\omega_0), \beta_y(\omega_0)$; nevertheless coupling cannot be neglected in describing the changing field polarization.

A statistical analysis of random mode coupling can be carried out starting from equations (2.20), assuming a large ensemble of macroscopically identical fibers. A first important statistical parameter of the fiber ensemble is the autocorrelation function of the random mode coupling process. If all the fibers are macroscopically uniform along the propagation axis and the core material is homogeneous, the random coupling is a stationary process and the autocorrelation function $K(z)$ does not depend on the coupling term under consideration. Therefore it can be written as

$$K(z) = \langle k_{12}(u)k_{21}^*(z+u) \rangle \qquad (2.21)$$

where $\langle \rangle$ indicate the ensemble average. Starting from the autocorrelation function, a characteristic coupling distance is the *correlation length*, L_c, of the random mode coupling. The correlation length is defined as the distance above which $K(z)$ vanishes.

It can be demonstrated [12] that the power coupled from one mode to the other is related to the Fourier transform of the function $K(z)$, called the *polarization holding function*. It is defined as

$$h_p(\xi) = \frac{1}{\pi^2} \int_{-\infty}^{+\infty} K(z) e^{-i\xi z} \, dz \qquad (2.22)$$

In particular, if the optical power P_o is launched into the fiber along the LP_{0y} mode, the average value $\langle P_{0x} \rangle$ of the power coupled to the orthogonal mode after propagation along a fiber of length z is given by

$$\frac{\langle P_{0x} \rangle}{P_o} = \frac{1}{2}[1 - e^{-2h_p(\Delta\beta)z}] \qquad (2.23)$$

where the polarization holding function is evaluated in correspondence of the deterministic birefringence parameters of the fiber $\Delta\beta = \beta_x(\omega_0) - \beta_y(\omega_0)$. The parameter $h_p(\Delta\beta)$, often called the *polarization holding parameter*, represents the inverse distance over which the power is scrambled between the two modes. Such a distance, L_h, often defines the characteristic length of the random mode coupling. For $\Delta\beta$ tending to infinity, $h_p(\Delta\beta)$ tends to zero, so one way to maintain the same state of polarization along a fiber link is to use high birefringence fibers [13]. This solution, however, is not practical due to the high cost of these fibers.

Principal States of Polarization
Polarization mode dispersion in the presence of random mode coupling can be effectively described by introducing the notions of the principal states of polarization (PSPs) and their differential group delay (DGD) [14, 15].

The input PSPs are two orthogonal states of polarization; they are defined in such a way that a narrowband pulse injected into the fiber along each PSP arrives at the fiber output along a fixed SOP that is independent of the pulse's frequency. The output SOPs corresponding to the input PSPs are called output PSPs. Moreover a pulse injected along a PSP does not broaden due to PMD and presents a group delay $\tau_i (i = 1, 2)$. In this formulation, the PMD is defined as the differential group delay between the PSPs, $\Delta\tau = \tau_1 - \tau_2$.

The PSPs can be evaluated starting from the expression of fiber transfer matrix given by equation (2.18). To derive the expression of the PSPs, the procedure introduced in [14] can be used. If the output field has to be independent of ω, its derivative with respect to ω has to be set to zero. The expression of this derivative can be obtained by equation (2.18); setting it to zero results in an eigenvalue equation for the matrix $[M_1]$, whose eigenvectors are the PSPs. In particular, the differential group delay (DGD) $\Delta\tau$ between the PSPs is given by

$$\Delta\tau = 2\sqrt{\left|\frac{dm_1}{d\omega}\right|^2 + \left|\frac{dm_2}{d\omega}\right|^2} \qquad (2.24)$$

The procedure adopted to obtain the PSPs ensures the independence of frequency only at the first order. An expression of the bandwidth in which the PSPs approximation is valid is obtained in [16]; in standard single-mode fibers this bandwidth is of the order of 100 GHz.

The evolution of field polarization in terms of PSPs can be easily described in the space of Stokes parameters, providing a simple and heuristic picture of the various phenomena.

The evolution of the SOP of a monochromatic field can be expressed in Stokes parameters, defined as $S_1 = |E_x|^2 - |E_y|^2$, $S_2 = E_x E_y^* + E_y E_x^*$, $S_3 = j(E_x E_y^* - E_y E_x^*)$, by the following equation [17]:

$$\frac{d\mathbf{S}}{d\omega} = \mathbf{\Gamma} \times \mathbf{S} \qquad (2.25)$$

where $\mathbf{S}$ is the Stokes vector corresponding to the output field and $\times$ indicates the vector product. The vector $\mathbf{\Gamma}$ is proportional to the Stokes vector of an output PSP (arbitrarily called *positive* PSP, while the other PSP is called *negative*), and its module is equal to the DGD $\Delta\tau$. If the vector $\mathbf{\Gamma}$ can be considered constant, varying the field frequency of a quantity $\Delta\omega$, the Poincaré sphere describes a rigid rotation around $\mathbf{\Gamma}$. It means that a point describing the output state of polarization moves on an arc of the circumference; this basic principle is adopted in the measurement of the PMD in the wavelength domain [17].

Another property of the PSPs, easily described in terms of the Stokes parameters is that the PMD of two concatenated fiber pieces can be obtained as a vector sum of the PMD of the two fiber pieces. In particular, the overall

PMD is given by [15]

$$\Delta\tau = \sqrt{\Delta\tau_1^2 + \Delta\tau_2^2 + 2\Delta\tau_1\Delta\tau_2\cos(2\theta)} \qquad (2.26)$$

where $\Delta\tau_1$ and $\Delta\tau_2$ are the DGD of the first and of the second fiber, respectively, and 2θ the angle between the Stokes vector of the output positive PSP of the first fiber and one of the input positive PSP of the second fiber.

Starting from equation (2.26), if we suppose to divide a fiber in many small sections, the PMD evolution can be seen as a random walk; thus the statistics of the PMD can be analytically evaluated. In particular, the probability distribution depends on the ratio between the link length and the characteristic length L_h.

A simple result can be found if the fiber is modeled as a structure with a deterministic birefringence, characterized by a constant DGD per unit length $\Delta\beta'$, plus a weak random mode coupling. In this case, if $z \ll L_h$, the random mode coupling has no effect, and the DGD at the link output is given by $\Delta\tau = \Delta\beta'z$.

On the other hand, if $z \gg L_h$, the PMD has a Maxwellian probability distribution that is written as

$$P(\Delta\tau)^2 = \frac{2\Delta\tau^2}{\sqrt{2\pi\Delta\beta'^6 z^3 L_h^3}} \exp\left(-\frac{\Delta\tau^2}{2\Delta\beta'^2 L_h z}\right), \qquad 0 \le \Delta\tau < \infty \qquad (2.27)$$

The PMD mean value and standard deviation are given by

$$\langle\Delta\tau\rangle = \sqrt{\frac{8zL_h}{3\pi}}\,\Delta\beta', \qquad \sigma_{\Delta\tau} = \sqrt{\frac{(3\pi-8)zL_h}{3\pi}}\,\Delta\beta' \qquad (2.28)$$

It is useful to note that the average PMD increases with the square root of the link length z. On this ground, when the link length is not fixed, the parameter quantifying the PMD is usually defined as $\langle\Delta\tau\rangle/\sqrt{z} = \sqrt{8L_h/3\pi}\,\Delta\beta'$, and it is measured in ps/$\sqrt{\text{km}}$.

It can be demonstrated that if a pulse signal, with a 3 dB width T_0, is launched into the fiber, at the output the pulse presents a 3-dB width T_1 given by [18]

$$\sqrt{T_1^2 - T_0^2} = \tfrac{1}{2}\Delta\tau|\sin(2\theta)|. \qquad (2.29)$$

To completely understand the meaning of equation (2.29), it is to be reminded that not only $\Delta\tau$ is different in fibers belonging to the same ensemble, it changes in the same fiber as a function of the frequency [15] and it fluctuates in times following the environmental conditions [19]; for example, in installed fiber cables $\Delta\tau$ varies in times of the order of 10 minutes. Thus the output pulse width also fluctuates.

Field Depolarization

The slow temporal fluctuation of the fiber Jones matrix induces a slow fluctuation of the polarization of the field at the fiber output and hence a fluctuation of the pulse width due to PMD. In other words, when a pulse is transmitted through the fiber, the light polarization slowly changes along the pulse profile.

Although this phenomenon is different from light depolarization such as can be observed in the sunlight, it can be described analogously by introducing the average Stokes parameters. Obviously in this case the average has to be realized in time, fixing an observation time interval T.

As in the case of standard depolarization, the sum of the squares of the average Stokes parameters is smaller than the field power. The difference between the two quantities increases with the light depolarization and can be used to define the polarization degree of the field as

$$\Delta = \frac{\sqrt{\langle s_1(z,t)\rangle_T^2 + \langle s_2(z,t)\rangle_T^2 + \langle s_2(z,t)\rangle_T^2}}{\langle P_0(z,t)\rangle_T} \tag{2.30}$$

where s_j indicates the jth Stokes parameter and $\langle\,\rangle_T$ the time average in the observation time interval T.

2.2 NONLINEAR EFFECTS IN OPTICAL FIBERS

Nonlinear effects in optical fibers are due either to changes in the refractive index with optical power or to scattering phenomena [20–22]. The power dependence of refractive index is responsible for the Kerr effect. Depending on the shape of the input signal, the Kerr nonlinearity manifests itself by different effects, such as self-phase modulation, cross-phase modulation, and four-wave mixing.

Scattering phenomena are responsible for Brillouin and Raman effects, in which part of the energy of the field is transferred to local phonons. In particular, in Brillouin scattering, acoustic phonons are involved, while in Raman scattering, optical phonons are generated. At high-power levels both of these processes can induce stimulated effects, and the intensity of the scattered light grows exponentially once the incident power exceeds a threshold value.

A fundamental difference between Brillouin and Raman scattering is that the Brillouin-generated phonons are coherent, giving rise to a macroscopic acoustic wave in the fiber; on the contrary, in the case of Raman scattering, the phonons are incoherent and no macroscopic wave is generated. Also the gain bandwidth and the direction of the scattered light are different. In the stimulated Brillouin scattering (SBS) the gain bandwidth is of the order of

20 MHz and the light is scattered in the backward direction, while in the stimulated Raman scattering (SRS) the bandwidth is of the order of some THz and the light is scattered in the forward direction.

2.2.1 Nonlinear Susceptibility

The general equation that describes the optical field evolution in a dielectric medium is given by [20]

$$\nabla^2 \mathbf{E} - \frac{1}{c^2} \frac{\partial^2 \mathbf{E}}{\partial t^2} = -\mu_0 \frac{\partial^2 \mathbf{P(E)}}{\partial t^2} \qquad (2.31)$$

where the polarization $\mathbf{P}$ characterizes the medium and it is a function of the electrical field. In the case of weak nonlinear behavior of the medium, the polarization can be expressed by a Taylor polynomial as

$$\mathbf{P} \approx \varepsilon_0 \{\chi^{(1)} \mathbf{E} + \chi^{(2)} : \mathbf{EE} + \chi^{(3)} \vdots \mathbf{EEE}\} \qquad (2.32)$$

where $\chi^{(1)}$ is the linear susceptibility, : represents the inner tensor product and the second- and third-order tensors $\chi^{(2)}$ and $\chi^{(3)}$ are responsible for the nonlinear behavior. In particular, $\chi^{(2)}$ is responsible for the second harmonic generation, while $\chi^{(3)}$ for the third-order harmonic generation and Kerr effect. In optical fibers the term $\chi^{(2)}$ can be neglected, since it is different from zero only for media without an inversion symmetry at the molecular level; this is not the case of the fiber that is composed by symmetric molecules (SiO_2). As a consequence the nonlinear behavior of the fiber is mainly due to the $\chi^{(3)}$ term of the susceptibility. In particular, the real part of $\chi^{(3)}$ is responsible for the Kerr effect, while the imaginary part for the Raman effect. Due to the presence of a coherent acoustic wave, the Brillouin scattering is very difficult to incorporate into an expression of $\chi^{(3)}$. To analyze SBS, the coupled equations system composed by equation (2.31) and by the propagation equation of the acoustic wave have to be solved [23].

2.2.2 Kerr Effect

The presence of $\chi^{(3)}$ implies that the refractive index depends on the field intensity $I = |\mathbf{E}|^2 / Z_I$, where $Z_I = \sqrt{\mu/\varepsilon}$ is the fiber electromagnetic impedance. In particular, the refraction index can be written as

$$r = r_0 + r_2 \frac{|\mathbf{E}|^2}{Z_I} = r_0 + r_2 I \qquad (2.33)$$

where r_0 and r_2 are the linear and the nonlinear refractive index, respectively. In a generic dielectric media with a third-order optical nonlinearity, the nonlinear index is related to all the components of the tensor $\chi^{(3)}$ and depends on the field polarization. This dependence in optical fibers will be described later and here we will assume a linearly polarized field. In this case only the component $\chi^{(3)}_{xxxx}$ of the $\chi^{(3)}$ tensor is involved and the nonlinear index is simply given by

$$r_2 = \frac{3}{8r_0} \, \text{Re}(\chi^{(3)}_{xxxx}) \qquad (2.34)$$

where Re( ) means the real part. Since $\chi^{(3)}_{xxxx}$ will be used only dealing with the Kerr nonlinearity, from now on, the real part of the third-order susceptibility will be simply indicated with $\chi^{(3)}_{xxxx}$. Typical values of r_2 and $\chi^{(3)}_{xxxx}$ are reported in Table 2.1.

The Kerr nonlinearity gives rise to different effects, depending on the shape of the field injected into the fiber. In the following, the main effects due to the Kerr nonlinearity will be reviewed.

When a composite optical signal (e.g., carrying information in optical transmission systems) feeds the fiber, all the signal frequency components interact, giving rise to a complicated behavior. However, the main phenomenological aspects can be often understood in terms of the elementary phenomena that are described in this Section.

Self-Phase Modulation

Due to r_2, the phase of a signal propagating through the fiber varies with the distance z according to the equation

$$\phi = (r_0 z + \phi_0) + \frac{2\pi}{\lambda} \, r_2 I(t) z \qquad (2.35)$$

where ϕ_0 is the initial phase. The first term in equation (2.35) represents the linear phase shift due to signal propagation; the second term represents the nonlinear phase shift. If the optical signal is intensity modulated, the nonlinear phase shift gives rise to a spurious phase modulation: This effect is called self-phase modulation (SPM).

In the simple case of sinusoidal intensity modulation, even the phase modulation is sinusoidal. The superposition of intensity and phase sinusoidal modulations causes an asymmetry in the sidebands of the output signal, revealing the presence of SPM.

If a more complex signal feeds the fiber, the time-dependent nonlinear phase shift causes a chirp in the transmitted field. In the absence of GVD, the presence of a chirp causes a nonlinear broadening of the signal spectrum. The spectrum broadening depends on the bandwidth and on the shape of the injected signal. In the particular case of a gaussian pulse of $1/e$ bandwidth B_0, the spectral broadening ΔB_0 caused by SPM can be simply evaluated [20] obtaining

$$\Delta B_o = 0.86 \, \frac{2\pi}{\lambda} \, r_2 I_p L B_o \qquad (2.36)$$

where I_p is the peak field intensity and L the fiber length.

It is to be noted that the SPM-induced chirp combines with the linear chirp generated by chromatic dispersion. The effect of the interaction between GVD and SPM depends on the sign of the dispersion β_2. If the fiber dispersion coefficient is positive (normal dispersion), the linear and nonlinear chirps have the same sign, while in the opposite case (anomalous dispersion) they have opposite signs. Thus we can expect that in the first case pulse broadening is enhanced by SPM while in the second case it is reduced.

Cross-Phase Modulation

When N signals having different carrier frequencies propagate into a fiber, the nonlinear phase evolution of the signal at frequency ω_i depends also on the power of the signals at frequencies different from ω_i according to the expression

$$\Delta\phi_i = \frac{2\pi r_2 z}{\lambda} \left[I_i(t) + 2 \sum_{i \neq j} I_j(t) \right] \qquad (2.37)$$

The first term in the square brackets represents the contribution of SPM, while the second term is the contribution from the cross-phase modulation (XPM). From equation (2.37) it is evident that the weight of the XPM is doubled respect to that one of SPM. XPM causes a further nonlinear chirp, thus interacting with fiber GVD as in the case of SPM.

From equation (2.37) it can be deduced that XPM is effective only when the interactive signals are superimposed in time. For example, if two pulses at different frequencies propagate through the fiber, they travel at different group velocities due to dispersion. In this condition XPM occurs only in the time intervals in which the pulses are superimposed. Thus increasing the GVD, the XPM efficiency decreases.

Four-Wave Mixing

When signals at different frequencies propagates through the fiber, besides XPM, another important effect occurs: four-wave mixing (FWM).

FWM is a parametric interaction among waves satisfying a particular phase relationship called *phase matching*. Different phenomena may be originated by FWM process depending on the relation among the interacting frequencies.

Let us assume that three linearly polarized monochromatic waves with angular frequencies $\omega_j (j = 1, 2, 3)$ propagate in the fiber. The overall third-order polarization vector **P** can be obtained by substituting the expression of the field in the third term of equation (2.32). By simplifying the resulting expression, it appears that the polarization vector has several components:

three components have the frequencies of the input fields, the others have an angular frequency ω_k given by

$$\omega_k = \omega_1 \pm \omega_2 \pm \omega_3 \tag{2.38}$$

Thus nonlinear interaction generates four new frequency components of the material polarization vector. If another field is yet present at the frequency ω_k, it is influenced by the nonlinear interaction between the fields at ω_j; this induces, for example, crosstalk in multiwavelength transmission. If no field is injected in the fiber at the frequency ω_k, a new field component is created at this frequency.

The expression of the amplitude of the third-order polarization at the frequency ω_k is

$$P_k = \frac{3\varepsilon_0}{4Z_I} \chi_{xxxx}^{(3)} \{|E_k|^2 E_k + 2(|E_1|^2 + |E_2|^2 + |E_3|^2)E_k$$
$$+ 2E_1 E_2 E_3 \, e^{i\Delta\theta_+} + 2E_1 E_2 E_3^* \, e^{i\Delta\theta_-} + \cdots\} \tag{2.39}$$

where

$$\Delta\theta_+ = (\beta(\omega_1) + \beta(\omega_2) + \beta(\omega_3) - \beta(\omega_k))z - (\omega_1 + \omega_2 + \omega_3 - \omega_k)t$$
$$\Delta\theta_- = (\beta(\omega_1) + \beta(\omega_2) - \beta(\omega_3) - \beta(\omega_k))z - (\omega_1 + \omega_2 - \omega_3 - \omega_k)t \tag{2.40}$$

In expression (2.39) the first term at the second member is responsible for SPM, the second for XPM, the third and fourth terms are responsible for FWM, and the other terms (represented by ... in the equation) represent degenerate FWM interaction (which will be described later) and the interaction among the various newly generated waves at frequency ω_k. Generally, this last interaction is negligible and will not be analyzed.

Effective interaction occurs when the infinitesimal contributions given at different times by the different fiber sections add in phase. This happens when the corresponding phase term $\Delta\theta$ vanishes: This condition is called *phase matching* between the interacting waves.

In the case of the wave at $\omega_k = \omega_1 + \omega_2 + \omega_3$, efficient generation can derive from the term $2E_1 E_2 E_3 \exp(i\Delta\theta_+)$. However, the space-dependent part of $\Delta\theta_+$ does not vanish in normal fibers, so generation at the frequency $\omega_k = \omega_1 + \omega_2 + \omega_3$ can be generally neglected.

In optical fibers, phase matching can be more easily obtained for the wave at $\omega_k = \omega_1 + \omega_2 - \omega_3$, by the term $2E_1 E_2 E_3^* \exp(i\Delta\theta_-)$. In this case the phase-matching condition is

$$\Delta\beta = \beta(\omega_1) + \beta(\omega_2) - \beta(\omega_3) - \beta(\omega_4) = 0 \tag{2.41}$$

and it requires that the fiber dispersion be negligible.

Besides the so-called nondegenerate configuration, in which the interacting waves have different frequencies, FWM can occur in a degenerate configuration when two of the interacting frequencies coincide. This case verifies, for example, when a couple of monochromatic waves at the angular frequencies ω_1 and ω_2 are injected into the fiber. At the fiber output two symmetric sidebands are created by FWM at the frequencies $2\omega_1 - \omega_2$ and $2\omega_2 - \omega_1$ by the interaction terms $2E_1^2 E_2^* \exp(i\Delta\theta_-)$ and $2E_2^2 E_1^* \exp(i\Delta\theta_-)$, which we did not explicitly report in equation (2.39) for brevity.

The power evolution along the fiber of an FWM-generated wave can be obtained by solving the coupled propagation equations of the four interacting waves. In the case where the power exchange among the interacting waves is very low (*not depleted pump condition*), these equations can be analytically solved. Assuming three waves at the fiber input, the power A_k^2 of the newly generated wave at the frequency $\omega_k = \omega_1 + \omega_2 - \omega_3$ is given by [21]

$$A_k^2(z) = 4\eta\gamma^2 d_e^2 L_e^2 A_1^2(z) A_2^2(z) A_3^2(z) e^{-\alpha z} \qquad (2.42)$$

where γ will be defined in equation (2.47).

In equation (2.42), $L_e = [1 - \exp(-\alpha z)]/\alpha$ is the effective interaction length and d_e the so-called degeneracy factor (equal to 3 if degenerate FWM is considered, 6 otherwise). The factor η is the FWM efficiency, and it can be written as

$$\eta = \frac{\alpha^2}{\alpha^2 + \Delta\beta^2} \left[1 + \frac{4e^{-\alpha z}\sin^2(\Delta\beta z/2)}{(1 - e^{-\alpha z})^2} \right] \qquad (2.43)$$

The dependence of the FWM efficiency on the fiber dispersion and on the frequency spacing among the input waves can be made clear by substituting into the expression (2.41) of the phase-matching parameter the expression (2.14) of the propagation constant of the mode. This obtains

$$\Delta\beta = \beta_2|\omega_1 - \omega_3||\omega_2 - \omega_3| \qquad (2.44)$$

In Fig. 2.5 the FWM efficiency is reported in the degenerate case ($\omega_1 = \omega_2$) versus the input signal spacing ($\omega_1 - \omega_3)/2\pi$ for two different values of the fiber dispersion β_2. Other parameters are a fiber loss equal to $0.25\,\text{dB/km}$ and z equal to $17\,\text{km}$, which almost coincides with the effective length.

It is useful to note that in the limit of no attenuation and good phase matching ($\alpha \to 0$ and $\Delta\beta z \ll 1$), equation (2.43) writes $\eta = 4/(\Delta\beta^2 z^2)$, clearly showing that the FWM efficiency is a decreasing function of the phase-matching parameter. Writing $(\omega_1 - \omega_2) = (\omega_2 - \omega_3) = 2\pi\Delta\nu$, $\eta = 1/(\pi^2\beta_2^2\Delta\nu^4 z^2)$ results. From this simple expression the detuning $\Delta\nu_{3\,\text{dB}}$ at which the FWM efficiency is decreased of $3\,\text{dB}$ due to lack of phase matching can be easily evaluated, obtaining

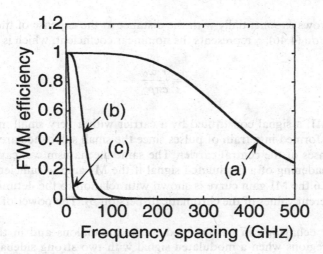

FIGURE 2.5 FWM efficiency in the degenerate case ($\omega_1 = \omega_2$) versus the frequency spacing ($\omega_1 - \omega_3)/2\pi$ for three different values of GVD: (a) $|\beta_2| = 0.01 \, \text{ps}^2/\text{km}$, (b) $|\beta_2| = 1 \, \text{ps}^2/\text{km}$, and (c) $|\beta_2| = 20 \, \text{ps}^2/\text{km}$.

$$\Delta \nu_{3\,\text{dB}} = \sqrt{\frac{\sqrt{2}}{\pi \beta_2 z}} \approx \sqrt{\frac{0.45}{\beta_2 z}} \qquad (2.45)$$

The value of $\Delta \nu_{3\,\text{dB}}$ is often called the *FWM bandwidth*. It is important to note that $\Delta \nu_{3\,\text{dB}}$ was obtained in the limit of $\Delta \beta z \ll 1$, so a rough approximation of the bandwidth in which the FWM efficiency is high can be often obtained starting from the expression of $\Delta \nu_{3\,\text{dB}}$ for an appropriate value of z. This can be the length over which good phase matching is preserved or the length between discontinuity points along the link. For example, in links adopting optical amplifiers, often the amplifiers spacing is used to evaluate $\Delta \nu_{3\,\text{dB}}$.

Modulation Instability
With the term modulation instability (MI) different phenomena are indicated, all generated by the interplay between Kerr effect and chromatic dispersion. All these phenomena generate the exponential growth or attenuation of the sidebands with respect to the carrier of a modulated signal propagating through the fiber [22].

The first form of MI can be observed in the anomalous dispersion region. Injecting in the fiber a strong CW signal with angular frequency ω_0 and power A^2 and a probe with an angular frequency satisfying the condition

$$\omega < \omega_c = \sqrt{\frac{4\gamma A^2}{\beta_2}} \qquad (2.46)$$

the probe grows exponentially with the distance at the expense of the signal at ω_0. In equation (4.46), γ represents the nonlinear coefficient, which is defined as

$$\gamma = \frac{r_2 \omega_0}{c \pi \rho_m^2} \tag{2.47}$$

Due to MI, a signal constituted by a carrier with a very small modulation can be transformed in a train of pulses since the small sidebands are amplified at the expenses of the central carrier. The same mechanism can cause a large spectral broadening of a modulated signal if the MI gain is sufficiently high.

In Fig. 2.6 the MI gain curve is shown with relation to the detuning $\omega - \omega_0$ for two different values of the chromatic dispersion β_2. The power of the CW is 1 mW.

Different behavior can be observed in the anomalous and in the normal dispersion regions when a modulated signal with two strong sidebands and a weak central carrier is injected into the fiber. Amplification of the central carrier will occur at the expense of the sidebands in these regions if the following conditions on the power A_s^2 of the sidebands are satisfied:

$$A_s^2 > \frac{|\beta_2| \Delta \omega^2}{2\gamma}, \qquad \beta_2 < 0$$

$$A_s^2 > \frac{|\beta_2| \Delta \omega^2}{6\gamma}, \qquad \beta_2 > 0 \tag{2.48}$$

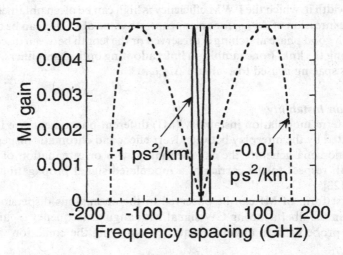

FIGURE 2.6 Gain curve of the modulation instability in the anomalous dispersion region versus the detuning $\omega - \omega_0$ for two different values of the chromatic dispersion β_2.

where $\Delta\omega$ is the modulation frequency, that is, the frequency spacing between a sideband and the carrier. In the case of normal dispersion the MI gain curve shows a peak whose frequency is

$$\Delta\omega_M = \sqrt{\frac{2\gamma A_s^2}{|\beta_2|}} \qquad (2.49)$$

In the case of anomalous dispersion the gain curve attains its maximum value for $\omega = \omega_c/\sqrt{2}$.

2.2.3 Brillouin Scattering

The Brillouin effect can be described as a three-wave interaction among the original optical wave, the acoustic wave, and the so-called Stokes wave. The three waves must satisfy the energy conservation law, which here we write as $\omega_o = \omega_s + 2\pi f_a$, where ω_o is the angular frequency of the injected optical wave, which is named the pump wave, ω_s the angular frequency of the Stokes wave, and f_a the frequency of the acoustic wave, which is about 11.1 GHz in conventional fibers. The maximum effect results if the wave vector obeys the conservation law. In a single-mode optical fiber, optical waves can propagate only along the direction of the fiber axis, so the conservation law is simply a scalar relationship which we write as $\beta_o = \beta_s + \beta_a$, where β_o, β_s, and β_a are the wave numbers of the original optical wave, the Stokes wave, and the acoustic wave, respectively. Since the acoustic wave velocity $v_a \approx 5960$ m/s is by far smaller than the light velocity, $|\beta_a| \approx 2\pi f_a/v_a > |\beta_o| \approx |\beta_s|$. In this case the momentum conservation law has the important consequence that no effect is generated if the Stokes and pump waves propagate in the same direction while maximum efficiency is obtained if the Stokes and the pump propagate in opposite directions. Moreover, due to energy conservation, the Stokes wave frequency is downshifted, with respect to the pump wave, by an amount equal to the acoustic frequency.

The Brillouin effect can be derived mathematically by combining the wave equation describing the propagation of a single-mode electromagnetic field through the fibers with that describing the propagation of elastic waves [23]. Applying the rotating wave and slowly varying approximations, the combined set of wave equations can be solved for a monochromatic optical field. Here the result is that the presence of the pump wave induces a Brillouin exponential gain G_B for a wave traveling in the opposite direction whose dependence on optical frequency, in the undepleted pump approximation, is expressed by

$$G_B(\omega) = \frac{g_B L_e}{\pi \rho_m^2} \frac{A_0^2}{1 + ((\omega - \omega_o + 2\pi f_a)/2\pi\Delta\nu_B)^2} \qquad (2.50)$$

The maximum gain at frequency $\omega_0 - 2\pi f_a$ is proportional to the pump power A_0^2 at the fiber input through the gain coefficient g_B which depends on the fiber material. The shape of the Brillouin gain curve is Lorentzian, and its full width at half maximum is given by $\Delta\nu_B$. It depends on λ^{-2}, and in conventional single-mode fibers, its value is about 16 MHz in the third window and about 23 MHz in the second window. Typical values of the parameters of the Brillouin gain curve are reported in Table 2.1.

If no Stokes wave is directly launched into the fiber, as in the case of a communication system, the presence of a sufficiently high Brillouin gain can amplify the spontaneous emission noise, which is always present in the medium, so as to generate a strong Stokes wave at the expense of the pump. Since the Brillouin gain is proportional to pump power, it becomes evident that with an increase of pump power the Stokes power at the fiber end increases exponentially and a threshold value can be defined for A_s^2. Above this threshold the increase of the input pump power is converted into Stokes power. This effect is shown in Fig. 2.7, where the optical powers exiting the fiber at transmitter and receiver sides are plotted with relation to the injected pump power. Conventionally, the Brillouin threshold A_{th}^2 is fixed to the fiber input power required to obtain a Stokes power at the fiber near end equal to the pump power at the fiber far end. For a fiber length of more than a few kilometers [24], the result is $(g_B A_{th}^2 L_e)/(\pi\rho_m^2) = 21$, obtaining a threshold power within $10 \div 15$ mW.

Previous results are obtained in the case of optical and acoustic monochromatic waves, and they can be applied with a good approximation as long as the Brillouin gain linewidth is substantially wider than the spectral width of the

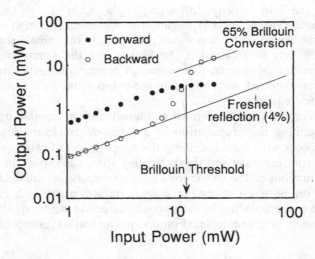

FIGURE 2.7 Optical power at the transmitter and at the receiver end of a 13.6-km-long single-mode fiber for $\lambda = 1.32\,\mu$m (from the paper of Cotter cited in [24]).

pump wave. Generally, this condition is not satisfied in communication systems. An analysis of the Brillouin effect in the presence of nonmonochromatic optical pump is rather difficult, but a simple approximate approach can be applied if the pump spectrum is much wider than the Brillouin gain linewidth [25]. Then the Brillouin gain curve $G_B(\omega)$ can be obtained by the convolution integral

$$G_B(\omega) = \frac{1}{A_s^2} \int_{-\infty}^{\infty} G_B^*(\xi) S_o(\omega - \xi) \, d\xi \qquad (2.51)$$

where the gain curve $G_B^*(\omega)$ is given by equation (2.50) and $S_o(\omega)$ is the pump optical spectrum.

2.2.4 Raman Scattering

Raman scattering is due to the interaction of photons with atomic nuclei through the transition of the nucleus to a vibrational excited level. In other words, it can be described as a scattering effect by incoherent optical phonons. If a sufficiently powerful optical wave (the pump) is launched into the fiber, a new field (the Stokes wave) is generated through stimulated scattering at the expense of the pump power. Since the spontaneous incoherent scattering is isotropous and the maximum gain of the stimulated scattering is obtained along the pump propagation direction, the Stokes wave propagates in the same direction of the pump wave.

The Raman gain curve for a typical single-mode fiber is shown in Fig. 2.8. Unlike Brillouin scattering, the Raman effect is extremely broadband and the

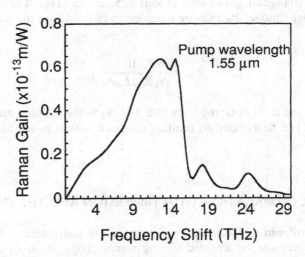

FIGURE 2.8 Raman gain in optical fibers at 1.55 μm (from the book of Agrawal cited in [20]).

Raman-Stokes wave is frequency downshifted of about 12 THz with respect to the pump (about 100 nm in the third window).

To analyze this effect quantitatively, the imaginary part of the third-order nonlinear dielectric susceptibility is introduced in the wave equation. By applying the rotating wave and slowly varying approximations, two coupled equations are obtained for both the pump and the Stokes wave and solved using suitable border conditions. In particular, a Raman threshold power is defined as in the Brillouin case. For a fiber longer than a few kilometers, the pump power at which a pump depletion of 3 dB occurs is given by [26] $(g_R A_{th}^2 L_e)/(b_p \pi \rho_m^2) = 16$, where g_R is the Raman gain coefficient and b_p is 1 if the pump and Stokes waves have the same polarization, and 2 if the polarizations are randomly scrambled such as occurs in a real fiber. The Raman threshold in conventional fibers is of the order of 500 mW for copolarized pump and Stokes wave (about 1W for random polarizations), making the Raman effect negligible for a single modulated signal.

However, because of the extremely wide band of the Raman gain curve, this effect is important for signals composed of numerous waves at different frequencies (a frequency multiplexed signal). In this case the signals at longer wavelengths are amplified by the Raman effect at the expense of those at shorter wavelengths, and a relevant degradation of the worst channel can occur even when the overall optical power is quite smaller than the Raman threshold.

A general result can be obtained by approximating the Raman gain curve versus the pump-Stokes separation as a triangle with a maximum at $\Delta\omega = \Omega_M$ and zero for $\Delta\omega < 0$ and $\Delta\omega > \Omega_M$ [27]. Since the worst channel is that at the lowest frequency ($\Delta\omega = 0$), the transmitted channels are all within the bandwidth of the triangular gain curve (about 125 nm, 15 THz). Thus the Raman power limit per channel P_R causing a depletion of 0.5 dB on the worst channel is given by

$$P_R = \frac{8.7 \cdot 10^{15}}{N_c B_{opt} L_{eff}} \tag{2.52}$$

where B_{opt} is the total optical bandwidth and N_c is the channel number. Thus Raman effect can be avoided by limiting the peak optical power along the link below P_R.

2.3 SIGNAL PROPAGATION WITH DISPERSION AND THE KERR EFFECT

Since both Brillouin and Raman nonlinearities are characterized by a threshold, these effects can be avoided during signal propagation by limiting the signal power. The behavior of Kerr effect is completely different, and it must be taken into account in analyzing propagation through long fiber links.

2.3.1   Nonlinear Schrödinger Equation

Whenever PMD can be neglected or we are not interested in the polarization evolution of the signal, the optical field can be assumed to be linearly polarized along the fiber. A scalar propagation equation can be derived from the wave equation (2.31) by introducing third-order susceptibility and applying slowly varying envelope and rotating wave approximations. The resulting equation, known as the nonlinear Schrödinger equation, is written as

$$\frac{\partial A}{\partial z} + \beta_1 \frac{\partial A}{\partial t} + \frac{i}{2} \beta_2 \frac{\partial^2 A}{\partial t^2} + \frac{\alpha}{2} A = i\gamma |A|^2 A$$

$$+ \frac{1}{6} \beta_3 \frac{\partial^2 A}{\partial t^3} - a_1 \frac{\partial(|A|^2 A)}{\partial t} - a_2 A \frac{\partial |A|^2}{\partial t} \tag{2.53}$$

where $A(z, t)$ is the field complex envelope ($|A|^2$ is measured in W) and

$$\beta_j = \frac{d^j \beta(\omega)}{d\omega^j} \bigg|_{\omega_0} \tag{2.54}$$

The term a_1 is responsible for the self-steepening effect of the pulse edge, and it is approximately given by $2\gamma/\omega_0$ [28], while the term a_2 takes into account the Raman effect and can be approximately written as $i\gamma T_R$, where T_R is related to the Raman gain curve. In optical communications the self-steepening effect can be neglected. Even the Raman effect can be avoided by limiting the signal power and the total signal bandwidth. As a consequence, for our purposes, both a_1 and a_2 can be eliminated by equation (2.53), obtaining an equation that takes into account only the linear effects (chromatic dispersion and attenuation) and Kerr nonlinearity.

Equation (2.53) in fact can be rewritten in simpler form by making some variable transformations. The first step is to adopt the reference frame that moves at the average group velocity. This can be done by the variable transformation

$$\bar{t} = \frac{t - z}{v_g} = t - \beta_1 z \tag{2.55}$$

The second step is to define a set of dimensionless variables by the transformation

$$U(z, t) = \frac{A(z, \bar{t})}{\sqrt{P_0}}, \quad \tau = \frac{\bar{t}}{T_0}, \quad \zeta = \frac{z|\beta_2|}{T_0^2} \tag{2.56}$$

where P_0 is the peak power of the signal at the fiber input.

Since Raman and self-steepening effects are neglected, by the transformations (2.55) and (2.56) the propagation equation (2.53) can be written as

$$\frac{\partial U}{\partial \zeta} + \frac{i}{2} \operatorname{sgn}(\beta_2) \frac{\partial^2 U}{\partial \zeta^2} + \frac{1}{2} \frac{L_D}{L_a} U = i \frac{L_D}{L_{NL}} |U|^2 U + \frac{1}{6} \frac{L_D}{L_D'} \frac{\partial^3 U}{\partial \tau^3} \qquad (2.57)$$

where

$$L_D = \frac{T_0^2}{|\beta_2|}, \quad L_{NL} = \frac{1}{\gamma P_0}, \quad L_D' = \frac{T_0^3}{|\beta_3|}, \quad L_a = \frac{1}{\alpha} \qquad (2.58)$$

are the dispersion length, the nonlinear length, the third-order dispersion length, and the attenuation length, respectively. The parameter T_0 is an arbitrary constant, which is generally set equal to some temporal characteristic value of the input signal, for example, the pulse 3 dB width (full width half maximum, or FWHM), the pulse rise time, and so on.

From (2.57) it can be seen that the propagation equation depends on four characteristic lengths. Intuitively each characteristic length can be considered as the fiber length over which the corresponding phenomenon generates visible effects.

Starting from the intuitive meanings of L_D and L_{NL}, linear propagation regime occurs when $L_D/L_{NL} \ll 1$. In this case signal evolution can be simply determined by adopting the Fourier transform method. In the opposite case, the nonlinear terms in equation (2.57) cannot be neglected.

As a first example of propagation in linear regime, let us analyze the evolution of a gaussian pulse with $1/e$ duration equal to T_0. If the fiber length L is shorter than L_D', the third-order chromatic dispersion can be neglected. In this case the output field has the same shape of the input one, even if its phase turns out to be time dependent and pulse duration is greater. The dependence on time of the phase can be characterized by the chirp $\delta\omega = -\partial\phi/\partial t$, which in this case is written as

$$\delta\omega = \frac{2\operatorname{sgn}(\beta_2)(z/L_D)}{1 + (z/L_D)^2} \frac{t}{T_0^2} \qquad (2.59)$$

The time broadening of the pulse is given by

$$\frac{T_{0,\text{out}}}{T_0} = \sqrt{1 + \left(\frac{z}{L_D}\right)^2} \qquad (2.60)$$

From equation (2.60) it is evident that the magnitude of the pulse broadening due to GVD can be measured by the ratio between the fiber length and the dispersion length. In particular, the dispersion length represents the distance at which a gaussian pulse exhibits a broadening factor equal to $\sqrt{2}$.

In the presence of an input chirp, the pulse feeding the fiber can be written as

$$U(0,\tau) = \exp\left[-(1+i\Psi)\frac{\tau^2}{2T_0^2}\right] \tag{2.61}$$

where Ψ is the chirp parameter and the $1/e$ spectral width of the pulse, is given by

$$B_0 = \frac{\sqrt{1+\Psi^2}}{T_0} \tag{2.62}$$

In this case the pulse at the fiber output is also a chirped gaussian pulse, and the temporal broadening factor is modified as

$$\frac{T_{0,\text{out}}}{T_0} = \sqrt{\left(1+\frac{\text{sgn}(\beta_2)\Psi z}{L_D}\right)^2 + \left(\frac{z}{L_D}\right)^2} \tag{2.63}$$

Equation (2.63) shows that in the presence of a chirp with $\Psi\beta_2 < 0$, the pulse can have a temporal narrowing. In particular, there is a value of the chirp parameter such that the input pulse width is reproduced at the fiber output.

The third-order chromatic dispersion must be taken into account when L is of the same order or greater than L_D'. In this case the pulse shape shows a temporal asymmetric modification. In particular, input gaussian pulses do not maintain their shapes. In this case an input chirp can compensate the broadening induced by the second-order chromatic dispersion but not that induced by the third-order dispersion.

The signal behavior gets very complex when both dispersion and Kerr nonlinearity are present. Neglecting the third-order dispersion, an interesting signal behavior can be observed when $L_D \sim L_{NL}$, that is, when both dispersion and the Kerr effect are significant.

In Section 2.2.2, in discussing SPM, it was shown that when a signal propagates in a nonlinear regime, a spurious phase modulation can arise if the signal is amplitude modulated. The resulting chirp, obtained by deriving equation (2.35), is always positive on the pulse leading edge (when the derivative of the pulse instantaneous power is positive) and negative on the pulse trailing edge (when the derivative is negative).

In the anomalous dispersion region, if opportune conditions are present, the chirp induced by the chromatic dispersion can be compensated by the chirp induced by the Kerr effect, which has opposite sign. This behavior has been widely studied for soliton signals, though a partial compensation of chromatic dispersion by means of the Kerr effect has also been observed for other signals [29, 30].

To better show the effect of the nonlinear compensation of the GVD, in Fig. 2.9 we report the evolution of a train of almost squared pulses

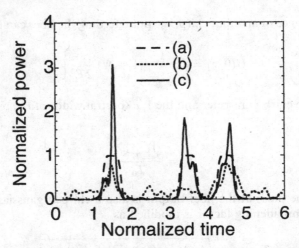

FIGURE 2.9 Evolution of a train supergaussian pulses ($m = 3$) in absence of loss, and third-order chromatic dispersion in a link 6000 km long with $\beta_2 = 1\,\text{ps}^2/\text{km}$. (a) The input signal, (b) the output signal in the absence of Kerr effect, and (c) the output signal in the presence of the Kerr effect, with $\gamma = 2.7\,\text{W}^{-1}\,\text{km}^{-1}$ and a peak power of 4 mW.

(supergaussian pulses with index three) in the absence of loss and third-order chromatic dispersion. The curves of the figure are obtained by numerically solving equation (2.57) using the method described in Appendix A1 and assuming $\alpha = 0$.

In the figure the fiber is 6000 km long, and its dispersion parameter is $\beta_2 = -1.28\,\text{ps}^2/\text{km}$ ($D = 1\,\text{ps/nm/km}$). Curve *a* refers to the input signal, *b* to the output signal in the absence of the Kerr effect, and *c* to the output signal in the presence of the Kerr effect with $\gamma = 2.7\,\text{W}^{-1}\,\text{km}^{-1}$ and $P_0 = 4\,\text{mW}$. In the absence of the Kerr effect, the signal shows a very strong time broadening due to chromatic dispersion. Conversely in the presence of Kerr nonlinearity, the output signal experiences almost no time broadening even if the pulse shape is distorted. Thus, in Fig. 2.9, a partial compensation of chromatic dispersion can be observed. To attain complete compensation, a particular pulse shape has to be selected, one called a *fundamental soliton* or simply a *soliton*.

2.3.2 Solitons in Optical Fibers

The word *soliton* was used for the first time to describe certain waves observed in water channels. In 1965 it was used to describe pulse envelopes in a dispersive nonlinear medium that behave like particles, since they maintained their shape also in the presence of perturbations. Soliton pulses in optical fiber were studied for the first time in 1973 by Hasegawa and Tapper and were experimentally observed in 1980 by Mollenauer, Stolen, and Gordon [31]. Up to the present day much research effort has been devoted to the study of solitons and their application in telecommunication systems because they have the peculiar behavior of preserving their shape during propagation. A detailed theoretical analysis of soliton propagation in optical communication links will be presented in Chapter 5. The purpose of this section is to introduce the intuition behind the soliton concept and to review some main characteristics.

In absence of loss and third-order chromatic dispersion, equation (2.57) obtains

$$\frac{\partial U}{\partial \zeta} + \frac{i}{2}\operatorname{sgn}(\beta_2)\frac{\partial^2 U}{\partial \tau^2} = iN_s|U|^2 U, \quad N_s = \frac{L_D}{L_{NL}} \tag{2.64}$$

where for pulse propagation, the dispersion length is evaluated by assuming that $T_0 = T_s$, with T_s the $1/e$ width of the transmitted pulse. Equation (2.64) can be solved for a generic input pulse by the so-called inverse scattering method [32] that transforms the nonlinear differential equation in a linear integral equation. Generally, the solution is not a trivial task, but a particularly simple case exists: The case where N_s is an integer number and β_2 is negative (propagation in the anomalous dispersion region of the fiber). In this case the solutions of the nonlinear Schrödinger equation are called solitons.

The general theory of solitons is outside the scope of this book; here will be discussed mainly the *fundamental* soliton solution corresponding to $N_s = 1$.

The expression of the fundamental soliton can be written as

$$U(\zeta, \tau) = \operatorname{sech}(\tau)\exp\left(\frac{i\zeta}{2}\right) \tag{2.65}$$

where sech(τ) is the secant hyperbolic function. It is a *bell-like pulse* whose 3-dB width T_F is approximately equal to $1.763\, T_s$. The amplitude shape of the fundamental soliton is shown in Fig. 2.10, where it is compared with a gaussian pulse of the same width.

By equation (2.65) the fundamental soliton preserves its shape during propagation, since the space variable x appears only in the expression of the soliton phase. This property derives from the interplay between GVD and the Kerr effect: The result obtained in evaluating the SPM-induced chirp by deriving equation (2.35) is exactly opposite that of the GVD-induced chirp.

When N_s is higher than one, the soliton solutions have a more complex shape. In the particular case where the initial pulse shape is a hyperbolic secant, that is,

$$U(0, \tau) = N_s \operatorname{sech}(\tau) \tag{2.66}$$

the soliton experiences a periodic evolution and reassumes its initial shape at multiple distances of the soliton period

$$z_0 = \frac{\pi}{2} L_D \tag{2.67}$$

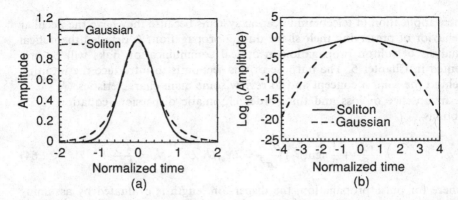

FIGURE 2.10 Pulse shape of a fundamental soliton (*continuous line*) and of a gaussian pulse (*dashed line*) with the same 3 dB width. In (a) the shapes are shown using a linear y scale; a logarithmic scale is instead used in figure (b) to evidence the different behavior of the tails of the two pulses.

The solutions are often called high-order solitons. The evolution of a fourth-order soliton ($N_s = 4$) is shown in Fig. 2.11.

Since N_s must have a fixed value, to obtain a soliton (fundamental or high order), the pulse launched into the fiber must have an hyperbolic secant shape, and its power and duration must be correlated. Using equations (2.58) and the definition of N_s, the following relation is obtained for the fundamental soliton

$$P_0 = \frac{|\beta_2|}{\gamma T_s^2} \approx \frac{3.11|\beta_2|}{\gamma T_F^2} \tag{2.68}$$

An important characteristic of the fundamental soliton is that it is a stable solution of the nonlinear Schrödinger equation. This means that a perturbed soliton recovers its initial shape in its evolution along a fiber.

A simple case showing soliton stability is when the pulse at the fiber input has the hyperbolic secant shape but not the soliton power [33]. The pulse can be expressed as

$$U(0, \tau) = (1 + \varepsilon)\text{sech}[(1 + \varepsilon)\tau] \tag{2.69}$$

where ε is a small quantity. In solving the propagation problem for this input pulse, it can be seen that it asymptotically tends to a fundamental soliton by adjusting its width if $|\varepsilon| < 0.5$. In particular, the value of $T_F(z)$ asymptotically tends to $1.763T_s/(1 + \varepsilon)$, and hence the correct relation among the soliton power and width is reached as well. As a result, if $\varepsilon < 0$, the pulse broadens; if $\varepsilon > 0$, the pulse narrows.

For the pulse fed to the fiber that is not hyperbolic secant in shape, both its width and its shape will change while the pulse tends to a soliton. The evolution

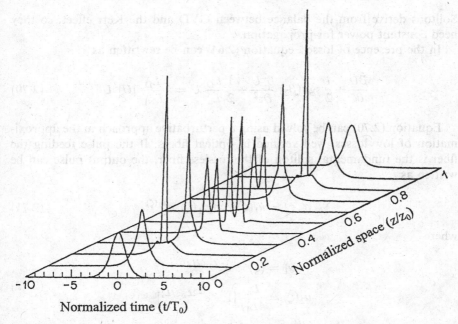

FIGURE 2.11 Periodic evolution of a third-order soliton (from the book of Agrawal cited in [20]).

producing this result is quite complex, and a part of the pulse power is lost in radiative waves in the process. The width and power of the soliton and the distance needed to complete the soliton formation will nevertheless depend on the initial pulse shape. In general, regular pulses are transformed in a soliton after a distance of a few soliton periods.

Ideal solitons are obtained by solving equation (2.64) and neglecting phenomena such as third-order dispersion, attenuation, and Brillouin scattering. However, soliton stability guarantees that as long as these phenomena can be viewed as small perturbations, solitons will exist.

In the following section we will briefly discuss the impact on soliton propagation of fiber attenuation, an initial spurious chirp, and the presence of other solitons in the fiber. The impact of the other nonlinear effects and of third-order dispersion are generally less important in this context. A complete discussion of the impact of other nonlinear effects can be found in [33], while a discussion of the effect of the third-order dispersion is presented in [34].

Solitons in Lossy Fibers

Since solitons are solutions of equation (2.64), under a mathematical point of view they exist only in a medium without losses. This can be easily understood:

Solitons derive from the balance between GVD and the Kerr effect, so they need constant power for propagation.

In the presence of losses, equation (2.64) can be rewritten as

$$\frac{\partial U}{\partial \zeta} + \frac{i}{2}\,\text{sgn}(\beta_2)\,\frac{\partial^2 U}{\partial \tau^2} + \frac{1}{2}\frac{L_D}{L_a}\,U = i\,\frac{L_D}{L_{NL}}\,|U|^2 U \qquad (2.70)$$

Equation (2.70) can be solved using a perturbative approach in the approximation of low losses, well verified in optical fibers. If the pulse feeding the fiber is the fundamental soliton of the lossless fiber, the output pulse can be written as

$$U(\zeta,\tau) = u_1(\zeta)\text{sech}[u_1(\zeta)]\,e^{iu_2(\zeta)} \qquad (2.71)$$

where

$$u_1 = (\zeta) = e^{-L_D\zeta/L_a}$$
$$u_2(\zeta) = \frac{L_a}{4L_D}\,[1 - e^{-2L_D\zeta/L_a}] \qquad (2.72)$$

The solution still obtains a hyperbolic secant pulse and reduces to the fundamental soliton for L_a tending to infinity, that is, for $\alpha = 0$. The pulsewidth increases with distance to compensate the decreasing of the pulse peak power and to maintain the relation between width and power that is characteristic to solitons. The 3-dB pulsewidth $T_F(\xi)$ is

$$T_F(\zeta) = T_F(0)\,e^{L_D\zeta/L_a} = T_F(0)\,e^{\alpha z} \qquad (2.73)$$

where $T_F(0)$ is the input pulse width.

Obviously the pulsewidth does not increase exponentially for arbitrarily long distances since at increasing distance the pulse peak power decreases to a linear propagation regime. When linear propagation is attained, the pulse broadening is mainly due to chromatic dispersion, which linearly increases $T_F(\zeta)$ with distance. The perturbative solution is expected to be accurate so long as the attenuation can be considered a weak perturbation to the propagation of the soliton, that is, provided that $z \ll L_a$.

In the opposite condition ($z \gg L_D$) the pulse broadening is linear. However, the broadening rate is lower than that expected if only GVD is present. This is a form of residual compensation due to the nonlinear effect.

Unwanted Chirp in Soliton Sources
Soliton sources can be affected by unwanted chirps as in the case of gain-switched semiconductor lasers [35]. The presence of a chirp is detrimental for soliton propagation because it disturbs the balance between the chirp due to

SPM and that generated by GVD. However, since the soliton is a stable solution of the propagation equation, it is expected that, within some tolerance, the input pulse will evolve into a soliton, eventually by losing some energy.

An input pulse that is linearly chirped can be written as

$$U(0, \tau) = \operatorname{sech}(\tau) \, e^{-i\Psi\tau^2/2} \tag{2.74}$$

The propagation equation can be solved with the border condition (2.74) obtaining that a limit value exists for the chirp parameter: $\Psi_{\lim} \approx 1.64$. For $\Psi < \Psi_{\lim}$ the propagating pulse evolves into a soliton loosing energy into the process; for $\Psi > \Psi_{\lim}$ the soliton is not formed and complex pulse evolution is observed. If signal propagation produces a soliton, the percentage of energy lost in the process increases as Ψ approximates $\Psi_{\lim}$; for example, the asymptotic soliton has only the 83% of the initial energy if $\Psi = 0.5$, while its energy decreases to the 62% of the initial value if $\Psi = 0.8$.

Soliton Interactions

Ideally the tails of a soliton pulse extend over all the time axis. Thus the distance between the soliton and any another pulse transmitted through the fiber should be as large as possible. However, this condition cannot be respected in transmission systems where the information is carried by a train of pulses. Thus it is important to study the evolution of two solitonlike pulses that are injected into the fiber with a given time separation [36].

The interplay between dispersion and the Kerr effect causes an interaction force between adjacent solitons, so soliton spacing has to be fixed above a critical value if the pulses are to reach the fiber end almost unperturbed. The interactions among the solitons and their implications for transmission systems adopting optical amplifiers will be discussed in detail in Chapter 5; in this section only some preliminary results will be presented.

For a pair of solitons injected into the fiber, the input normalized envelope can be written as

$$U(0, \tau) = \operatorname{sech}(\tau - q_0) + \xi \operatorname{sech}(\tau + q_0) \, e^{i\phi} \tag{2.75}$$

where $2q_0$ is the initial normalized soliton spacing ξ is the ratio between the amplitudes of the two solitons, and ϕ is the relative phase.

The propagation equation can be solved by the boundary condition (2.75): It can be found that the two pulses propagating into the fiber interact, experiencing a complex evolution. The type of interaction critically depends on the relative amplitude and phase. For example, in the case of $\xi = 1$ and $\phi = 0$ (two in-phase solitons with equal energy), the soliton interaction is periodical, with period ζ_p. In the first half of the period the solitons reduce their distance up to the point of collapse. In the second part they separate again and reach their initial distance at the end of the period. A simple expression of the period in the case $q_0 \gg 1$ based on the perturbation theory [37] is obtained as

$$\zeta_p = \frac{\pi}{2} e^{q_0} \qquad (2.76)$$

From expression (2.76) it is evident that the period of the interaction increases with the initial separation q_0, exponentially tending to infinity.

If initially the solitons do not have the same amplitude, the interacting behavior may be similar, but the point of collapse is not reached. Rather, on increasing ξ, the soliton separation experiences smaller and smaller fluctuations. For example, the separation does not vary more than 10% over the period if $\xi = 1.1$ and $q_0 > 4$.

The evolution is even more different if the solitons are not in phase at the fiber input.

2.4 Polarization Evolution in the Presence of the Kerr Effect

Until now we have neglected the field polarization. When both polarization modes are considered, the nonlinear phase variation of one component depends also on the power of the other.

The nonlinear field propagation in a birefringent fiber with random mode coupling can be described by two coupled nonlinear Schrödinger equations [20, 38]:

$$\frac{\partial A_x}{\partial z} + \beta'_x \frac{\partial A_x}{\partial t} + \frac{i}{2} \beta_2 \frac{\partial^2 A_x}{\partial t^2} + \frac{\alpha_x}{2} A_x - \frac{1}{6} \beta_3 \frac{\partial^3 A_x}{\partial t^3} + i\kappa(z) A_y \exp(-i\Delta\beta z)$$

$$= i\gamma(|A_x|^2 + \tfrac{2}{3}|A_y|^2) A_x + \frac{i\gamma}{3} A_x^* A_y^2 \exp(-2i\Delta\beta z)$$

$$\frac{\partial A_y}{\partial z} + \beta'_y \frac{\partial A_y}{\partial t} + \frac{i}{2} \beta_2 \frac{\partial^2 A_y}{\partial t^2} + \frac{\alpha_y}{2} A_y - \frac{1}{6} \beta_3 \frac{\partial^3 A_y}{\partial t^3} - i\kappa^*(z) A_x \exp(+i\Delta\beta z)$$

$$= i\gamma(|A_y|^2 + \tfrac{2}{3}|A_x|^2) A_y + \frac{i\gamma}{3} A_y^* A_x^2 \exp(+2i\Delta\beta z) \qquad (2.77)$$

where $\kappa(z)$ is the random coupling coefficient, $\Delta\beta = \beta_x(\omega_0) - \beta_y(\omega_0)$ is the birefringence parameter and α_x, α_y are the losses for the two LP_0 modes. The average loss is given by $\alpha = (\alpha_x + \alpha_y)/2$, while $\Delta\alpha = (\alpha_x - \alpha_y)$ is the polarization dependent loss (PDL).

Equations (2.77) are worthy of some comments. First of all, XPM occurs not only between signals at different optical frequencies but also between the different polarization components of the same signal due to the term $2i\gamma|A_y|^2 A_2/3$ in the first equation and to the term $2i\gamma|A_x|^2 A_y/3$ in the second. Moreover another interaction arises among the polarization components not of the XPM type. It is given by the term $i\gamma A_y^* A_y^2 \exp(-2i\Delta\beta z)/3$ in the first equation and by an analogous term in the second; this interaction depends on the fiber birefringence.

Due to the complex nonlinear coupling between the polarization modes, a description of field propagation that takes into account polarization evolution is much more complex that the scalar model provided by equation (2.57). First it is useful to determine the framework in which the scalar model can be adopted even in the presence of birefringence and PMD.

Comparing the result obtained by the numerical solutions of equations (2.57) and (2.77), yields that the scalar model can be adopted where it is not important to include the field SOP. In general, the scalar model given an overestimation of the nonlinear effect, and using this model is equivalent to assuming parallel and linearly polarized fields. Generally this overestimation is small, but it can be significant in certain situations such as when considering FWM among signals at different angular frequencies [39]. In the limit case of two waves with orthogonal polarization no FWM product exists, contrary to what is foreseen by the scalar theory.

Various attempts have been made to take into account the weaker effect of fiber nonlinearity due to birefringence and PMD within a scalar model. One simple way is to introduce a phenomenological constant smaller than one in front of the nonlinear term, as done in [40]. In this reference this constant was deduced by the interpolation of some experimental results and set to $\frac{1}{2}$.

However, in analyzing the performances of optical transmission systems, this phenomenological method is not generally used: A worst-case analysis is preferred, mainly because in many cases the worst case is unlikely but not impossible.

In this book we will adopt the worst-case approach, following much of the technical literature. We will use equations (2.77) only where polarization evolution is important while in the other cases we will adopt the scalar model given by equation (2.57).

Another consideration is that since equations (2.77) are stochastic equations due to the random mode coupling term, they present complex problems, both from analytical and numerical points of view [41–44]. Simulations of signal propagation in the presence of random mode coupling are provided in Appendix A1. In this section two conditions that enable equations (2.77) to be analytically simplified are described.

2.4.1 Propagation with Negligible Random Mode Coupling

The major way in which equations (2.77) can be analytically simplified is if the random mode coupling can be neglected and equations (2.77) treated as deterministic equations. In fact this condition can occur if the polarization evolution is largely determined by a high fiber birefringence or if random mode coupling is prevented by a fiber structure that removes the polarization mode degeneracy.

Once the random mode coupling is neglected by setting $\kappa(z) = 0$, there are two conditions by which equations (2.77) can be notably simplified. The first case is the low birefringence condition, where $\Delta\beta T_0^2 \ll \beta_2$. Since the effect of the birefringence is low in the dispersion length, the term $\exp(-2i\Delta\beta z)$ can be set to one. The second case is the high birefringence condition, where $\Delta\beta T_0^2 \gg \beta_2$. Here the exponential term, $\exp(-2i\Delta\beta z)$, fluctuates rapidly in the dispersion length, so its average effect tends to vanish. Neglecting this term and third-order dispersion, the propagation equations can be written as

$$\left(\frac{\partial A_x}{\partial z} + \frac{\beta_x'}{2}\frac{\partial A_x}{\partial t}\right) + \frac{i}{2}\beta_2\frac{\partial^2 A_x}{\partial t^2} + \frac{\alpha_x}{2}A_x = i\gamma(|A_x|^2 + \tfrac{2}{3}|A_y|^2)A_x$$

$$\left(\frac{\partial A_y}{\partial z} + \frac{\beta_y'}{2}\frac{\partial A_y}{\partial t}\right) + \frac{i}{2}\beta_2\frac{\partial^2 A_y}{\partial t^2} + \frac{\alpha_y}{2}A_y = i\gamma(|A_y|^2 + \tfrac{2}{3}|A_x|^2)A_y$$

(2.78)

Since the high birefringence approximation is generally satisfied by optical fibers for telecommunications, equations (2.78) can be used both where the effects of random mode coupling can be neglected and where they are removed as in the case of polarization maintaining fibers.

2.4.2   Propagation with Negligible PMD

A second way in which the vectorial propagation equations can be simplified and transformed into deterministic equations is that of strong random mode coupling in the absence of PMD and PDL. This condition can be analytically expressed as follows:

- $\Delta\beta' = 0$ (absence of PMD),
- $\alpha_x = \alpha_y$ (absence of polarization dependent losses),
- $L_h \ll L_{NL}$ (the characteristic length of the random mode coupling is much shorter than the nonlinear length).

Here, if the fiber length is at least of the order of the nonlinear length, the output polarization is independent of the input polarization, and the field can have every polarization state with the same probability. Exploiting this property, the propagation equations (2.77) can be averaged over the Poincaré sphere, and the resulting equations, called Manakov equations, are [45]

$$\frac{\partial A_x}{\partial z} + \frac{i}{2}\beta_2\frac{\partial^2 A_x}{\partial t^2} + \frac{\alpha}{2}A_x - \frac{1}{6}\beta_3\frac{\partial^3 A_x}{\partial t^3} = \gamma\tfrac{8}{9}(|A_x|^2 + |A_y|^2)A_x$$

$$\frac{\partial A_y}{\partial z} + \frac{i}{2}\beta_2\frac{\partial^2 A_y}{\partial t^2} + \frac{\alpha}{2}A_y - \frac{1}{6}\beta_3\frac{\partial^3 A_y}{\partial t^3} = i\gamma\tfrac{8}{9}(|A_y|^2 + |A_x|^2)A_y$$

(2.79)

Even in the case $\Delta\beta' \neq 0$, equations (2.79) can be used in any instance where the polarization evolution is largely driven by random mode coupling and the PMD has little relevant effect. This is generally the case of optical fibers for telecommunications. The condition of weak PMD can be analytically expressed by $\Delta\tau \ll T$, where T is the time duration of the propagating pulse and $\Delta\tau$ the delay between the polarization states.

The Manakov equations are quite useful in the analysis of soliton propagation, as shown in Section 2.4.4.

2.4.3 Self-Trapping Effect

The presence of PMD and random mode coupling induce a depolarization of the field propagating along an optical fiber. This depolarization can be observed in a linear regime, as described in Section 2.1.3. Due to the Kerr-induced interaction between the field polarization components, the field depolarization in a nonlinear propagation regime is different from that observed in a linear regime. In particular, in the anomalous dispersion region, the so-called self-trapping effect can take place.

Self-trapping occurs when, due to the Kerr nonlinearity, an attractive force arises between the polarization components of a pulse propagating along the optical fiber, suppressing or at least reducing the temporal broadening due to PMD [41].

An example of self-trapping is reported in Fig. 2.12, where we consider a train of almost squared pulses that are transmitted through 5000 km of lossless fiber with $\beta_2 = -1.28\,\mathrm{ps^2/km}$ and a constant DGD per unitary length $\Delta\beta' = 0.025\,\mathrm{ps/km}$. Random mode coupling is neglected. The normalized output field components, $|U_1|$ and $|U_2|$, are shown with relation to the normalized time (t/T, where $T = 200\,\mathrm{ps}$ is the pulse duration). The input signal was launched with a linear input SOP presenting an angle of $\pi/4$ with respect to the x axis. In Fig. 2.12a the output signal is assumed to have linear propagation and both the broadening due to the GVD and splitting due to the PMD are shown. Figure 2.12b shows the output signal for an input peak power equal to 0.53 mW: Nonlinear compensation of the GVD arises due to propagation in the anomalous dispersion region. Above a certain power threshold, self-trapping occurs, as shown in Fig. 2.12c, for an input power of 0.72 mW.

Self-trapping can also be detected in the presence of random birefringence. The example in Fig. 2.13 shows the behavior of a train of NRZ pulses at the output of a 1000-km-long link in the absence of losses. The following different cases are considered: a the presence of PMD only with a value of $1\,\mathrm{ps}/\sqrt{\mathrm{km}}$ ($L_c = 500\,\mathrm{m}$), b the presence of PMD and GVD with $\beta_2 = -1.28\,\mathrm{ps^2/km}$, and c the presence of PMD, GVD, and the Kerr effect ($\gamma = 2.7\,\mathrm{W\,km^{-1}}$) with an input peak power of 1 mW.

Comparing the different curves, we see that the interplay between GVD and PMD induces a pulse broadening that can be reduced by the self-trapping.

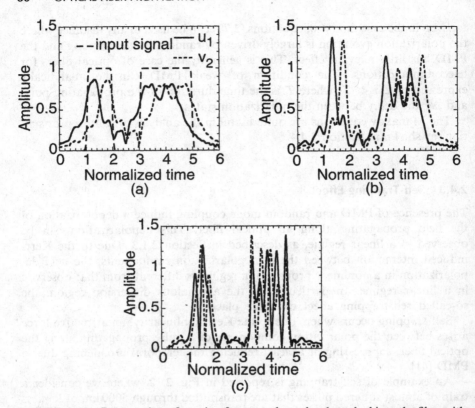

FIGURE 2.12 Propagation of a train of rectangular pulses launched into the fiber with a linear SOP at $\pi/4$ with respect to the x-axis. Normalized output field components, $|U_1|$ and $|U_2|$, versus the normalized time for a pulse width of 200 ps. In (a) the output signal is reported in linear propagation regime; (b) shows the output signal for an input peak power equal to 0.53 mW; and (c) for an input power of 0.72 mW.

Moreover, from the analysis of the polarization degree of the light, it can be seen that pulse depolarization is also reduced if nonlinear propagation occurs.

2.4.4 Soliton Propagation in Fibers Showing Birefringence and Random Mode Coupling

In the presence of birefringence but absence of random mode coupling, the propagation through the fiber can be described by equations (2.78). In this case stable solitons can propagate due to the self-trapping effect, as demonstrated in [42,43].

Self-trapping is effective in maintaining soliton stability even in the presence of PMD as shown in [44]. If self-trapping occurs, the soliton not only maintain

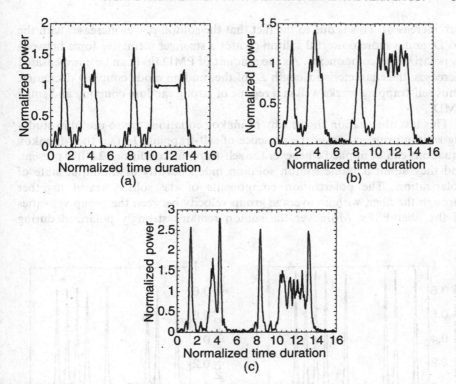

FIGURE 2.13 Behavior of a 10-Gbit/s-NRZ signal at the output of a 1000-km-long link in the absence of loss. Case (a) refers to the presence of only PMD with a value of $1 \, ps/\sqrt{km}$ ($L_c = 500 \, m$), case (b) to the presence of PMD and GVD with $\beta_2 = -1.28 \, ps^2/km$, and (c) to the presence of PMD, GVD, and the Kerr effect ($\gamma = 2.7 \, (W \, km)^{-1}$) with an input peak power of 1 mW.

its width but also remains highly polarized despite any polarization changes during propagation.

If self-trapping does not occur, stable solitons cannot exist in the fiber. The presence of both birefringence and random coupling will cause the soliton pulses to split into different pulses. This effect can be observed by simulating both low and high random mode coupling between polarization modes.

The condition for effectively raising self-trapping during soliton propagation can be deduced by numerically solving equations (2.77) [43]. The result verifies that self-trapping will occur under the following condition:

$$\Delta\beta'\sqrt{L_h} < 0.3\sqrt{\beta_2} \tag{2.80}$$

From equation (2.80) it can be seen that as the value of β_2 increases, the amount of PMD the soliton can compensate increases (the maximum value of

$\Delta\beta'$ increases). This is due to the fact that the soliton power increases with the GVD, and a more powerful soliton creates a stronger attractive force between its polarization components. As the amount of PMD that can be compensated increases, the characteristic length L_h of the random mode coupling decreases. Thus self-trapping works well in a regime of strong random coupling and small PMD.

This last observation shows why Manakov equations are so useful in studying soliton propagation in the presence of self-trapping. In fact the Manakov equations can be applied whenever the self-trapping condition (2.80) is present, and they admit a stable soliton solution independently of the input state of polarization. The polarization components of the soliton travel together through the fiber, with an average group velocity between the group velocities of the fiber PSPs. Moreover the soliton remains strongly polarized during

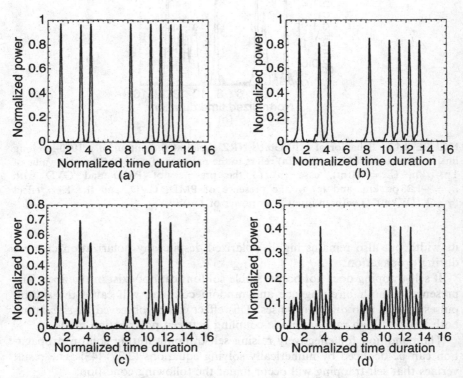

FIGURE 2.14 Propagation of soliton signals in link 1000 km long in the absence of loss for different values of PMD. The pulses have a 3 dB width of 20 ps and a time distance $\Delta T = 100$ ps, corresponding to a bit-rate of 10 Gbit/s. The fiber parameters are $\beta_2 = -1.28$ ps^2/km and $\gamma = 2.7$ (W km)$^{-1}$. Case (a) refers to a PMD of 0.3 ps/$\sqrt{\text{km}}$), case (b) to a PMD of 1 ps/$\sqrt{\text{km}}$, case (c) to a PMD of 2.3 ps/$\sqrt{\text{km}}$, and case (d) to a PMD of 3.3 ps/$\sqrt{\text{km}}$.

propagation. Although the soliton polarization changes in the different sections of the fiber, it remains always the same over all the pulse profile [45]. Thus, averaging over the pulse duration, the soliton has a polarization degree Δ equal to one. On the other hand, in the case of nonsoliton pulses, the pulse polarization changes along the pulse profile after a sufficiently long propagation distance, as noted in Section 2.1.3.

The peak power of the soliton resulting from the solution of the Manakov equations is equal to

$$P_0 = \frac{9}{8} \frac{|\beta_2|}{\gamma t_s^2} \tag{2.81}$$

From equation (2.81) it can be noted that the presence of random mode coupling slightly increases the soliton power with respect to the scalar case. The self-trapping effect for soliton signals is graphically shown in Fig. 2.14 where we plot the signal at the output of a 1000-km-long link in the absence of loss for different values of PMD. The pulses have a 3 dB width of 20 ps and a time distance of 100 ps. The fiber parameters are $\beta_2 = -1.28 \, \text{ps}^2/\text{km}$ and $\gamma = 2.7 \, (\text{W km})^{-1}$. The different cases refer to *a* a PMD of $0.3 \, \text{ps}/\sqrt{\text{km}}$, *b* a PMD of $1 \, \text{ps}/\sqrt{\text{km}}$, *c* a PMD of $2.3 \, \text{ps}/\sqrt{\text{km}}$, and *d* a PMD of $3.3 \, \text{ps}/\sqrt{\text{km}}$.

REFERENCES

1. The possibility of realizing low-loss silica fibers, whose loss is mainly limited by Rayleigh scattering, was first proposed in the fundamental paper
   K. C. Kao and G. A. Ockam, Dielectric fiber surface waveguides for optical frequencies, *Proceedings IEE* 113: 1151–1158, 1966.
   The first low-loss fiber operating in the first-transmission window (with an attenuation of 4 dB/km at a wavelength of 0.87 μm) was realized at Corning Glass and presented by
   R. D. Maurer, First european electro-optics market and technology, Geneva, September 12–15, 1972.

2. The first fiber operating in the second- and third-transmission windows was proposed by
   M. Horiguchi and H. Osanai, Spectral losses of low-OH-content optical fibers, *Electronics Letters* 12: 310–312, 1976.

3. One of the first papers on dispersion-shifted fibers is
   L. G. Cohen, C. Lin, and W. G. French, Tailoring zero chromatic dispersion into the 1.5–1.6 μm low-loss spectral region of single-mode fibres, *Electronic Letters* 15: 334–335, 1979.
   For a good review paper in the structure and fabrication of dispersion-shifted fibers, see
   B. J. Ainslie and C. R. Day, A review of single-mode fibers with modified dispersion characteristics, *IEEE-Journal of Lightwave Technology* 4: 967–979, 1989.

4. An analysis of the propagation in dielectric waveguides starting from the Maxwell equations can be found in the classic book
A. W. Snyder and J. D. Love, *Optical Waveguide Theory*, Chapman and Hall, London, 1983.
For a more specific analysis of optical fibers, see
H. G. Unger, *Planar Optical Waveguides and Fibers*, Clarendon Press, Oxford, 1977.

5. A complete analysis of fiber propagation based on geometric optics can be found in
J. Arnaud, *Beams and Fiber Optics*, Academic Press, New York, 1976.

6. D. Gloge, Weakly guiding fibers, *Applied Optics* 10: 2252–2258, 1971.

7. For a detailed analysis of the LP approximation and of its relation with the theory yielding to HE and EH modes, see
D. Marcuse, *Light Transmission Optics*, Van Nostrand-Reinhold, New York, 1972.

8. The characteristic parameters of the optical fibers can be defined in different, equivalent ways. Fundamental references, adopted when needed in this books, are the ITU-T recommendations Rec G.651, Rec G.652, Rec G.653.

9. It is beyond the scope of this book to provide a detailed analysis of the various dispersion contributions. For complete expressions of the dispersion coefficient D, see, for example,
M. J. Adams, *An Introduction to Optical Waveguides*, Wiley, New York, 1981.
L. B. Jeunhomme, *Single-Mode Fiber Optics*, Dekker, New York, 1981.

10. One of the first analysis concerning the polarisation evolution along an optical fiber can be found in
A. Simon, R. Ulrich, Evolution of polarisation along a single mode fiber, *Applied Physics Letters* 31: 517–520, 1977.
For a complete description of random mode coupling, see the following papers:
M. Moniere and L. Jeunhomme, Polarization mode coupling in long single-mode fibers, *Optical and Quantum Electronics* 12: 449–461, 1980.
I. P. Kaminow, Polarization in optical fibers, *IEEE-Journal on Quantum Electronics* 17: pp. 15–21, 1981.
A detailed analytical theory is presented in ref. [12].

11. For this decomposition of the propagation matrix of an optical fiber, see, for example,
F. Curti, B. Daino, G. De Marchis, and F. Matera, Statistical treatment of the evolution of the principal states of polarization in single mode fibers, *IEEE-Journal of Lightwave Technology* 8: 1162–1166, 1990.

12. An analytical demonstration of this property in the context of a general detailed description of the coupled mode theory, both in the deterministic and in the random case, can be found in
D. Marcuse, *Theory of Dielectric Optical Waveguides*, Academic Press, New York, 1974, ch. 5.

13. This technique is described in
S. C. Rasleigh, W. K. Burns, R. P. Moeller, and R. Ulrich, Polarization holding in birefringent single-mode fibers, *Optics Letters* 7: 40–42, 1982.

14. The principal states of polarization were first introduced in
C. D. Poole and R. E. Wagner, Phenomenological approach to polarization dispersion in single mode fibers, *Electronics Letters* 22: 1029–1030, 1986.

15. The properties of the principal state of polarization hold only if the optical spectrum of the transmitted signal is not excessively wide; an extensive study has been carried out to determine the validity limits of this representation. A paper in which experimental results concerning undersea transmission are reported is the following one:

C. D. Poole, N. S. Bergano, R. E. Wagner, and H. J. Schutte, Polarization dispersion and principal states in a 147 km undersea lightwave cable, *IEEE-Journal of Lightwave Technology* 6: 1185–1190, 1988.

Results for terrestrial cables are presented in

F. Curti, B. Daino, Q. Mao, F. Matera, and C. G. Someda, Concatenation of polarisation dispersion in single mode fibers, *Electronics Letters* 25: 290–292, 1989. A theoretical analysis of a single fiber and a fiber cascade can be found, for example, in

F. Curti, B. Daino, G. De Marchis, and F. Matera, Statistical treatment of the evolution of the principal states of polarization in single mode fibers, *IEEE-Journal of Lightwave Technology* 8: 1162–1166, 1990.

A rigorous and complete investigation of all the PSP's properties can be found in the papers

C. D. Poole and G. J. Foschini, *IEEE-Journal of Lightwave Technology* 9: 1439 (1991).

F. Matera and C. G. Someda, Random birefringence and polarization dispersion in long single-mode fibers, in

C. G. Someda and G. Stegeman (eds.), *Anisotropic and Nonlinear Optical Waveguides*, Elsevier Science, Amsterdam, 1992.

16. See, for example, S. Betti, F., Curti, B. Daino, G. De Marchis, E. Iannone, and F. Matera, Evolution of the bandwidth of the principal states of polarisation in single mode fibers, *Optics Letters* 16: 467–469, 1991.

17. D. Andresciani, F. Curti, B. Daino, and F. Matera, Measurement of the group-delay difference between the PSP on a low-birefringence terrestrial fiber cable, *Optics Letters* 12: 144–146, 1987.

18. G. De Marchis and E. Iannone, Polarization dispersion in single-mode optical fibers: A simpler formulation based on pulse envelope propagation, *Microwave and Optical Technology Letters* 4: 75–77, 1991.

19. C. De Angelis, A. Galtarossa, G. Gianello, F. Matera, and M. Schiano, Time evolution of polarization mode dispersion in long terrestrial links, *IEEE-Journal of Lightwave Technology* 10: 552–555, 1992.

20. Nonlinear processes arising in fiber propagation are analyzed in the books:

G. P. Agrawal, *Nonlinear Fibre Optics*, Academic Press, New York, 1989, in which an exhaustive description of the Brillouin, Raman and Kerr effects in optical fibers is shown.

A. Yariv, *Quantum Electronics*, 3rd ed., Wiley, New York, 1989, where the nonlinear effects are analyzed from a macroscopic point of view as well as in relation to the microscopic behaviour.

An analysis of nonlinear optical effects from a microscopic point of view can be found in

M. Schubert and B. Wilhelmi, *Nonlinear Optics and Quantum Electronics*, Wiley, New York, 1986.

21. For an analysis of the efficiency of four-wave mixing, see, for example,
N. Shibata, R. P. Braun, and R. G. Waarts, Phase-mismatch dependence of
efficiency of wave generation through four-wave mixing in a single-mode optical
fiber, *IEEE-Journal of Quantum Electronics* 23: 1205–1210, 1987.
R. W. Tkach, A. R. Chraplyvy, F. Forghieri, A. H. Gnauck, and R. M. Derosier,
Four-photon mixing and high-speed WDM systems, *IEEE-Journal of Lightwave
Technology* 13: 841–849, 1995.

22. For the phenomenon of the Modulation Instability in the anomalous dispersion
region, see
D. Anderson and M. Lisak, Modulation instability of coherent optical-fibre trans-
mission signals, *Optics Letters* 9: 468–470, 1984.
G. P. Agrawal, Modulation instability induced by cross phase modulation, *Physical
Review Letters* 59: 880–883, 1987.
For a discussion of the effect in the normal dispersion region, see
A. Barthelemy and R. La Fuente, Unusual modulation instability in fibers with
normal dispersion, *Optical Communications* 73: 409–412, 1989.
A further analysis on the spatial instability effects can be found in
G. Cappellini and S. Trillo, Third order three-wave mixing in single-mode fibers:
Exact solutions and spatial instability effects, *Journal of Optical Society of America*
B 8: 824–838, 1991.

23. The coupled equations describing the Brillouin scattering are derived in
N. M. Kroll, Excitation of hypersonic vibrations by means of photoelastic coupling
of high-intensity light waves to elastic waves, *Journal of Applied Physics* 36: 34–43,
1965.
A more complex approach based on an analysis of the medium dynamic using the
continuity equation, the Navier-Stokes equation, and the energy balance equation
is reported in
D. Pohl and W. Kaiser, Time-resolved investigation of stimulated Brillouin scattering in transparent and absorbing media: Determination of phonon lifetimes,
Physical Review B 1: 31–43, 1970.

24. A definition of the Brillouin threshold can be found, for example, in
D. Cotter, Stimulated Brillouin scattering in monomode optical fibre, *Journal of
Optical Communications* 4: 10–19, 1983.
A. Cosentino and E. Iannone, SBS threshold dependence on line coding in phase
modulated coherent optical systems, *Electronics Letters* 25: 1459–1460, 1989.

25. Brillouin scattering in the case of a wide linewidth pump, starting from the con-
volution formula (3.30), is analyzed by
M. Denariez and G. Bret, Investigation of Rayleigh wings and Brillouin stimulated
scattering in liquids, *Physical Review* 171: 160–170, 1968.
An analysis of the potentialities and the limits of the use of such equations in the
performance evaluation of transmission systems is reported in
S. Betti, G. De Marchis, and E. Iannone, *Coherent Optical Communications
Systems*, Wiley, New York, 1995, sec. 7.5.

26. For a definition of the Raman threshold, see, for example,
J. Auyeung and A. Yariv, Spontaneous and stimulated Raman scattering in long
low loss fibers, *IEEE-Journal of Quantum Electronics* 14: 347–413, 1978.
Y. Ohmori, Y. Sasaki, M. Kawachi, and T. Edahiro, Fibre-length dependence of

critical power for stimulated Raman scattering, *Electronics Letters* 17: 593–594, 1981.

27. A. R. Chraplyvy, Optical power limits in multi-channel wavelength-division-multi-plexed systems due to stimulated Raman scattering, *Electronics Letters* 20: 58–59, 1984.

28. G. P. Agrawal, *Nonlinear Fibre Optics*, Academic Press, New York, 1989, sec. 2.3.

29. The nonlinear compensation of chromatic dispersion for a train of squared pulses (an NRZ signal) was studied for the first time in
J. P. Hamaide and P. Emplit, Limitations in long haul IM/DD optical fiber systems caused by chromatic dispersion and nonlinear Kerr effect, *Electronics Letters* 26: 1451–1453, 1990.
A similar effect was found for a phase modulated signal in
E. Iannone, F. S. Locati, F. Matera, M. Romagnoli, and M. Settembre, Nonlinear evolution of ASK and PSK signals in repeaterless fiber links, *Electronics Letters* 28: 1902–1903, 1992.
A comparison between the cases of intensity- and phase-modulated signals is reported in
F. Matera and M. Settembre, Nonlinear compensation of chromatic dispersion for phase- and intensity-modulated signals in the presence of amplified spontaneous emission noise, *Optics Letter* 19: 1198–1200, 1994.
Nonlinear compensation of chromatic dispersion for nonsoliton signals is also experimentally demonstrated in
A. H. Gnauk, R. W. Tkach, and M. Mazurczyk, Interplay of chirp and self phase modulation in dispersion-limited optical transmission systems, *Proceedings of ECOC'93*, pp. 105–108, Montreux, Switzerland, September 12–16, 1993.

30. D Andeson, Variational approach to nonlinear pulse propagation in optical fibers, *Physical Review* A 727: 3135–3145, 1983.
M. Florjanczyk and R. Tremblay, RMS width of pulses in nonlinear dispersive fibers: Pulses of arbitrary initial form with chirp, *IEEE-Journal of Lightwave Technology* 13: 1801–1806, 1995.

31. Solitons in optical fibers were introduced by
A. Hasegawa and F. Tapper, Transmission of stationary nonlinear optical pulses in dispersive dielectric fibers I. anomalous dispersion, *Applied Physics Letters* 23: 142–144, 1973.
The first observation of such optical pulses is presented in
L. F. Mollenauer, R. H. Stolen, and J. P. Gordon, Experimental observation of picosecond pulse narrowing and solitons in optical fibers, *Physical Review Letters* 45: 1095–1098, 1980.

32. For a rigorous derivation of all the soliton solutions from the nonlinear propagation equation, see, for example,
G. L. Lamb, *Element of Soliton Theory*, Wiley, New York, 1980.
R. K. Dodd, J. C. Eilbeck, J. D. Gibbon, and H. C. Morris, *Solitons and Nonlinear Wave Equations*, Academic Press, New York, 1984.

33. Most of the discussion of the soliton stability in this section was inspired by
G. P. Agrawal, *Nonlinear Fibre Optics*, Academic Press, New York, 1989, ch. 5.

34. Y. Kodama, M. Romagnoli, S. Wabnitz, and M. Midrio, Role of third order dispersion on soliton instabilities and interactions in optical fibers, *Optics Letters* 19: 165–167, 1994.

35. For experiments in which the transmitted soliton train is obtained by gain switching a semiconductor laser and the obtained pulses are processed to eliminate the unwanted chirp, see
E. Yamada, K. Suzuki, and M. Nakazawa, 10 Gbit/s single-pass soliton transmission over 1000 km, *Electronics Letters* 27: 1289–1291, 1991.
K. Iwatsuki, K. Suzuki, S. Nishi, and M. Saruwatari, 40 Gbit/s optical soliton transmission over 65 km, *Electronics Letters* 28: 1821–1822, 1992.
For ways to avoid the unwanted chirp, such as external modulation, see
L. F. Mollenauer, M. J. Neubelt, M. Haner, E. Lichtman, S. G. Evangelides, and B. M. Nyman, Demonstration of error-free soliton transmission at 2.5 Gbit/s over more than 14000 km, *Electronics Letters* 27: 2055–2056, 1991.

36. Soliton interaction has been widely studied, both for theoretical interest in the behavior of this particlelike pulse and for its practical importance. See, for example,
J. P. Gordon, Interaction forces among solitons in optical fibers, *Optics Letters* 8: 596–598, 1983.
For the perturbative approach, see
V. I. Karpman and V. V. Solov'ev, A perturbational approach to the two-solitons system, *Physica* 3D: 487–502, 1981.
For experimental results, see
F. M. Mitschke and L. F. Mollenauer, Experimental observation of interaction forces between solitons in optical fibers, *Optics Letters* 12: 355–357, 1987.

37. For an exposition of the theory of small perturbations in the case of solitons, see, for example,
V. I. Karpman, Perturbation theory for solitons, *Soviet Physics*, JETP, 46: 281–291, 1977.
D. J. Kaup, Perturbation theory for solitons in optical fibers, *Physical Review* A 42: 5689–5694, 1990.
An expansion to the second order of the perturbative theory can be found in
D. J. Kaup, Second order perturbations for solitons in optical fibers, *Physical Review* A, 44: 4582–4590, 1991.

38. For a mathematical derivation of the coupled nonlinear Schrödinger equations that describe propagation in the presence of birefrigence and the Kerr effect, see, for example,
Y. R. Shen, *Principle of Nonlinear Optics*, Wiley, New York, 1984.

39. K. Inoue, Polarization effect on four-wave mixing efficiency in a single-mode fiber, *IEEE-Journal of Quantum Electronics* 28: 883–891, 1992.

40. A. R. Chraplyvy, Limitations on lightwave communications imposed by optical-fiber nonlinearities, *IEEE-Journal of Lightwave Technology* 8: 1548–1557, 1990.

41. For an example of self-trapping in the case of the propagation of a squared pulses train (NRZ signal), see
F. Matera and M. Settembre, Compensation of the polarization mode dispersion by means of the kerr effect for non return to zero signals, *Optics Letters* 20: 28–30, 1995.

42. L. F. Mollenauer, K. Smith, J. P. Gordon, and C. R. Menyuk, Resistance of solitons to the effects of polarization dispersion in optical fibers, *Optics Letters* 9: 1218–1221, 1989.

43. C. R. Menyuk, Stability of solitons in birefringent optical fibers. II. Arbitrary amplitudes, *Journal of Optical Society of America* B 5: 392–402, 1988.

44. The experimental demonstration of stable soliton propgation in the presence of random mode coupling is reported in
    L. F. Mollenauer, K. Smith, J. P. Gordon, and C. R. Menyuk, Resistance of solitons to the effect of polarization dispersion in optical fibers, *Optics Letters* 14: 1219–1221, 1989.
    See also
    P. K. Wai, C. R. Menyuk, and H. H. Chen, Stability of solitons in randomly varying birefringent fibers, *Optics Letters* 16: 1231–1233, 1991.

45. The Manakov equations were reported in the papers
    S. V. Manakov, On the theory of two-dimensional self-focusing of electromagnetic waves, *Zh. Eksp. Teor. Fiz.* 65: 505–516, 1973, or S. V. Manakov, *Sov. Phys. JEPT* 38: 248–253, 1974.
    A fundamental work in which the Manakov equations are derived is
    P. K. Way and C. R. Menyuk, Polarization mode dispersion, decorrelation, and diffusion in optical fibers with randomly varying birefringence, *IEEE-Journal of Lightwave Technology* 14: 148–157, 1996.
    Other works in which the Manakov equations are considered are in the papers
    P. K. Wai, C. R. Menyuk, and H. H. Chen, Stability of solitons in randomly varying birefringent fibers, *Optics Letters* 16: 1231–1233, 1991.
    S. G. Evangelides, L. F. Mollenauer, J. P. Gordon, and N. S. Bergano, Polarization multiplexing with solitons, *IEEE-Journal of Lightwave Technology* 10: 28–35, 1992.

CHAPTER THREE

Optical Amplifiers

Optical amplifiers are key devices in the design of high-capacity optical networks. Their fundamental characteristics will be reviewed in this chapter. In its simplest form an optical amplifier consists of a two-level system where population inversion is induced by the presence of a resonant optical pump [1]. Of course no real optical amplifier can be accurately modeled as a two-level system, but consideration of this model is useful in pointing out in a simple way the most important characteristics of real amplifiers.

More specific information will be provided about erbium doped fiber amplifiers (EDFAs) and semiconductor optical amplifiers (SOAs). These optical amplifiers are widely used both in communication links and systems for optical routing and switching.

In the case of doped fiber amplifiers, since early 1960s, when the first $1.06\,\mu m$ laser [2] and amplifier [3] were realized by Nd^{3+} doped silica fibers, silica glass has been known to be a host for rare earth ions. The amazing acceleration toward the development of this technology followed the first high-gain low-noise amplifier in the third transmission window, realized by doping a silica fiber with Er^{3+} ions [4]. Indeed, among the main fiber optical amplifier types such as Brillouin amplifiers [5] and Raman amplifiers [6], only erbium-doped amplifiers have been developed at an industrial level, and commercial devices are nowadays available. For this reason only this type of amplifier will be analyzed in this chapter, though the discussed physical phenomena are shared by all the rare-earth-doped fiber amplifiers [7].

Semiconductor amplifiers, on the other hand, relay on the same principle as semiconductor lasers. Although the strong nonlinearity of semiconductor amplifiers degrades the performances of transmission systems, the state-of-the-art semiconductor devices seem to be the most interesting amplifiers for transmission in the second transmission window (around $1.3\,\mu m$). Their nonlinear behavior can be exploited to obtain devices for all optical processing of optical signals. For example, all optical TDM demultiplexers [8], wavelength converters [9], and logical gates [10] have been proposed based on different nonlinearities in semiconductor amplifiers.

3.1 CONTINUOUS WAVE BEHAVIOR OF THE TWO-LEVEL AMPLIFIER

A block diagram of an ideal optical amplifier is shown in Fig. 3.1. Amplification in an inverted medium takes place due to stimulated emission inside the medium. However, spontaneous emission always occurs, even during the amplification process. This simple observation anticipates the fact that the propagation equation of an optical field into an amplifying medium must be composed of two parts: coherent amplification and incoherent photon production caused by spontaneous emission. The spontaneously emitted photons are amplified besides the signal photons so that, at the amplifier output, an amplified spontaneous emission (ASE) noise is present. Since spontaneous emission always takes place, ASE noise is unavoidable and does not depend on the amplifier temperature. This is one of the most important differences between optical and electrical amplifiers where amplifier noise is of thermal origin and can be reduced by lowering the amplifier temperature.

The characteristics of the ASE noise can be deduced by a complete quantum model of the two-level amplifier so that this effect can be phenomenologically added to the classical Maxwell equations.

Assuming the signal field to be linearly polarized and perfectly monochromatic, the propagation equation for the signal envelope $A(z)$ can be obtained from the Maxwell equations, including noise, by the slowly varying, rotating wave approximation:

$$\frac{dA}{dz} = [\tfrac{1}{2}g(\nu, P) + ir(\nu)]A + n(t, z) \tag{3.1}$$

where $g(\nu, P)$ is the local gain, which in general depends on the signal power

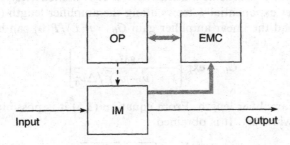

FIGURE 3.1 Block diagram of a generic optical amplifier; IM = inverted medium, OP = optical pump, EMC = electronic monitoring and control circuit.

$P(z) = |A(z)|^2$ due to gain saturation, $r(\nu)$ the refraction index, and $n(t, z)$ the ASE noise process.

The local gain, assuming an homogeneously broadened system [1], can be written as

$$g(\nu, P) = \frac{g_0}{1 + \dfrac{(\nu - \nu_0)^2}{\Delta \nu_0^2} + \dfrac{P}{P_{sat}}} \quad (3.2)$$

where g_0 is the unsaturated local gain, ν_0 the resonance frequency, $\Delta \nu_0$ the 3-dB local gain bandwidth, and P_{sat} the local saturation power. It must be emphasized that $\Delta \nu_0$ and P_{sat} refer to the local gain; the macroscopic amplifier bandwidth and saturation power have different values that will be evaluated in the following.

The noise process can be approximated as a gaussian, zero-mean, broadband noise, whose autocorrelation function can be written as

$$\langle n(t, z) n^*(t + \Delta t, z + \Delta z) \rangle = h\nu \left(\frac{\mu_1}{\mu_1 - \mu_0} \right) g\delta(\Delta z)\delta(\Delta t) \quad (3.3)$$

where g is the local gain whose expression is given in equation (3.2). The atomic population of the fundamental level is indicated with μ_0 and that of the upper level with μ_1. From equation (3.2) it is evident that at a given frequency, the noise power depends on two factors, the amplifier gain G and the so-called inversion factor $\mu_1/(\mu_1 - \mu_0)$. In particular, the locally generated noise power increases linearly both with G and with the inversion factor.

To describe the average signal evolution, equation (3.1) can be averaged to eliminate the noise term and added to its conjugate equation. The result is an equation for the average signal power, which is given by

$$\frac{dP}{dz} = g(\nu, P)P \quad (3.4)$$

Equation (3.4) can be simply solved assuming $P(0) \ll P_{sat}$. In this case the local gain can be approximately assumed to be independent of the signal power, the amplifier works as a linear device. It is immediately obtained that the signal power exponentially grows along the amplifier length (as shown by equation 3.4) and the linear amplifier gain $G_0 = P(L)/P(0)$ can be written as

$$G_0 = \exp\left[\frac{g_0 L}{1 + (\nu - \nu_0)^2/\Delta \nu_0^2} \right] \quad (3.5)$$

where L is the amplifier length. From equation (3.5) it is possible to evaluate the 3-dB bandwidth B_o. It is obtained

$$B_o = \Delta \nu_0 \sqrt{\left(\frac{\ln 2}{g_0 L - \ln 2} \right)} \quad (3.6)$$

From equation (3.6) it is clear that the macroscopic bandwidth of the amplifier B_o is smaller than the microscopic gain bandwidth $\Delta\nu_0$. This is an intuitive result if the amplifier is seen as a cascade of small amplifier sections of bandwidth $\Delta\nu_0$, since a cascade of filters has a smaller bandwidth than a single filter of the cascade.

In the general case equation (3.4) can be integrated by variable separation obtaining an implicit equation for $P(z)$ as a function of the input signal power $P(0)$. Exploiting the definition of G_0 and $G = P(L)/P(0)$, and assuming $\nu = \nu_0$, the general solution of equation (3.4) can be written as

$$(1 - G)\,\frac{P(0)}{P_{sat}} = \ln\left(\frac{G}{G_0}\right) \tag{3.7}$$

In Fig. 3.2 the amplifier gain is plotted versus the input power for a linear gain $G_0 = 20\,\mathrm{dB}$ and a microscopic saturation power $P_{sat} = 10\,\mathrm{mW}$. From the figure it is clear that the input power at which the gain is reduced by 3 dB, that is, the macroscopic input saturation power P_s^{in}, does not coincide with P_{sat}. Indeed, it is obtained by substituting G with $G_0/2$ in equation (3.7):

$$P_s^{in} = \frac{2\ln(2)P_{sat}}{(G_0 - 2)} \tag{3.8}$$

Hence, the output saturation power is $P_s^{out} = G_0 P_s^{in}/2$.

Once the output power is evaluated, the output phase $\phi(z)$ is simply obtained by averaging equation (3.1) and substituting $A(z)$ with $P(z)^{1/2}\,e^{i\phi(z)}$.

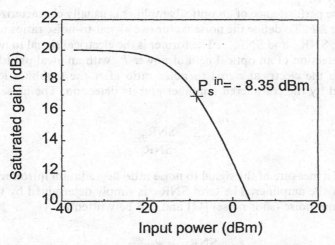

FIGURE 3.2   Amplifier gain versus the input power ($G_0 = 20\,\mathrm{dB}$, $P_{sat} = 10\,\mathrm{mW}$).

The noise power at the amplifier output in a given bandwidth can be evaluated starting from the expression (3.3) of the noise autocorrelation function. Equation (3.3) holds for a rigorously white noise, whose power spectral density is constant over all the frequency axis and whose overall power is infinite. This is why the equal time autocorrelation function is infinite. Without defining in detail the shape of the filter needed to measure the noise power in a bandwidth B_n, it can be noted that the filtering does not affect the signal, which is perfectly monochromatic, but it limits the value of the equal time noise autocorrelation function. This value can be assumed to be equal to

$$\langle n(t,z)n^*(t,z+\Delta z)\rangle = h\nu\left(\frac{\mu_1}{\mu_1-\mu_0}\right)g(\omega,z)B_n\delta(\Delta z) \qquad (3.9)$$

where the indirect dependence of the local gain g on the coordinate z due to the dependence of g on $P(z)$ is evidenced.

It is worth noting that in practice the noise spectrum of amplifiers does not extend over an infinite bandwidth. Thus, if a filter is not present at the amplifier output, B_n represents the real amplifier noise bandwidth.

Starting from the formal solution of equation (3.1) and from the expression of the noise correlation function, the field envelope variance σ_n^2 at the amplifier output is given by

$$\sigma_n^2 = S_{sp}B_n = h\nu\left(\frac{\mu_1}{\mu_1-\mu_0}\right)(G-1)B_n \qquad (3.10)$$

where the ASE noise spectral density, which is assumed to be constant and obtained by dividing the noise power by the bandwidth, is indicated by S_{sp}. The factor $\mu_1/(\mu_1-\mu_0)$, called the amplifier inversion factor, is generally indicated by F.

The noise performance of an optical amplifier is usually characterized by the noise factor NF. To define the noise factor two signal-to-noise ratios have to be introduced: SNR_{in} and SNR_{out}. The former is the electrical signal to noise ratio after the detection of an optical field of power P_0 with an ideal photodetector; the latter is the electrical signal to noise ratio after the amplification of the optical field by the considered amplifier and its detection. The noise factor is defined as

$$NF = \frac{SNR_{out}}{SNR_{in}} \qquad (3.11)$$

Thus NF is a measure of the signal to noise ratio degradation introduced by the presence of the amplifier. The term SNR_{in} is simply determined by the detection quantum noise (shot noise) [11] and can be written as

$$SNR_{in} = \frac{P_0}{2h\nu B_e} \qquad (3.12)$$

The electrical bandwidth of the photodetector is indicated by B_e, and its optical bandwidth by B_o. To evaluate SNR_{out}, it is useful to write the optical signal $A(L)$ at the amplifier output as

$$A(L) = \sqrt{GP_o}\, e^{i\phi} + n_{tot}(t) \tag{3.13}$$

where the average envelope $\sqrt{GP_o}\, e^{i\phi}$ (generally called the signal) has been separated by the fluctuations, indicated by $n_{tot}(t)$. After detection three electrical noise components are present: the shot noise $c_{sh}(t)$ which is proportional to the overall detected power, a component $c_{nn}(t)$ which is proportional to the square of the ASE noise $n_{tot}(t)$, and a component $c_{sn}(t)$ which is proportional to the beating between the signal and the ASE due to the detection process. Thus, indicating by R_p the detection responsivity, the photocurrent $c(t)$ can be written as

$$c(t) = R_p G P_o + c_q(t) + c_{sn}(t) + c_{nn}(t) \tag{3.14}$$

The power of the different noise terms is given by the following equations in which each term has the same index of the related photocurrent:

$$
\begin{aligned}
\sigma_q^2 &= 2qR_pGP_oB_e \\
\sigma_{sn}^2(t) &= 4R_p^2GP_oS_{sp}B_e \\
\sigma_{nn}^2(t) &= 3R_p^2\sigma_n^4\frac{B_e}{B_o} = 3R_p^2S_{sp}^2B_eB_o
\end{aligned}
\tag{3.15}
$$

By equations (3.13) and (3.14), SNR_{out} can be written as

$$SNR_{out} = \frac{R_p^2G^2P_0^2}{2qR_pGP_oB_e + 4R_p^2GP_oS_{sp}B_e + 3R_p^2S_{sp}^2B_eB_o} \tag{3.16}$$

From (3.16) and (3.12) the amplifier noise factor can be easily evaluated.

If we are considering a low noise amplifier, similar to that used in communication systems, the signal power impinging the photodetector is by far larger than the optical noise power and the shot noise power. As a consequence the electrical noise due to the signal-ASE beating is dominant and the other noise terms can be neglected in equation (3.16). In this case the noise factor has a quite simple expression: Replacing the S_{sp}, expression (3.10) is rewritten as

$$NF = 2\left(\frac{\mu_1}{\mu_1 - \mu_0}\right)\frac{G-1}{G} \approx 2\left(\frac{\mu_1}{\mu_1 - \mu_0}\right) = 2F \tag{3.17}$$

where the approximation holds when the gain is much higher than one. In the case of an ideal amplifier the population of the ground level μ_0 is equal to zero, thus realizing complete population inversion. In this case, by equation (3.17), $NF = 2$, which is the minimum attainable value of the noise factor.

3.2   TIME-DEPENDENT BEHAVIOR OF THE TWO-LEVEL AMPLIFIER

If the input signal variation in time intervals of the order of the lifetime of the excited atoms is not negligible, the amplifier model described in the previous section does not hold, since the amplifier gain saturation does not instantaneously respond to the fluctuations of the signal power.

When the signal bandwidth is by far larger than the spontaneous decay rate, the amplifier gain becomes almost constant and is determined only by the average signal power. In the intermediate regime, a dynamic amplifier model is needed.

In the simple case of a two-level amplifier, three equations are needed to describe the dynamic amplifier behavior [12]. In general, they are stochastic equations, since both the optical field and the atomic populations are affected by random fluctuations. However, the noise properties of the amplifier does not change dramatically in a time-dependent regime, while the corresponding stochastic equations are quite different to solve. Thus in this section the behavior of a noiseless amplifier will be considered.

The first equation is the rate equation of the atoms in the upper level which is written as

$$\frac{d\mu_1}{dt} = I - \frac{\mu_1}{\tau} - g \frac{|A|^2}{h\nu} \qquad (3.18)$$

At the right-hand side of equation (3.18), three terms are present: I is the pumping rate, the second term represents the contribution of spontaneous emission (τ is the spontaneous lifetime of the atomic system), and the third term represents the stimulated emission rate which depends on the density of photons $|A|^2/h\nu$.

The second equation is the equation relating the local gain g to the atomic populations; it is written as

$$g = a(\mu_1 - \mu_0) \qquad (3.19)$$

where a is the differential gain.

The third equation is the propagation equation for the optical field in the amplifier. This equation, obtained by the Maxwell equations using the rotating wave and slowly varying envelope approximations, is written as

$$\frac{\partial A}{\partial z} + \frac{1}{v_g} \frac{\partial A}{\partial t} = \left(\frac{g}{2} + ir \right) A \qquad (3.20)$$

where v_g is the wave group velocity. It is to be noted that due to the dependence of μ_1 on time and length, the local gain g is a function of t and z.

Equations (3.18), (3.19), and (3.20) can be rewritten in a simpler form by eliminating the atomic populations substituting in equation (3.20) the time variable t, with the reduced time $\theta = t - z/g_g$ and expanding the complex field envelope as $A(z, t) = \sqrt{P(z, t)}\, e^{i\phi(z,t)}$. Then, substituting equation (3.19) into equation (3.18) and defining the linear local gain as $g_0 = a\tau I - a\mu_0$, the following equations are obtained:

$$\frac{\partial g}{\partial \theta} = \frac{g_0 - g}{\tau} - g\frac{|A|^2}{E_{\text{sat}}} \tag{3.21}$$

$$\frac{\partial P}{\partial z} = gP \tag{3.22}$$

$$\frac{\partial \phi}{\partial z} = -r \tag{3.23}$$

where the saturation energy (not to be confused with the different saturation powers introduced in the previous section) is given by $E_{\text{sat}} = h\nu/a$. The physical meanings of E_{sat} is evident from equation (3.21): It is the pulse energy above which the amplifier is heavily saturated. The relation between E_{sat} and the local saturation power P_{sat} can be easily obtained by the model detailed in this section, which is $P_{\text{sat}} = E_{\text{sat}}/\tau$.

Equations from (3.21) to (3.23) can be simplified by introducing the integrated local gain

$$h(\theta) = \int_0^L g(z, \theta)\, d\theta \tag{3.24}$$

Equations (3.22) and (3.23) can be immediately integrated over the amplifier length, obtaining

$$P(L, \theta) = P(0, \theta)\, e^{h(\theta)} \tag{3.25}$$

$$\phi(L, \theta) = \phi(0, \theta) - rL \tag{3.26}$$

where $P(0, \theta)$ is the input power and $\phi(0, \theta)$ the input phase. Thus the amplifier behavior is completely described by the integrated gain $h(\theta)$. A simple equation for the integrated gain can be obtained by integrating equation (3.21) with respect to z. Exploiting equation (3.22) to substitute $g|A|^2$ with $\partial P/\partial z$ results in

$$\frac{dh}{d\theta} = \frac{g_0 L - h}{\tau} - \frac{P(0, \theta)}{E_{\text{sat}}}[e^{h(\theta)} - 1] \tag{3.27}$$

By numerically solving equation (3.27), for example, with a Runge-Kutta algorithm, and exploiting (3.25) and (3.26) the output field can be evaluated.

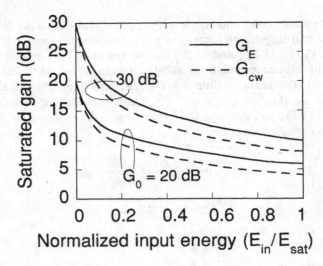

FIGURE 3.3 Saturated gain versus the ratio among the input and the saturation energy [12]. The energy gain when amplifying a pulse train is indicated by G_E and the power gain in CW conditions by G_{cw}. The linear power gain is indicated by G_0.

From the analysis described in this section, it is clear that, feeding the amplifier with a pulse whose duration is of the order of τ, the overall amplifier gain G is a function of time. A gain-related quantity of practical interest when amplifying pulses is the energy gain. Indicating with $E(0)$ and $E(L)$ the input and output pulse energy respectively, the energy gain G_E is given by

$$G_E = \frac{E(L)}{E(0)} = \frac{1}{E(0)} \int_{-\infty}^{\infty} P(0, \theta) \, e^{h(\theta)} \, d\theta \qquad (3.28)$$

Figure 3.3 shows the energy gain behavior of a two-level amplifier in continuous wave and pulse operation. In particular, the continuous wave gain G_{cw} and the energy gain G_E are shown versus the ratio $E(0)/E_{sat}$, respectively.

3.3 SEMICONDUCTOR OPTICAL AMPLIFIERS

A simplified structure of a semiconductor amplifier is shown in Fig. 3.4, with the band-gap behavior along the growth direction [13]. The amplifier is essentially constituted by a p-i-n heterojunction diode whose intrinsic layer the (i-layer) is formed by a direct gap semiconductor material. In the case of semiconductor amplifiers used in telecommunication systems, typical active materials are $In_xGa_{1-x}As$ and $In_xGa_{1-x}As_yP_{1-y}$: These materials allow spontaneous emission in the second- and third-transmission windows to be obtained with an opportune choice of x and y.

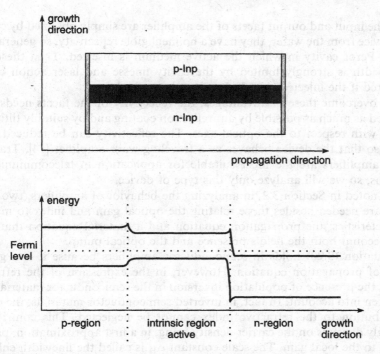

FIGURE 3.4 General scheme of a semiconductor amplifier and band-gap behavior along the growth direction.

When an electrical current is injected by directly polarizing the diode, the injected carriers are trapped in the intrinsic region (i-region), while band gap is smaller than those of the n- and p-regions. Thus, above a threshold value of the injected current, the population inversion generates an optical gain.

Carrier and photon confinement in the growth direction is obtained, in the structure of Fig. 3.4, by the heterojunction; no lateral confinement is present. This device requires a very high injection current to obtain a reasonable gain with great problems related to heat dissipation. To obtain an effective amplifier, lateral confinement has to be provided. The different structure that have been proposed can be divided in two categories: gain-guided and index-guided amplifiers [13]. In gain-guided amplifiers population inversion is generated only at the center of the i-region, injecting the current by means of a thin electrode. In this way the carriers are confined by the current injection while the photons are confined by the gain profile in the active region. In the index-guided amplifiers the active region is surrounded by a passive semiconductor with a different diffraction index. The obtained optical waveguide allows photon confinement. Carrier confinement is achieved by the same method in gain-guided devices and may even be enhanced by the guiding structure of the amplifier.

If the input and output facets of the amplifier are simply obtained by cutting the device from the wafer, they have a nonnegligible reflectivity, so generating a Fabry-Perot cavity in which the active medium is inserted. Then the device bandwidth is strongly limited by the cavity finesse and laser action can be triggered if the injected electrical current is too high.

To overcome these disadvantages, the reflectivity of the facets needs to be reduced as much as possible by antireflection coating and by suitably tilting the facets with respect to the optical axes. The reflectivity can be reduced up to 10^{-5}, so that the device behaves as a traveling wave amplifier [14]. Traveling wave amplifiers are the most suitable for application in telecommunication systems, so we will analyze only this type of device.

As noted in Section 3.2, in analyzing the behavior of amplifiers, two equations are needed besides those relating the optical gain and index to material characteristics: the propagation equation and a material equation that takes into account both the field's presence and the optical pump.

Equation (3.20) holds in semiconductor amplifiers because it is a general form of propagation equation. However, in the expression of the refraction index, the presence of population inversion in the semiconductor material must be taken into account. In fact, in inverted semiconductor materials, the carrier contribution to the refractive index cannot be neglected. This contribution strongly depends on the carrier density being, in a first approximation, proportional to the local gain. The scale constant α_H is called the linewidth enhancement factor, since the same constant has a major role in determining the linewidth of semiconductor lasers. Taking into account this phenomenon, the refraction index can be written as

$$r = r_1 - \alpha_H g \tag{3.29}$$

with r_1 being the lattice contribution.

Since g is a function of the injected optical power, gain saturation induces a nonlinear phase shift of the injected pulses, known as self-phase modulation, which is similar to the phenomenon taking place in optical fibers [15]. As far as the material equation is concerned, it can be written as

$$\frac{d\mu}{dt} = D_D \nabla^2 \mu + \frac{C}{qV_A} - R_R(\mu) \tag{3.30}$$

where $\mu(x, y, z, t)$ is the carrier density which depends both on the spatial coordinates and on time, D_D is the carrier diffusion coefficient, and $R_R(\mu)$ is the recombination rate. The injected current is indicated by C and V_A is the active region volume.

The first term at the right-hand side of equation (3.30) takes into account the carrier diffusion. The diffusion length, which is the maximum length over which diffusion is effective, is of the order of 50 to 100 µm. Since the device length is generally between 400 and 1000 µm, the diffusion has a negligible effect along this direction. On the other hand, the transversal dimensions of the active region are 0.1–0.2 µm and 1–2 µm, respectively: On the transverse plane diffusion is very efficient, and the carrier density is almost independent of x and y. Thus, assuming μ to be independent of x and y, the diffusion term can be neglected.

As far as the recombination term is considered, there are four major contributions to be taken into account:

1. *Nonradiative recombination by traps or surface phenomena.* This is a single particle transition giving a recombination rate proportional to μ by the scale constant A_R [16].

2. *Spontaneous interband recombination.* This is a two-particle transition giving a recombination rate proportional to μ^2 by the scale constant B_R.

3. *Auger recombination.* This is a four-particle transition (three electrons and one hole, two electrons and two holes, etc.) in which the energy released during the electron-hole recombination is transferred to another carrier (electron or hole). This carrier is excited at a higher energy level in its energy band and then relaxes back to thermal equilibrium by phonon emission. From the above description it is clear that different types of Auger recombination exist, depending on the involved particles, and that the evaluation of the Auger recombination rate is quite complex [16]. In a first approximation the Auger rate is proportional to μ^3 by the scale constant C_R.

4. *Stimulated emission.* In evaluating the stimulated emission rate, the band structure of the semiconductor has to be taken into account [16]. In a first approximation the stimulated emission rate can be written as a $(\mu - \mu_0)|A|^2/h\nu$, where a is the differential gain and μ_0 is the transparency carrier density.

Taking into account all the recombination phenomena, the recombination term can be written as

$$R_R(\mu) = A_R\mu + B_R\mu^2 + C_R\mu^3 + \frac{a(\mu - \mu_0)|A|^2}{h\nu} \qquad (3.31)$$

Equation (3.31) can be rewritten defining a decay time $\tau(\mu)$ as

$$\tau(\mu) = \frac{1}{A_R + B_R\mu + C_R\mu^2} \qquad (3.32)$$

Finally, eliminating the diffusion term and exploiting the expression of the recombination rate, the material equation can be written as

$$\frac{d\mu}{dt} = \frac{C}{qV_A} - \frac{\mu}{\tau(\mu)} - a(\mu - \mu_0)\frac{|A|^2}{h\nu} \qquad (3.33)$$

TABLE 3.1 Typical device parameters of a semiconductor amplifier.

Device length	L	500 µm
Active region volume	V_A	100 µm³
Driving current	C	200 mA
Transparency current	C_0	30 mA
Unsaturated device gain	G_0	28 dB
Local saturation power	P_{sat}	10 dBm
Linewidth enhancement factor	α_H	5
Carrier decay time	τ	200 ps
Noise factor	NF	9 dB

If the decay time variation can be neglected during the operation of the amplifier, equation (3.33) is formally identical to the equation obtained by substituting (3.19) into (3.18). Since this approximation is often well respected in practice, the semiconductor optical amplifier behaves approximately as a two-level system.

In fact the set of approximations adopted to obtain equations (3.33) and (3.29) are equivalent to approximating the carrier population inside the active zone as a population of noninteracting two-level systems with homogeneous lineshape broadening. Therefore, with the correct expression adopted for the refraction index, the model detailed in Sections 3.1 and 3.2 can be fully applied.

Typical values of the parameters of traveling wave semiconductor amplifiers are summarized in Table 3.1. From the table can be seen that the carrier spontaneous lifetime τ is of the order of the bit time of high-speed optical transmissions. Thus, when semiconductor amplifiers are used in a communication link, the amplifier gain varies in time following the time behavior of the input optical signal. This effect causes pulse distortion and nonlinear crosstalk in frequency multiplexed systems and is one of the main disadvantages of semiconductor optical amplifiers.

Semiconductor amplifiers also have a wide bandwidth, which is mainly determined by the dependence of the local gain on the wavelength. A typical value of the 3-dB bandwidth for a semiconductor optical amplifiers is 50 nm (about 6.2 THz at $\lambda = 1.55$ µm).

3.4 ERBIUM-DOPED FIBER AMPLIFIERS

The erbium-doped fiber amplifier is based on a single-mode optical fiber suitably doped with erbium ions that constitute optically active elements [17]. The energy-level scheme of the erbium ion in glass is shown in Fig. 3.5, where the energy levels are indicated by the dominant Russel-Saunders terms. For each energy level $^S L_J$, both the wavelength of the ground state absorption (GSA) and the wavelength of the excited state absorption (ESA) is indicated.

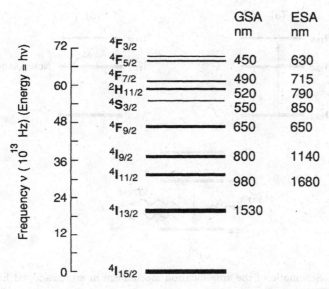

FIGURE 3.5   Energy levels for Er^{3+} ions in silica glass. Energy levels are indicated by the dominant Russel-Saunders term. GSA = ground state absorption, ESA = excited state absorption.

The GSA is related to the transitions from the ground-state ($^4I_{15/2}$ level) to the considered ^SL_J level, while the ESA is referred to the transition from the metastable laser state ($^4I_{13/2}$ level) to the considered ^SL_J level.

Because of the introduction of Er^{3+} ions, the amplifier operation can be described by assuming a three-level system, as illustrated in Fig. 3.6 for a 0.98-µm pump. A pump photon is absorbed by a ground-state erbium ion that jumps to a higher energy level (in the reported example the $^4I_{11/2}$ state), and the excited ion jumps in its turn to the metastable $^4I_{13/2}$ level through nonradiative decay. Once in the metastable state, a radiative decay toward the ground-state leads to the emission of a photon at the 1.530-µm wavelength. In this last transition a stimulated emission process can occur, generating, in the presence of suitable conditions, either amplification or lasing action.

To design reliable and feasible optical amplifiers, feasible pumping frequencies must be determined from the absorption spectrum, taking into account the wavelengths of existing laser sources. From Fig. 3.5 two absorption wavelengths, at 800 nm and 980 nm, result that are interesting as pump wavelengths because they correspond to possible emission frequencies of GaAs and AlGaAs semiconductor lasers. Moreover the wide absorption spectrum of around 1.5 µm permits to an Er^{3+} device to be pumped directly to the metastable state by InGaAsP semiconductor lasers emitting at 1.48 µm, obtaining a high quantum efficiency.

Among the considered pumping bands, that around 800 nm is affected by

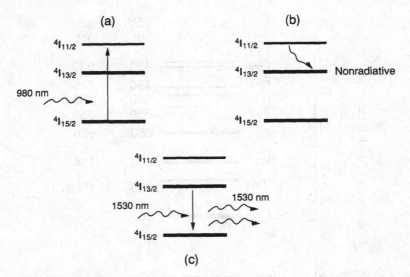

FIGURE 3.6 Schematic of the amplification mechanism in erbium-doped fibers for a 980-nm pump.

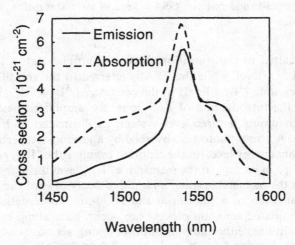

FIGURE 3.7 Absorption coefficient versus wavelength for Er^{3+} ions in silica glass.

the phenomenon of ESA, schematically depicted in Fig. 3.8. Once a pump photon is absorbed to excite an ion on the pump level, the ion relaxes on a metastable state where, instead of causing stimulated emission by decaying toward the ground-state, another pump photon is absorbed so as to make the ion jump to a different excited state. This phenomenon strongly reduces the pumping efficiency at 800 nm so that it is preferable to adopt a pump at 980 nm or direct pumping at 1.48 μm.

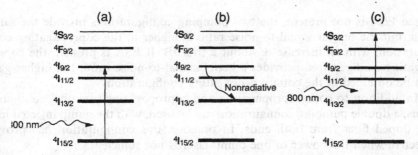

FIGURE 3.8 Schematic for the excited-state absorption mechanism for an 800-nm pump.

Even if at a first glance resonant pumping at 1.48 μm seems to produce a two-level dynamics among the Er^{3+} ions, it should be noted that the peaks in the absorption and stimulated emission cross-section spectra are at slightly different frequencies, leading to a three-level system dynamics (see Fig. 3.7). If it is more convenient to use either the 980-nm or 1.48-μm pumping wavelength is not simple to state, since the pumping efficiencies are comparable and commercial devices are available for both pumping wavelengths.

The structure of an erbium-doped fiber amplifier (EDFA) that can provide a small signal gain in excess of 30 dB is shown in Fig. 3.9. Besides the doped fiber and the pump laser, the amplifier is constituted by a power supply line, a pump-driving circuit, a monitoring circuit, and an electrical monitoring line. Moreover an optical tunable filter is usually placed at the amplifier output.

The pump can be launched into the doped fiber in either the copropagating or counterpropagating direction with respect to the signal. The performance of EDFA has been evaluated for both pumping configurations by accurate physical models [18] that take into account the presence of ESA as well.

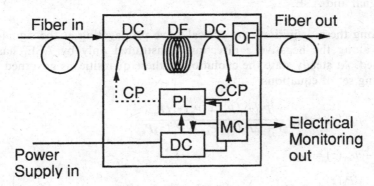

FIGURE 3.9 Block scheme of an erbium-doped fiber amplifier. DF = doped fiber, DC = directional coupler, MC = monitoring circuit, OF = optical tunable filter, CP = copropagating pumping configuration, CCP = contropropagating pumping configuration, PL = pump laser.

When ESA is not present, the two pumping configurations provide the same gain, but the output signal-to-noise ratio is higher in the copropagating configuration, with an increase of about 2 to 3 dB. If ESA is present, the copropagating configuration provides a better signal-to-noise ratio, but higher gain can be obtained in the counterpropagating configuration.

In addition to counterpropagating and copropagating pumping configurations, a double pumping configuration can be used, with the pump injected into the doped fiber from both ends. In practice, this configuration has proved effective when the power of one pump laser is not sufficient.

To provide some insight on the physical behavior of an EDFA, a simple performance evaluation model is presented here, by which the main features of the device will be described [18, 19]. In this model the EDFA is assumed to be a three-level one-dimension system in which the lifetime of the pumping state is much shorter than the other times of interest so that the population of this state can be considered negligible. Several assumptions can be made to get a simpler analysis: The transverse doping shape and the dependence of the optical fields on the transverse coordinates are not taken into account; the guiding function of the fiber and consequent effects due to chromatic dispersion and polarization evolution can be neglected; both ESA and ion-ion interaction are not considered. In addition any emission of photons at the pump wavelength can be neglected.

Representing by z the spatial coordinate along the doped fiber axis, the physical variables describing the amplifier behavior are as follows:

- Population $\mu_{gr}(z)$ of the ground state.
- Population $\mu_{me}(z)$ of the metastable state.
- Power $P_p(z)$ of the pump wave at the wavelength λ_p (frequency ν_p).
- Power $P_s(z)$ of the optical wave at the signal wavelength λ_s (frequency ν_s), propagating in the positive z direction, which is constituted by the actual signal and ASE.

Among these definitions the optical wave propagating at the signal wavelength along the negative z direction, constituted only by ASE, has been neglected. At steady state the evolution of these quantities is governed by the following set of equations:

$$\mu_{me}(z) = \mu_{tot} \frac{P_s(z)/P_{ss} + P_p(z)/P_{sp}}{1 + 2P_s(z)/P_{ss} + P_p(z)P_{sp}} \tag{3.34}$$

$$\mu_{me}(z) + \mu_{gr}(z) = \mu_{tot} \tag{3.35}$$

$$\frac{dP_s(z)}{dz} = 2\pi\Gamma_s \left[s_{se}\mu_{me}(z)P_s(z) + \frac{s_{se}}{\Gamma_s}\mu_{me}(z)P_n - s_{sa}\mu_{gr}(z)P_s(z) \right] \tag{3.36}$$

$$\frac{dP_p(z)}{dz} = \pm 2\pi\Gamma_p s_p \mu_{gr}(z)P_p(z) \tag{3.37}$$

Equation (3.34) represents the steady state value of the population of a three-level system in the presence of pumping and signal waves, where the population of the upper level is set to zero. The overall density of erbium ions is indicated with μ_{tot}, while the saturation effects due to pump and signal are taken into account by means of the local pump saturation power P_{sp} and the local signal saturation power P_{ss}. The saturation powers can be evaluated starting from the doped fiber parameters, obtaining $P_{sp} = \pi \rho_c^2 h\nu_p/(s_p\tau_{sp})$ and $P_{ss} = \pi \rho_c^2 \nu_s/(s_{se}\tau_{ss})$, where h is the Planck constant and τ_{sp}, τ_{ss} are the spontaneous emission lifetimes of the metastable state for the pump and the signal respectively. The emission and absorption cross sections at the signal wavelength are indicated by s_{se} and s_{sa}, respectively, and the fiber core radius by ρ_c. These equations are somewhat similar to the equation $P_{sat} = h\nu/(a\tau)$ introduced in Section 3.2 for a two-level amplifier.

Equation (3.35) states that all the ions that are not on the metastable state are on the ground-state. Equation (3.36) gives the propagation for the signal wave. The three terms at the right-hand side refer to phenomena involving photons at the signal wavelength: The first one represents the stimulated emission, the second the spontaneous emission, and the third the absorption. Since the first term is proportional to the occupation number of the metastable state $\mu_{me}(z)$, and the third to the occupation number of the ground-state $\mu_{gr}(z)$, the signal wave undergoes a net amplification if $s_{se}\mu_{me} > s_{sa}\mu_{gr}$; otherwise, it becomes attenuated. The coupling between the signal mode and the core of the erbium-doped fiber is given by Γ_s which represents the overlapping integral between the normalized transverse signal mode and the erbium concentration in the doped fiber transverse section.

The term P_n represents spontaneous emission, which turns out to be a random process with photons spontaneously emitted at random intervals. Since it is important to determine the value of the signal-to-noise ratio at the amplifier output, P_n can be substituted with its average value. Based on the results of Section 3.1, it is evident that $P_n = 2h\nu_s B$, where B is the optical signal bandwidth.

Equation (3.37) represents the evolution equation of the pump wave: The minus sign refers to the copropagating pump, and the plus sign to the counterpropagating pump. The parameter s_p is the cross section of the pump's absorption and Γ_p the overlapping integral of the normalized transverse pump mode and the erbium concentration in a doped fiber transverse section.

Numerical values of the parameters in equations (3.34) to (3.37) are reported in Table 3.2 for a low-noise in-line amplifier. If the EDFA is designed to operate with a high output power, the saturation powers will be higher, but often at the expense of an increase in noise power. By equation (3.36) two important parameters of the doped fiber amplifier can be introduced: the local gain per unit length $g_m(z)$ and the attenuation per unit length $\alpha_m(z)$. They are written as

$$g_m(z) = 2\pi s_{se}\mu_{me}(z)$$
$$\alpha_m(z) = 2\pi s_{sa}\mu_{gr}(z)$$

(3.38)

TABLE 3.2 Typical parameters of an EDFA amplifier for low-noise signal amplification.

Core area	$\pi\rho^2$	$13\,\mu m^2$
Erbium concentration	μ_{tot}	$5.4 \times 10^{24}\,m^{-3}$
Signal overlapping integral	Γ_s	0.4
Pump overlapping integral	γ_p	0.4
Signal emission cross section	s_{se}	$5.3 \times 10^{-25}\,m^2$
Signal absorption cross section	s_{sa}	$3.5 \times 10^{-25}\,m^2$
Pump absorption cross section	s_p	$3.2 \times 10^{-25}\,m^2$
Signal local saturation power	P_{ss}	$1.3\,mW$
Pump local saturation power	P_{sp}	$1.6\,mW$

According to the definitions above, the overall amplifier gain G can be evaluated by setting $P_n = 0$ into equation (3.36), that is, assuming a noiseless amplifier. Solving the amplifier equations and evaluating the ratio between the output and input signals obtains

$$G = \frac{P_s(L)}{P_s(0)} = \exp\left(\int_0^L [g_m(z) - \alpha_m(z)]\, dz\right) \qquad (3.39)$$

where L is the doped fiber length. This equation is absolutely equivalent to equation (3.5) derived for the two-level amplifier, taking into account the expression of net gain of the two-level amplifier given by equation (3.2).

Equations from (3.34) to (3.37) must be numerically solved to evaluate the performance of the amplifier. An example of the performance of a typical EDFA for signal amplification, whose parameters are given in Table 3.2, is given in Figs. 3.10, 3.11, and 3.12. In particular, Fig. 3.10 shows the gain G as a function of the pump power for different values of the fiber length. The pump saturation effect occurs for input pump power values in the range 1 to 6 mW. In Fig. 3.11 the gain is shown with relation to the doped fiber length for different pump power values. The curves allow a comparison of maximum given fiber lengths: It is evident that the optimum length is of the order of a few tens of meters and increases with the pump power. Figure 3.12 shows the gain saturation induced by the output power at the signal wavelength.

Regarding the gain saturation in EDFAs, it must be noted that an excited ion's lifetime is of the order of 11 ms [19]. Thus in optical communication systems the amplifier saturation is determined only by the average signal power, and no signal distortion is induced by the gain dynamic.

In many applications it is important to analyze the spectral behavior of the EDFA gain. Depending on the fiber parameters, the 3-dB bandwidth of the EDFA in a small-signal amplification regime ranges from 25 to 44 nm, which corresponds to a frequency bandwidth from about 3000 to 5300 GHz at the wavelength of interest [20]. Such a wide bandwidth is another important feature of the EDFA.

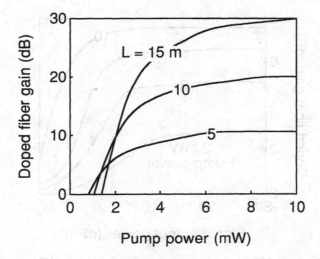

FIGURE 3.10   Doped fiber gain versus pump power for different fiber lengths.

The noise performances of EDFA can be analyzed by the amplifier noise factor. In practice the EDFA's performance is not far from that of an ideal amplifier ($\mu_0 = 0$ implying that $NF = 2$), since NF is of the order of 2.6 to 3 when the pump wavelength is 980 nm, whereas for a pump at 1.48 μm, NF is about 3 to 4.

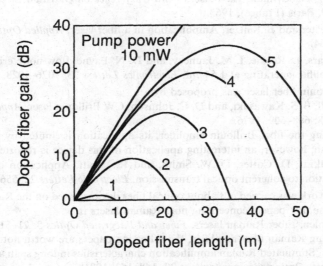

FIGURE 3.11   Doped fiber gain versus fiber length for different pump powers.

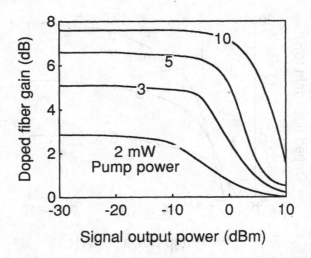

FIGURE 3.12   Doped fiber gain versus output signal power for different pump powers.

REFERENCES

1. The model of a two levels optical system can be found in all the textbooks dealing the interaction between the electromagnetic radiation and the matter. See, for example,
    A. Yariv, *Quantum Optics*, 3rd ed., Wiley, New York, 1989, ch. 8.
    D. Marcuse, *Principles of Quantum Electronics*, Academic Press, New York, 1980, ch. 5.

2. E. Snitzer, Neodymium glass laser, *Third Conference on Quantum Electronics*, pp. 999–1019, Paris (France), 1963.

3. C. J. Koester and E. Snitzer, Amplification in a fiber laser, *Applied Optics* 3: 1182–1186, 1964.

4. R. J. Mears, L. Reekie, I. M. Jauncey, and D. N. Payne, Low noise erbium doper fiber amplifier operating at 1.54 μm, *Electronics Letters* 23: 1026–1028, 1987.

5. The Brillouin fiber laser was proposed by
    K. O. Hill, B. S. Kawasaki, and D. C. Johnson, CW Brillouin laser, *Applied Physics Letters* 28: 608–609, 1976.
    Concerning the fiber Brillouin amplifier, its application is limited by its reduced bandwidth; however, an interesting application of this device is reported by
    C. G. Atkins, D. Cotter, D. W. Smith, and R. Wyatt, Application of Brillouin amplification to coherent optical transmission, *Electronic Letters* 22: 556–558, 1986.

6. A lot of work was carried out about optical fiber devices based on the Raman effect. One of the first papers concerning fiber Raman lasers is
    R. H. Stolen, Fiber Raman lasers, *Fiber and Integrated Optics* 3: 21–51, 1980.
    Concerning Raman fiber amplifier, the following papers are worth noting:
    M. Ikeda, Stimulated Raman amplification characteristics in long span single-mode silica fibers, *Optical Communications* 39: 148–152, 1981

Y. Aoki, Properties of fiber Raman amplifiers and their applicability to digital optical communication systems, *IEEE-Journal of Lightwave Technology* 6: 1225–1239, 1988.

T. Nakashima, S. Sekai, M. Nakazawa, and Y. Negishi, Theoretical limit of repeater spacing in an optical transmission line utilizing Raman amplification, *IEEE-Journal of Lightwave Technology*, 4: 1267–1272, 1986.

7. An exhaustive review of the material characteristics of erbium doped fibers and of the atomic phenomena governing the behavior of these devices is presented in
W. J. Miniscalco, Erbium doped glasses for fiber amplifiers at 1500 nm, *IEEE-Journal of Lightwave Technology* 9: 234–250, 1991.

8. Good examples of the numerous experiments dealing with demultiplexing of very high-speed time division multiplexed optical signals exploiting semiconductor optical amplifiers nonlinearities are
D. M. Patrick and R. J. Manning, 20 Gbit/s all-optical clock recovery using semiconductor nonlinearity, *Electronics Letters* 30: 151–153, 1994.
T. Morioka, H. Takara, S. Kawanishi, K. Uchiyama, and M. Saruwatari, Polarization independent all-optical demultiplexing up to 200 Gbit/s using four-wave mixing in a semiconductor laser amplifier, *Proceedings of OFC'96*, pp. 131–132, San Jose, CA, February 25–March 1, 1996.

9. An example of wavelength converter based on fast gain saturation in a semiconductor amplifier is reported, along with device performances and possible system applications, in
B. Glance et al., High performance optical wavelength shifters, *Electronics Letters* 28: 1714–1715, 1992.
Index saturation has been exploited to obtain wavelength converters based on semiconductor amplifiers. In particular, two semiconductor amplifiers are generally put in the branches of an interferometer so that signal induced index saturation changes the interference condition at the device output. This modulates the probe field at a different frequency. Such devices have been realized with different techniques, see, for example,
T. Durhuus et al., Penalty free all-optical wavelength conversion by SOAs in Mach-Zehnder configuration, in *Proceedings of ECOC '93*, paper TuC5.2, Montreaux, Switzerland, 1993.
K. E. Stubkjaer et al., Optical wavelength converters, in *Proceedings of ECOC'94*, vol. 2, pp. 635–642, Florence, 1994.
Finally, even four-wave mixing in semiconductor optical amplifiers has been exploited to obtain all-optical wavelength conversion. For example see
A. D'Ottavi, E. Iannone, A. Mecozzi, S. Scotti, P. Spano, R. Dall'Ara, J. Eckner, and G. Guekos, Efficiency and noise performances of wavelength converters based on FWM in semiconductor optical amplifiers, *IEEE-Photonics Technology Letters* 31: 1995.

10. All-optical AND and XOR gates have been realized by using four-wave mixing in semiconductor optical amplifiers, see, for example,
D. Nesset, N. C. Tatham, L. D. Westbrook, and D. Cotter, Ultrafast all-optical AND gate using four wave mixing in a semiconductor laser amplifier, *Proceedings of ECOC'94*, pp. 529–532, Florence, September 25–29, 1994.
A. D'Ottavi, E. Iannone, and S. Scotti, Address recognition in all-optical packet

switching by FWM in semiconductor amplifiers, *Microwave and Optical Technology Letters* 10: 228–230, 1995.

11. The detection noise is present, even if an ideal detector is assumed, due to the quantum nature of the optical field. For a quantum discussion of this issue, see, for example,

    Y. Yamamoto and H. A. Haus, Preparation, measurement and information capacity of optical quantum states, *Review of Modern Physics* 58: 1001–1020, 1986. A semiclassical approach to the problem is reported in all the books on optical communications or quantum optics. See, for example,

    G. P. Agrawal, *Fiber-Optic Communication Systems*, Wiley, New York, 1992, ch. 4.

    S. Betti, G. De Marchis, and E. Iannone, *Coherent Optical Communications Systems*, Wiley, New York, 1995, ch. 2.

12. The dynamic behavior of a two-level amplifier has been studied mainly in relation to semiconductor optical amplifiers when the active semiconductor is modeled as a collection of noninteracting two-level systems. See the original paper in which the dynamic behavior of semiconductor amplifiers was analyzed using this approach:

    G. P. Agrawal and N. A. Olsson, Self-phase modulation and spectral broadening of optical pulses in semiconductor laser amplifiers, *IEEE-Journal of Quantum Electronics* 25: 2297–2306, 1989.

    See also the book

    G. P. Agrawal, *Fiber-Optic Communication Systems*, Wiley, New York, 1992, ch. 8.

13. The structure of semiconductors amplifiers is similar to that of semiconductor lasers but for the antireflection coating on the amplifiers facets. Different possible structures of semiconductors lasers for optical communications, along with a review of the adopted semiconductor materials, are presented in

    G. P. Agrawal and N. K. Dutta, *Long Wavelength Semiconductor Lasers*, Van Nostrand Reinholds, New York, 1986.

    Specific reviews of the semiconductor amplifier are

    M. J. O'Mahony, Semiconductor laser optical amplifiers for use in future fiber systems, *IEEE-Journal of Lightwave Technology* 6: 531–544, 1988.

    T. Saito and T. Mukai, Recent progress in semiconductor laser amplifiers, *IEEE-Journal of Lightwave Technology* 6: 1656–1654, 1988.

    T. Saito and T. Mukai, Traveling-wave semiconductor amplifiers, in *Coherence, Amplification, and Quantum Effects in Semiconductor Lasers*, Y. Yamamoto (ed.), Wiley, New York, 1991.

14. See, for example,

    P. Doussiere et al., Polarization independent 1550 nm semiconductor optical amplifier packaged module with 29 dB fiber to fiber gain, *Proceedings of Optical Amplifiers and Their Applications*, pp. 119–122, Davos, June 15–17, 1995.

    C. H. Holtmann, P. A. Besse, and H. Melchior, Polarization resolved, complete noise characterization of bulk ridge-waveguide semiconductor optical amplifiers, *Proceedings of Optical Amplifiers and Their Applications*, pp. 115–118, Davos, June 15–17, 1995.

15. G. P. Agrawal and N. A. Olsson, Self-phase modulation and spectral broadening of optical pulses in semiconductor laser amplifiers, *IEEE-Journal of Quantum Electronics* 25: 2297–2306, 1989.

16. G. P. Agrawal and N. K. Dutta, *Long Wavelength Semiconductor Lasers*, Van Nostrand Reinholds, New York, 1986, ch. 3.

17. For a complete description of the state of the art on EDFAs, see
E. Desurvire, *Erbium Doped Fiber Amplifiers, Principles and Applications*, Wiley, New York, 1994.
W. J. Miniscalco, Erbium doped glasses for fiber amplifiers at 1500 nm, *IEEE-Journal of Lightwave Technology* 9: 234–250, 1991.
More complex amplifier structures and a discussion on their performances are reported in
J.-M. P. Delavaux and J. A. Nagel, Multi-stage erbium doped fiber amplifier design, *IEEE-Journal of Lightwave Technology* 13: 703–720, 1995.

18. The equations governing the behavior of an Er^{3+}-doped fiber amplifier are derived, for example, in
E. Desurvire and R. J. Simpson, Amplification of spontaneous emission in erbium-doped single-mode fiber amplifiers, *IEEE-Journal of Lightwave Technology* 7: 835–845, 1989.
An accurate model for the analysis of different pumping configurations as well as the excited state absorption phenomenon are reported in
M. Montecchi, A. Mecozzi, M. Settembre, M. Tamburrini, and L. Di Gaspare, Gain and noise in rare-earth-doped optical fibers, *Journal of Optical Society of America* B 8: 134–141, 1991.
P. R. Morkel and L. I. Laming, Theoretical modeling of erbium doped fiber amplifiers with excited state absorption, *Optics Letters* 14: 1062–1064, 1989.

19. An accurate analysis of the dynamic of ion populations is reported in the book of Desurvire cited in ref. [17]; an analysis for transient effect in doped fiber amplifiers is also presented in
C. R. Giles, E. Desurvire, and J. R. Simpson, Transient gain and crosstalk in erbium doped fiber amplifiers, *Optics Letters* 14: 880–882, 1989.

20. On the bandwidth of erbium-doped fiber amplifiers, see, for example, ref. [17] and
P. C. Becker, A. Lidgard, J. R. Simpson, and N. A. Olsson, Erbium doped fiber amplifier pumped in the 950–1000 nm region. *IEEE-Photonics Technology Letters* 2: 35–37, 1990.
J. M. Armitage, Spectral dependence of the small signal gain around 1.5 µm in erbium doped silica fiber amplifiers, *IEEE-Journal of Quantum Electronics* 26: 223–225, 1990.

CHAPTER FOUR

Optical Transmission Systems

This chapter introduces the main types of optical transmission systems and analyzes their performances in cases where fiber propagation can be assumed to be linear. There are two major ways of classifying optical transmission systems. One way is by the adopted modulation format and the other by the receiver scheme. All the modulation formats used in radio communications (amplitude, phase, and frequency modulation), can be also adopted in optical communications [1]. Additionally in the optical case even polarization modulation can be used if the polarization fluctuations at the receiver input are compensated.

Considering the detection scheme, optical transmission systems are further divided into two categories: *direct detection* and *coherent* transmission systems. The structure of a direct-detection system is simple. At the transmitter the information bearing signal is obtained by modulating an optical source (generally a semiconductor laser) either directly or by an external modulator; at the receiver the optical signal is directly detected by one or more photodiodes so as to obtain baseband electrical currents; these currents are suitably processed to estimate the transmitted message.

Several modulation formats are possible for direct-detection systems: even phase modulation can be adopted if an optical interferometer is used at the receiver to transform phase modulation into amplitude modulation before the direct detection takes place. In practice, much attention has been devoted to intensity modulation–direct detection (IM-DD) systems [2] because of their simple structure. In fact, angular modulation does not significantly improve system performance and the required receiver structure is quite complex.

Besides the conventional IM-DD systems, there has been some interest shown in polarization-modulation systems [3] whose modulation format allows very efficient multilevel systems to be realized [4].

In the field of optical communications the term *coherent* is used with a different meaning with respect to that in radio frequency communication systems. An optical system is *coherent* if the coherence properties of the transmitted optical field are somehow exploited at the receiver. Generally, at the receiver the incoming field is mixed with the field emitted by a laser, the local oscillator. The resulting wave is detected by one or more photodiodes so that the electrical currents carry information about amplitude, phase, frequency, and polarization of the received optical signal.

If the electrical currents are baseband signals, the optical receiver is called a *homodyne*; if instead they are at radio or microwave frequency, the receiver is called a *heterodyne*. In a heterodyne receiver, electronic demodulation is required to obtain the baseband signal from which the transmitted message is estimated. The electronic demodulation can be either synchronous, based on electronic phase lock loop (PLL), or asynchronous using, for example, square law devices. A large number of different coherent systems have been theoretically and experimentally studied for amplitude, frequency, phase modulation, and adopting different receivers [5].

Optical multiplexing is another important issue to include in a review of optical transmission techniques. Optical bandwidth is much wider than the electrical one, and efficient use of the optical channel call for optical multiplexing. Two multiplexing techniques can be considered: frequency division multiplexing [6] (usually called wavelength division multiplexing, WDM) and optical time division multiplexing (OTDM) [7]. The first WDM commercial systems are appearing, and it looks like the use of WDM will become widespread in the near future. The OTDM-related technology is less mature, and its application in optical networks is perhaps farther away in time. However, OTDM has great potentiality, and considerable research effort is directed to make OTDM feasible and reliable as required by practical applications.

4.1 BASIC SCHEMES OF OPTICAL TRANSMISSION SYSTEMS

4.1.1 Intensity-Modulation Direct-Detection Systems

The block diagram of an IM-DD system is shown in Fig. 4.1. The intensity of the transmitting laser is modulated by direct modulation or by an external modulator so that the emitted electrical field can be written as

$$\mathbf{E} = A\sqrt{m(t)}e^{i[\omega_0 t + \phi]}\mathbf{x} \qquad (4.1)$$

where A is the field amplitude, $\mathbf{x}$ is the unit vector in the field direction, and the field phase ϕ can depend on time if, for example, a spurious chirp is introduced by the transmitter. To simplify the notation, in this chapter we normalize the electrical field so that $|\mathbf{E}|^2$ is the optical power measured in [W]. Therefore the field $\mathbf{E}$ and the envelope A are both measured in [$W^{1/2}$].

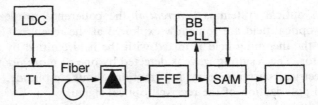

FIGURE 4.1 Block diagram of an IM-DD transmission system adopting direct modulation. TL = transmission laser, LDC = laser driving circuit, EFE = electrical front end, BB PLL = baseband PLL, SAM = sampler, DD = decision device. If external modulation is adopted, an external modulator is present at the output of the transmitting laser.

The modulation function $m(t)$ is given by

$$m(t) = \sum_{k=1}^{N_b} b_k g(t - kT) \tag{4.2}$$

where $b_k = 0, 1$ is the transmitted bit, T the bit time, and $g(t)$ is the transmitted pulse. The number of transmitted bits is N_b, which can be assumed to be infinite for statistical purposes.

In the case where $g(t)$ is equal to one in the bit interval and zero elsewhere, the adopted transmission format is called non return to zero (NRZ). In this case, if two consecutive marks are transmitted, the optical signals does not return to zero between the two pulses. In the other cases, where different pulses are used, the transmission format is called return to zero (RZ). In RZ signals the ratio between the adopted pulse width and the bit interval is called the *duty cycle*; the duty cycle is between zero and one, in RZ signals but equal to one in NRZ signals.

The spectral width of the transmitted signals depends on the pulses used: It is narrowest in the case of NRZ transmission and increases by decreasing the duty cycle. The power spectral density $S_o(\nu)$ of the optical signal (4.1) is given by

$$S_o(\nu) = \frac{A^2}{2} S_b(\nu - \nu_0) + \frac{A^2}{2} S_b(\nu + \nu_0) \tag{4.3}$$

where $S_b(\nu)$ is the baseband power spectral density and $\nu_0 = \omega_0/2\pi$ is the optical carrier frequency. The baseband spectrum $S_b(\nu)$ is the spectrum of the modulating signal $\sqrt{m(t)}$ and depends both on the transmitted message and on the shape of the modulating pulse. Limiting our attention to the case in which uncorrelated bits are transmitted, $S_b(\nu)$ can be written as [1]

$$S_b(\nu) = \frac{\sigma_m^2}{T} Y(\nu)^2 + \frac{\mu_m^2}{T^2} \sum_{k=-\infty}^{\infty} \left| Y\left(\frac{k}{T}\right) \right|^2 \delta\left(\nu - \frac{k}{T}\right) \tag{4.4}$$

where $Y(\nu)$ is the Fourier transform of the pulse $\sqrt{g}(t)$ and $\delta(\ )$ is the Dirac distribution. The standard deviation and the average of the transmitted bit stream b_k are indicated by σ_m and μ_m and are evaluated as

$$\mu_m = \sum_{k=0}^{1} b_k \pi_k = \pi_1, \quad \sigma_m^2 = \sum_{k=0}^{1} (b_k - \mu_m)^2 \pi_k \qquad (4.5)$$

where π_1 is the probability of transmitting a mark and π_0 that of transmitting a space.

In the important case of equiprobable bits and NRZ transmission, the expression of the baseband spectrum of the signal gets particularly simple. In this case $\mu_m = \frac{1}{2}$, $\sigma_m = \frac{1}{2}$ and the Fourier transform of $\sqrt{g(t)}$, which in this case is equal to the modulating pulse itself, is given by

$$Y(\nu) = T \frac{\sin(\pi\nu T)}{\pi\nu T} e^{-i\pi\nu T} \qquad (4.6)$$

From equation (4.6) it is clear that $Y(k/T)$ is zero but for $k = 0$; thus the only Dirac pulse in the spectrum is located in the origin. Substituting the values of μ_m and σ_m and equation (4.6) in equation (4.4) obtains

$$S_b(\nu) = \frac{TA^2}{2} \frac{\sin^2(\pi\nu T)}{(\pi\nu T)^2} + \frac{A^2}{2} \delta(\nu) \qquad (4.7)$$

The plot of the spectrum of equation (4.7) is shown in Fig. 4.2. Generally the bandwidth of the NRZ signal is considered the distance between the first zeros

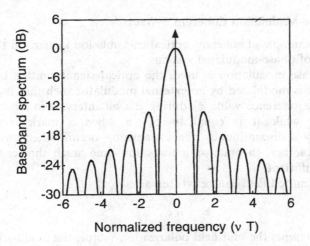

FIGURE 4.2 Plot of the baseband power spectral density of an IM-DD NRZ signal. A logarithmic scale is used on the ordinate axis to show the ripple in the spectrum tails.

of the spectrum: This is equal to $2R$, with $R = 1/T$ the bit rate. About 97% of the transmitted power is contained within this bandwidth.

The transmitted field is coupled to the fiber, and it is attenuated during propagation toward the fiber far end where it is detected by a photodiode. In IM-DD systems, both PIN and APD photodiodes can be used.

After amplification and possible equalization, assuming ideal fiber propagation and a perfectly linear processing at the receiver, the electrical current at the sampler input can be expressed as

$$c(t) = [R_p A^2 m(t) M_p + n_s(t)] \otimes h_R(t) + n_t(t) \qquad (4.8)$$

where $c(t)$ is the photocurrent generated by the received optical signal, A^2 the peak optical power, $h_R(t)$ the receiver transfer function, and $\otimes$ indicates the convolution operation. The photodiode responsivity is given by R_p, and the mean value of the photodiode gain is given by M_p; $M_p = 1$ for a PIN photodetector.

Two noise contributions have been considered in equation (4.8): the composition of shot noise and multiplication noise $n_s(t)$ [8] and the thermal noise due to the front-end electronics $n_t(t)$ [9]. Background noise and the photodiode's dark current noise have been neglected [10].

The baseband current given by equation (4.8) is sampled in the middle of each bit interval to obtain the decision variable $C_k = c(t_k)$, where t_k is the kth sampling instant.

The decision variable is compared with a threshold to estimate the transmitted bit. If C_k exceeds the threshold, a mark is estimated; otherwise, a space is assumed.

4.1.2 Phase-Modulation Coherent Systems

We give an example of coherent optical transmission systems in the following description of phase-modulated systems.

When phase modulation is used, the optical signal emitted by the transmitting laser is modulated by an external modulator such that the field phase is equal to a reference value ϕ_0 during the bit intervals in which a space is transmitted, while it is equal to $\phi_0 + \pi$ when a mark is transmitted. Theoretically a discontinuous phase transition occurs when switching from one bit to another. In practice, a transition time much shorter than the bit interval is sufficient.

The transmitted field can be written as

$$\mathbf{E} = A\, e^{i[\omega t + \pi m(t) + \phi]} \mathbf{x} \qquad (4.9)$$

where $\mathbf{x}$ represents the unit field polarization vector, the modulation function $m(t)$ is given by equation (4.2), and the pulse $g(t)$ is equal to one for $t \in (0, T)$ and zero elsewhere.

The spectrum of the signal (4.9) can be evaluated by noting that the signal can be written as

$$\mathbf{E} = Am'(t)e^{i[\omega t + \phi]}\mathbf{x} \tag{4.10}$$

where $m'(t)$ is given by equation (4.2) but with $b_k = -1$ or $b_k = 1$ and assuming rectangular pulses with duty cycle equal to one. Thus applying equation (4.4) obtains

$$S_b(\nu) = TA^2 \frac{\sin^2(\pi\nu T)}{(\pi\nu T)^2} \tag{4.11}$$

The transmitted field is coupled to the fiber and is attenuated during propagation toward the fiber far end where it is detected by a coherent receiver [5]. We describe here two particular realizations of a coherent receiver: heterodyne and homodyne synchronous receivers.

PSK Homodyne Receiver

The basic block diagram of a PSK homodyne receiver is shown in Fig. 4.3. A polarization controller is used to match the receiver signal to the local oscillator polarization state. Then the signal is mixed with the local oscillator field that has exactly the same optical frequency as the signal. The mixing is realized by an unbalanced directional coupler. The resulting field is detected by a photodiode.

The baseband electrical current $c(t)$ is expressed by [12–15]

$$c(t) = 2R_p A_s A_{1o} \alpha_{dc} \sqrt{1 - \alpha_{dc}^2} \cos[\Delta\phi + \pi m(t)] + n_q(t) + n_t(t) \tag{4.12}$$

with A_s, A_{1o} as the signal and local oscillator amplitudes, respectively, $\Delta\phi$ the phase difference between the signal and the local oscillator field, and α_{dc} the power splitting ratio of the directional coupler. The shot noise and the receiver thermal noise are indicated, respectively, by $n_q(t)$ and $n_t(t)$.

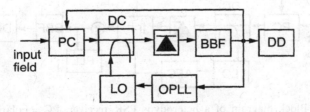

FIGURE 4.3 Block diagram of a homodine PSK receiver. PC = polarization controller, DC = directional coupler, LO = local oscillator, BBF = baseband filter, OPLL = optical phase lock loop, DD = decision device.

As in the case of the IM-DD receiver, the baseband current is sampled in the middle of each bit interval to obtain the decision variable C_k. The decision variable is then compared with zero to estimate the transmitted bit: If $C_k > 0$, a mark is estimated; otherwise, a space is assumed. From equation (4.12) it is clear that to correctly estimate the transmitted bit, the phase difference $\Delta\phi$ must be as near to zero as possible, that is, the local oscillator must be phase locked to the received signal. This condition is realized by the optical PLL as indicated in Fig. 4.3. An analysis of the structure and performance of optical PLLs is beyond of the scope of this book; a reference list on this complex argument is given in [11].

PSK Heterodyne Receiver

The basic block diagram of a PSK heterodyne receiver is shown in Fig. 4.4. A polarization controller is used to match the received signal to the local oscillator polarization state. Then the signal is mixed with the local oscillator field that has a frequency detuned from the signal frequency; the mixing is realized by an unbalanced directional coupler. The resulting field is detected by a photodiode.

The electrical current $c(t)$ at the photodetector output is expressed by

$$c(t) = 2R_p A_s A_{lo}\alpha_{dc}\sqrt{1 - \alpha_{dc}^2}\,\cos[2\pi f_{IF}t + \Delta\phi + \pi m(t)] + n_q(t) + n_t(t) \quad (4.13)$$

with f_{IF} being the so-called intermediate frequency, that is, the difference between the signal and the local oscillator frequencies. If the intermediate frequency is much greater than the bit rate, the current $c(t)$ has a bandwidth approximately equal to $2R$ centered around the frequency f_{IF}. Generally, f_{IF} is in the microwave or in the millimeters wave region, so the signal $c(t)$ can be demodulated by a standard synchronous radio receiver based on an electrical PLL.

It is important to note that no optical phase locking is needed in the heterodyne receiver, although the local oscillator must be frequency locked to the

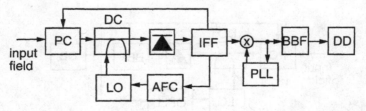

FIGURE 4.4 Block diagram of a heterodyne PSK receiver. PC = polarization controller, DC = directional coupler, AFC = automatic frequency control, LO = local oscillator, IFF = itermediate frequency filter, BBF = baseband filter, DD = decision device.

signal to ensure the stability of the intermediate frequency. This is realized by a frequency stabilization loop that is generally simpler to realize than an optical PLL.

4.1.3 Multiplexed Systems

Since the bandwidth of optical transmission systems is much larger than that of electrical systems, optical multiplexing and demultiplexing is required to fully exploit the potentialities of fiber systems. Two kinds of optical multiplexing concern us here: optical time division multiplexing (OTDM) and wavelength division multiplexing (WDM).

OTDM Systems

In general, the message transmitted over an OTDM channel is divided into frames. The simpler OTDM frame, which is shown in Fig. 4.5, is divided into an header and a payload. The header contains control and synchronization information, while each bit of the payload is associated to a different tributary channel. If the bit rate of each tributary channel is R, the bit rate of an OTDM channel multiplexing N_T tributary channels with an header N_H bit long is given by $(N_T + N_H)R$.

More complex frame structures can be adopted, for example, without a fixed relation among bits in the payload and tributary channels; however, a detailed description of all the possible realizations of TDM is beyond the scope of this chapter. The simple frame structure of Fig. 4.5 is sufficient to describe OTDM systems and to understand the problems that are present in their implementation. An accurate and complete description of the TDM technique can be found in several books on communication systems and networks [16].

The block scheme of an electrooptic OTDM multiplexer is shown in Fig. 4.6. A high-speed optical source provides a train of short pulses, one for each OTDM frame. The pulses feed a multiplexer composed of $N_T + N_H$ branches. On the generic branch the pulse stream is delayed and modulated by a standard electrooptic modulator; then the pulses are combined to form the OTDM frame. The electrooptic modulators work at the speed of a single tributary channel.

This multiplexer structure is simple, but its optical loss cannot be lower than $10 \log_{10}(N_T + N_H)$ dB, and very accurate control of the delay along each

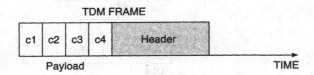

TDM FRAME

Payload   TIME

FIGURE 4.5   Scheme of a simple TDM frame. The slots assigned to the four tributary channels are indicated by $c1$, $c2$, $c3$, and $c4$.

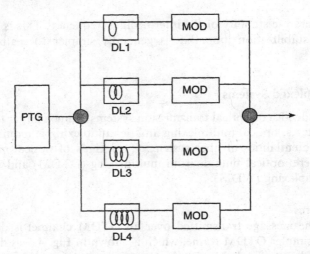

FIGURE 4.6 Example of an OTDM multiplexer. PTG = pulse train generator, MOD = modulator, DL = delay line.

branch is necessary. A possible scheme of an OTDM demultiplexer is shown in Fig. 4.7. The incoming OTDM signal is split into $N_T + 1$ branches. The signal on one branch feeds an optical device able to recover the frame clock, that is, a pulse stream presenting a pulse at the beginning of each OTDM frame. The frame clock is split by an $1 \times N_T$ splitter and delayed on each output branch of

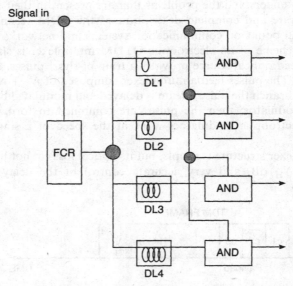

FIGURE 4.7 Example of an OTDM demultiplexer. FCR = frame clock recovery, DL = delay line.

the splitter so as to constitute the reference clock for each tributary channel. The reference clock selects a tributary channel on each signal branch by means of N_T optical AND gates.

The critical issue in this OTDM demultiplexer is the implementation of the AND gates and the frame clock extractor. For the AND gates, different implementations have been proposed, based on second harmonic generation in materials as KTP [17] and on four-wave mixing in an optical fiber [18] or in a semiconductor optical amplifier [19].

In the case of the frame clock extractor, its structure mainly depends on structure of the frame header. For example, if the header starts with a reference pulse more powerful than the other pulses of the frame, it can be insulated by an optical nonlinear gate that extracts the frame clock.

Besides the demultiplexer described in Fig. 4.7, other kinds of demultiplexers for OTDM signals have been proposed and experimentally demonstrated. Among them, one based on a nonlinear optical loop mirror [20] seems to be quite promising.

WDM Systems

The WDM signal is obtained by simply assigning to each tributary signal a different frequency band and transmitting all the signals on the same fiber. The resulting spectrum of the WDM signal would appear as shown in Fig. 4.8. Since the spectra tails of each transmitted channel theoretically occupy all the frequency axis, the different WDM channels can experience crosstalk. To limit this effect, a guard bandwidth between adjacent channels is placed in the WDM spectrum as shown in the figure.

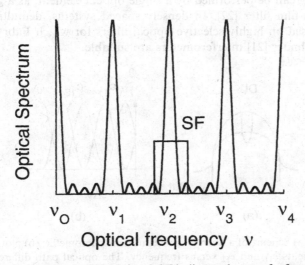

FIGURE 4.8 Spectrum of a WDM signal. SF indicates the transfer function of an ideal optical selection filter placed in front of the receiver.

Optical multiplexing of N_T signals at different frequencies is carried out in two ways that differ according to the multiplexing device: by passive star couplers or by frequency selective devices.

Using a star coupler, a simple multiplexer structure can be designed that is independent of the frequencies of the tributary channels but experiences an unavoidable internal loss of $10 \log_{10}(N_T)$ dB. Frequency selective elements can be used to reduce the internal loss up to negligible values (equal to zero in the case of ideal devices), but the multiplexer structure becomes more complex [21]. An example of a frequency selective multiplexer based on a cascade of Mach-Zehnder interferometers is shown in Fig. 4.9. The scheme of a fiber Mach-Zehnder interferometer is given in Fig. 4.9a, and the periodic behavior of the transfer matrix elements c_{11} and c_{12} is shown in Fig. 4.9b, where the frequency spacing between the maxima of the two transfer matrix elements is given by $\Delta \nu = \Delta L c/2$, with ΔL being the optical path difference between the interferometer branches and c the speed of light in the fiber. As the figure shows, if two carriers with optical frequencies ν_1 and $\nu_2 = \nu_1 + \Delta \nu$ feed the two inputs, they are multipexed at the output. Considering the periodic characteristic of the elements of the interferometer transfer matrix, a WDM multiplexer can be realized, as shown in Fig. 4.10, by a cascade of interferometers. This structure can also be used as a WDM demultiplexer, since all the tributary signals are splitted at the N_s inputs if the multiplexed signal feeds the output.

The structure of frequency selective WDM multiplexers/demultiplexers heavily depends on the frequency spacing between adjacent WDM channels. In Fig. 4.11 different multiplexer technologies are considered with the range of channel spacing for which they can be used. In the case of wide spacing WDM, demultiplexing can be performed by a single optical element, as a grating or a dielectric thin-film filter [22]. In densely spaced systems, demultiplexers are essentially based on highly selective optical filters for which Fabry-Perot [23] and Mach-Zehnder [21] interferometers are suitable.

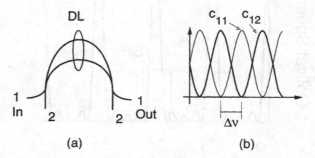

FIGURE 4.9 (a) Scheme of a fiber Mach-Zehnder interferometer; (b) plot of the transfer matrix elements c_{11} and c_{12} versus frequency. The optical path difference between the interferometer branches (DL) is indicated as ΔL and the frequency period of the transfer matrix elements as $\Delta \nu$.

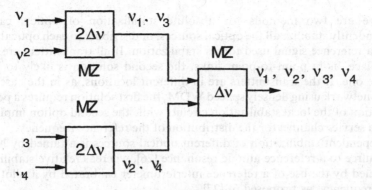

FIGURE 4.10 Block diagram of a WDM multiplexer build by a cascade of Mach-Zehnder interferometers (MZ). The characteristic frequency of each interferometer is reported inside its corresponding block as a function of the output WDM comb spacing $\Delta\nu$.

Integrated optics grating filters, based on waveguides like those realized in DFB lasers, have been also proposed: They are much more frequency selective than microoptic gratings [24]. The performance of integrated gratings are highlighted in Fig. 4.11 by the different shading. Optical filters based on liquid crystals have been proposed as well for applications in densely spaced WDM systems. Their performances are not reported in Fig. 4.11, but their potentialities are discussed in [25].

In densely spaced WDM systems, the absolute frequency of the optical carriers must be carefully controlled, typically with an accuracy of the order of a small fraction of the bit-rate. This is mainly due to the high frequency selectivity of the optical filters required for demultiplexing.

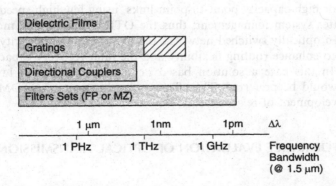

FIGURE 4.11 Relationship between multiplexer technology and attainable channel spacing in WDM transmission.

There are two methods for absolute stabilization of optical carriers: Independently stabilize all the optical sources, or distribute to each optical transmitter a reference signal used for its stabilization. If all transmitters are at the same place, as in point-to-point links, the second solution is likely to be the suitable one. If the transmitters are in different locations, as in the case of an optical network using densely spaced WDM, the first solution requires a periodic calibration of the local stabilization circuits while the second option implies the use of a service channel for the distribution of the reference frequency.

Independent stabilization of different optical sources is attained by locking each source to a reference atomic resonance [26], whereas relative stabilization is attained by the use of a reference interferometer calibrated by a pilot stabilizing frequency, as proposed in [27].

OTDM versus WDM Technique

The state of the art of WDM permits an easier use of the wide bandwidth of fiber optic systems. Multiplexers and demultiplexers are at the industrial development stage, while point-to-point WDM systems are already on the market. Besides technological maturity, an important advantage promoting the rapid distribution of WDM systems is that wavelength division can be used not only for multiplexing but also for network functions such as routing and switching [28].

The OTDM technology is less mature. The spectral efficiency of OTDM systems would be greater exploiting all the available optical bandwidth with a single OTDM channel due to the absence of the guard bandwidth. Since the bandwidth of an optical link is very wide (at least a few THz), this solution seems not possible. However, two other solutions could be considered: transmitting few high-speed OTDM channels at different frequencies or a large number of lower-speed channels by electrical TDM. In Section 4.2.3 it will be seen that the crosstalk induced penalty depends on the relative channel spacing $\Delta\nu/R$. Therefore the spectral efficiency of a WDM system does not depend on the bit-rate of the single channel, and the choice between WDM systems based on OTDM or on electrical TDM channels depends on other factors. For high-capacity point-to-point links, using few high-speed channels allows better system management; thus the OTDM may be the more viable solution. In optically switched networks the high WDM granularity could be exploited to enhance routing flexibility and the fault recovery capacity of the network. In this case a solution based on a large number of lower-speed channels would be preferred. The effective development of OTDM depends on the development of related technology.

4.2 PERFORMANCE EVALUATION OF OPTICAL TRANSMISSION SYSTEMS

To analyze the performance of optical transmission systems, a generalized model of the transmission system can be useful. Different performance

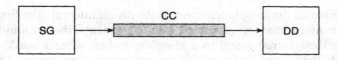

FIGURE 4.12   Block diagram of a generic transmission system: SG = signal generator, CC = communication channel, DD = decision device.

evaluation parameters could then be used to describe the impact of the various transmission impairments.

A general diagram of a transmission system is shown in Fig. 4.12. The signal generator encodes the transmission message in a baseband signal, whose expression is given in equation (4.2). The communication channel transmits the baseband signal from the signal generator to the decision device. The modulator, the transmission medium, and the receiver are all enclosed within the communication channel. The signal $r(t)$ at the decision device input is a random signal due to different noise disturbances. By separating from the average signal its randomness and neglecting, for sake of simplicity, the time delay due to signal propagation along the communication channel, this random process can be written as

$$r(t) = \sum_{j=1}^{N_b} s_j(t) + n(t, m) \tag{4.14}$$

The deterministic term in equation (4.14) can be divided by different contributions $s_j(t)$ that are zero out of the jth-bit interval. However, each contribution depends not only on the bit transmitted in the same bit interval but also on all of the transmitted message. This effect is known as the *pattern effect*. In practice, the dependence of the signal in the kth-bit interval on distant bits becomes weaker and weaker. Thus, in general, a finite channel memory can be defined so that the signal received in $(t_0, t_0 + T]$ depends only on the signal transmitted in $(t_0 - MT, t_0 + T + MT]$, M being an integer number. The difference $d_j(t) = s_j(t)/\sqrt{P_0} - g(t - jT)$ between the average received signal normalized to the received power P_0 and the transmitted signal is known as *signal distortion*.

The random term $n(t, m)$, which is the noise term, generally depends not only on the channel but even on the transmitted message $m(t)$. The decision device infers the transmitted message by sampling $r(t)$ at the center of each bit interval and comparing each sample with a decision threshold. If the sample is lower than the threshold, a space is assumed; otherwise, it is a mark.

A simple performance evaluation parameter for the communication system is the so-called eye-opening penalty (EOP). The eye-opening penalty is defined as

$$EOP = \frac{\min[s_j(t_j, b = 1)] - \max[s_j(t_j, b = 0)]}{\sqrt{P_0}\, g(T/2)} \tag{4.15}$$

where t_j represents the sampling instant of the jth-bit interval. The first addendum in the numerator of equation (4.15) represents the minimum value assumed by the received signal in a sampling instant when a mark is transmitted, while the second addendum is the maximum value assumed in correspondence of a space. Thus the eye-opening penalty is a measure of the reduction of the distance between the mark and the space due to distortion. Since only the average received signal appears in the definition of the EOP, it does not depend on noise, and in measuring the EOP in real systems, the noise contribution must be eliminated by an opportune average of the output signal.

A performance evaluation parameter that takes into account the presence of noise and is widely used is the average amplitude signal-to-noise ratio (ASN). This parameter is sometimes called the Q factor. However, we will adopt a different definition for the Q factor in the following. The ASN is defined as

$$\text{ASN} = \frac{1}{N_b} \sum_{j=0}^{N_b} (2b_j - 1) \frac{|s_j(t_j)|}{\sigma_n(t_j)} = \sum_{k=1}^{PP} \Pi_k (2b_k - 1) \frac{|s_k(t_k)|}{\sigma_n(t_k)} \qquad (4.16)$$

The variance of the random variable $n(t_j, m)$ is indicated by $\sigma_n^2(t_j)$, and the factor $(2b_j - 1)$ ensures that the contributions arising from bit intervals in which a mark is transmitted are added and those corresponding to a space are subtracted. The second expression of the ASN in equation (4.16) holds when a finite channel memory can be identified. In this equation $PP = 2^{2M+1}$ is the number of possible patterns of $2M + 1$ bits, while $s_k(t)$, b_k, and t_k are the signal, the transmitted bit, and the sampling instant of the mid-bit interval of the kth pattern, respectively. Finally the probability of occurrence of the kth pattern is indicated with Π_k. The expression of ASN is greatly simplified if no pattern effect is present ($s_j(b)$ depends only on b_j) and if the noise variance does not depend on the transmitted message. In this case, assuming equiprobable bits, ASN is given by

$$\text{ASN} = \frac{|s_1(t_j)| - |s_0(t_j)|}{2\sigma_n(t_j)} \qquad (4.17)$$

The most important performance evaluation parameter for a digital transmission system is the error probability. To determine the probability of an error occurring, a statistical analysis is made of the random variable $r(t_j)$, for example, finding the probability density function (PDF) $p(r, j)$. Then, assuming that the jth bit is a mark, the probability of error on the jth bit is

$$P_e(j) = \int_{-\infty}^{r_{th}} p(r, j) \, dr \qquad (4.18)$$

where r_{th} is the decision threshold. An analogous equation holds if a space is transmitted. The overall error probability P_e is the average of $P_e(j)$ over the transmitted message. This average can be expressed by a simple equation in the case where the channel has a finite memory, obtaining

$$P_e = \sum_{k=1}^{PP} \Pi_k P_e(k) \tag{4.19}$$

where $P_e(k)$ is the error probability for the bit at the center of the kth pattern.

In some important cases the characteristic function $G(\rho, j)$ of $r(t_j)$ is found instead of the PDF. In this book the characteristic function will be defined as the bilateral Laplace transform of the PDF (Appendix A2, equation A2.15); however, all of the following equations can be easily modified if the Fourier transform is used [30]. Starting from $G(\rho, j)$, the error probability $P_e(j)$ can be evaluated by the Cauchy formula [31]

$$P_e(j) = \frac{1}{2\pi i} \int_{\varepsilon-i\infty}^{\varepsilon+i\infty} \frac{G(\rho, j)}{\rho} e^{(2b_j-1)\rho r_{th}} \, d\rho \tag{4.20}$$

where ε is a positive real number inside the convergence strip of $G(\rho, j)$. The integral (4.20) can often be solved exactly by the residue theorem. When this is not possible, various asymptotic approximations have been proposed, such as those based on the Chernov bound or on the steepest descendent method (also called the saddle point approximation) [32]. Some of these methods are summarized in Appendix A2.

In general, the ASN is not sufficient to evaluate the error probability. However, in some cases a simple relation exists between P_e and ASN. The most important is the case of the gaussian channel. The gaussian channel is a channel with no distortion and in which the noise term is a gaussian wideband white process, independent of the transmitted signal. Then the error probability is derived directly as

$$P_e = \frac{1}{4} \operatorname{erfc}\left(\frac{|s_1 - r_{th}|}{\sqrt{2}\sigma_n}\right) + \frac{1}{4} \operatorname{erfc}\left(\frac{|s_0 - r_{th}|}{\sqrt{2}\sigma_n}\right) \tag{4.21}$$

where s_0 and s_1 are the average samples as a space or a mark is transmitted. The threshold r_{th} must be optimized to find the minimum error probability, since the optimum threshold is $r_{th} = 0.5(s_1 + s_0)$, equation (4.21) can be rewritten as

$$P_e = \frac{1}{2} \operatorname{erfc}\left(\frac{|s_1 - s_0|}{2\sqrt{2}\sigma_n}\right) = \frac{1}{2} \operatorname{erfc}\left(\frac{\text{ASN}}{\sqrt{2}}\right) \tag{4.22}$$

From equation (4.22) it derives that, the error probability is less than 10^{-9}, if the ASN factor is greater than 6.

Equation (4.22) can be used to obtain a rough evaluation of the error probability, starting with the knowledge of ASN, even if the communication channel is more complex than a simple gaussian channel [29]. However, a *widespread* use of equation (4.22) can lead to inaccurate results. The classical case is when $r(t_j)$ is the square of a gaussian variable. In this case the adoption of the so-called gaussian approximation (equation 4.22) introduces an error of 3 dB in evaluating the ASN factor needed to attain an error probability of 10^{-9}.

A performance evaluation parameter that is more accurate than ASN in representing the impact of pattern effects on the error probability is the Q factor. The Q factor is defined as

$$\frac{1}{2} \operatorname{erfc}\left(\frac{Q}{\sqrt{2}}\right) = \frac{1}{2N_b} \sum_{j=1}^{N_b} \operatorname{erfc}\left(\frac{|s_j - r_{th}|}{\sqrt{2}\sigma_n(j)}\right) \tag{4.23}$$

where the decision threshold r_{th} has to be optimized to attain a minimum value at the right-hand side of equation (4.23). In the case of a finite channel memory, the Q factor can be written as

$$Q = \sqrt{2}\operatorname{erfc}^{(-1)}\left[\sum_{k=1}^{PP} \Pi_k \operatorname{erfc}\left(\frac{|s_k - r_{th}|}{\sqrt{2}\sigma_n(k)}\right)\right] \tag{4.24}$$

where $\operatorname{erfc}^{(-1)}(\ )$ indicates the inverse of the complementary error function.

In the communication channel affected by gaussian noise, the right-hand side of equation (4.23) is equal to the error probability. In particular, in the absence of pattern effects and if $s_0 = 0$, this reduces to equation (4.22), and the Q factor coincides with ASN. Thus in the case an error probability of 10^{-9} is attained for a Q factor greater than 6.

If the decision variable PDF is not gaussian, the right-hand side of equation (4.23) is not the error probability, and the optimum threshold evaluated by maximizing the Q factor will not coincide with that obtained by minimizing the real error probability.

4.2.1 Performances of Direct-Detection Systems

In this section we present a simple model for the analysis of the performance of IM-DD systems. We will analyze the effects of important noise sources as well as of pulse broadening and time jitter.

The electrical current in front of the decision device in an IM-DD system is given by equation (4.8). An analysis of system performance can be carried out by assuming the use of an APD: The PIN is obtained as a special case where $M_p = 1$.

Equation (4.8) can be rewritten by separating the signal by the noise. Assuming that the receiver does not introduce distortion and taking into account that the APD gain is a random variable, we obtain

$$c(t) = [R_p A^2 m(t) M_p + n_s(t)] \otimes h_R(t) + n_t(t) \tag{4.25}$$

The term $n_s(t)$ is not gaussian, and its power spectral density is as large as the APD bandwidth [33]. If the overall receiver bandwidth (of the order of the bit rate) is much smaller than the APD bandwidth, the central limit theorem can be applied, and the term $n_s(t) \otimes h_R(t)$ can be modeled as gaussian noise [34]. Under this hypothesis the IM-DD system appears as a gaussian channel with a signal-depending noise variance. In particular, the overall noise variance can be written, in a first approximation, as

$$\sigma_n^2(b_j) = \sigma_s^2(b_j) + \sigma_t^2 = qR_p A^2 b_j M_p^{2+x_a} B_R + \frac{4k_B \Theta}{R_c} F_e B_R \qquad (4.26)$$

where the first addendum σ_s^2 is the filtered multiplication noise variance and the second the thermal noise variance σ_t^2. Further in the equation q is the electron charge, x_a is a parameter that characterizes the APD, and B_R is the electrical receiver bandwidth which is of the order of the bit-rate. The absolute temperature is indicated by Θ, the Boltzman constant by k_B, the receiver load resistance by R_c, and the electrical front-end noise factor by F_e.

As shown by equations (4.22) and (4.23), the Q factor of a gaussian channel without pattern effects coincides with the ASN; exploiting this property in this case obtains

$$Q = \text{ASN} = \frac{R_p A^2 M_p}{\sqrt{\sigma_s^2(1) + \sigma_t^2}} = \frac{R_p A^2 M_p}{\sqrt{2\left(qR_p A^2 M_p^{2+x_a} + \frac{4k_B \Theta}{R_c} F_e\right) B_R}} \qquad (4.27)$$

The error probability can be expressed by equation (4.21), which holds for a generic gaussian channel, substituting equation (4.26) for the noise variance.

To evaluate the performances of IM-DD systems using the simple model presented in this section, the APD gain and the decision threshold must be optimized to attain the lowest error probability. The optimum value of the APD gain can be obtained by maximizing the Q factor. Starting from equation (4.27), this optimization obtains

$$M_p = \left(\frac{8k_B \Theta F_e}{qR_p A^2 R_c}\right)^{1/(2+x_a)} \qquad (4.28)$$

The optimization of the threshold can be carried out by minimizing the error probability given by equation (4.21), which yields

$$r_{th} = \frac{R_p A^2 M_p \sigma_n(0)}{\sigma_n(0) + \sigma_n(1)} \qquad (4.29)$$

The parameters of a typical IM-DD receiver for high-speed optical signals are reported in Table 4.1, the error probability is shown in Fig. 4.13 as a

TABLE 4.1   Parameters of a sample IM–DD receiver for 10 Gbit/s signals.

Parameter	Value	Unit
Bit-rate, R	10	Gbit/s
Photodiode responsivity, R_p	0.9	A/W
Absolute temperature, Θ	300	K
Equivalent load resistance, R_c	100	Ω
APD noise parameter, x_a	0.8	—
Front end noise factor, F_e	6	dB

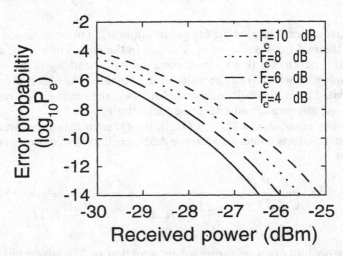

FIGURE 4.13 Error probability versus the received power for an IM-DD system for different values of the electrical noise factor. The system parameters are reported in Table 4.1.

function of the optical power at the receiver input for different values of the front-end noise factor and for a bit-rate $R = 10$ Gbit/s.

Examples of performance evaluation of IM-DD systems in the presence of noise disturbances are presented below. Two important phenomena will be analyzed: pulse broadening during propagation and time jitter in the received pulses. Both are important in practical systems. Pulse broadening is caused mainly by fiber dispersion and is one of the main impairments in standard IM-DD systems. Jitter can arise in IM-DD systems in the transmitter, and it is critical in soliton systems where it rises from the interaction of solitons with the amplifier noise, as shown in Chapter 5.

IM-DD Systems Performance in the Presence of Pulse Broadening

Pulse broadening during propagation along the fiber introduces a deterministic distortion in the transmitted signal. As this phenomenon occurs, the pulse partially occupies the contiguous bit intervals giving rise to the so-called inter-symbol interference. The pattern effect that arises in this case can be limited to the memory of one bit ($M = 1$) just as a small pulse broadening occurs. Assuming RZ transmission using gaussian pulses with a standard deviation σ_T, the received pulses always have a gaussian shape but with a standard deviation $\sigma'_T = \sigma_T + \Delta\sigma$, with $\Delta\sigma$ being the broadening factor. The overall error probability can be expressed by equation (4.19) with $PP = 8$ to represent all the patterns of three bits and $\Pi_k = \frac{1}{8}$. The conditional error probability $P_e(k)$ is given by

$$
P_e(k) = \frac{1}{2} \operatorname{erfc} \left[\frac{|R_p P_o(k) M_p - r_{th}|}{\sqrt{2 \left(q R_p A^2(k) M_p^{2+x_a} + \dfrac{4 k_B \Theta}{R_c} F_e \right) B_R}} \right]
\tag{4.30}
$$

where $P_o(k)$ is the optical power detected at the central bit of the kth pattern, and it can be evaluated by the broadening factor. Optimization of the decision threshold can proceed starting with equation (4.30).

The obtained error probability is reported in Fig. 4.14 as a function of the received optical power for different values of the pulse broadening normalized to the bit-rate ($\Delta\sigma R$). The error probability evaluated starting from the ASN by equation (4.22) is also reported in Fig. 4.14 for comparison.

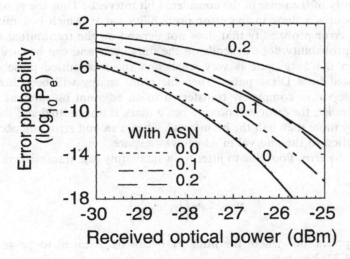

FIGURE 4.14 Error probability versus the received power for an IM-DD system in the presence of time pulse broadening for different values of the pulse broadening normalized to the bit duration. The error probability curves evaluated by the gaussian approximation starting from the ASN factor are reported for comparison.

In this case the approximation of the error probability by the ASN factor and equation (4.22) is acceptable only if the pulse is moderately broadened. If $\Delta\sigma R = 0.1$, the pulse broadening introduces a penalty of about 1.7 dB in the system sensitivity (received power for $P_e = 10^{-9}$). In this case the error caused by the use of equation (4.22) and ASN is about 0.6 dB. If the normalized pulse broadening is far below 10%, the ASN approximation is acceptable; if $\Delta\sigma R > 0.1$, this approximation yields serious errors.

On the other hand, since we have a gaussian channel, an exact estimate of the error probability via equation (4.22) can be made by using the Q factor instead of the ASN.

IM-DD Systems Performances in the Presence of Time Jitter

In the presence of jitter, the modulating function $m(t)$ in equation (4.2) is written as

$$m(t) = \sum_{k=-\infty}^{\infty} b_k g(t - kT - \tau_k) \tag{4.31}$$

where the random variables τ_k represent the jitter; these variables are assumed to be independent and with the same PDF, indicated with $p_j(\tau)$.

Due to the jitter, energy from a bit is transferred to adjacent bit intervals, thus increasing the error probability. When almost all of the pulse's energy is transferred to a bit interval where the transmission is zero, the error occurs independently of the noise in the considered bit intervals. Thus the presence of jitter introduces a floor in the error probability curve, which is a minimum achievable error probability that does not depend on the transmitted energy. The error probability floor depends on the duty cycle and can be easily evaluated when the duty cycle is very small and the transmitted pulse can be approximated by a Dirac pulse. Then the pulse energy will be either in its own bit interval or completely transferred to an adjacent bit interval. In the absence of noise, the error will occur when a mark is transmitted and the pulse is shifted by more than half the bit interval. Then a second error will occur as a pulse is shifted to the interval in which was a space.

In sum, the error floor due to jitter for a small duty rate transmission can be written as

$$P_e = \frac{3}{4}\Pr\left\{|\tau_k| > \frac{T}{2}\right\} \tag{4.32}$$

in those applications where the jitter PDF can be assumed to be gaussian, equation (4.32) becomes

$$P_e = \frac{3}{4}\operatorname{erfc}\left(\frac{T}{2\sqrt{2}\,\sigma_j}\right) \tag{4.33}$$

where σ_j is the standard deviation of the jitter. It can be observed that by equation (4.33) an error floor of 10^{-9} is achieved if $\sigma_j/T \approx 0.08$. This result is influenced by the assumption of *zero* pulse duration; if a small but nonzero pulse duration is assumed, this value is higher.

When the duty rate is not so small, jitter can induce pattern effects. A limited memory may be introduced since a bit is influenced by its nearest neighbors. Equiprobable are the patterns of three bits, and they can be used for the evaluation of the error probability, $PP = 8$ and $\Pi_k = \frac{1}{8}$ for $k = 1, \ldots, 8$. Equation (4.19) can be used in calculating the pattern error probabilities $P_e(k)$.

Let us indicate with b_{-1}, b_0, and b_1 the three bits of a generic pattern and with τ_{-1}, τ_0, and τ_1 the corresponding jitter variables. The optical average power P_o detected in the central bit interval of the pattern is given by

$$P_o(\tau_{-1}, \tau_0, \tau_1) = \frac{A^2}{T} \sum_{v=-1}^{1} b_v \int_{-T/2}^{T/2} g(t - vT - \tau_v)\, dt \qquad (4.34)$$

where the dependence of the optical power on the jitter variables is explicit. Substituting the average power given by equation (4.34) to A^2 in equation (4.27), a jitter dependent ASN is obtained, from which the pattern error probability conditioned to the jitter variables can be found by the formula

$$P_e(k, \tau_{-1}, \tau_0, \tau_1) = \frac{1}{2} \operatorname{erfc}\left[\frac{|\mathrm{ASN}(k, \tau_{-1}, \tau_0, \tau_1) - \mathrm{ASN}_{th}|}{\sqrt{2}} \right] \qquad (4.35)$$

where $\mathrm{ASN}_{th} = r_{th}/\sigma_n(b_0)$ takes into account the decision threshold.

From (4.35) the overall pattern error probability can be obtained by saturation with respect to the jitter variables; thus

$$P_e(k) = \int_{-\infty}^{\infty} \int_{-\infty}^{\infty} \int_{-\infty}^{\infty} P_e(k, \tau_{-1}, \tau_0, \tau_1) p_j(\tau_{-1}) p_j(\tau_0) p_j(\tau_1)\, d\tau_{-1}\, d\tau_0\, d\tau_1 \qquad (4.36)$$

In general, it is difficult to apply equations (4.34) through (4.36) directly. However, there is at least one interesting case in which the integration can be done by an asymptotic approximation that treats the erfc( ) function as an exponential (i.e., for high error probability, the rule of thumb is to obtain the solution for $P_e < 10^{-4}$). This is possible when the jitter PDF is gaussian (with zero mean and standard deviation σ_j) and the received pulses are rectangular; that is, $g(t)$ is equal to A for $|t| < t_g/2$ and zero elsewhere, with $t_g \leq T$.

These results are shown in Fig. 4.15 and 4.16. In particular, in Fig. 4.15 the error probability is shown with relation to the ideal ASN (ASN of the system with no jitter) for a duty cycle $t_g/T = 0.5$ and different values of the standard deviation of the jitter. In the figure the jitter-induced error probability floor is evident. The abrupt slope change of the error probability curves, passing from an almost ideal behavior to the floor within about 2.5 dB, depends on the pulse

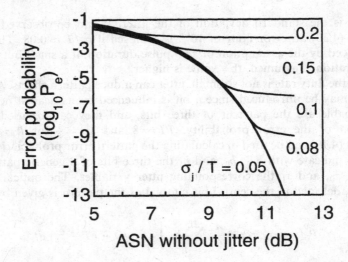

FIGURE 4.15 Error probability in the presence of gaussian jitter versus the ideal ASN (ASN of the system with no jitter) for a duty cycle $t_g/T = 0.5$ and different values of the jitter standard deviation.

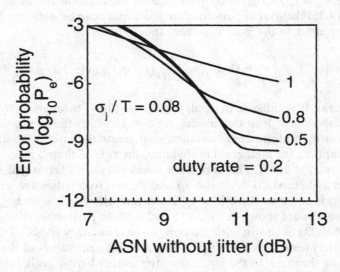

FIGURE 4.16 Error probability in the presence of gaussian jitter versus the ideal ASN fixing a jitter standard deviation ($\sigma_j = 0.08\,T$) and for different values of the duty cycle.

shape and on the duty cycle. In the case of gaussian pulses, for example, the curves behavior is smoother.

In Fig. 4.16 the error probability is shown with relation to the ideal ASN for different values of the duty cycle, fixing the standard deviation of jitter at $\sigma_j = 0.08\,\mathrm{T}$. Increasing the duty cycle produces smoother curves, but the performance gets worst, as can be expected by the fact that wider the pulse, greater the energy transferred to the adjacent bit interval.

Using equation (4.22), an approximation of the average ASN in the presence of jitter can be made:

$$ASN = \int_{-\infty}^{\infty} \int_{-\infty}^{\infty} \int_{-\infty}^{\infty} ASN(\tau_{-1}, \tau_0, \tau_1) p_j(\tau_{-1}) p_j(\tau_0) p_j(\tau_1)\, d\tau_{-1}\, d\tau_0\, d\tau_1 \quad (4.37)$$

where the jitter conditioned value of the ASN can be derived from equation (4.16) by using the three-bit patterns.

The exact value of the error probability and the value obtained by combining equation (4.22) and the ASN are compared in Fig. 4.17. The result, if time jitter is present, indicates that the ASN is not sufficient to evaluate the system's performance. Indeed, when transmission errors are mainly determined by time jitter, an high value of ASN does not ensure a low error probability. This is due to the fact that the ASN is basically the average value of the square signal-to-noise ratio. As a simple example, consider the case where the duty cycle is very small (short pulses in a long bit interval) and the ASN is quite high. Moreover assume that due to the time jitter, a pulse is translated out of its bit interval

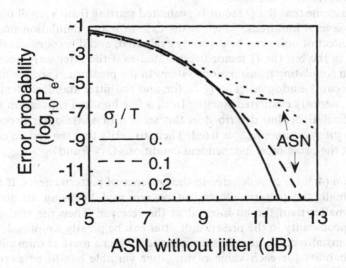

FIGURE 4.17 Comparison between the exact error probability and the error probability evaluated by means of equation (4.22) and the ASN in the presence of jitter. The conditions are the same as those of Fig. 4.15.

with a probability equal to 10^{-6}. The overall error probability is only 10^{-6}, but the ASN value remains very high because only one term every 10^{6} is affected by the jitter on the average.

This result is important when the system performance is evaluated via a simulation program. In this case it is easy to approximate equation (4.37) by an average over the entire transmitted message, thus obtaining the ASN.

As far as the use of the Q factor and equation (4.22) are concerned, the accuracy of the approximation depends on how the Q factor is defined in the presence of jitter. If the Q factor is defined as $Q = \sqrt{2}\,\mathrm{erfc}^{-1}(2P_e)$, where P_e is the exact error probability, obviously equation (4.22) can be used without error.

In a simulation program in which a bit-dependent error probability is evaluated, the average error probability is obtained by averaging over the transmitted message and then the Q factor is calculated. However, this algorithm requires a sufficiently long message to correctly estimate the average with respect to the jitter variables. This condition is difficult to obtain when simulating optical transmission systems because of the low required error probability.

An example may better elucidate the situation. Let us assume an RZ system with a very small duty cycle so that the transmitted pulses can be assumed to be Dirac pulses. If the system has a very high signal-to-noise ratio and the probability that the pulse is shifted out of the bit interval due to jitter is 10^{-8}, then the error probability is about 10^{-8}. Now, to correctly estimate the Q factor by a Monte Carlo method, more than 10^{9} bits have to be considered. Such a high number of simulation runs is not feasible; thus the Q factor in this situation cannot be correctly evaluated.

Let us assume that the Q factor is evaluated starting from a small number of bits (e.g., several hundreds, which is the case in many simulation programs). Then the effect of noise can be correctly evaluated, and if no considerable jitter is present in the bit, the Q factor is estimated as if the jitter were not present.

Thus, in simulating transmission systems in the presence of sizable jitter, this procedure can be adopted. The Q factor and the jitter standard deviation σ_j are simultaneously evaluated starting from a few hundred simulation runs. By the analytical algorithm described in this section or some other procedure, a maximum jitter variance σ_{max} is fixed. Then, to verify that the system can work effectively, the two almost independent conditions $Q > 6$ and $\sigma_j > \sigma_{max}$ must be verified.

Equation (4.36) is complex due to the presence of pattern effects. If the ratio σ_j/T is small, the pattern effects can be removed by using an acceptance window smaller than the bit-interval at the receiver. Then the evaluation of the error probability in the presence of jitter can be greatly simplified. When a space is transmitted, the jitter has no influence; when a mark is transmitted, the error probability for each value of the jitter variable has to be averaged to obtain the final error probability. In this case the equations for the gaussian channel case can be applied, substituting the received power with the power measured in the receiver acceptance window.

4.2.2 Performance of Coherent Systems

In the case of a PSK homodyne receiver, the photocurrent in front of the decision device is given by equation (4.12). The ASN factor for the PSK homodyne receiver can be evaluated assuming perfect phase matching between the signal and the local oscillator and the use of a PIN photodiode. This last option is preferred because the optical power at the photodetector is so high that an APD, where internal gain is present together with multiplication noise, is not required. In the absence of multiplication noise, the detection noise reduces to the shot noise, as indicated by equation (4.12). Moreover the shot noise PDF can be approximated with a gaussian PDF, since the number of photons impinging the photodiode is very high due to the local oscillator. The PSK homodyne systems is a gaussian channel without pattern effects, so the ASN and the Q factor coincide. Assuming that the receiver bandwidth exactly equals the bit-rate,

$$Q = \text{ASN} = \sqrt{\frac{4R_p^2 A_s^2 A_{lo}^2 \alpha_{dc} \sqrt{1 - \alpha_{dc}^2}}{qR_p A_{lo}^2 \sqrt{1 - \alpha_{dc}^2}\, R + \dfrac{4k_B \Theta}{R_c} F_e}} \tag{4.38}$$

By increasing the local oscillator power, it is possible to make the thermal noise negligible with respect to the shot noise. Then the receiver operates at the quantum limit. In this case, assuming that α_{dc} is near to one and $R_p = q/h\nu$, with ν representing the optical frequency,

$$Q = \text{ASN} = \sqrt{\frac{4A_s^2}{h\nu R}} = \sqrt{4E_s} \tag{4.39}$$

where E_s represents the number of signal photons received in the bit interval.

The optimum threshold is zero, since the average signals in correspondence of a mark and a space have the same module and opposite signs and the noise variance does not depends on the signal. Thus the overall error probability is given by equation (4.22) which can be written as

$$P_e = \frac{1}{2}\,\text{erfc}\left(\frac{Q}{\sqrt{2}}\right) = \frac{1}{2}\,\text{erfc}(\sqrt{2E_s}) \tag{4.40}$$

The performance of a PSK heterodyne receiver can be derived by an assumption analogous to that used for the homodyne receiver. In particular, the expression of Q at the quantum limit is given by

$$Q = \text{ASN} = \sqrt{\frac{2A_s^2}{h\nu R}} = \sqrt{2E_s} \tag{4.41}$$

It is worth noting that the heterodyne receiver suffers a 3-dB penalty in the Q factor with respect to the homodyne receiver. This is due to the fact that the noise bandwidth of the heterodyne receiver coincides with the intermediate frequency bandwidth, which is of the order of $2R$ (it is assumed to be exactly equal to $2R$ in equation 4.41). On the other hand, the noise bandwidth of the homodyne receiver is R. Since the shot noise spectral density is equal for the two receivers, the heterodyne receiver is affected by a greater noise power.

The error probability for the heterodyne receiver is given by

$$P_e = \frac{1}{2}\operatorname{erfc}\left(\frac{Q}{\sqrt{2}}\right) = \frac{1}{2}\operatorname{erfc}(\sqrt{E_s}) \qquad (4.42)$$

4.2.3 Crosstalk in IM-DD WDM Systems

In Sections 4.2.1 and 4.2.2 single-channel systems were considered. If WDM is adopted, besides the receiver noise and other performance-limiting phenomena discussed in Sections 4.2.1 and 4.2.2, crosstalk can arise among the WDM channels.

Figure 4.8 shows the positive frequency part of the optical spectrum of an IM-DD WDM signal; the transfer function of the optical filter operating as a demultiplexer is shown in the figure as well. A disturbance arises in detecting the selected channel due the presence of the tails of other channels in the filter bandwidth. The crosstalk causes a sensitivity penalty that increases with decreasing channel spacing.

In this section the effect of crosstalk is analyzed for an IM-DD WDM system in order to show a way to study this phenomenon. A detailed analysis of crosstalk in coherent systems is provided in a paper by Kazowski cited in [6].

After optical filtering in front of the receiver, it can be assumed that only two interfering channels contribute to the optical signal besides the selected channel. Under this assumption the received signal can be written as

$$\mathbf{E} = \sum_{h=-1}^{1} A_h[\sqrt{m_h(t)}\, e^{i(h\Delta\omega t + \psi_h)}] \otimes h(t)\xi_h e^{i\omega t} \qquad (4.43)$$

where A_h, ψ_h, $m_h(t)$, and ξ_h are, respectively, amplitude, phase, modulating function, and polarization unit vector of the hth channel and $\Delta\omega$ is the channel spacing. The convolution operation is indicated by $\otimes$, the selected channel has been labeled with the index 0, and $h(t)$ represents the pulse response of the optical filter. After detection of the optical signal by a photodiode and filtering by a baseband filter, the electrical signal can be written as

$$c(t) = R_p|\mathbf{E}|^2 \otimes h_R(t) + n(t) \qquad (4.44)$$

where $h_R(t)$ is the electrical filter pulse response. The noise process $n(t)$ at the electrical filter output can be assumed to be gaussian, as explained in Section 4.2.1.

In general, the messages transmitted on the interfering channels are not synchronous with the message on the considered channel. However, the cross-talk effect is maximum in the synchronous case, in which the bit intervals begin at the same instants on the three channels. Thus synchronous messages can be assumed. If b_h ($h = -1, 0, 1$) denotes the bit transmitted on the hth channel in the considered bit interval, the error probability P_e can be expressed as a function of the error probability $P_e(b_{-1}, b_0, b_1)$ conditioned to the bit transmitted on the three channels. In particular, assuming equiprobable bits,

$$P_e = \frac{1}{8} \sum_{b_{-1}, b_0, b_1 = 0}^{1} P_e(b_{-1}, b_0, b_1) \qquad (4.45)$$

The conditional error probability $P_e(b_{-1}, b_0, b_1)$ can be simply evaluated by the equation

$$P_e(b_{-1}, b_0, b_1) = \frac{1}{2} \operatorname{erfc}\left[\frac{|C(b_{-1}, b_0, b_1) - r_{\text{th}}|}{\sqrt{2}\sigma(b_{-1}, b_0, b_1)}\right] \qquad (4.46)$$

once the average value of the decision variable $C(b_{-1}, b_0, b_1)$ and its standard deviation $\sigma(b_{-1}, b_0, b_1)$ are known.

The conditional error probability depends on different parameters. Regarding the relative phases of the optical carriers ψ_h ($h = -1, 0, 1$), an evaluation of the crosstalk-induced penalty for different values of ψ_h shows that this dependence is negligible; thus in-phase optical carriers can be assumed. However, the dependence on relative polarizations of the transmitted channels is important. In fact, due to the nonlinear photodiode characteristics, beating products arise if the channels are equally polarized but not if adjacent channels have orthogonal polarizations.

To obtain results useful for system design, the penalties relative both to the best and to the worst cases are evaluated and compared in this section. To obtain the maximum penalty, the scalar products $\xi_1 \cdot \xi_0^*$, $\xi_1 \cdot \xi_{-1}^*$, and $\xi_{-1} \cdot \xi_0^*$ are assumed to equal a unit, which is equivalent to supposing that all the channels feeding the receiver have the same polarization. The penalty in the best case is evaluated assuming that the scalar products $\xi_1 \cdot \xi_0^*$, $\xi_1 \cdot \xi_{-1}^*$, and $\xi_{-1} \cdot \xi_0^*$ are equal to zero, which is the same thing as assuming that the polarization state of the selected channel is orthogonal with respect to those of the interfering channels.

Numerical results can be obtained by inserting into equations (4.45) and (4.46) the shapes of the optical and electrical filters. The crosstalk-induced penalty has been evaluated for a Fabry-Perot filter with a 3-dB optical band-width equal to B_o and a free-spectral-range equal to $200 B_o$, which is the same as assuming a finesse equal to 100. The electrical filter is a second-order Butterworth filter with a bandwidth equal to the bit-rate R.

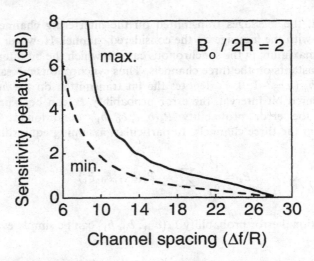

FIGURE 4.18 Minimum and maximum sensitivity penalty due to channel crosstalk in a WDM system versus the normalized channel spacing.

The penalties for the best and the worst case are shown in Fig. 4.18, relative to an optical filter bandwidth equal to four times the bit-rate. For a 1-dB maximum sensitivity penalty, the channel spacing must be wider than 20 R. The maximum penalty is shown in Fig. 4.19 for different values of the ratio B_o/R.

4.3 TRANSMISSION SYSTEMS ADOPTING OPTICAL AMPLIFIERS

4.3.1 Design of Optically Amplified Systems

As shown in Chapter 3, the noise performance of an optical amplifier is characterized by the inversion factor F. Assuming that the input signal is affected only by the shot noise and that $G \gg 1$, the optical signal-to-noise ratio SNR_o at the amplifer output is

$$SNR_o = \frac{A^2}{h\nu F B_o} \tag{4.47}$$

where A^2 is the input optical power, ν the optical frequency, h the Planck constant, and B_o the optical bandwidth.

As shown in Fig. 4.20, an amplified fiber link is a cascade of optical amplifiers and passive fiber sections; in-line filters and other optical devices can be present as well at the amplifiers locations, at the transmitter output, or in front of the receiver.

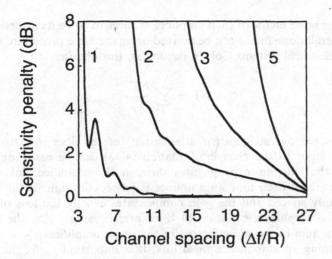

FIGURE 4.19 Maximum sensitivity penalty due to channel crosstalk versus the normalized channel spacing for different values of the optical bandwidth normalized to twice the bit-rate $(B_o/2R)$.

To carry out the performance analysis of the amplified system, the signal-to-noise ratio has to be calculated at the receiver side. Thus it is fundamental to evaluate the noise factor for the whole link.

For an exact evaluation the overall noise factor, the shot noise has to be taken into account [35]. However, the optical power is generally sufficiently high to assume that the main noise contribution for each amplifier is due to the beat between spontaneous emission and signal wave at the amplifier output. In

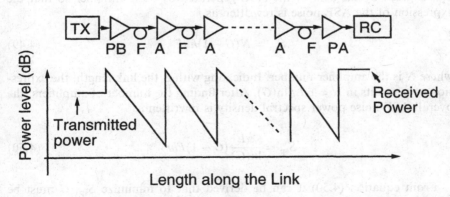

FIGURE 4.20 Basic block scheme of a transmission system adopting optical amplifiers and power level diagram along the link. TX = transmitter, PB = power booster, F = fiber link, A = amplifier, PA = receiver preamplifier, RC = receiver.

this case the noise factor of each amplifier reduces to twice its inversion factor, and the overall noise factor can be derived using the same procedure as that for amplified electrical systems [36]. In particular, this obtains

$$NF = \sum_j \frac{NF_j}{\Gamma_j} \qquad (4.48)$$

where Γ_j is the overall gain (or attenuation) of the fiber link from the jth amplifier output to the receiver. Equation (4.48) can be easily obtained by observing that the noise propagates through the amplified link, and new ASE is generated each time an amplifier is passed through. If the amplifiers are uniformly spaced and the gain compensates exactly the loss of the fiber between two amplifiers, equation (4.48) expresses simply that the link noise factor is the sum of the noise factors of the single amplifiers.

In designing an amplified optical link, it is important to fix the optimum value for amplifier spacing. First of all the conditions under which such optimization is performed must be defined. In long fiber links that are not based on soliton transmission, the effects due to nonlinear phenomena arising during propagation have to be reduced as much as possible. Since the impact of fiber nonlinearities depends essentially on the average optical power along the link, it is reasonable to perform optimization for a given value of the average power. This value is to be chosen to minimize the penalty induced by nonlinear effects. With this assumption, since the signal power at the receiver end is proportional to the average power in the link, the maximum signal-to-noise ratio at the receiver is obtained minimizing the ASE power. For the sake of simplicity, it is useful to assume identical amplifiers with gain G, which exactly compensates the loss of a fiber section. This yields $Ge^{-\alpha L_A} = 1$, where α is the fiber loss (m^{-1}) and L_A the amplifier spacing. The ASE power spectral density at the receiver, S_{sp}, is given by the sum of the ASE generated by each amplifier so that the expression of the ASE noise is rewritten as

$$S_{sp} = N(G - 1)h\nu F \qquad (4.49)$$

where N is the amplifier number. Indicating with L the link length, the expression of G results in $N = \alpha L / \ln(G)$. After finding the number of amplifiers, the overall ASE noise power spectral density is rewritten as

$$S_{sp} = \frac{\alpha L}{\ln(G)}(G - 1)Fh\nu \qquad (4.50)$$

From equation (4.50) it can be derived that to minimize S_{sp}, G must be minimized. To maintain the constraint $Ge^{-\alpha L_A} = 1$, L_A must be reduced while decreasing G up to the limit of distributed amplification. Obviously this is only a theoretical solution, since a long chain of doped fibers implies a number of application problems. Anyway, as equation (4.50) shows, if lumped amplifiers are used with a spacing of 30 to 50 km, the increase in the ASE power spectral density is of the order of a few dBs with the usual system parameters, which are acceptable in practice.

4.3.2 Performances of Optically Amplified Systems with Ideal Fiber Propagation

Optically Amplified IM-DD Systems

ASE noise is the main source of performance limitation for amplified IM-DD systems if fiber dispersion and nonlinearities can be neglected [37]. Indeed, in the presence of an optical amplifier in front of the photodetector, the received optical power is sufficiently high to adopt a PIN photodiode and to neglect the electronic thermal noise. Because here the propagation along the link is ideal, the signal waveform is not distorted. To further simplify the performance evaluation, it is assumed that the optical bandwidth B_o is determined by an optical filter placed in front of the receiver and that the electrical part of the receiver behaves as an electrical integrator. Then, the arbitrarily polarized optical field at the receiver input can be expressed as

$$\mathbf{E} = A\sqrt{m(t)}\,\boldsymbol{\xi}e^{i(\omega t + \phi)} + \mathbf{n}(t)e^{i\omega t} \tag{4.51}$$

where the modulating function $m(t)$ is given by equation (4.2) and $\boldsymbol{\xi}$ is the incoming field polarization unit vector. The amplifiers spontaneous emission is represented by the complex vector $\mathbf{n}(t)$ whose four quadratures are independent, bandlimited, gaussian, random processes with zero mean, and power spectral density S_{sp} within the considered bandwidth.

The optical field expressed by equation (4.51) is detected by the PIN, the resulting electrical signal is amplified, filtered by an electrical front end, and finally sampled at the center of each bit interval. The photocurrent $c(t)$ generated by the photodiode can be written as

$$c(t) = \frac{R_p}{T}\int_{-T/2}^{T/2}|\mathbf{E}(t-t')|^2\,dt' = \frac{R_p}{T}\int_{-T/2}^{T/2}\sum_{j=1}^{4}e_j^2(t-t')\,dt' \tag{4.52}$$

where $e_j(t)(j=1,2,3,4)$ are the four quadratures of the received field. Because of the presence of ASE, $e_j(t)$ are gaussian, bandlimited, independent random processes with bandwidth B_o. The decision variable C_k is obtained by the photocurrent $c(t)$ by sampling at the center of the kth bit interval.

Since the quadratures $e_j(t)$ are bandlimited processes, the sampling theorem can be applied to equation (4.52). Thus the decision variable C_k can be rewritten as

$$C_k = c(t_k) = \frac{R_p}{m}\sum_{w=-m/2}^{m/2-1}\left(\sum_{j=1}^{4}e_j^2(t_k - wT_o)\right) \tag{4.53}$$

where T_o is the sampling interval and m is assumed even. If m is odd, equation (4.53) has to be trivially modified.

The number of samples per bit is assumed to be equal for all the bits of the message. This is the same as assuming that $B_o/(2R)$ is an integer number, which is equal to m in this case. The model can be easily generalized to the case where $B_o/(2R)$ is a rational number, but it implies a more complex formalism without changing the results.

To determine the error probability, it is necessary to evaluate the probability density function of the decision variable. This can be conveniently evaluated through its characteristic function.

The variable C_k is given by the sum of m samples of the square module of the photocurrent: Since the sampling rate has been chosen equal to the Nyquist rate and the optical channel transfer function has been assumed to have an ideal square shape, the samples are statistically independent.

The single photocurrent sample is an hermitian quadratic form of the independent random gaussian variables $e_j(t_k - wT_o)$ $(j = 1, 2, 3, 4)$; its characteristic function $G(\xi/b_k)$, conditioned to the transmitted bit b_k, can be analytically calculated with the algorithm detailed in Appendix A2 [38], to obtain

$$G(\xi/b_k) = \frac{\exp\{-\xi[A^2 b_k/(1 + 2S_{sp}B_o\xi)]\}}{(1 + 2S_{sp}B_o\xi)^2} \tag{4.54}$$

Since the different degrees of freedom of the optical signal are independent random variables, the characteristic function $G_k(\xi/b_k)$ of C_k can be derived by raising $G(\xi/b_k)$ to the power m. Considering the scale factor R_p/m and assuming that $\zeta = 2S_{sp}B_o\xi$, it comes out that

$$G_k(\zeta/b_k) = \frac{\exp\{-\text{SNR}_o[mb_k\zeta/(1 + \zeta)]\}}{(1 + \zeta)^{2m}} \tag{4.55}$$

where $\text{SNR}_o = A^2/S_{sp}B_o$ is the optical signal-to-noise ratio.

The characteristic function of the decision variable depends on only two parameters: the optical signal-to-noise ratio SNR_o and the degrees of freedom of the optical signal, which is the ratio m between optical bandwidth B_o and electrical bandwidth $2R$.

Besides the SNR_o, the noise performances of the systems can be analyzed by the ASN or the Q factor at the receiver. The ASN can be evaluated by neglecting the receiver noise and assuming a square transfer function for both the optical and the electrical filter. The electrical bandwidth is assumed to be equal to R and the optical bandwidth to mR. Since no memory is present, the Q factor can be evaluated starting with equation (4.24) for $PP = 2$. Optimizing the threshold obtains [37]

$$Q = \frac{A^2}{\sqrt{2A^2 S_{sp}R + S_{sp}^2 R^2 m} + \sqrt{S_{sp}^2 R^2 m}} \tag{4.56}$$

Starting from the expression of the characteristic function, the conditional error probability can be evaluated by the Cauchy formula reported in equation (4.20). In the case $b_k = 0$, the Cauchy formula can be solved in closed form by the residue theorem, obtaining

$$P_e(0) = \exp\left\{-\frac{r_{th}}{2R_p S_{sp} B_o}\right\} \sum_{j=0}^{m-1} \frac{1}{j!} \left(\frac{r_{th}}{2R_p S_{sp} B_o}\right)^j \tag{4.57}$$

When $b_k = 1$ a closed form, a solution is not possible. However, the saddle-point approximation can be used to obtain a very good approximation of the error probability (see Appendix A2). It yields [39]

$$P_e(1) = \frac{(\sqrt{r_{th}/R_p A^2})^m \exp\{-r_{th}/2R_p S_{sp} B_o(1 - \sqrt{2R_p A^2/r_{th}})^2\}}{(\sqrt{R_p A^2/r_{th}} - 1)\sqrt{4\pi(r_{th}/R_p S_{sp} B_o \sqrt{r_{th}/R_p A^2})}} \tag{4.58}$$

Finally the error probability can be evaluated, in the absence of pattern effect, by averaging the error probabilities with $b_k = 0$ and $b_k = 1$ and numerically optimizing the decision threshold r_{th}. In Fig. 4.21 the error probability is shown versus the optical signal-to-noise ratio, calculated in the electrical bandwidth, SNR_e for different values of m. The optical signal-to-noise ratio in the electrical bandwidth is defined as

$$SNR_e = \frac{A^2}{RS_{sp}} = mSNR_o \tag{4.59}$$

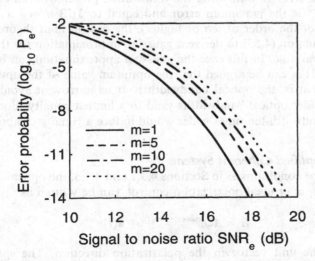

FIGURE 4.21 Error probability versus the signal-to-noise ratio in the electrical bandwidth for an IM-DD system adopting optical amplifiers. Different values of the ratio m between the optical and the electrical bandwidth are considered.

This parameter, instead of SNR_o, has been chosen for the error probability plot because SNR_e and m are independent variables for the given bit-rate (the first depends on A_s^2 and S_{sp}, and the second on B_o). On the other hand, SNR_o and m are correlated, since both depend on the optical bandwidth B_o. As a consequence, for a given value of SNR_o, the error probability becomes a decreasing function of m, as can be deduced from the data in Fig. 4.21. However, this does not mean that an increase in the bandwidth of the optical filter leads to an improvement in the system performance, for if the optical bandwidth is increased, the same value of SNR_o can be maintained only increasing the received signal power. Moreover the SNR_e can be directly evaluated through the overall link noise factor, obtaining, if the ASE-signal beating is the dominant noise term,

$$SNR_e = \frac{A^2}{(G-1)RNFh\nu} \tag{4.60}$$

Moreover the Q factor (or the ASN) can be directly evaluated starting from SNR_e by the equation

$$Q = \frac{SNR_e}{\sqrt{2\,SNR_e + m} + \sqrt{m}} \tag{4.61}$$

Since the distribution of the decision variable is not gaussian, the evaluation of the error probability by the Q factor and equation (4.22) implies a significant error. In cases where m is small the decision variable PDF is quite different from a gaussian PDF, and the approximation (4.20) leads to a sizable error. In particular, the error in estimating the transmitted power with an error probability of 10^{-9} is the maximum error and equal to $3\,dB$ if $m = 1$. For high values of m (of the order of ten or higher), the central limit theorem can be applied to equation (4.53) to derive a gaussian approximation for the PDF of the decision variable. In this case the gaussian approximation can be used.

By Fig. 4.21 it can be argued that the optimum value of the optical bandwidth is R; that is, the optical bandwidth is at its narrowest avoiding signal distortion. Wider optical bandwidths yield to a limited penalty: For example, an optical bandwidth ten times wider would induce a penalty of about $2.1\,dB$.

Optically Amplified Coherent Systems

Using the same conditions as in Sections 4.3.2 and 4.3.3, the optical field at the receiver input, after ideal polarization control, can be written as

$$E = A\xi e^{i[\omega t + \phi + \pi m(t)]} + \mathbf{n}(t)e^{i\omega t} \tag{4.62}$$

where ξ is the unit vector in the polarization direction. The spontaneous emission of the amplifiers is represented by the complex vector $\mathbf{n}(t)$, whose four quadratures can be considered independent, bandlimited, gaussian random processes with zero mean and power spectral density $S_{sp}/2$ within the bandwidth. The factor $\frac{1}{2}$ in the power spectral density is related to the fact that the ASE power is uniformly distributed between orthogonal polarization states. After mixing with a linearly polarized local oscillator, optical-to-electrical conversion and filtering by an ideal bandpass filter of bandwidth $2R$, the intermediate frequency current $c(t)$ is given by

$$c(t) = 2R_p A A_{1o} \cos[\omega_{IF} t + \Delta\phi + \pi m(t)] + n_{tot}(t) \qquad (4.63)$$

where $\Delta\phi = \omega_{IF} = 0$ for homodyne detection, and the noise term $n_{tot}(t)$ can be written as

$$n_{tot}(t) = 2R_p A_{1o}[n_{Ix}(t) \otimes h_{el}(t)] \cos(\omega_{IF} t) + n_s(t) \qquad (4.64)$$

where $n_{Ix}(t)$ is the in-quadrature term of the x polarization of the ASE noise, $h_{el}(t)$ the electrical filter pulse response (IF or baseband for heterodyne or homodyne detection), and $n_s(t)$ the shot noise at the filter output. Therefore the process $n_{tot}(t)$ is a zero mean, gaussian process whose variance σ_{tot}^2 is given by

$$\sigma_{tot}^2 = R_p^2 A_{1o}^2 B_e \frac{S_{sp}}{2} + q R_p A_{1o}^2 = R_p A_{1o}^2 B_e \left(\frac{NGF}{2} + 1\right) \qquad (4.65)$$

In the third member of equation (4.65) it is assumed that all the m amplifiers of gain G are equal and that their gains counterbalance exactly the losses of the fiber section. With these assumptions, S_{sp} can be expressed by means of equation (4.49). From (4.65) it is obtained that the sensitivity penalty P_{en} of the coherent receiver due to ASE noise can be expressed as

$$P_{en} = 1 + \frac{NFG}{2} \qquad (4.66)$$

Of course the presence of a sensitivity penalty does not mean that the amplifier is useless as shown in the following examples.

The first example concerns a 4.8 Gbit/s heterodyne PSK long haul system: without optical amplification, the maximum span is ~260 km because of the Kerr induced penalty, even if a dispersion shifted fiber in the anomalous dispersion region is used [38]. If optical amplifiers are used, the transmitted optical power can be much lower and long distances can be covered without any significant nonlinear effect arising along propagation. As an example, this is possible for a 4.8 Gbit/s PSK system using a dispersion shifted fiber in the anomalous dispersion region with the parameters reported in Table 2.1 and erbium doped fiber amplifiers whose parameters are reported in Table 3.2. An amplifier can be placed every 50 km and the gain is adjusted to balance exactly the fiber loss. In this case, a link length of 700 km can be achieved without any considerable nonlinear phenomena in fiber propagation (as demonstrated by numerical solution of the nonlinear Shrödinger equation considering the actual noise), if the optical power is kept sufficiently low that the receiver operates at its sensitivity limit, which is obviously affected by ASE noise.

The second example is the use of optical amplifiers in a passive star optical network: every branch of the optical star is supposed 40 km long and an amplifier is placed at the output of the star center on each branch; the other relevant parameters are PSK modulation, transmission in the third window, a fiber attenuation of 0.25 dB/km, an amplifier gain of 18 dB and a transmitted power of 1 mW. In the absence of amplification, the receiver sensitivity is 20 photons/bit that, at a bit rate of 1 Gbit/s, corresponds to $\sim 0.26\,\mu$W therefore, the maximum number of users is given by $N_U = A_t P_o / S_r \sim 38$, where S_r is the receiver sensitivity. Using optical amplifiers, the receiver sensitivity changes to ~ 56 photons/bit, for an amplifier inversion factor of 1.8: therefore, the maximum number of users is given by $N_U = A_t P_o G / S_r \sim 1370$ with a sensible improvement in comparison with the previous case.

REFERENCES

1. Numerous books on telecommunications describe the main modulation formats and introduce the analytical methods for performance evaluation of transmission systems. The interested reader can refer, for instance, to

   J. G. Proakis, *Digital Communications*, 2d ed., McGraw-Hill, New York, 1989.

   S. Benedetto, E. Biglieri, and V. Castellani, *Digital Transmission Theory*, Prentice Hall, Englewood Cliffs, NJ, 1987.

2. Among the vast literature produced on IM-DD systems, a complete presentation of unrepeatered systems comprising structure, performance evaluation, and several engineering problems can be found in

   M. J. Howes and D. V. Morgan, eds., *Optical Fiber Communications*, Wiley, New York, 1980.

   More recent results and indications about the impact of new technologies on IM-DD systems can be found in Technical staff of CSELT, *Fiber Optics Communication Handbook*, 2d ed., F. Tosco (ed.), TAB Books, McGraw-Hill, New York, 1990, part IV.

3. The structure and the performances of direct-detection polarization modulated systems are presented in

   S. Betti, G. De Marchis, and E. Iannone, Polarization modulated direct detection optical transmission systems, *IEEE-Journal of Lightwave Technology*, 10: 1985–1997, 1992.

   Experimental implementations are presented in

   K. Fukuchi, S. Yamazaki, T. Ono, and M. Rangaraj, Polarization shift keying-direct detection (PolSK-DD) scheme for fiber nonlinear effect insensitive communication system, *Proceedings of ECOC'92*, pp. 169–172, Berlin (Germany) September 27–October 1, 1992.

   Polarization modulated systems adopting coherent detection are introduced in

   S. Betti, F. Curti, G. De Marchis, and E. Iannone, Phase noise and polarization state insensitive coherent optical systems, *IEEE-Journal of Lightwave Technology*, 5: 561–572, 1987.

An experimental realization of this kind of transmission system is reported in
S. Benedetto, R. Paoletti, P. Poggiolini, C. Barry, A., Djupsjobacka, and
B. Lagerstrom, Coherent and direct detection polarization modulation system experiments, *Proceedings of ECOC'94*, pp. 67–71, Florence, September 25–29, 1994.

4. Multilevel polarization modulation is described, for example, in
S. Betti, F. Curti, G. De Marchis, and E. Iannone, Multilevel coherent optical systems based on Stokes parameters modulation, *IEEE-Journal of Lightwave Technology*, 8: 1127–1136, 1990;
See also the first paper of ref. [3].

5. A complete review of different coherent optical transmission systems is presented in
S. Betti, G. De Marchis, and E. Iannone, *Coherent Optical Communications Systems*, Wiley, New York, 1995.

6. A review on WDM techniques with some orientation to network problems is given in
C. A. Brackett, Dense wavelength division multiplexing: Principles and applications, *IEEE-Journal on Selected Areas in Communications*, 2: 669–672, 1990.
Crosstalk and intermodulation in WDM systems are treated in an extensive number of papers. For linear crosstalk, see, for example,
L. G. Kazovsky and J. L. Gimlet, Sensitivity penalty in multichannel coherent optical communications, *IEEE-Journal of Lightwave Technology*, 6: 1353–1365, 1988.
For nonlinear crosstalk, see, for example,
R. W. Tkach, A. R. Chraplyvy, F. Forghieri, A. H. Gnauck, and R. M. Derosier, Four-photon mixing and high speed WDM systems, *IEEE-Journal of Lightwave Technology*, 13: 841–849, 1995.
Analyses of problems arising in amplified WDM systems can be found in several recent papers. See, for example,
G. Jacobsen, Multichannel system design using optical preamplifiers and accounting for the effects of phase noise, amplifier noise and receiver noise, *IEEE-Journal of Lightwave Technology*, 10: 367–376, 1992.
A. E. Willner and S.-M. Hwang, Transmission of many WDM channels through a cascade of EDFA's in long-distance links and ring networks, *IEEE-Journal of Lightwave Technology*, 13: 802–817, 1995.

7. One of the early experiments on time division multiplexing of optical signals using conventional IM-DD transmission and all-optical processing is reported in
P. R. Prucnal, M. A. Santoro, S. K. Sehgal, and I. P. Kaminow, TDMA fibre-optic network with optical processing, *Electronics Letters*, 22: 1218–1219, 1986.
Recent studies of all-optical time division multiplexing and switching of intensity modulated signals can be found in
D. M. Spirit, D. E. Wickens, T. Widdowson, G. R. Walker, D. L. Williams, and L. C. Blank, 137 km, 4×5 Gbit/s optical time division multiplexed unrepeated system with distributed erbium fiber preamplifier, *Electronics Letters*, 28: 1218–1219, 1992.
Y. Shimazu and M. Tsukada, Ultrafast photonic ATM switch with optical output buffers, *IEEE-Journal of Lightwave Technology*, 10: 265–271, 1992.
The potential impressive capacity of OTDM systems is demonstrated, for example, in
S. Kawanishi, H. Takara, O. Kamatani, and T. Morioka, 100 Gbit/s, 500 km, optical transmission experiment, *Electronics Letters*, 31: 737–741, 1995.

S. Kawanishi, H. Takara, T. Morioka, O. Kamatani, and M. Saruwatari, 200 Gbit/s, 100 km, transmission using supercontinuum pulses with prescaled PLL timing extraction and all-optical demultiplexing, *Electronics Letters*, 31: 816–820, 1995.

An experiment on optical time division multiplexing using solitons is reported in

C. E. Soccolich, M. N. Islam, B. J. Hong, M. Chbat, and J. R. Sauer, Application of ultrafast gates to a soliton ring network, *Conference on Nonlinear Guided-Wave Phenomena*, pp. 366–369, Cambridge, England, September 2–4, 1991.

8. The ultimate cause of shot noise is the quantum nature of the electromagnetic field. This basis of shot noise is discussed in

J. H. Shapiro, Quantum noise and excess noise in optical homodyne and heterodyne receivers, *IEEE-Journal of Quantum Electronics*, 21: 237–250, 1985.

A semiclassical discussion of shot noise can be found in all the books on optical communications; see, for example, those of ref. [2].

Multiplication noise is generated by the random nature of the internal gain of an APD photodiode. A discussion of multiplication noise useful for optical systems performance evaluation can be found, for example, in

Technical staff of CSELT, *Fiber Optics Communication Handbook*, 2d ed., F. Tosco (ed.), TAB Books, McGraw-Hill, New York, 1990, sec. 6.2, pp. 544–552.

9. The thermal noise introduced by the receivers' front end depends essentially on the electronic front end structure. A classic analysis of this noise source is found in

R. G. Smith and S. D. Personik, Receiver design for optical fiber communication systems, in *Semiconductor Devices for Optical Communications*, H. Kressel (ed.), Springer, New York, 1980, ch. 4.

More recent results for wide bandwidth receivers can be found in

J. M. Golio, *Microwave MESFET & HEMT*, Arthec House, Boston, 1991.

10. For an introduction to the impact of these noise sources on the performances of optical transmission systems, see, for example,

M. J. Howes and D. V. Morgan, eds., *Optical Fiber Communications*, Wiley, New York, 1980.

S. Betti, G. De Marchis, and E. Iannone, *Coherent Optical Communications Systems*, Wiley, New York, 1995, ch. 5, pp. 193–210.

11. Many different optical PLL structures have been introduced and studied for use in homodyne PSK receivers. The balanced loop was first proposed for the suppression of laser intensity noise in homodyne receivers by

G. L. Abbas, V. W. S. Chan, and T. K. Lee, Local oscillator excess noise suppression for homodyne and heterodyne detection, *Optics Letters*, 8: 419–421, 1983.

An analysis of the loop performance assuming linear loop approximation is reported in

L. G. Kazovsky, Balanced phase locked loop for optical homodyne receivers: Performance analysis, design considerations and laser linewidth requirements, *IEEE-Journal of Lightwave Technology*, 4: 182–195, 1986.

The decision-driven PLL was first introduced for homodyne optical receivers in

L. G. Kazovsky, Decision-driven phase-locked loop for optical homodyne receivers: Performance analysis and laser linewidth requirements, *IEEE-Journal of Lightwave Technology*, 3: 1238–1247, 1985.

The Costas loop for optical communications is analyzed in

T. G. Hodgkinson, Costas loop analysis for coherent optical receivers, *Electronics Letters*, 22: 394–396, 1986.

12. Besides direct-detection and coherent heterodyne receivers introduced in the papers listed in [3] even coherent homodyne detection is possible for polarization modulated signals, as reported in
S. Betti, G. De Marchis, E. Iannone, and P. Lazzaro, Homodyne optical coherent systems based on polarization modulation, *IEEE-Journal of Lightwave Technology*, 9: 1314–1320, 1991.

13. Optical polarization controller have been studied mainly for application in coherent systems, in which the polarization of the local oscillator has to be matched with that of the received field. Two review papers on polarization controllers are
T. Okoshi, Polarization-state control schemes for heterodyne and homodyne optical fiber communications, *IEEE-Journal of Lightwave Technology*, vol 3, n.6, pp. 1232–1236, 1985.
N. G. Walker and G. R. Walker, Polarization control for coherent communications, *IEEE-Journal of Lightwave Technology*, 8: 438–458, 1990.

14. Besides the polarimeter considered in the text, other possible polarimeter structures are described, for example, in
R. M. A. Azzam, I. M. Elminyavi, and A. M. El-Sabe, General analysis and optimization of the four detector polarimeter, *Journal of Optical Society of America*, A5: 681–689, 1988.
T. Pkaar, A. C. van Bochove, M. O. van Deventer, H. J. Fraukene, and F. H. Groen, Fast complete polarimeter for optical fibers, in *Proceedings of EFOC/LAN'89*, pp. 206–209, Amsterdam, June 12–16, 1989.

15. S. Betti, G. De Marchis, and E. Iannone, Polarization modulated direct detection optical transmission systems, *IEEE-Journal of Lightwave Technology*, 10: 1985–1997, 1992.

16. TDM in electrical communication systems is described and analyzed in several classic books on communications. For example,
A. Bruce Carlson, *Communication Systems*, 2d ed., McGraw-Hill, New York, 1975.
A. S. Tanenbaum, *Computer Networks*, Prentice Hall, Englewood Cliffs, NJ, 1989.
Electrical TDM in networks adopting optical transmission and in particular synchronous digital hierarchy (SDH) are described, for example, in
B. Gi Lee, M. Kang, and J. Lee, *Broadband Telecommunications Technology*, Arthec House, Boston, 1993.

17. H. Lin and J. S. Smith, Optical time-division demultiplexing using second-order-optical nonlinear effects, *Applied Physics Letters*, 59: 2802–2804, 1991.

18. P. A. Andrekson, N. A. Olsson, J. R. Simpson, T. Tanbuk-Ek, R. A. Logan, and M. Haner, 16 Gbit/s all-optical demultiplexing using four-wave mixing, *Electronics Letters*, 27: 992–995, 1991.

19. A first proposal for an AND gate based on FWM in semiconductor amplifiers is reported in
R. Shnabel, W. Pieper, R. Ludwig, and H. G. Weber, All optical AND gate using femtosecond non-linear gain dynamics in semiconductor laser amplifiers, *Proceedings of ECOC'93*, pp. 133–136, Montreaux, Switzerland, September 12–16, 1993.

A different AND gate based on the same principle but able to process signals at the same wavelength is reported in

D. Nesset, M. C. Tatham, L. D. Westbrook, and D. Cotter, Ultrafast all optical AND gate for signals at the same wavelength using four wave mixing in semiconductor laser amplifiers, *Proceedings of ECOC'94*, pp. 529–532, Florence, September 25–29, 1994.

Nondegenerate four-wave mixing in semiconductor amplifiers has been also proposed to implement other functions useful in OTDM systems. See, for example,

M. Jinno, J. B. Schlager, and D. L. Franzen, Optical sampling using nondegenerate four-wave mixing in a semiconductor laser amplifier, *Electronics Letters*, 30: 1489–1491, 1994.

A. D'Ottavi, E. Iannone, and S. Scotti, Address recognition in all-optical packet switching by FWM in semiconductor amplifiers, *Microwave and Optical Technology Letters*, 10: 228–230, 1995.

20. The nonlinear loop mirror could be a key technology to realize OTDM systems. The use of nonlinear loop mirror as clock recovery circuit was introduced in
K. Smith and J. K. Lucek, All-optical clock recovery using a mode-locked laser, *Electronics Letters*, 28: 1814–1816, 1992.
Very high-speed demultiplexing was demonstrated in
I. Glesk, J. P. Sokoloff, and P. R. Prucnal, Demonstration of all-optical demultiplexing of TDM data at 250 Gbit/s, *Electronics Letters*, 30: 339–341, 1994.
The use of nonlinear loop mirror as frame clock recovery circuit is suggested in
J. K. Lucek and K. Smith, All-optical clock multiplication and division using an optically mode-locked laser, *Proceedings of ECOC'95*, pp. 941–944, Brussels, Belgium, September 17–21, 1995.

21. Periodic filters based on Mach-Zehnder interferometers, for multiplexing and demultiplexing optical signals, were first developed for millimeter wave applications; an analysis of millimeter wave devices can be found, for example, in
H. Kumazawa and I. Ohtomo, 30-GHz band periodic branching filter using travelling wave resonator for satellite applications, *IEEE-Transaction on Microwave Theory and Techniques*, 25: 683–689, 1977.
Experimental results on using a periodic filter in a WDM optical system are reported in the second paper in ref. [3] and in
N. Takato et al., Guided wave multi/demultiplexer for optical FDM, *Proceedings of ECOC'86*, pp. 443–447, Barcellona, (Spain), September 22–25, 1986.
H. Toba, K. Oda, K. Nosu, and N. Takato, 16-channels, optical FDM distribution/transmission experiment utilizing multichannel frequency stabilizer and waveguide frequency selection filter, *Electronics Letters*, 25: 574–575, 1989.

22. A classic review on widely spaced WDM in which this technology is highlighted is
H. Ishio, J. Minowa, and K. Nosu, Review and status of wavelength division multiplexing technology and its applications, *IEEE-Journal of Lightwave Technology*, 2: 448–463, 1984.

23. Among the first papers on the application of miniature Fabry–Perot interferometers to WDM systems, the following ones are recalled:
J. Stone and L. W. Stulz, Pigtailed high-finesse fibre Fabry–Perot interferometers with large, medium and small free spectral ranges, *Electronics Letters*, 23: 781–783, 1987.

J. S. Harper, P. A. Rosher, S. Fenning, and S. R. Mallinson, Application of miniature micromachined Fabry–Perot interferometer to optical fiber WDM system, *Electronics Letters*, 25: 1065–1066, 1989.
A possible application of these devices to optical networks is shown in
I. P. Kaminow, P. P. Iannone, J. Stone, and L. W. Stulz, FDMA-FSK star network with a tunable optical filter as demultiplexer, *IEEE-Journal of Lightwave Technology*, 6: 1406–1416, 1988.
High-performance miniaturized Fabry–Perot interferometers are reported in
J. Stone and L. W. Stulz, High performance fibre Fabry–Perot filters, *Electronics Letters*, 27: 2239–2240, 1991.

24. Integrated optics diffraction gratings are described and analyzed in
T. Numai, S. Murata, and I. Mito, Tunable wavelength filters using $\lambda/4$-shifted waveguide grating resonators, *Applied Physics Letters*, 53: 83–85, 1988.
I. M. I. Habbab, S. L. Woodward, and L. J. Cimini, Jr., DBR-based tunable optical filter, *IEEE-Photonics Technology Letters*, 2: 337–339, 1990.
W. P. Huang and J. Hong, A coupled-waveguide grating resonator filter, *IEEE-Photonics Technology Letters*, 4: 884–886, 1992.

25. A notable example of high-selectivity optical filter based on liquid crystals is reported in
K. Hirabayashi, H. Tsuda, and T. Kurokawa, Tunable wavelength-selective liquid crystal filters for 600-channel FDM system, *IEEE-Photonics Technology Letters*, 4: 597–599, 1992.

26. Experimental results on the absolute frequency stabilization of optical carriers are reported in
B. Villeneuve, N. Cyr, and M. Tetu, Precise optical heterodyne beat frequency from laser diodes locked to atomic resonances, *Electronics Letters*, 23: 1082–1084, 1987.
T. Yanagawa, S. Saito, S. Machida, Y. Yamamoto, and N. Noguchi, Frequency stabilization of an InGaAsP distributed feedback laser to an NH_3 absorption line at 1515 nm with an external frequency modulator, *Applied Physics Letters*, 47: 1036–1038, 1985.
S. Sudo, Y. Sakai, H. Yasaka, and T. Ikegami, Frequency stabilized DFB laser module using 1.53159 µm absorption line of C_2H_2, *IEEE-Photonics Technology Letters*, 1: 281–284, 1989.

27. F. Favre and D. Le Guen, High frequency stability of laser diode for heterodyne communication systems, *Electronics Letters*, 16: 709–710, 1980.

28. Large research and industrial projects are underway on introducing all-optical transport networks based on WDM technology in geographical regions and large metropolitan areas. Examples are the RACE MWTN project (ended in 1994) and the METON project in Europe, and the MONET project in the United States. A large amount of research has been also devoted to local area networks based on WDM technology. The wide range of activity in this fields is found in
IEEE-Journal of Lightwave Technology. Special issue on Broad-Band Optical Networks, 11: 667–679, 1993.
Among the numerous other interesting papers are the following which give special attention to the optical transport network:
K.-I. Sato, S. Okamoto, and H. Hadama, Network performance and integrity enhancement with optical path layer technologies, *IEEE-Journal of Selected Areas in Communications*, 12: 159–170, 1994.

A. Watanabe, S. Okamoto, and K.-I. Sato, Optical path cross-connect node architecture with high modularity for photonic transport networks, *IEICE Transaction on Communications*, E77-B: 1220–1229, 1994.

E. Iannone and R. Sabella, Performance evaluation of an optical multi-carrier network using wavelength converters based on FWM in semiconductor optical amplifiers, *IEEE-Journal of Lightwave Technology*, 13: 312–324, 1995.

29. The gaussian approximation for the evaluation of the error probability when the decision variable PDF is not known has been widely used. See, for example,

    C. J. Anderson and J. A. Lyle, Technique for evaluating system performance using Q factor in numerical simulations exhibiting intersymbol interference, *Electronics Letters*, 30: 71–72, 1991.

    N. S. Bergano, F. W. Kerfoot, and C. R. Davidson, Margin measurements in optical amplifier systems, *IEEE-Photonics Technology Letters*, 3: 304–306, 1993.

30. For a derivation of the error probability from the Fourier transform of the PDF, see, for example,

    J. E. Mazo and J. Salz, Probability of error for quadratic detectors, *Bell System Technical Journal* (November): 2165–2186, 1965.

    J. G. Proakis, *Digital Communications*, 2d ed., McGraw-Hill, New York, 1989, app. 4B, pp. 344–349.

31. The expression of error probability by the Cauchy integral and a review of exact solutions and possible approximations of this integral are found in

    S. Betti, G. De Marchis, and E. Iannone, *Coherent Optical Communications Systems*, Wiley, New York, 1995, app. C, pp. 523–529.

32. The Chernov bound was originally introduced in

    H. Chernov, A measure of asymptotic efficiency of tests of a hypothesis based on the sum of observations, *Mathematical Statistics Annals*, 23: 493–507, 1952.

    The steepest descendent approximation is a classic method of mathematical physics essentially due to Debye. A complete and rigorous review of this method can be found in

    P. M. Morse and H. Feshbach, *Methods of Theoretical Physics*, Part I, McGraw-Hill, New York, 1935.

    A simple description of the two considered approximations can be found in ref. [31], and a discussion of the relation between them can be found in

    K. Shumacher and J. J. O'Reilly, Relationship between the saddle point approximation and the modified Chernov bound, *IEEE-Transactions on Communications*, 38: 270–272, 1990.

33. M. J. McIntyre, The distribution of gain in uniformly multiplying avalanche photodiodes, *IEEE-Transaction on Electron Devices*, 19: 703–712, 1972.

34. The first paper introducing a detailed analysis of IM-DD systems based on the gaussian approximation was

    S. D. Personick, Receiver design for digital fibre optic communication systems, parts I and II, *Bell System Technical Journal*, 52: 843–886, 1973.

    This approximation is used in some classic system analyses such as those contained in the books

    M. J. Howes and D. V. Morgan, eds., *Optical Fibre Communications*, Wiley, New York, 1980.

    Technical staff of CSELT, *Fibre Optic Communication Handbook*, 2d ed., F. Tosco (ed.), TAB Books, McGraw-Hill, New York, 1990, ch. 4.

An analysis of the accuracy and the limits of this approximation, where results are compared with results obtained adopting more rigorous approaches, is carried out in

S. D. Personik, P. Balaban, J. H. Bobasin, and P. R. Kumar, A detailed comparison of four approaches to the calculation of the sensitivity of optical fibre systems receivers, *IEEE-Transaction on Communications*, 25: 541–548, 1977.

35. A suggestion that a complete quantomechanical evaluation be made of the noise factor of an amplifier cascade is given in
K. Kikuchi, Generalized formula for optical amplifier noise and its application to erbium doped fiber amplifiers, *Electronics Letters*, 26: 1851–1856, 1990.
Starting with the results in this paper, a complete expression of the noise factor can be derived by considering the fibers in the amplifier link as passive attenuators contributing quantum noise.

36. The overall noise factor of an amplified electrical link is evaluated in all the classic books on electrical communications. See, for example,
J. M. Wozencraft and I. M. Jacobs, *Principles of Communication Engineering*, Wiley, New York, 1965.

37. D. Marcuse, Derivation of analytical expressions for the bit-error probability in lightwave systems with optical amplifiers, *IEEE-Journal of Lightwave Technology*, 8: 1816–1823, 1990.
For a more complete analysis, taking into account even the receiver noise, see
D. Marcuse, Calculation of bit-error probability for a lightwave system with optical amplifiers and post-detection gaussian noise, *IEEE-Journal of Lightwave Technology*, 9: 505–513, 1991.

38. S. Betti, G. De Marchis, and E. Iannone, *Coherent Optical Communications Systems*, Wiley, New York, 1995.

39. J. Salz, Coherent lightwave communications, *AT&T Technical Journal*, 64: 2153–2209, 1985.

CHAPTER FIVE

Soliton Optical Communications

When in the early 1973 Hasegawa and Tappert [1] proposed the use of solitons for optical communications, five years had to come before the first generation of fiber-optic-based lightwave systems could be deployed [2]. Researchers were facing two problems at that time, the attenuation of optical fibers and the fiber dispersion. The idea of Hasegawa and Tappert was to use Kerr nonlinearity in compensating for fiber dispersion in the wavelength region approximating 1.55 nm where the fiber loss was lowest. Their work was stimulated by the earlier work of Zakharov and Shabat [3] who showed in a seminal paper that the nonlinear Schrödinger (NLS) equation, which describes the propagation of an optical pulse in a dispersive medium in the presence of Kerr nonlinearity, belongs to a class of equations that can be integrated by the inverse scattering method [4] and admit soliton solutions. The solitary pulse that Hasegawa and Tappert proposed to use is indeed the simplest among the soliton solutions discovered by Zakharov and Shabat, which is referred to as the fundamental soliton solution. Obviously solitons were a solution to the problem of pulse spreading caused by fiber dispersion, but they could do nothing against the attenuation that a pulse traveling in a fiber experiences due to waveguide loss. Unfortunately, when the use of soliton was proposed, there were no practical methods to compensate for fiber loss; therefore the soliton literature remained nothing more than an elegant mathematical curiosity.

When in 1980 Mollenauer, Stolen, and Gordon [5] reported the first experimental evidence of solitons in optical fibers, the interest in the field of soliton transmission was renewed. In 1983 Hasegawa [6] proposed the use of Raman amplification to compensate for fiber loss, and the first successful experiment was reported in 1988 by Mollenauer and Smith [7]. Two years before the experiment of Mollenauer and Smith, in 1986 Gordon and Haus wrote a paper that is a milestone for the later development of the field of soliton optical communication [8]. They predicted that the transmission of a signal made of optical solitons could not be extended an unlimited distance in the presence of fiber loss compensated by optical amplification. Optical gain, for quantum mechanical reasons related to the preservation of the commutation brackets of the field operator across the amplifier [9], require that optical gain is associated with spontaneous emission noise. The coupling of the spontaneous emission noise with the soliton spectrum causes a random walk of the soliton center frequency. Since the group velocity of a soliton is proportional to its center frequency because of the group velocity dispersion of the fiber, a random walk of the soliton centre frequency produces a random walk of the soliton velocity. The arrival time of the soliton then is a random variable whose variance is proportional to the cube of the propagation distance. This is the so-called Gordon-Haus effect.

The Gordon-Haus effect was discovered assuming that fiber loss is compensated by distributed Raman gain. During the same years, however, a new type of optical amplifiers was being developed, the erbium-doped fiber amplifiers (EDFA) [10]. The development of the EDFA was a major breakthrough in the field of optical communications. In particular, the invention of the new device drastically changed the perspective on soliton communications. Lumped EDFAs are in fact much more practical than distributed Raman amplifiers. Because of the low gain of Raman amplifiers, kilometers of fibers are required to get some dB of amplification, so Raman amplifiers were mainly proposed as distributed amplifiers, able to continuously compensate for fiber loss along the line. Lumped amplification is more practical for a number of reasons but mainly because it is easier to design a device made of a few meters than a device made of kilometers of fiber. The first successful demonstration of soliton transmission with EDFAs was reported by Nakazawa, Kimura, and Suzuki in 1989 [11]. A number of theoretical (and conceptual) issues had to be solved, however. How can pulses that are attenuated and periodically amplified still be considered solitons? To this question, three research groups gave the same answer almost simultaneously, in three independent papers [12, 13, 14]. If the loss-gain period is much shorter than the length over which the soliton wavevector gives a change of 2π to the phase of the soliton, then the averaged dynamics over the loss-gain period follows, to a first-order approximation, a lossless NLS that therefore presents soliton solutions. The requirement that the loss-gain period be shorter than the nonlinear length limits the maximum compatible amplifier spacing for soliton transmission. Averaged soliton transmission made possible, in 1991, the transmission of Gbit/s soliton optical signals over transoceanic distances [15]; the limit was set only by the Gordon-Haus effect.

It was then quite obvious that researchers had to focus their effort on ways to extend the limit set by the Gordon-Haus effect. At the end of 1991 two research groups, one at MIT and the other at AT&T, Bell Labs., came out independently with a method of extending the limit set by the Gordon-Haus effect by using passive frequency filtering [16, 17]. Using filters, Mollenauer et al. were able, in a recirculating loop experiment, to extend the propagation distance and the bit-rate well beyond the Gordon-Haus limit [18]. The filtering method does not require any additional device to make it compatible with wavelength division multiplexing (WDM) systems. As predicted in one of the two papers that originally proposed the filtering control [16], the limit to the maximum propagation distance with filtering control was set by the extra amplifier gain required to compensate for the filter loss on the spectral wings of the soliton. This extra gain opens a window of positive gain at the center of the filter passband, leading to the generation of a continuous wave radiation seeded by the spontaneous emission noise of the amplifiers. The amplified spontaneous emission noise grows exponentially with distance leading, eventually, to unacceptable transmission errors in the detection of zeros. This problem was beautifully solved by Mollenauer, Gordon, and Evangelides [20]. The solution to this problem was continuously sliding the filter center frequency with propagation distance: The soliton follows the sliding; the linear noise does not. The soliton line with sliding filters is a line that is transparent to signal and opaque to the noise, and this property is possible because the line is intrinsically nonlinear. By using sliding filters, the Mollenauer group was able to greatly extend the error-free transmission distance [21]. The latest published results (August 1996) are 6×10 Gbit/s channels at 11.4 Mm and 7×10 Gbit/s channels at 9.4 Mm [22]. At the same time when this filter scheme was proposed, Nakazawa, Yamada, Kubota, and Suzuki proposed a method for controlling soliton transmission based on a time modulator [23]. This method is not immediately compatible with WDM. Very recently, however, the same group has reported high bit-rate transmission of a WDM signal by demultiplexing the signals before the retiming stage, using three modulators to retime the three channels independently, and multiplexing them again before retransmission [24].

Most of the recent progress in soliton transmission has been in the development of systems mathematically described by dynamics equations that a suitable *linear* transformation converts into a slightly perturbed NLS. In a more novel recent approach, however, KDD researchers have proposed an alternative to long-distance transmission whereby dispersion compensating devices, which may be either dispersion compensating fibers or fiber gratings, are inserted along the line with a spacing of the same order of the soliton period. If the linear devices do not compensate for the line dispersion totally, the remaining dispersion is compensated by the nonlinearity, and stable pulses can propagate [60]. The path-averaged dispersion of such schemes can be made very low, and this greatly reduces the Gordon-Haus jitter compared to conventional soliton systems. More recently the capability of this scheme was shown in combination with WDM to transmit 3 channels, 20 Gbit/s each, over transoceanic distances (7000 km) [26]. This scheme cannot be strictly considered a "perturbed" soliton transmission scheme. The pulses are more gaussians than sechlike, and they undergo a large amount of reshaping during the dispersion period. However, some of the properties of solitons are preserved, like their robustness to perturbations. The KDD scheme is a valuable alternative to sliding filters for the next generation of soliton-based undersea transmission systems.

In this chapter we will give a brief account of the theory of optical soliton communication. The analysis, for space constraints, is sometimes not rigorous and necessarily not complete. The reader interested in the more mathematical aspects of the theory is advised to refer to the recent book of Hasegawa and Kodama [27] and to the very comprehensive reference list therein. Also Haus and Wong gave an excellent review of the latest developments in the soliton field in a paper recently published in the *Review of Modern Physics* [28]. An even more recent comprehensive review is given by Essiambre and Agrawal [29].

5.1 BASIC EQUATIONS

5.1.1 Averaged Equation

We showed in Chapter 2 that in averaging the fast polarization dynamics (having a characteristic length much shorter than the characteristic length of the nonlinearity), one ends up, in the lossless case, with a Manakov equation, which we can rewrite, neglecting third-order dispersion, as [30]

$$\frac{\partial U_1}{\partial z} = -i\,\frac{\beta_2}{2}\,\frac{\partial^2 U_1}{\partial t^2} + i\,\frac{8}{9}\,\gamma(|U_1|^2 + |U_2|^2)U_1 \tag{5.1}$$

$$\frac{\partial U_2}{\partial z} = -i\,\frac{\beta_2}{2}\,\frac{\partial^2 U_2}{\partial t^2} + i\,\frac{8}{9}\,\gamma(|U_1|^2 + |U_2|^2)U_2 \tag{5.2}$$

Here U_1 and U_2 are related to the two linearly polarized components of the electric field by a linear, unitary, polarization transformation that depends on the position along the fiber and on the frequency. The linear transformation that brings the U_1 and U_2 to the real field that propagates along the fiber may be assumed to be approximately constant within the bandwidth of the principal states of polarization.

Assume for the moment that the bandwidth of the field is less than the bandwidth of the principal states of polarization. If the field is launched with a fixed state of polarization, equations (5.1) and (5.2) do not provide any mechanism of transfer of power between the polarization state of the field and the orthogonal polarization state. Even in the presence of disturbances produced, for instance, by the ASE noise of the in-line amplifiers, experiments have shown that the field stays copolarized in single-channel transmission over distances of the order of thousands of kilometers [30]. For this reason the analysis is greatly simplified if one uses the nonlinear Schrödinger equation for a single polarization, which, for a lossy fiber where loss is periodically compensated by lumped amplifiers, is

$$\frac{\partial U}{\partial z} = f(z)U + i\,\frac{|\beta_2|}{2}\,\frac{\partial^2 U}{\partial t^2} + i\gamma|U|^2 U \tag{5.3}$$

where $\gamma = 2\pi n_2/(\lambda A_{\text{eff}})$ is the nonlinear coefficient, n_2 the Kerr coefficient of the fiber, λ the free-space wavelength and $A_{\text{eff}} = \pi\rho_c^2$ the effective area of the fiber. In this equation and in those following, to avoid the introduction everywhere of the factor 8/9 that accounts for the polarization averaging, we assume that 8/9 has been included in the Kerr coefficient of the fiber. Since we are interested in bright soliton solutions, which exist only in the anomalous dispersion region of the fiber, we assume a negative group velocity dispersion β_2. The function $f(z)$ accounts for the linear loss-gain cycle. If we assume that amplifiers are periodically spaced with period L_A, $f(z)$ is equal to $-\alpha/2$, where α is the loss per unit length of the transmission fiber, in the regions of propagation through the lossy fiber and equal to $g/2$, where g is the gain per unit length of the erbium fiber within the EDFAs. Introducing in equation (5.3) the transformation

$$U(z) = F(z)V(z) \tag{5.4}$$

where

$$F(z) = \exp\left[\int_0^z f(z')\,dz'\right] \tag{5.5}$$

obtains

$$\frac{\partial V}{\partial z} = i\,\frac{|\beta_2|}{2}\,\frac{\partial^2 V}{\partial t^2} + i\gamma|F(z)|^2|V|^2 V \tag{5.6}$$

If the amplifier length is much shorter than the amplifier spacing, then the function $|F(z)|^2$ is well approximated by a periodic, pointwise, discontinuous function with period L_A. Within the period $0 \leq z < L_A$, $|F(z)|^2$ assumes the values

$$|F(z)|^2 = \exp(-\alpha z) \tag{5.7}$$

The function $|F(z)|^2$ may be expanded as a spatial Fourier series

$$|F(z)|^2 = \sum_{n=-\infty}^{\infty} \exp\left(in\,\frac{2\pi}{L_A}\,z\right)f_n \tag{5.8}$$

where

$$f_n = \frac{1 - \exp(-\alpha L_A)}{\alpha L_A + in2\pi} \tag{5.9}$$

The coefficient of the z independent term is

$$f_0 = \frac{1 - \exp(-\alpha L_A)}{\alpha L_A} \tag{5.10}$$

The terms that oscillate with spatial period L_A/n are formally of the order L_A/n. So in the perturbative expansion in L_A,

$$V = V_0 + L_A V_1 + \cdots \qquad (5.11)$$

The equation for the lowest order term V_0 is

$$\frac{\partial V_0}{\partial z} = i \frac{|\beta_2|}{2} \frac{|\partial^2 V_0|}{\partial t^2} + i\gamma \frac{1 - \exp(-\alpha L_A)}{\alpha L_A} |V_0|^2 V_0 \qquad (5.12)$$

On physical grounds, keeping only the term V_0 of the expansion, which is equivalent to neglecting the spatially ocillating terms in $|F(z)|^2$, gives an excellent approximation of the exact solution V if the length scale over which the nonlinear and the dispersive terms gives significant changes to the pulse is much longer than the amplifier spacing. We can reasonably assume that V does not change much within the amplifier period; hence the oscillating terms of $|F(z)|^2$ in the product $|F(z)|^2 |V|^2 V$ average to zero. We will see that for the fundamental soliton, this condition holds for L_A much shorter than eight times the soliton period, the characteristic distance for the soliton dynamics which will be defined later.

We have already shown that the average field V_0 evolves according to an NLS, formally similar to the NLS for lossless propagation with a rescaled nonlinear coefficient that accounts for the periodic variation of the power along the path. It is worth noting that $V \sim V_0$ is not the field evolving within the fiber. The field within the fiber is U, which is related to V by equation (5.4). Since $F(z)$, as given by equation (5.5), is a periodic function of period L_A and $F(0) = 1$, we have $F(nL_A) = 1$; hence V coincides with U at the fiber input and after each amplifier.

5.1.2 Amplifier Noise

Assuming that the conditions for the validity of the average NLS are met, equation (5.12) can be seen as the equation describing the evolution of the field propagating along the line on a grid having grid points at the amplifier outputs. Solving equation (5.12), the minimum incremental step having physical meaning is then L_A, which corresponds to the differential dz in a conventional differential equation [12]. With this in mind, one may easily understand how it is possible to add the effect of the amplifier noise into equation (5.12). The amplified sponetaneous emission (ASE) noise added at each amplifier is well described, semiclassically, by a while noise term of zero average and gaussian distribution. The statistical properties of the ASE noise term are

$$\langle N_i(t) \rangle = 0 \qquad (5.13)$$

$$\langle N_i(t) N_j(t') \rangle = 0 \qquad (5.14)$$

$$\langle N_i(t) N_j^*(t') \rangle = \hbar \omega_0 F(G - 1)\delta_{i,j}\delta(t - t') \qquad (5.15)$$

In the expressions above, $\hbar = h/2\pi$ is the reduced Planck constant and ω_0 is the angular carrier frequency of the soliton, G is the amplifier gain, and F is the inversion factor of the amplifier (see Chapter 3) [31]. For complete inversion, $F = 1$. The Krönecker delta appears in equation (5.15) because the white noise added by each amplifier is independent to that added by the others, as obvious. Equation (5.14) is the mathematical expression of ASE's property of having no preferential phase. One may see this by expanding, into equation (5.14), N_i in its real and imaginary parts and setting to zero the real part of the average of the product. One finds that the variance of the real part of N_i is equal to the variance of the imaginary part and that the correlation function of real and imaginary parts is zero. These two properties together are equivalent to the phase-independence of the ASE noise.

Adding white noise terms accounting for the ASE noise of the amplifiers, and suitably averaged over the amplifier spacing which, we recall is the smallest dz of the equation having a correspondence to a physical length, equation (5.12) becomes

$$\frac{\partial V_0}{\partial z} = i\,\frac{|\beta_2|}{2}\,\frac{\partial^2 V_0}{\partial t^2} + i\gamma\,\frac{1 - \exp(-\alpha L_A)}{\alpha L_A}\,|V_0|^2 V_0 + N(t,z) \qquad (5.16)$$

where the statistical properties of the continuous noise terms are

$$\langle N(t,z) \rangle = 0 \qquad (5.17)$$

$$\langle N(t,z) N_j(t',z') \rangle = 0 \qquad (5.18)$$

$$\langle N(t,z) N_j^*(t',z') \rangle = \hbar\omega_0 F\,\frac{(G-1)}{L_A}\,\delta(z-z')\delta(t-t') \qquad (5.19)$$

One can immediately verify that

$$N_i(t) \equiv \int_z^{z+L_A} N(t,z')\,dz' \qquad (5.20)$$

where the integral is extended from $z' \geq z$ up to $z' < z + L_A$, excluding the upper extreme, and z is the position of the ith amplifier. The equivalence equation (5.20) should be intended in the sense that the statistical properties of the left-hand side are the same as those of the right-hand side.

5.1.3   Soliton Units

Let us rescale equation (5.16) by the following transformations:

$$\tau = \frac{t}{T_s} \qquad (5.21)$$

$$\zeta = \frac{z}{L_D} \qquad (5.22)$$

$$q = \frac{V_0}{V_{nor}} \tag{5.23}$$

$$P_O = \frac{|\beta_2|}{\gamma T_s^2} \tag{5.24}$$

where

$$L_D = \frac{T_s^2}{|\beta_2|} \tag{5.25}$$

$$V_{nor} = P_k^{1/2} = \left[\frac{1 - \exp(-\alpha L_A)}{\alpha L_A} \gamma L_D \right]^{-1/2}$$

$$= \left[\frac{G-1}{G \ln G} \gamma L_D \right]^{-1/2} = \left[\frac{G \ln G}{G-1} P_O \right]^{1/2} \tag{5.26}$$

Equation (5.16) becomes

$$\frac{\partial q}{\partial \zeta} = \frac{i}{2} \frac{\partial^2 q}{\partial \tau^2} + i|q|^2 q + n(\tau, \zeta) \tag{5.27}$$

The dimensionless units defined by equations (5.21)–(5.26) are called soliton units because they are particularly useful for the analysis of the soliton solution of the NLS. The quantity L_D is the soliton length that in Chapter 2 has been called dispersion length, V_{nor} is a normalization field, and T_s is a (so far arbitrary) normalization time. In equation (5.26) we have used the fact that $G = \exp(\alpha L_A)$, which derives from the assumption that the amplifier gain should compensate for the line loss. In equation (5.27) we have defined

$$n(\tau, \zeta) = \frac{L_D}{v_{nor}} N(t, z) \tag{5.28}$$

The statistical properties of n are the same of N for the first two, equations (5.17) and (5.18), while equation (5.19) becomes

$$\langle n(\tau, \zeta) n^*(\tau', \zeta') \rangle = \frac{L_D^2}{v_{nor}^2} \frac{(G-1)}{L_A} \hbar \omega_0 \delta(z - z') \delta(t - t')$$

$$= \frac{L_D^2}{v_{nor}^2} \frac{(G-1)}{L_A} \hbar \omega_0 \frac{\delta(\zeta - \zeta')}{z_0} \frac{\delta(\tau - \tau')}{T_0}$$

$$= \frac{L_D^2 \gamma}{T_s} \frac{1 - \exp(-\alpha L_A)}{\alpha L_A} \frac{(G-1)}{L_A} \hbar \omega_0$$

$$\times \delta(\zeta - \zeta') \delta(\tau - \tau') \tag{5.29}$$

We finally get

$$\langle n(\tau, \zeta) \rangle = 0 \tag{5.30}$$

$$\langle n(\tau, \zeta) n(\tau', \zeta') \rangle = 0 \tag{5.31}$$

$$\langle n(\tau, \zeta) n^*(\tau', \zeta') \rangle = \langle n \rangle \delta(\zeta - \zeta') \delta(\tau - \tau') \tag{5.32}$$

where we have added the mean and the other second-order correlation function of n for completeness, and

$$\langle n \rangle = \hbar \omega_0 F \frac{L_D^2 \gamma}{T_s} \frac{1 - \exp(-\alpha L_A)}{\alpha L_A} \frac{(G-1)}{L_A} \tag{5.33}$$

If we define

$$f = \frac{(G-1)^2}{G(\ln G)^2} \tag{5.34}$$

we get

$$\langle n \rangle = \hbar \omega_0 F \alpha \frac{\gamma L_D^2}{T_s} f \tag{5.35}$$

The ASE noise with lumped amplifier is enhanced by the factor $f > 1$ with respect to the case in which the loss is compensated by distributed gain if the comparison is made for *the same average power* or, equivalently, for the same pulsewidth. The parameter f can be decomposed as the product of two terms, $f = f_1 f_2$. The first one $f_1 = (G-1)/(\alpha L_A) = (G-1)/\ln G$ is the ratio of the gain required to overcome the loss with lumped and distributed amplification. The second $f_2 = [1 - \exp(-\alpha L_A)]/(\alpha L_A) = (1 - G^{-1})/\ln G$ is the ratio of the output power of the amplifiers to the average power along the line. The first parameter f_1 is larger than one, it is an enhancement factor, which is obvious since more gain, hence more ASE, is required to overcome loss when the gain is lumped. The second parameter f_2 is less than one. This parameter accounts for the fact that we are comparing lumped and distributed amplification with the same average power. The comparison between lumped and distributed amplification is made, for the lumped case, at the point between two amplifiers where the power is equal to the average power. So, before making the comparison, the ASE noise after one amplifier should be reduced by the loss that the ASE experiences before reaching this point. This loss is equal to f_2. The product f between f_1 and f_2 is, however, larger than one and tends to unity for the amplifier spacing L_A approaching zero.

In Fig. 5.1 we show the enhancement factor f versus the amplifier gain in dB. The enhancement factor does not exceed 2 for a gain less than 12.95 dB. Assuming a fiber loss of 0.24 dB/km, this gain corresponds to amplifier spacing

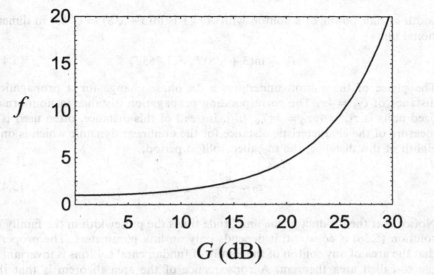

FIGURE 5.1 ASE enhancement factor f versus amplifier gain G in dB.

of 54 km. With 30-km amplifier spacing, the enhancement factor with the same value of fiber loss is $f = 1.25$, which is very close to the unit value holding for distributed amplification.

5.2 SOLITON PERTURBATION THEORY

5.2.1 Fundamental Soliton Solution

Without the noise term $n(\tau, \zeta) = 0$, the fundamental soliton solution of equation (5.27) is

$$q_0(\tau, \zeta) = A(\zeta)\mathrm{sech}\{A(\zeta)[\tau - T(\zeta)]\}\exp[-i\Omega(\zeta)\tau + i\phi(\zeta)] \qquad (5.36)$$

where

$$A(\zeta) = A \qquad (5.37)$$

$$\Omega(\zeta) = \Omega \qquad (5.38)$$

$$\phi(\zeta) = \phi + \tfrac{1}{2}[A^2 - \Omega^2]\zeta \qquad (5.39)$$

$$T(\zeta) = T - \Omega\zeta \qquad (5.40)$$

where the constants A, Ω, ϕ, and T are the values of $A(\zeta)$, $\Omega(\zeta)$, $\phi(\zeta)$, and $T(\zeta)$ at the origin $\zeta = 0$, which we set at the fiber input. In soliton units the full width at half power of a soliton with $A = 1$ is $\ln(3 + \sqrt{8}) \approx 1.763$, in dimensional units

$$T_F = \ln(3 + \sqrt{8})T_s \approx 1.763T_s \tag{5.41}$$

The phase of the soliton undergoes a 2π phase change for a propagation distance of $\zeta_{2\pi} = 4\pi$. The corresponding propagation distance in nonnormalized units is $z_{2\pi} = 4\pi z_0 = 4\pi T_0^2/|\beta_2|$. Instead of this distance, often used is a measure of the characteristic distance for the nonlinear dynamic which is one-eighth of this distance, the so-called soliton period,

$$z_0 = \frac{\pi T_s^2}{2|\beta_2|} = \frac{\pi}{2}L_D \tag{5.42}$$

Notice that the product of the amplitude times the pulsewidth in the family of solution (5.36) is constant; it depends only on link parameters. The property that the area of any soliton of the family of fundamental solitons is invariant is the so-called area theorem. A consequence of the area theorem is that the soliton energy, which is proportional to the amplitude squared times the pulsewidth, is proportional to the amplitude. In the following, when we discuss the soliton amplitude, we infer also the soliton energy, since the two quantities are proportional.

In real-world units, the soliton solution (5.36) is

$$V_0(t, z) = A \operatorname{sech}\left(\frac{t - t_p + |\beta_2|\Delta\omega_0 z}{T_s}\right)$$

$$\times \exp\left(-i\Delta\omega_0 t + i\frac{|\beta_2|}{2T_s^2}z - i\frac{|\beta_2|\Delta\omega_0^2}{2}z\right) \tag{5.43}$$

where t_p is the timing of the soliton peak at $z = 0$, $\Delta\omega_0 = \Omega/T_s$ is the angular frequency shift of the soliton from the angular frequency of the carrier ω_0, and

$$A = \left[\frac{G\ln(G)|\beta_2|}{(G-1)\gamma}\right]^{1/2}\frac{1}{T_s} \tag{5.44}$$

is the soliton's amplitude. The area of the pulse is

$$\text{Area} = \int dt|V_0| = \pi\left[\frac{G\ln(G)|\beta_2|}{(G-1)\gamma}\right]^{1/2} \tag{5.45}$$

and the pulse energy

$$\mathcal{E} = \int dt|V_0|^2 = \left[\frac{2G\ln(G)|\beta_2|}{(G-1)\gamma T_s}\right] \tag{5.46}$$

The amplitude of the pulse $A \propto T_s^{-1}$ and $\mathcal{E} \propto A^2 T_s$, so $\mathcal{E} \propto T_s^{-1}$. The area of the pulse, proportional to AT_0 is independent of T_0, and this is the area theorem, which we noted above.

The spectrum of the soliton is easily evaluated from (5.43). At $z = 0$ it is

$$V_0(\Delta\omega, 0) = \int_{-\infty}^{\infty} dt \, \exp(i\Delta\omega t) V_0(t, 0) = \pi A T_s \, \text{sech}\left(\frac{\pi}{2} \Delta\omega T_s\right) \qquad (5.47)$$

A useful measure of the spectral extension of a soliton is the so-called time-bandwidth product, defined as the product between T_F and the full-width at half maximum of the power spectral density of the soliton $\Delta\nu_F$. Straightforward algebra gives

$$\Delta\nu_F T_F = \frac{\ln^2(3 + \sqrt{8})}{\pi^2} = 0.315 \qquad (5.48)$$

5.2.2 Soliton Perturbation Theory

Let us make equation (5.27) more general by including other sources of perturbation besides the noise caused by spontaneous emission,

$$\frac{\partial q}{\partial \zeta} = \frac{i}{2} \frac{\partial^2 q}{\partial \tau^2} + i|q|^2 q + R \qquad (5.49)$$

where R can be function of τ, ζ, and of q, q^* and their derivatives. The special case $R = n(\tau, \zeta)$ corresponds to a soliton propagation perturbed only by the ASE noise of the in-line amplifiers. Equation (5.36) is the solution of (5.49) with $R = 0$. We assume that the perturbation due to R is small, and its effect on the soliton dynamics can be analyzed to the first order. Aim of this section is to arrive at an equation for the adiabatic motion of the soliton parameters, A, ϕ, Ω, and T. Although the final equations are simple, and all terms can be given a direct physical meaning, their derivation requires some work. The reader who is not interested in a detailed mathematical derivation should proceed directly to equations (5.91)–(5.94).

Let us first consider the case $\Omega = 0$. We can assume the following form for the solution of (5.27):

$$q(\tau, \zeta) = q_0(\tau, \zeta) + \delta q(\tau, \zeta) \exp\left(\frac{i}{2} A^2 \zeta\right) \qquad (5.50)$$

Inserting equation (5.50) into equation (5.27), we get, to first order,

$$\frac{\partial \delta q}{\partial \zeta} = \mathcal{L}(\delta q) + R \qquad (5.51)$$

where

$$\mathcal{L}(\delta q) = i\left(\frac{1}{2}\frac{\partial^2}{\partial \tau^2} + 2|q_0|^2\right)\delta q + iq_0^2\delta q^* = \frac{i}{2}A^2\delta q \tag{5.52}$$

In terms of the actual physics here, the displacement of the soliton from the soliton solution (5.27) involves a combination of contributions having very different physical meanings. If the soliton timing T changes into $T + \delta T$, it perturbs q and contributes to δq, and so do displacements of the soliton's frequency, amplitude, and phase, which change Ω, A, or ϕ into $\Omega + \delta\Omega$, $A + \delta A$, and $\phi + \delta\phi$. These changes are very different from the generic changes of q which cannot be reduced to changes in any of the soliton parameters. To sort out the various contributions, for δq we can write, without loss of generality, the following expansion [32]:

$$\delta q = \frac{\partial q_0}{\partial A}\delta A + \frac{\partial q_0}{\partial \phi}\delta\phi + \frac{\partial q_0}{\partial \Omega}\delta\Omega + \frac{\partial q_0}{\partial T}\delta T + \delta q_c(\tau, \zeta) \tag{5.53}$$

The rest of the above expansion, $\delta q_c(\tau, \zeta)$, is any perturbation of the solitons that cannot be reduced to a change of soliton parameters, an excitation of the soliton "continuum" [33, 34]. Let us calculate the partial derivatives of q_0 with respect of A, ϕ, Ω, and T

$$\frac{\partial q_0}{\partial A} = \frac{1}{A}\{1 - A(\tau - T)\tanh[A(\tau - T)]\}q_0 \tag{5.54}$$

$$\frac{\partial q_0}{\partial \phi} = iq_0 \tag{5.55}$$

$$\frac{\partial q_0}{\partial \Omega} = -i\tau q_0 \tag{5.56}$$

$$\frac{\partial q_0}{\partial T} = A\tanh[A(\tau - T)]q_0 \tag{5.57}$$

The above derivatives can be expressed as a linear combination of four functions:

$$f_A = \frac{1}{A}\{1 - A(\tau - T)\tanh[A(\tau - T)]\}q_0 \tag{5.58}$$

$$f_\phi = iq_0 \tag{5.59}$$

$$f_\Omega = -i(\tau - T)q_0 \tag{5.60}$$

$$f_T = A\tanh[A(\tau - T)]q_0 \tag{5.61}$$

They are generalized eigenfunctions of the operator $\mathcal{L}$ with eigenvalue zero,

$$\mathcal{L}(f_A) = Af_\phi \tag{5.62}$$

$$\mathcal{L}(f_\phi) = 0 \tag{5.63}$$

$$\mathcal{L}(f_\Omega) = -f_T \tag{5.64}$$

$$\mathcal{L}(f_T) = 0 \tag{5.65}$$

The moduli of the four functions $|f_j|$ for $A = 1$ are shown versus $\tau - T$ in Fig. 5.2. Since they coincide with the derivatives of the soliton with respect to one of the four parameters for $T = 0$, their physical meaning is obvious: At $T = 0$, $f_j \Delta j$ is the change of the soliton field caused by a displacement Δj of the parameter labeled by j ($j = A, \phi, \Omega$, and T).

Using definitions (5.54)–(5.57), we can rewrite equation (5.53) as

$$\delta q = f_A \delta A(\zeta) + f_\phi \delta \phi(\zeta) + (f_\Omega - Tf_\phi)\delta \Omega(\zeta) + f_T \delta T(\zeta) + \delta q_c(\tau, \zeta) \tag{5.66}$$

Now we define the inner product in the space of functions δq as

$$\langle f|g \rangle = \mathrm{Re} \int d\tau \, f^* g \tag{5.67}$$

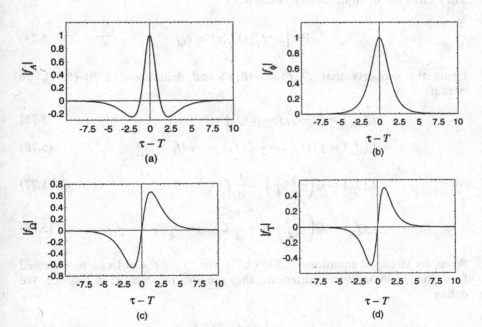

FIGURE 5.2 Moduli of the four functions $|f_j|$, for $A = 1$, versus $\tau - T$.

and the adjoint operator of $\mathcal{L}$, $\underline{\mathcal{L}}$ as

$$\text{Re} \int d\tau \, f^* \mathcal{L}(g) = \text{Re} \int d\tau \underline{\mathcal{L}}(f)^* g \tag{5.68}$$

By direct substitution and integration by parts, it can be shown that the adjoint of $\mathcal{L}$ defined by equation (5.52) is

$$\underline{\mathcal{L}}(\delta q) = -i \left(\frac{1}{2} \frac{\partial^2}{\partial \tau^2} + 2|q_0|^2 \right) \delta q + i q_0^2 \delta q^* + \frac{i}{2} A^2 \delta q \tag{5.69}$$

We notice that $\underline{\mathcal{L}}(if) = -i\mathcal{L}(f)$. Let us define then the adjoint set of the f_j:

$$\underline{f}_A = -if_\phi \tag{5.70}$$

$$\underline{f}_\phi = if_A \tag{5.71}$$

$$\underline{f}_\Omega = -i\frac{f_T}{A} \tag{5.72}$$

$$\underline{f}_T = i\frac{f_\Omega}{A} \tag{5.73}$$

They meet the orthonormality condition

$$\text{Re} \int d\tau \underline{f}_j^*(\tau) f_i(\tau) = \delta_{i,j} \tag{5.74}$$

Using the property that $\underline{\mathcal{L}}(if) = -i\mathcal{L}(f)$ and definitions (5.70)–(5.73), we obtain

$$\underline{\mathcal{L}}(\underline{f}_A) = \underline{\mathcal{L}}(-if_\phi) = i\underline{\mathcal{L}}(f_\phi) = 0 \tag{5.75}$$

$$\underline{\mathcal{L}}(\underline{f}_\phi) = \underline{\mathcal{L}}(if_A) = -i\underline{\mathcal{L}}(f_A) = -iAf_\phi = A\underline{f}_A \tag{5.76}$$

$$\underline{\mathcal{L}}(\underline{f}_\Omega) = \underline{\mathcal{L}}\left(-i\frac{f_T}{A}\right) = \frac{i}{A}\underline{\mathcal{L}}(f_T) = 0 \tag{5.77}$$

$$\underline{\mathcal{L}}(\underline{f}_T) = \underline{\mathcal{L}}\left(i\frac{f_\Omega}{A}\right) = -\frac{i}{A}\underline{\mathcal{L}}(f_\Omega) = \frac{i}{A}f_T = -\underline{f}_\Omega \tag{5.78}$$

Since, by virtue of equations (5.70)–(5.73), the $|\underline{f}_j|$ for $A = 1$ can be obtained from the $|f_j|$ by index permutations, they are also represented in Fig. 5.2. We define

$$F_j(\zeta) = \int d\tau \, \underline{f}_j(\tau) \delta q_c(\tau, \zeta) \tag{5.79}$$

Then, neglecting the derivatives of $A(\zeta)$, $\theta(\zeta)$, and $T(\zeta)$ with respect to ζ, which means that we assume that those quantities are slowly varying with respect to the soliton dynamics, we write

$$\frac{\partial F_j(\zeta)}{\partial \zeta} = \int d\tau \, \underline{f}_j(\tau) \mathcal{L}[\delta q_c(\tau, \zeta)] = \int d\tau \underline{\mathcal{L}}[\underline{f}_j(\tau)] \delta q_c(\tau, \zeta) \qquad (5.80)$$

At $\zeta = 0$ we decompose δq such that the continuum is orthogonal to the f_j,

$$F(0) = \int d\tau \, \underline{f}_j(\tau) \delta q_c(\tau, 0) = 0 \qquad (5.81)$$

Since $\underline{\mathcal{L}} \, \underline{f}_j(\tau)$ is always proportional to one of the $\underline{f}_h(\tau)$, by equations (5.80) and (5.81) the continuum will stay orthogonal for all $\zeta \geq 0$. This is an important result. The continuum, which is generated by the perturbation R, does not couple with the soliton to the first order. The back-coupling of the continuum onto the soliton is a second-order effect [32, 33].

Inserting equation (5.66) into equation (5.51), we get

$$f_A \frac{d\delta A(\zeta)}{d\zeta} + f_\phi \frac{d\delta\phi(\zeta)}{d\zeta} + (f_\Omega - Tf_\phi) \frac{d\delta\Omega(\zeta)}{d\zeta} + f_T \frac{d\delta T(\zeta)}{d\zeta}$$
$$+ \frac{\partial \delta q_c(\tau, \zeta)}{\partial \zeta} = \mathcal{L}(\delta u) + R \qquad (5.82)$$

Let us project equation (5.82) over f_ϕ by multiplying both sides of (5.82) by $\underline{f}_\phi$, integrating over T, and taking the real part. Using the orthonormality relations (5.74), we get

$$\frac{d\delta\phi(\zeta)}{d\zeta} - T \frac{d\delta\Omega(\zeta)}{d\zeta} = \text{Re} \int d\tau \, \underline{f}_\phi^* \mathcal{L}(\delta q) + \text{Re} \int d\tau \, \underline{f}_\phi^* R \qquad (5.83)$$

Using the definition of adjoint operator, we get

$$\frac{d\delta\phi(\zeta)}{d\zeta} - T \frac{\partial\delta\Omega(\zeta)}{\partial\zeta} = \text{Re} \int d\tau \underline{\mathcal{L}}(\underline{f}_\phi)^* \delta q + \text{Re} \int d\tau\tau \, \underline{f}_\phi^* R \qquad (5.84)$$

Using equation (5.76) and definition (5.66), we get

$$\frac{d\delta\phi(\zeta)}{d\zeta} = T \frac{d\delta\Omega(\zeta)}{d\zeta} + A\delta A + A \, \text{Re} \int d\tau \, \underline{f}_A^* \delta q_c + \text{Re} \int d\tau \, \underline{f}_\phi^* R \qquad (5.85)$$

that is, by equation (5.79),

$$\frac{d\delta\phi(\zeta)}{d\zeta} = T \frac{\partial\delta\Omega(\zeta)}{\partial\zeta} + A\delta A + \text{Re} \int d\tau \, \underline{f}_\phi^* R \qquad (5.86)$$

The equations for the other perturbations can be obtained analogously. We give just the final result, repeating equation (5.86) for the sake of completeness:

$$\frac{d\delta A}{d\zeta} = \text{Re} \int d\tau \, \underline{f}_A^* R \tag{5.87}$$

$$\frac{d\delta\phi}{d\zeta} = A\delta A + T \frac{d\delta\Omega(\zeta)}{d\zeta} + \text{Re} \int d\tau \, \underline{f}_\phi^* R \tag{5.88}$$

$$\frac{d\delta\Omega}{d\zeta} = \text{Re} \int d\tau \, \underline{f}_\Omega^* R \tag{5.89}$$

$$\frac{d\delta T}{d\zeta} = -\delta\Omega + \text{Re} \int d\tau \, \underline{f}_T^* R \tag{5.90}$$

These equations were obtained under the assumption that observable changes of the soliton are small. Conceptually one might extend the above analysis to large variations as follows: Assume an unperturbed soliton at input. At $\Delta\zeta \ll 1$ the soliton observable amplitude, phase, frequency, and timing will change slightly by the perturbation effect. One can then take the soliton with the modified observables as the new unperturbed solution, again applying first-order perturbation theory. This procedure does not need to be applied literally; it just tells us that large changes in the soliton observables can be analyzed by extending equations (5.87)–(5.90) to large signals if the rate of change is small with respect to eight times the soliton period which, as we have discussed above, is the length scale of the unperturbed soliton dynamics. In doing so, equations (5.87) to (5.90) are first extended to $\Omega \neq 0$. This can be easily done, since the soliton wavevector is $(A^2 - \Omega^2)/2$; for $\Omega \neq 0$ the phase ϕ changes into $\phi - \Omega^2\zeta/2$, so to the right-hand side of (5.88) is added the term $-\Omega^2/2$. Also $T(\zeta)$ changes into $T(\zeta) - \Omega\zeta$, so the term $-\Omega$ is set at the right-hand side of (5.90). Finally we add to the equation for the phase the term $A^2/2$ which does not appear in (5.88), since in the definition of the field perturbation $\delta q(\tau, \zeta)$, equation (5.50), we have subtracted the phase term $\exp(iA^2\zeta/2)$. Therefore the generalized equations of the first-order perturbation theory are

$$\frac{dA}{d\zeta} = \text{Re} \int d\tau \, \underline{f}_A^* R \tag{5.91}$$

$$\frac{d\phi}{d\zeta} = \frac{1}{2}(A^2 - \Omega^2) + T \frac{d\Omega(\zeta)}{d\zeta} + \text{Re} \int d\tau \, \underline{f}_\phi^* R \tag{5.92}$$

$$\frac{d\Omega}{d\zeta} = \text{Re} \int d\tau \, \underline{f}_\Omega^* R \tag{5.93}$$

$$\frac{dT}{d\zeta} = -\Omega + \text{Re} \int d\tau \, \underline{f}_T^* R \tag{5.94}$$

Equations (5.87)–(5.90) may be easily obtained by linearizing equations (5.91)–(5.94) around A, ϕ, $\Omega = 0$, and T.

The physical meaning of the different terms at right-hand side of equations (5.91)–(5.94) are as follows: The integrals

$$\text{Re} \int d\tau \, \underline{f}_j^* R \tag{5.95}$$

are the effect of perturbation R, which is mathematically represented by a projection over the pertinent adjoint function f_j^*, defined by equations (5.70)–(5.23) and (5.58)–(5.61), on a particular soliton parameter. Without perturbation R, or if the perturbation R is orthogonal to the respective adjoint function, the amplitude A and the frequency Ω would be in constant motion. The phase ϕ has free motion because the soliton wavevector $(A^2 - \Omega^2)/2$ is unperturbed; see equation (5.36) and (5.39). The timing T also experiences unperturbed motion if $\Omega \neq 0$ because of the functional form of the soliton solutions (5.36) and (5.40) in turn caused by the nonzero group velocity dispersion of the fiber, making the group velocity proportional to the central frequency of the soliton. Since with anomalous dispersion the velocity of a pulse is proportional to its frequency, as the frequency of the soliton decreases, its timing increases (i.e., the soliton is delayed). Finally, one may understand the reasons for the appearance of the term $T d\Omega/d\zeta$ at right side of the phase equation (5.92) by considering a periodic stream of solitons separated by T and uniformly accelerated, that is, having a constant $d\Omega/d\zeta = \Omega' \neq 0$. The periodic stream of solitons is represented in the Fourier space as a sequence of lines, spaced in normalized units by $\Delta\Omega = 2\pi/T$. At $\zeta = 0$ all the solitons are assumed to be in phase. The spectrum of the sequence is the symmetric, with one line at the center. A soliton can be frequency shifted by the action of a gain, which depends linearly on frequency; it is zero at the center of the spectrum, positive for the lines at one side of the spectrum, and negative (i.e., loss) for those at the other side. A frequency-dependent gain simply changes the amplitude of the lines, therefore changing the position of the center of the spectral evelope, not the position. At $\zeta_1 = \pi/(T\Omega')$ the center of the envelope of the spectrum shifts by $\Omega'\zeta_1| = \pi/T$, half the line spacing. The spectral lines are now still symmetrically distributed with respect to the center of the envelope, but there is no line at the center. This spectrum is that of a soliton stream π which is out of phase. Using (5.92), the term $T d\Omega/d\zeta = T\Omega'$ at right-hand side of the dynamical equation for the phase gives just the π phase shift when integrated over ζ_1, resulting in $T\Omega'\zeta_1 = \pi$. This phase change is applied to each element of the sequence of solitons spaced by T, and it produces a change in the spectrum of the sequence from the in-phase stream to the out-of-phase stream of solitons. This is consistent with the simple physical picture give earlier.

5.3 GORDON-HAUS EFFECT AND ITS CONTROL

5.3.1 Gordon-Haus Effect

Assume that the soliton propagation is perturbed by the ASE noise of the in-line amplifiers only, $R = n(\tau, \zeta)$. Assume also that the perturbation due to noise is small and $\Omega = 0$ so that equations (5.87)–(5.90) can be used instead of equations (5.91)–(5.94). Let us further assume that $T = 0$. The statistical properties of the noise terms

$$n_i(\zeta) = \int d\tau \, \underline{f_i}(\tau) n(\tau, \zeta) \tag{5.96}$$

characterize the statistical properties of the perturbations of the soliton observables. This statistical properties are easily evaluated with the help of equations (5.30)–(5.32)

$$\langle n_i(\zeta) \rangle = \mathrm{Re} \int d\tau \, f_i(\tau)^* \langle n(\tau, \zeta) \rangle = 0 \tag{5.97}$$

$$\langle n_i(\zeta) n_j(\zeta') \rangle = \left\langle \mathrm{Re} \int d\tau \, \underline{f_i}(\tau)^* n(\tau, \zeta) \, \mathrm{Re} \int d\tau' \, \underline{f_j}(\tau')^* n(\tau', \zeta') \right\rangle \tag{5.98}$$

Notice that $\mathrm{Re} \, \underline{f_i}(\tau)^* n(\tau, \zeta)$ is the component of the phasor representing $n(\tau, \zeta)$ along the direction of the phasor representing $\underline{f_i}(\tau)$, so equation (5.98) becomes for $i = j$,

$$\langle n_i(\zeta) n_i(\zeta') \rangle = \frac{1}{2} \int d\tau \int d\tau' \, \underline{f_i}(\tau)^* \, \underline{f_i}(\tau') \langle n(\tau, \zeta)^* (\tau', \zeta') \rangle$$

$$= \frac{\langle n \rangle}{2} \int d\tau |\, \underline{f_i}(\tau)|^2 \delta(\zeta - \zeta') \tag{5.99}$$

Represented by $\frac{1}{2}$ is the fact that the noise in-phase with $\underline{f_j}$ is one-half of the total due to the phase indpendence of the ASE noise. It can be verified at once that because the $\underline{f_i}$ have different parity, or they are in quadrature in the complex plane, $\langle n_i(\zeta) n_j(\zeta') \rangle = 0$ for $i \neq j$. Equation (5.98) then becomes

$$\langle n_i(\zeta) n_j(\zeta') \rangle = n_i \delta(\zeta - \zeta') \delta_{i,j} \tag{5.100}$$

where

$$n_i = \frac{\langle n \rangle}{2} \int d\tau |\, \underline{f_i}(\tau)|^2 \tag{5.101}$$

By definitions (5.70)–(5.73) we get

$$n_A = \langle n \rangle A, \tag{5.102}$$

$$n_\phi = \frac{\langle n \rangle}{3A} \left(1 + \frac{\pi^2}{12}\right) \tag{5.103}$$

$$n_\Omega = \frac{\langle n \rangle}{3} A \tag{5.104}$$

$$n_T = \frac{\langle n \rangle}{A^3} \frac{\pi^2}{12} \tag{5.105}$$

The correlation function is proportional to a Dirac delta function in space. This is the mathematical consequence, in the average model, of the statistical independence of the ASE noise added to the signal by different amplifiers. Solution of equations (5.87)–(5.90) with $\delta A(0) = \delta\phi(0) = \delta\Omega(0) = \delta T(0) = 0$ can be found immediately:

$$\delta A(\zeta) = \int_0^\zeta d\zeta' n_A(\zeta') \tag{5.106}$$

$$\delta\phi(\zeta) = A \int_0^\zeta d\zeta' \delta A(\zeta') + \int_0^\zeta d\zeta' n_\phi(\zeta') \tag{5.107}$$

$$\delta\Omega(\zeta) = \int_0^\zeta d\zeta' n_\Omega(\zeta') \tag{5.108}$$

$$\delta T(\zeta) = -\int_0^\zeta d\zeta' \delta\Omega(\zeta) + \int_0^\zeta d\zeta n_T(\zeta') \tag{5.109}$$

We prefer to keep the explicit dependence on A, instead of setting $A = 1$, to maintain in the equations the explicit dependence of the statistical properties of the soliton perturbations on soliton amplitude (and pulsewidth).

Let us calculate first the correlation function of the amplitude fluctuations [35]

$$\langle \delta A(\zeta) \delta A(\zeta') \rangle = \int_0^\zeta d\zeta_1 \int_0^{\zeta'} d\zeta_2 \langle n_A(\zeta_1) n_A(\zeta_2) \rangle$$

$$= n_A \int_0^\zeta d\zeta_1 \int_0^{\zeta'} d\zeta_2 \delta(\zeta_1 - \zeta_2)$$

$$= n_A \min(\zeta, \zeta') \tag{5.110}$$

where $\min(\zeta, \zeta')$ is the minimum of ζ and ζ'. Analogously we find that

$$\langle \delta\Omega(\zeta) \delta\Omega(\zeta') \rangle = n_\Omega \min(\zeta, \zeta') \tag{5.111}$$

In particular, for $\zeta = \zeta'$ we have the variance

$$\langle \delta A(\zeta)^2 \rangle = n_A \zeta \tag{5.112}$$

$$\langle \delta\Omega(\zeta)^2 \rangle = n_\Omega \zeta \tag{5.113}$$

The variance of the amplitude and frequency fluctuations grows proportionally to propagation distance ζ. The variance of the phase and timing is also easily evaluated (the correlation function for $\zeta \neq \zeta'$ in not relevant). For the phase

$$\langle \delta\phi(\zeta)^2 \rangle = A \int_0^\zeta d\zeta_1 \int_0^\zeta d\zeta_2 \langle \delta A(\zeta_1) \delta A(\zeta_2) \rangle + \int_0^\zeta d\zeta_1 \int_0^\zeta d\zeta_2 \langle n_\phi(\zeta_1) n_\phi(\zeta_2) \rangle$$

$$= 2An_A \int_0^\zeta d\zeta_1 \int_0^{\zeta_1} d\zeta_2 \zeta_2 + n_\phi \zeta$$

$$= n_A A \frac{\zeta^3}{3} + n_\phi \zeta \tag{5.114}$$

Analogously we find for the timing fluctuations

$$\langle \delta T(\zeta)^2 \rangle = n_\Omega \frac{\zeta^3}{3} + n_T \zeta \tag{5.115}$$

Let us concentrate on the result on timing fluctuations. We have found that the variance of timing fluctuations grows, asymptotically, proportionally to the cube of propatation distance ζ. This rapid growth of timing fluctuations is the consequence of the frequency, hence velocity, fluctuations induced by the ASE noise on the soliton. This effect, discovered by Gordon and Haus in 1986 [8] and named after them, has important consequences for the potential use of optical solitons in long-distance transmission systems.

In real-world units the variance of timing fluctuations is

$$\langle \delta t(z)^2 \rangle = \frac{\hbar\omega_0 \gamma F\alpha f |\beta_2|}{9T_s} z^3 + \frac{\pi^2 \hbar\omega_0 \gamma T_s^3 F\alpha f}{12|\beta_2|} z \tag{5.116}$$

Let us discuss the physical meaning of the expression above. The second term at right-hand side of (5.116) is the effect of the direct coupling of the ASE noise with the soliton timing, and it grows proportionally to the linear power of propagation distance. It is usually negligible. The first term is the mathematical expression of the Gordon-Haus effect, it is the effect of the random walk of the soliton center frequency caused by the coupling of the ASE noise with soliton spectrum. The soliton center frequency is proportional to the soliton group velocity, because of the nonzero group velocity dispersion of the line; hence a random walk of the soliton center frequency produces a random walk of the soliton velocity. The random walk of the soliton velocity in turn causes the growth of the variance of the arrival time of the solitons proportionally to the cube of propagation distance. This growth is much faster than that caused by the direct coupling of the ASE noise with timing, which is instead linear with distance.

The Gordon-Haus effect poses a severe limit to the maximum distance that can be reached with a soliton-based optical transmission system. It sets a limit to the maximum product bit-rate times transmission distance that can be achieved. This may be easily understood by dividing by T_0^2 (remember that T_0 is related to the full width at half the power of the soliton by $T_{FWHM} = 1.763 T_0$) both sides of (5.116),

$$\frac{\langle \delta t(z)^2 \rangle}{T_s^2} = \frac{\hbar \omega_0 \gamma F \alpha f |\beta_2|}{9} \left(\frac{z}{T_s} \right)^3 + \frac{\pi^2 \hbar \omega_0 \gamma T_s F \alpha f}{12 |\beta_2|} z \qquad (5.117)$$

As we mentioned above, the second term at right-hand side of equation (5.117) is usually negligible; let us then focus our attention on the first. To avoid large errors caused by timing jitter, $\langle \delta t(z)^2 \rangle / T_0^2$ should be less than a value that depends on the ratio of the pulsewidth to the bit period. The ratio of the pulsewidth to the bit-period is usually set by the requirement of avoiding significant soliton-soliton interaction; therefore also the maximum value of $\langle \delta t(z)^2 \rangle / T_0^2$ is fixed. Hence (5.117) gives a limit to z / T_0, which is the ratio of the transmission distance to the bit-period or, equivalently, to the maximum bit-rate distance product.

5.3.2 Soliton Transmission Control

The limit imposed by the Gordon-Haus effect can be extended by using methods to control soliton propagation. A unique feature of solitons compared to linear waves is their roboustness, which makes them able to sustain the action of controlling elements without being distroyed, at least if the action of the controlling elements is weak enough that their effect is small within the characteristic length of the soliton dynamics, eitht times the soliton period. If this condition is met, the action of the controlling elements can be analyzed by first-order perturbation theory.

The two methods that have been proposed first, and successfully experimentally tested, will be described below. One is based on the periodical insertion along the line of passive frequency filters, the other on active modulators. We will study the soliton control based on filters first.

Filter Control

The idea of the filtering control of soliton transmission is simple [16, 50]. We have seen that the Gordon-Haus effect is the random walk that the soliton center frequency undertakes when the soliton propagates in the presence of ASE noise. This effect produces the growth in the variance of the timing fluctuations proportionally to the cube of the propagation distance, which severely limits the practical applications of optical solitons as information carriers. So it might come naturally to search for ways to prevent fluctuations of the soliton center frequency or, at least, to look for devices able to provide a restoring force for the frequency fluctuations. Weak band-pass filters inserted along the line are good candidates. Their use is especially effective because of the peculiar robustness of the fundamental NLS soliton against small perturbations.

To understand how filtering control works, assume simple parabolic filters to fix ideas, since any shallow filter can be approximated to first order by the parabolic filter having the same curvature at pass-band center. Assume that the center of the filter pass-band coincides with the center frequency of the soliton. When the ASE noise causes the soliton center frequency to shift from the center of the filter pass-band, the soliton experiences an asymmetric in frequency loss that pushes it back to the center of the filter profile. As anticipated above, one may quantitatively characterize the effect of this "restoring force" that the soliton experiences in the frequency domain by using soliton perturbation theory.

The spectral profile of the simplest Fabry-Perot filter is

$$H(\omega) = \frac{1 - R_f}{1 - R_f \exp[i(\omega - \omega_0)t_d]} \tag{5.118}$$

where t_d is the filter delay defined as $t_d = 2d/v_g$, d is the distance between the two mirrors of power reflectivity R_f, ω is the optical frequency, and ω_0 is the center frequency of the soliton that coincides with the center frequency of one of the filter pass-band. If a Fabry-Perot filter is inserted periodically along the line with period L_f, the equivalent averaged filter profile is

$$H_{av}(\omega) = \frac{1}{L_f} \ln\left\{\frac{1 - R_f}{1 - R_f \exp[i(\omega - \omega_0)t_d]}\right\} \tag{5.119}$$

Equation (5.119) can be written by separating the real and imaginary parts as

$$H_{av}(\omega) = h(\omega) + i\varphi(\omega) \tag{5.120}$$

where

$$h(\omega) = \frac{1}{2L_f} \ln\left\{\frac{1}{1 + F \sin^2[(\omega - \omega_0)t_d/2]}\right\} \tag{5.121}$$

$$F_r = \frac{4R_f}{(1 - R_f)^2}, \tag{5.122}$$

$$\varphi(\omega) = \frac{1}{L_f} \arctan\left\{\frac{R_f \sin[(\omega - \omega_0)t_d]}{1 - R_f \cos[(\omega - \omega_0)t_d]}\right\} \tag{5.123}$$

By expanding the phase profile as a power series of $(\omega - \omega_0)$, one finds that the filter contributes to the odd terms of the line dispersion: The first-order term is a simple group velocity delay term, the term proportional to $(\omega - \omega_0)^3$ is a third-order dispersion term which adds to the third-order dispersion of the line

[36]. For shallow filters and for fiber dispersion sufficiently large, the phase contribution of the filter is negligible. For analyzing the effect of shallow filters on soliton propagation, it might be enough expanding $h(\omega)$ to second order in $(\omega - \omega_0)$ as

$$h(\omega) = -\frac{F_r t_d^2}{8L_f} (\omega - \omega_0)^2 \tag{5.124}$$

By using the transformation $-i(\omega - \omega_0) \rightarrow \partial/\partial t$, the averaged filter action given by (5.124) becomes in the time domain

$$h\left(\frac{\partial}{\partial t}\right) = \left(\frac{F_r t_d^2}{8L_f}\right) \frac{\partial^2}{\partial t^2} \tag{5.125}$$

Equation (5.12) extended to include the effect of in-line filtering becomes

$$\frac{\partial V_0}{\partial z} = \frac{\delta g}{2} V_0 + \left(\frac{F_r t_d^2}{8L_f}\right) \frac{\partial^2 V_0}{\partial t^2} + i \frac{|\beta_2|}{2} \frac{\partial^2 V_0}{\partial t^2}$$
$$+ i\gamma \frac{1 - \exp(-\alpha L_A)}{\alpha L_A} |V_0|^2 V_0 \tag{5.126}$$

Because of filtering, the soliton experiences larger loss on the wings of its spectrum than on the center of the filter profile. For the gain being equal to loss across the soliton spectrum, the amplifiers must provide extra gain at the center of the filter profile to compensate for the extra loss in the wings due to filtering. The excess gain of the amplifiers has been included in equation (5.126) by the term $\delta g V_0/2$. In soliton units, and with the ASE noise term added, equation (5.126) becomes (see equation (5.27))

$$\frac{\partial q}{\partial \zeta} = \frac{i}{2} \frac{\partial^2 q}{\partial \tau^2} + i|q|^2 q + \frac{\Delta \alpha}{2} q + \frac{\eta}{2} \frac{\partial^2 q}{\partial \tau^2} + n(\tau, \zeta) \tag{5.127}$$

where

$$\Delta \alpha = \delta g z_0 \tag{5.128}$$

$$\eta = \left(\frac{F t_d^2}{4L_f |\beta_2|}\right) \tag{5.129}$$

We may use the perturbative approach with

$$R = \frac{\Delta \alpha}{2} q + \frac{\eta}{2} \frac{\partial^2 q}{\partial \tau^2} + n(\tau, \zeta) \tag{5.130}$$

Using equations (5.91)–(5.94) we get after some algebra [16, 50]

$$\frac{dA}{d\zeta} = \Delta\alpha A - \eta\left(\Omega^2 + \frac{A^2}{3}\right)A + n_A(\zeta) \tag{5.131}$$

$$\frac{d\phi}{d\zeta} = \tfrac{1}{2}(A^2 - \Omega^2) + T\frac{d\Omega}{d\zeta} + n_\phi(\zeta) \tag{5.132}$$

$$\frac{d\Omega}{d\zeta} = -\tfrac{2}{3}\eta A^2\Omega + n_\Omega(\zeta) \tag{5.133}$$

$$\frac{dT}{d\zeta} = -\Omega + n_T(\zeta) \tag{5.134}$$

Neglecting the noise terms, the steady state of (5.131) and (5.133) is

$$\Omega_s = 0 \tag{5.135}$$

$$A_s = \sqrt{\frac{3\Delta\alpha}{\eta}} \tag{5.136}$$

and T_s arbitrary. Within our definitions, the phase ϕ is not stationary; it grows linearly with ζ, $\phi = A_s^2\zeta/2$. This is the consequence of including in the definition of the phase the contribution of the soliton wavevector $(A^2 - \Omega^2)/2$ (see equations (5.36) and (5.39)). A steady state amplitude $A_s = 1$ implies the excess gain $\Delta\alpha = \eta/3$. Linearizing around the steady state the above equations, assuming $T_s = 0$ obtains

$$\frac{d\delta A}{d\zeta} = -\beta\delta A + n_A(\zeta) \tag{5.137}$$

$$\frac{d\delta\phi}{d\zeta} = A_s\delta A + n_\phi(\zeta) \tag{5.138}$$

$$\frac{d\delta\Omega}{d\zeta} = -\beta\delta\Omega + n_\Omega(\zeta) \tag{5.139}$$

$$\frac{dT}{d\zeta} = -\Omega + n_T(\zeta) \tag{5.140}$$

In (5.138) we have defined the damping strength for amplitude and frequency fluctuations as

$$\beta = \tfrac{2}{3}\eta A_s^2 \tag{5.141}$$

and we have subtracted from the phase the contribution of the soliton wave-vector at $\Omega_s = 0$ by defining $\delta\phi = \phi - A_s^2\zeta/2$. The solution for δA with $\delta A(0) = 0$ is

$$\delta A(\zeta) = \int_0^\zeta d\zeta' \exp[-\beta(\zeta - \zeta')]n_A(\zeta') \tag{5.142}$$

and the two-point correlation function

$$\langle\delta A(\zeta)\delta A(\zeta')\rangle = \frac{n_A}{2\beta}\{\exp[-\beta|\zeta - \zeta'|] - \exp[-\beta(\zeta + \zeta')]\} \tag{5.143}$$

The solution of equation (5.138) with $\delta\phi(0) = 0$ is

$$\delta\phi(\zeta) = A_s\int_0^\zeta d\zeta'\delta A(\zeta') + \int_0^\zeta d\zeta'n_\phi(\zeta') \tag{5.144}$$

The variance of the phase fluctuations is

$$
\begin{aligned}
\langle(\delta\phi(\zeta))^2\rangle &= A_s^2\int_0^\zeta d\zeta_1\int_0^\zeta d\zeta_2\langle\delta A(\zeta_1)\delta A(\zeta_2)\rangle \\
&\quad + \int_0^\zeta d\zeta_1\int_0^\zeta d\zeta_2\langle n_\phi(\zeta_1)n_\phi(\zeta_2)\rangle \\
&= \frac{A_s^2 n_A}{\beta}\int_0^\zeta d\zeta_1\int_0^{\zeta_1} d\zeta_2\{\exp[-\beta(\zeta_1 - \zeta_2)] \\
&\quad - \exp[-\nu(\zeta_1 + \zeta_2)]\} + n_\phi\zeta
\end{aligned} \tag{5.145}
$$

which, performing the integrals, becomes

$$
\langle(\delta\phi(\zeta))^2\rangle = \frac{A_s^2 n_A}{2\beta^3}[-\exp(-2\beta\zeta) + 4\exp(-\beta\zeta) - 3 + 2\beta\zeta] \\
+ n_\phi\zeta \tag{5.146}
$$

Analogously, we get for the frequency fluctuations,

$$\langle\delta\Omega(\zeta)\delta\Omega(\zeta')\rangle = \frac{n_\Omega}{2\beta}\{\exp[-\beta|\zeta - \zeta'|] - \exp[-\beta(\zeta + \zeta')]\} \tag{5.147}$$

and for the timing fluctuations,

$$\langle(\delta T(\zeta))^2\rangle = \frac{n_\Omega}{2\beta^3}[-\exp(-2\beta\zeta) + 4\exp(-\beta\zeta) - 3 + 2\beta\zeta] + n_T\zeta \tag{5.148}$$

The variance of the amplitude and frequency fluctuations is also obtained by setting $\zeta = \zeta'$ in (5.143) and (5.147),

$$\langle \delta A(\zeta)^2 \rangle = \frac{n_A}{2\beta} \left[1 - \exp(-2\beta\zeta)\right] \tag{5.149}$$

$$\langle \delta\Omega(\zeta)^2 \rangle = \frac{n_\Omega}{2\beta} \left[1 - \exp(-2\beta\zeta)\right] \tag{5.150}$$

First, we notice that both frequency and amplitude fluctuations are bounded at steady state. The simple picture we have given above gives the reason for the boundness of the frequency fluctuations: When the soliton undertakes any fluctuation of its center frequency, it experiences an asymmetric, in frequency, loss that pushes it back at the center of the filter profile.

The effect of filters on frequency fluctuations does not depend on any particular scaling of the soliton parameters. An optical pulse such as a gaussian pulse, which is not a soliton but capable of regenerating, by some mechanism, the radiation lost in the wings of the spectrum, would behave similarly if filtered. An example of such a pulse is a gaussian pulse running in an active mode-locked laser. In these lasers the filtering action is obtained by the frequency-dependent gain of the active medium. The wings of the pulse spectrum are regenerated by the action of an active amplitude modulator inserted inside the cavity, which continuously pumps energy from the central region of the pulse spectrum to its wings through the generation of modulation sidebands [40]. In actively mode-locked lasers, the pulse is trapped within the active medium bandwidth even if it is not a soliton.

On the other hand, the effect on amplitude fluctuation is indirect and it is the consequence of the peculiar scaling of the soliton parameters. By the area theorem, the product amplitude times pulsewidth is constant. Shorter solitons have larger amplitude (hence energy). Assume that a soliton, intially at steady state, undertakes a positive amplitude fluctuation. By the area theorem, as the soliton becomes shorter, hence wider in frequency, its loss grows because the loss is higher in the wings of filter profile. So the amplitude is pushed back to the equilibrium value. The same occurs for negative fluctuations. Thus the filters have the beneficial effect of stabilizing amplitude fluctuations of the soliton in the presence of any disturbance and, in particular, in the presence of the ASE noise of the in-line amplifiers.

Equation (5.148) gives the quantitative measure of the benefit of filters in extending the limit imposed by the Gordon-Haus effect. Without filters, the varance of the timing fluctuations grows proportionally to the cube of the transmission distance. With filters there is still a monotonic growth of the variance of the timing fluctuations, but this growth is only linear with propagation distance. One may easily find the expression for the timing fluctuations in real-world units. We just give here the expression for the asymptotic part which depends on the frequency fluctuations

$$\langle \delta T(\zeta)^2 \rangle = \frac{\langle \delta(z)^2 \rangle}{T_s^2} \sim \frac{9 n_\Omega}{4 \eta^2} \frac{z}{L_D} = \frac{3}{4} \hbar \omega_0 F \alpha \frac{T_s}{|\beta_2|} \gamma f \left(\frac{4 L_f |\beta_2|}{F_r t_d^2} \right)^2 z \quad (5.151)$$

Without control the relative timing fluctuations are inversely proportional to the cube of the pulsewidth (see equation 5.117), but with filters the relative timing fluctuations are proportional to the pulsewidth [37, 38]. The different qualitative behavior of soliton transmission with filtering control is because the controlling action of the filters of the fixed bandwidth is larger for shorter pulses, of the larger bandwidth.

With filtering control, the limit to the transmission distance is set by the stability of the soliton background [16]. The problem arises because of the excess gain δg required for compensating the loss that the soliton experiences in the wings of the spectrum due to filtering. Whenever a zero is transmitted, the ASE radiation generated by the in-line amplifiers experiences a positive gain, which is amplified. To analyze this process on a quantitative basis, we may neglect in (5.127) the nonlinear term because of the low intensity of the ASE radiation, obtaining the linear equation

$$\frac{\partial q}{\partial \zeta} = \frac{i}{2} \frac{\partial^2 q}{\partial \tau^2} + \frac{\delta \alpha}{2} q + \frac{\eta}{2} \frac{\partial^2 q}{\partial \tau^2} + n(\tau, \zeta) \quad (5.152)$$

This equation can be solved by transformation into the Fourier domain. Defining the Fourier transform pair as

$$q = \int \frac{d\Omega}{2\pi} \exp(-i\Omega\tau) \tilde{q}(\Omega) \quad (5.153)$$

$$\tilde{q} = \int d\tau \exp(i\Omega\tau) q \quad (5.154)$$

equation (5.152) becomes

$$\frac{\partial \tilde{q}}{\partial \zeta} = -\frac{i}{2} \Omega^2 \tilde{q} + \frac{\Delta\alpha}{2} \tilde{q} - \frac{\eta}{2} \Omega^2 \tilde{q} + \tilde{n}(\Omega, \zeta) \quad (5.155)$$

where the statistical properties of the Fourier transform of the white noise term $n(\tau, \zeta)$ are

$$\langle \tilde{n}(\Omega, \zeta) \rangle = 0 \quad (5.156)$$

$$\langle \tilde{n}(\Omega, \zeta) \tilde{n}(\Omega', \zeta') \rangle = 0 \quad (5.157)$$

$$\langle \tilde{n}^*(\Omega, \zeta) \tilde{n}(\Omega'\zeta') \rangle = 2\pi \langle n \rangle \delta(0\Omega' - \Omega) \delta(\zeta - \zeta') \quad (5.158)$$

Equation (5.155) can be further simplified by noting that the term $-i\Omega^2\tilde{q}/2$ gives only a ζ-dependent phase shift; one can get rid of by the transformation $\tilde{q} \rightarrow \tilde{q}\exp(-i\Omega^2\zeta/2)$, which leaves the intensity unaltered. Equation (5.155) is then equivalent to

$$\frac{\partial\tilde{q}}{\partial\zeta} = \frac{\Delta\alpha}{2}\tilde{q} - \frac{\eta}{2}\Omega^2\tilde{q} + \tilde{n}(\Omega,\zeta) \tag{5.159}$$

Solution of (5.159) with $\tilde{q}(\Omega,0) = 0$ is

$$\tilde{q}(\Omega,\zeta) = \int_0^\zeta d\zeta' \exp\left[\frac{\Delta\alpha - \eta\Omega^2}{2}(\zeta - \zeta')\right]\tilde{n}(\Omega,\zeta) \tag{5.160}$$

The correlation function is

$$\langle \tilde{q}^*(\Omega,\zeta)\tilde{q}(\Omega',\zeta)\rangle = 2\pi S(\omega)\delta(\Omega - \Omega') \tag{5.161}$$

where

$$S(\Omega) = \langle n\rangle \int_0^\zeta d\zeta' \exp[(\Delta\alpha - \eta\Omega^2)\zeta'] \tag{5.162}$$

By Fourier inversion we get

$$\langle |q(\tau,\zeta)|^2\rangle = \int \frac{d\Omega}{2\pi}\exp(i\Omega\tau)\int \frac{d\Omega'}{2\pi}\exp(-i\Omega'\tau)\langle \tilde{q}^*(\Omega,\zeta)\tilde{q}(\Omega',\zeta)\rangle$$

$$= \int \frac{d\Omega}{2\pi}S(\Omega) \tag{5.163}$$

The spectrum of the ASE noise on zeros can be explicitly calculated with the help of (5.162),

$$S(\Omega) = \langle n\rangle\zeta \frac{\exp[(\Delta\alpha - \eta\Omega^2)\zeta] - 1}{(\Delta\alpha - \eta\Omega^2)\zeta} \tag{5.164}$$

Compared with the simple linear growth of the ASE power $\langle n\rangle\zeta$ that would take place without filters, in the presence of filtering the ASE power become narrowband and its peak power grows exponentially as the ASE field propagates down the fiber. If we define the normalized spectrum $S_{nor}(\Omega)$ as the ratio of the ASE noise back ground with filtering and without filtering, we have

$$S_{nor}(\Omega) = \frac{\exp[(\Delta\alpha - \eta\Omega^2)\zeta] - 1}{(\Delta\alpha - \eta\Omega^2)\zeta} \tag{5.165}$$

According to equation (5.163), the total ASE power is given by the integral of the ASE noise spectrum. To evaluate the ASE power, it is convenient to use (5.163) where the ASE spectrum is given in the integral form (5.162), and invert the order of the two integrals. We get

$$\langle |q(\tau,\zeta)|^2 \rangle = \int \frac{d\Omega}{2\pi} \, S(\Omega) = \langle n \rangle \int_0^\zeta d\zeta' \, \frac{1}{2\sqrt{\pi\eta\zeta'}} \exp(\Delta\alpha\zeta')$$

$$= \frac{\langle n \rangle}{\sqrt{\pi\eta\Delta\alpha}} \int_0^{\Delta\alpha\zeta} dY \exp(Y^2)$$

$$= \frac{\langle n \rangle}{\sqrt{\pi\eta\Delta\alpha}} \exp(\Delta\alpha\zeta) D(\sqrt{\Delta\alpha\zeta}) \qquad (5.166)$$

where $D(x) = \exp(-x^2) \int_0^x dt \exp(t^2)$ is the Dawson integral, related to the imaginary error function by $D(x) = -i(\sqrt{\pi}/2)\exp(-x^2)\mathrm{erf}(ix)$. By using the condition that comes from the requirement that the net gain of the soliton is zero, $\Delta\alpha = \eta/3$, equation (5.166) becomes

$$\langle |q(\tau,\zeta)|^2 \rangle = \frac{\langle n \rangle}{\eta} \sqrt{\frac{3}{\pi}} \exp\left(\frac{\eta}{3}\zeta\right) D\left(\sqrt{\frac{\eta}{3}}\zeta\right) \qquad (5.167)$$

Using the asymptotic expansion of the Dawson integral $D(x) \sim 1/(2x)$, we obtain the approximate expression

$$\langle |q(\tau,\zeta)|^2 \rangle \sim \frac{3\langle n \rangle}{2\eta\sqrt{\pi\eta\zeta}} \exp\left(\frac{\eta}{3}\zeta\right) \qquad (5.168)$$

The ASE energy within a time interval T_{int} is given by $\langle |q(\tau,\zeta)|^2 \rangle T_{int}$.

In Fig. 5.3 we show the ASE background $\langle |q(\tau,\zeta)|^2 \rangle / \langle n \rangle$ with relation to ζ. The solid line was obtained with expression (5.167), the dashed line by approximation (5.168).

According to (5.168), the power of the ASE background integrated over frequency is proportional to $\exp(\eta\zeta/3)$. The damping of the frequency fluctuations is instead proportional to $\exp[-(4/3)\eta\zeta]$; its characteristic length is shorter by a factor 4 than the characterizing length of the growth of the ASE background. This makes the use of frequency filters for the control of the Gordon-Haus effect effective, at least down to a propagation distance where the growth of the ASE background does not completely overhelm the soliton stream.

Modulation Control

Before the use of filters was proposed to extend the Gordon-Haus effect, Nakazawa proposed the use of active amplitude modulators to periodically retime the soliton stream during propagation [23]. As we will see below, soliton

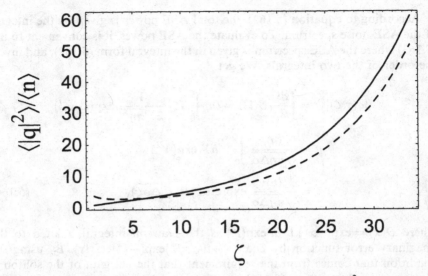

FIGURE 5.3 Amplified spontaneous emission background $\langle |q(\tau,\zeta)|^2\rangle/\langle n\rangle$ versus ζ. Solid line was obtained by expression (5.167); dashed line by approximation (5.168).

propagation in the presence of active amplitude modulators is unstable. However, soliton propagation with in-line modulators and filters is stable provided that a stability condition is met. In the first experiment of Nakazawa, a band-pass filter was also used in the recirculating loop, and its action was essential for the success of the experiment, although its role was not recognized before the work [39]. For these reasons we will consider below the case in which both modulators and filters are inserted along the line.

The transfer function of the modulator depends on the type of modulators. The time-dependent loss of a traveling-wave electroabsorption modulator, for instance, is

$$M(t) = \exp\{\Gamma_M L_M [\cos(\Omega_M t) - 1]\}, \tag{5.169}$$

where $\Gamma_M[\cos(\Omega_M t) - 1]$ is the absorption coefficient per unit length, Ω_M is the angular frequency of the modulator, and L_M is the length of the modulator waveguide. As done with the filters, let us define the averaged action of the modulator per unit length as

$$m(t) = \frac{1}{2L_m} \ln M(t) = \frac{\Gamma_M L_M}{2L_A} [\cos(\Omega_M t) - 1] \tag{5.170}$$

where L_m is the spacing between modulators. Let us assume shallow modulators so that their response can be approximated to second order in t:

$$m(t) \sim -\frac{\Gamma_M L_M}{4L_A} \Omega_M t^2 = -\frac{m''}{2} t^2 \tag{5.171}$$

Equation (5.126) extended to include the effect of modulators and in-line filtering becomes

$$\frac{\partial V_0}{\partial z} = \frac{\delta g}{2} V_0 + \left(\frac{F_r t_d^2}{8 L_f} \right) \frac{\partial^2 V_0}{\partial t^2} - \frac{m''}{2} t^2 V_0$$

$$+ i \frac{|\beta_2|}{2} \frac{\partial^2 v_0}{\partial t^2} + i\gamma \frac{1 - \exp(-\alpha L_A)}{\alpha L_A} |V_0|^2 V_0 \qquad (5.172)$$

where the excess gain δg compensates for the extra loss caused by both filtering and modulators. In soliton units, and with the ASE noise term added, (5.172) becomes

$$\frac{\partial q}{\partial \zeta} = \frac{i}{2} \frac{\partial^2 q}{\partial \tau^2} + i|q|^2 q + \frac{\Delta \alpha}{2} q + \frac{\eta}{2} \frac{\partial^2 q}{\partial \tau^2} - \frac{\mu}{2} \tau^2 q + n(\tau, \zeta) \qquad (5.173)$$

where the symbols are the same as those of equation (5.127) and

$$\mu = m'' T_s^2 z_0 \qquad (5.174)$$

Let us use the perturbative equations (5.91)–(5.94) with

$$R = \frac{\Delta \alpha}{2} q + \frac{\eta}{2} \frac{\partial^2 q}{\partial \tau^2} - \frac{\mu}{2} \tau^2 + n(\tau, \zeta) \qquad (5.175)$$

From equations (5.91)–(5.94) we get, after some algebra,

$$\frac{dA}{d\zeta} = \Delta \alpha A - \eta \left(\Omega^2 + \frac{A^2}{3} \right) A - \mu \left(T^2 + \frac{\pi^2}{12} \frac{1}{A^2} \right) A + n_A(\zeta) \qquad (5.176)$$

$$\frac{d\phi}{d\zeta} = \frac{1}{2} (A^2 - \Omega^2) + T \frac{d\Omega}{d\zeta} + n_\phi(\zeta) \qquad (5.177)$$

$$\frac{d\Omega}{d\zeta} = -\frac{2}{3} \eta A^2 \Omega + n_\Omega(\zeta) \qquad (5.178)$$

$$\frac{dT}{d\zeta} = -\Omega - \mu \frac{\pi^2}{6} \frac{T}{A^2} + n_T(\zeta) \qquad (5.179)$$

To obtain the above equations, we have used

$$\int d\tau \, \tau^2 A^2 \, \mathrm{sech}^2 [A(\tau - T)] = \frac{1}{A} \frac{\pi^2}{6} + 2T^2 A \qquad (5.180)$$

$$\int d\tau \, \tau^2 (\tau - T) A^2 \, \mathrm{sech}^2 [A(\tau - T)] = -\frac{T}{A^2} \frac{\pi^2}{3} \qquad (5.181)$$

The modulators add a restoring force directly on the timing fluctuations by a similar mechanism to that which induces a restoring force on frequency fluctuations with filters. We have seen that filters control the soliton frequency and hence the soliton velocity. Modulators, instead, act directly on timing. When a soliton advances in time, it experiences a loss that is asymmetric in time and pushes the soliton back to the center of the modulation profile and the opposite if the soliton is late. Also the restoring force on timing, like the restoring force on frequency, does not require any particular scaling of the soliton parameter to hold true. The same modulator action is effective for the gaussian pulses running within an actively mode-locked laser cavity, where the pulses are also trapped at the center of the modulator profile.

The amplitude modulator further affects the amplitude fluctuations of the soliton. This effect depends on the scaling of the soliton parameters, as we discuss below. By inspection of (5.176), we first notice the similarity between the term describing the loss of filter and the modulator. This is the consequence of the filter and the modulator both being considered in the parabolic approximation, in frequency and time, respectively, and of the property of being the spectrum of a hyperbolic secant pulse that remains a hyperbolic secant

$$\int d\tau \, \text{sech}(A\tau)\exp(i\Omega\tau) = \pi \, \text{sech}\left(\frac{\pi}{2A}\Omega\right) \qquad (5.182)$$

The coefficient $\pi^2/12$ of the modulator's loss term at $T = 0$ instead of a $\frac{1}{3}$ in front of the filler's loss term at $\Omega = 0$ is due to the coefficient $\pi/2$ which has scaled the Ω axis in the Fourier transform of the hyperbolic secant. Besides this different coefficient, there is a qualitative difference between the loss term of the filter at $\Omega = 0$ and the loss term of the modulator at $T = 0$ that has important practical consequences. The filter loss is proportional to A^2, and the modulator loss is proportional to A^{-2}. Let us see the reason for this different dependence.

The loss caused by a parabolic frequency filter on a pulse is proportional to the pulse bandwidth squared; the loss caused by a modulator is proportional to its pulsewidth square. Pulses of larger pulsewidth (bandwidth) experience larger loss in the wings of the modulator (filter). At this point, the scaling of the soliton parameters comes into play. The amplitude of the soliton is inversely proportional to its pulsewidth and therefore proportional to its bandwidth. So loss is proportional to the amplitude squared with filters but inversely proportional to the amplitude squared with modulators.

The consequence of the dependence of loss on the soliton amplitude is that modulators have a different effect than filters on amplitude fluctuations. We have already seen that filters stabilize the amplitude fluctuations of the soliton, since higher loss is felt by solitons of higher energy, and vice versa. However, modulators make them unstable because lower loss is felt by solitons of higher energy, and vice versa. Consider·the dynamic response of a soliton to an amplitude fluctuation in the presence of amplitude modulators, the equivalent for a filter is shown for comparison in parenthesis. A positive amplitude fluctuation produces the decrease (increase) of the soliton pulsewidth (bandwidth), therefore the reduction (increase) of its loss. Conversely, a negative amplitude fluctuation produces the increase (reduction) of the soliton pulsewidth (bandwidth), therefore the increase (decrease) of its loss. While filters establish a negative, stable, feedback loop in the amplitude fluctuations, modulators establish a positive, unstable, one.

The above qualitative discussion has immediate correspondence with the stability analysis of equations (5.176)–(5.179). The steady state of equations (5.176), (5.178), and (5.179) are

$$\Delta \alpha = \eta \frac{A_s^2}{3} + \mu \frac{\pi^2}{12} \frac{1}{A_s^2} \tag{5.183}$$

$$\Omega_s = 0 \tag{5.184}$$

$$T_s = 0 \tag{5.185}$$

For $A_s = 1$, the excess gain must be $\Delta \alpha = \eta/3 + \mu \pi^2/12$. Linearizing (5.176)–(5.179) about the steady state (the phase change is defined as $\delta \phi = \phi - A_s^2 \zeta/2$, as in the case of simple filtering), we obtain

$$\frac{d\delta A}{d\zeta} = -(\beta - \beta_T)\delta A + n_A(\zeta) \tag{5.186}$$

$$\frac{d\delta \phi}{d\zeta} = A_s \delta A + n_\phi(\zeta) \tag{5.187}$$

$$\frac{d\delta \Omega}{d\zeta} = -\beta \delta \Omega + n_\Omega(\zeta) \tag{5.188}$$

$$\frac{d\delta T}{d\zeta} = -\delta \Omega - \beta_T \delta T + n_T(\zeta) \tag{5.189}$$

where β is given by equation (5.141) and

$$\beta_T = \frac{\pi^2}{6} \mu \frac{1}{A_s^2}. \tag{5.190}$$

First, we notice that as anticipated, there is a minimum filter strength for the stability of amplitude fluctuations. For the stability of (5.186), it is required that

$$\beta > \beta_T \tag{5.191}$$

Equation (5.191) makes it explicit that solitons are stable if the stabilizing action of the filters is larger than the opposite action of modulators. If this condition is not met, amplitude fluctuations grow exponentially with propagation distance.

The correlation function of frequency fluctuations is the same as that for simple filtering, equation (5.147). By following the same procedure used for filtering, we obtain after some algebra the variance of the timing fluctuations,

$$\langle \delta T(\zeta)^2 \rangle = \frac{n_T}{2\beta_T} [1 - \exp(-2\beta_T \zeta)] + n_\Omega \left\{ \frac{1}{4\beta\beta_T(\beta + \beta_T)} \right.$$
$$\left. - \frac{\exp(-2\beta\zeta)}{4\beta(\beta - \beta_T)^2} - \frac{\exp(-2\beta_T \zeta)}{4\beta_T(\beta - \beta_T)^2} + \frac{\exp[-(\beta + \beta_T)\zeta]}{(\beta - \beta_T)^2(\beta_T + \beta)} \right\} \quad (5.192)$$

Asymptotically for $\zeta \to \infty$ the timing fluctuations are bounded,

$$\langle \delta T(\zeta)^2 \rangle_{\zeta \to \infty} = \frac{n_T}{2\beta_T} + \frac{n_\Omega}{4\beta\beta_T(\beta + \beta_T)} \quad (5.193)$$

This is an expected result because both timing and frequency (velocity) fluctuations are controlled, which is the reason for the very long transmission distance that can be achieved with the combined use of filtering and modulators. In-line filtering, besides ensuring the stabilty of amplitude fluctuations, is then also essential for the boundeness of the timing fluctuations. Without filters the uncontrolled frequency fluctuations experience linear growth as the distance of the timing jitter increases, and similarly for the case of transmission with frequency filtering and without modulators. This may be directly checked for taking the limit for $\beta \to 0$ in equation (5.192), which corresponds to having no filtering control,

$$\langle \delta T(\zeta)^2 \rangle_{\beta \to 0} = \frac{n_T}{2\beta_T} [1 - \exp(-2\beta_T \zeta)]$$
$$+ \frac{n_\Omega}{4\beta_T^3} [-\exp(-2\beta_T \zeta) + 4\exp(-\beta_T \zeta) - 3 + 2\beta_T \zeta] \quad (5.194)$$

A comparison between the growth with distance of the timing jitter in the three cases of uncontrolled transmission, filtering control, and combined modulation-filtering control is provided in Fig. 5.4, where in a double logarithmic scale the variance of the timing fluctuations normalized to $\langle n \rangle$ is shown in relation to the normalized distance ζ. The solid line corresponds to uncontrolled transmission. The cubic growth with distance typical of the Gordon-Haus effect is clearly in evidence. The dashed line refers to in-line filtering only, with $\eta = 0.4$. Asymptotically the variance grows linearly with distance. The dot-dashed line refers to the case where synchronous modulators with $\mu = 0.1$ are added to the filters with $\eta = 0.4$ along the line, showing saturation of the timing fluctuations to a finite value.

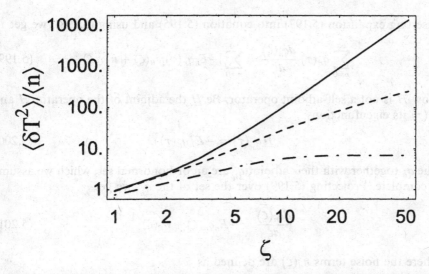

FIGURE 5.4 Variance of the time fluctuations, normalized to $\langle n \rangle$, versus distance, in soliton units. Solid line: no in-line control; dashed line: with in-line filters ($\eta = 0.4$); dot-dashed line: with both filters ($\eta = 0.4$) and modulators ($\mu = 0.1$).

The study of the stability of the zeros of equation (5.173) is not as simple as that of filtering. One could, however, give an analytic criterion for the stability of zeros [41]. This task may be accomplished by neglecting the nonlinear term, which gives a good approximation because the total power of zeros must be kept low to have good transmission quality. Equation (5.173) becomes

$$\frac{\partial q}{\partial \zeta} = Hq + n(\tau, \zeta) \tag{5.195}$$

where

$$Hq = \frac{i}{2} \frac{\partial^2 q}{\partial \tau^2} + \frac{\Delta \alpha}{2} q + \frac{\eta}{2} \frac{\partial^2 q}{\partial \tau^2} - \frac{\mu}{2} \tau^2 q \tag{5.196}$$

We can expand (5.195) the following way:

$$q(\tau, \zeta) = \sum_{n=0}^{\infty} q_n(\tau) c_n(\zeta) \tag{5.197}$$

where the $q_n(\tau)$ are eigenfunctions of the operator H associated to the eigenvalues E_n, which in general takes complex values

$$Hq_n(\tau) = -E_n q_n(\tau) \tag{5.198}$$

Inserting expansion (5.197) into equation (5.195) and using (5.198), we get

$$\sum_{n=0}^{\infty} q_n(\tau) \frac{dc_n(\zeta)}{d\zeta} = \sum_{n=0}^{\infty} [-E_n q_n(\tau)] c_n(\zeta) + n(\tau, \zeta) \qquad (5.199)$$

Now H is not a self-adjoint operator. Be $\underline{H}$ the adjoint of the operator H and $\underline{q}_n(\tau)$ its eigenfunctions:

$$\underline{H}\underline{q}_n(\tau) = -E_n^* q_n(\tau) \qquad (5.200)$$

The q_n together with their adjoint $\underline{q}_n$ are an orthonormal set, which we assume is complete. Projecting (5.199) over the set of the $\underline{q}_n$, we get

$$\frac{dc_n(\zeta)}{d\zeta} = -E_n c_n(\zeta) + n_n(\zeta) \qquad (5.201)$$

where the noise terms $n_n(\zeta)$ are defined as

$$n_n(\zeta) = \int d\tau \, \underline{q}_n^*(\tau) n(\tau, \zeta) \qquad (5.202)$$

For the stability of zeros, all eigenvalues E_n should have positive real part. This, according to equation (5.201), corresponds to a decay of the temporal "modes" excited by the ASE noise. The operator H is similar to that encountered in quantum mechanics in the analysis of a particle in a harmonic potential. In the place of the particle mass, there is a complex number, and this is why H is not self-adjoint. The eigenfunctions are still, however, hermite-gaussian functions, although complex. The eigenfunction associated with the eigenvalue with the smallest real part is a gaussian,

$$q_0(\tau) = \mathcal{N}_0 \exp\left(-\frac{K_0 \tau^2}{2}\right) \qquad (5.203)$$

where $\mathcal{N}_0$ is a normalization constant. The corresponding eigenvalue equation is

$$\frac{i}{2} \frac{\partial^2 q_0}{\partial \tau^2} + \frac{\Delta\alpha}{2} q_0 + \frac{\eta}{2} \frac{\partial^2 q_0}{\partial \tau^2} - \frac{\mu}{2} \tau^2 q_0 = -E_0 q_0 \qquad (5.204)$$

Substituting (5.203) in the above equation, we obtain

$$(-K_0 + K_0^2 \tau^2) \frac{i + \eta}{2} + \frac{\Delta\alpha}{2} - \frac{\mu}{2} + E_0 = 0 \qquad (5.205)$$

that is

$$K_0 = \sqrt{\frac{\mu}{i + \eta}} \qquad (5.206)$$

$$2E_0 = -\Delta\alpha + \sqrt{\mu}\sqrt{i + \eta} \qquad (5.207)$$

Using equation (5.183) and condition $\text{Re}E_0 > 0$ for stability, we get

$$\eta \frac{A_s^2}{3} + \mu \frac{\pi^2}{12} \frac{1}{A_s^2} < \sqrt{\mu} \, \text{Re}(\sqrt{i + \eta}) \qquad (5.208)$$

This equation gives the condition for the stability of zeros. It should be met simultaneously with condition (5.191), which gives the condition for the stability of the amplitude fluctuations on solitons, that is, for the stability of the "ones" of the transmission code. The two conditions to be met simultaneously are then

$$\sqrt{\mu} \, \text{Re}(\sqrt{i + \eta}) > \eta \frac{A_s^2}{3} + \mu \frac{\pi^2}{12} \frac{1}{A_s^2}, \qquad \text{zeros stable} \qquad (5.209)$$

$$\frac{\pi^2}{6} \mu \frac{1}{A_s^2} < \frac{2}{3} \mu A_s^2, \qquad \text{ones stable} \qquad (5.210)$$

Assume that $\eta \ll 1$ and $\eta \ll 1$. Under this condition $\text{Re}(\sqrt{i + \eta}) \sim \text{Re}(\sqrt{i}) = \sqrt{2}/2$. Equation (5.209) becomes, to leading order,

$$\frac{\sqrt{2}}{2} \sqrt{\mu} > \frac{A_s^2}{3} \eta \qquad (5.211)$$

In a $\mu - \eta$ plane equation (5.211) is met by the internal points of a semi-parabola having its vertex in the origin and axis coinciding with the μ axis. In the same plane the points that meet (5.210) are those above a straight line passing through the origin. By simple geometrical considerations, one may easily understand that there always exist points that simultaneously meet (5.209) and (5.210) for μ and η sufficiently small. Both μ and η are inversely proportional to $|\beta_2|$ when transformed back to real-world units. Thus small μ and η means large dispersion. It makes good sense that the stability of zeros is enhanced for large dispersion. The radiation generated by the ASE noise in those regions where the modulator opens a window of positive gain, that is, close to $T = 0$, is dispersed by β_2 into regions where it experiences positive loss, and it is absorbed. If both conditions for the stability of ones and zeros are met, transmission up to unlimited distance might be possible, maybe only limited by "tunnelling" of ones into zeros, and vice-versa [23].

Equations (5.209) and (5.210) can also be exactly solved for μ. One gets

$$\mu > \left[\frac{6}{\pi^2} A_s^2 r(\eta) - \sqrt{s(\eta)} \right]^2 . \qquad (5.212)$$

$$\mu < \frac{4}{\pi^2} A_s^4 \eta \qquad (5.213)$$

where

$$s(\eta) = \left[\frac{6}{\pi^2} A_s^2 r(\eta)\right]^2 - \frac{4}{\pi^2} A_s^4 \eta, \tag{5.214}$$

$$r(\eta) = \text{Re}(\sqrt{i + \eta})$$

$$= (1 + \eta^2)^{1/4} \cos\left[\frac{\pi}{4} - \frac{1}{2} \arctan(\eta)\right] \tag{5.215}$$

The right-hand side of (5.212) gives real values for s positive to zero. For $s = 0$, it is immediate to show by using (5.214) that (5.212) gives $\mu = (4/\pi^2)A_s^4\eta$. This means that the extreme of the curve defining the region of validity of (5.212) is on the line defining the region where (5.213) is met. The region of validity of both equations, (5.212) and (5.213), is a closed region in the $\mu - \eta$ plane. The largest values of μ and η compatible with stability of both zeros and one are set by condition $s = 0$, and for $A_s = 1$ they are $\eta \simeq 1.536$ and $\mu \simeq 0.6225$. The region of the (μ, η) plane where the loop is stable, obtained by equations (5.212)–(5.215), is shown in Fig. 5.5.

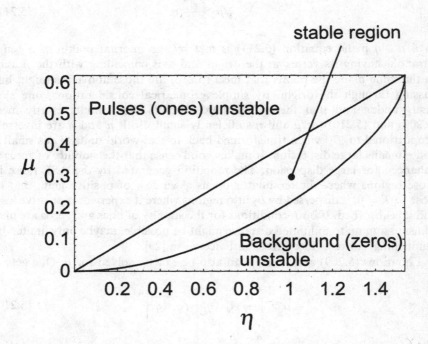

FIGURE 5.5 Region in the (η, μ) plane where the soliton storage loop is stable.

5.3.3 Sliding Filters

We showed in Section 5.3.2 that a major problem of soliton control by filtering is the instability of the soliton background (see Fig. 5.3). The narrowband radiation generated close to the center of the filter profile grows exponentially with transmission distance and eventually overwhelms the soliton bit-stream. A brilliant solution to this problem has been proposed by Mollenauer, Gordon, and Evangelides [20] and successfully implemented in both the recirculating loop [22] and in straight-line transmission [42] experiments.

The idea behind sliding filters is simple. The filter center frequency is changed linearly with propagation distance. The soliton is dragged by sliding because of its capacity of readjusting its spectrum and following the filter sliding. The linear ASE cannot; therefore the linear ASE generated at a given distance at the center of the filter profile is then absorbed at longer distances because of the filter center frequency shift. Alternatively, instead of shifting the filter center frequency, one can instead keep fixed the center frequency of the filters and shift the carrier frequency of the field (soliton plus radiation) in front of the filters by acoustooptic modulators. Once again the linear radiation is absorbed while solitons, because of their particlelike nature, reach a dynamic equilibrium between the action of modulators shifting their carrier frequency and that of filters pushing them back to the center of the filter profile [43].

The averaged action of the filter with a center frequency linearly dependent on z is (see equation 5.121)

$$h(\omega) = \frac{1}{2L_f} \ln\left\{ \frac{1}{1 + F_r \sin^2[(\omega - \omega_0 - \omega_0'z)t_d/2]} \right\} \tag{5.216}$$

with F given by equation (5.122). As for fixed filters, we will neglect the effect of the filter phase and expand $h(\omega)$ to second order in $(\omega - \omega_0)$:

$$h(\omega) = -\frac{F_r t_d^2}{8L_f}(\omega - \omega_0 - \omega_0'z)^2 \tag{5.217}$$

With transformation $-i(\omega - \omega_0) \rightarrow \partial/\partial t$, we obtain in the time domain

$$h\left(\frac{\partial}{\partial t}\right) = -\left(\frac{F_r t_d^2}{8L_f}\right)\left(i\frac{\partial}{\partial t} - \omega_0'z\right)^2 \tag{5.218}$$

Equation (5.126) with sliding filters becomes

$$\frac{\partial V_0}{\partial z} = \frac{\delta g}{2}V_0 - \left(\frac{F_r t_d^2}{8L_f}\right)\left(i\frac{\partial}{\partial t} - \omega_0'z\right)^2 V_0 + i\frac{|\beta_2|\partial^2 V_0}{\partial t^2}$$
$$+ i\gamma \frac{1 - \exp(-\alpha L_A)}{\alpha L_A}|V_0|^2 V_0 \tag{5.219}$$

and transformed into soliton units and with the ASE noise term added (see equations 5.127–5.129)

$$\frac{\partial q}{\partial \zeta} = \frac{i}{2}\frac{\partial^2 q}{\partial \tau^2} + i|q|^2 q + \frac{\Delta\alpha}{2}q - \frac{\eta}{2}\left(i\frac{\partial}{\partial \tau} - \omega_f'\zeta\right)^2 q + n(\tau,\zeta) \qquad (5.220)$$

where

$$\omega_f' = \omega_0' z_0 T_s \qquad (5.221)$$

One can still use the perturbation equations (5.91)–(5.94) provided that $\omega_f'\zeta \ll 1$, the condition required by the adiabatic hypothesis. The equations for the soliton perturbations are

$$\frac{dA}{d\zeta} = \Delta\alpha A - \eta\left[(\Omega - \omega_f'\zeta)^2 + \frac{A^2}{3}\right]A + n_A(\zeta) \qquad (5.222)$$

$$\frac{d\phi}{d\zeta} = \frac{1}{2}(A^2 - \Omega^2) + T\frac{d\Omega}{d\zeta} + n_\phi(\zeta) \qquad (5.223)$$

$$\frac{d\Omega}{d\zeta} = -\frac{2}{3}\eta A^2(\Omega - \omega_f'\zeta) + n_\Omega(\zeta) \qquad (5.224)$$

$$\frac{dT}{d\zeta} = -\Omega + n_T(\zeta) \qquad (5.225)$$

The equation for the amplitude A, equation (5.222), and that for the frequency Ω, equation (5.224), are stationary for

$$\Omega(\zeta) = \omega_f'\zeta + \Delta\Omega \qquad (5.226)$$

$$\Delta\Omega = -\frac{3\omega_f'}{2\eta A_s^2} \qquad (5.227)$$

$$\Delta\alpha = \eta(\Delta\Omega^2 + \frac{1}{3}A_s^2) \qquad (5.228)$$

The equation for the timing is not stationary; the pulse undertakes a continuous acceleration which, being deterministic, does not affect the transmission characteristics

$$T_a(\zeta) = -\omega_f'\frac{\zeta^2}{2} - \Delta\Omega\zeta \qquad (5.229)$$

The phase also evolves with ζ,

$$\phi(\zeta) = -\frac{A_s^2}{2}\zeta + \frac{\Delta\Omega^2}{2}\zeta + \frac{\omega_f'}{4}\zeta^2 + T\omega_f'\zeta \qquad (5.230)$$

Because of the $T\omega'_f\zeta$, the phase undergoes a linear sweep that explicitly depends on timing. This means that two solitons initially separated by T continuously change their relative phase. We will see below that in-phase solitons attract and out-of-phase solitons repel. Two solitons initially in phase become out-of-phase at $\zeta = \pi/(T\omega'_f)$. If the length scale of this phase evolution is short enough, the soliton interaction can be expected to significantly reduce, and under some conditions it can average to zero. When the Gordon-Haus jitter is controlled, soliton interaction sets the limits to the minimum interval between adjacent solitons of a given pulsewidth and therefore sets the limit to the maximum bit-rate. As we will see in more detail later in Section 5.4.1 devoted to soliton-soliton interaction, besides the reduction of the ASE noise to zero, the other main advantage of the sliding filter transmission scheme is the reduction of soliton-soliton interaction.

Let us analyze the growth of the ASE background with sliding filters. Equation (5.220), when the nonlinear term is neglected, becomes

$$\frac{\partial q}{\partial \zeta} = \frac{i}{2}\frac{\partial^2 q}{\partial \tau^2} + \frac{\Delta\alpha}{2}\,q - \frac{\eta}{2}\left(i\frac{\partial}{\partial\tau} - \omega'_f\zeta\right)^2 q + n(\tau,\zeta) \qquad (5.231)$$

In the Fourier domain this equation takes the form

$$\frac{\partial\tilde{q}}{\partial\zeta} = -\frac{i}{2}\,\omega^2\tilde{q} + \frac{\Delta\alpha}{2}\,\tilde{q} - \frac{\eta}{2}\,(\omega - \omega'_f\zeta)^2\tilde{q} + n(\omega,\zeta) \qquad (5.232)$$

Getting rid of the term $-i\omega^2\tilde{q}$ by transformation $\tilde{q} \to \tilde{q}\exp(-i\omega^2\zeta)$ leaves (5.232) unaltered, since the noise source is phase independent; hence $\tilde{n}(\omega,\zeta)$ has the same statistical properties as $\tilde{n}(\omega,\zeta)\exp(-i\omega^2\zeta)$. Equation (5.232) then can be simplified as

$$\frac{\partial\tilde{q}}{\partial\zeta} = \frac{\Delta\alpha}{2}\,\tilde{q} - \frac{\eta}{2}\,(\omega - \omega'_f\zeta)^2\tilde{q} + \tilde{n}(\omega,\zeta) \qquad (5.233)$$

The solution of this equation is

$$\tilde{q}(\omega,\zeta) = \int_0^\zeta d\zeta'\tilde{n}(\omega,\zeta')\exp\left[\frac{\Delta\alpha}{2}\,(\zeta - \zeta') - \frac{\eta}{2}\int_{\zeta'}^\zeta d\zeta''(\omega - \omega'_f\zeta'')^2\right] \qquad (5.234)$$

and the power spectrum at ζ,

$$\langle\tilde{q}^*(\omega,\zeta)\tilde{q}(\omega',\zeta)\rangle = 2\pi\langle n\rangle\delta(\omega - \omega')\zeta N(\omega,\zeta) \qquad (5.235)$$

where

$$N(\omega,\zeta) = \frac{1}{\zeta}\int_0^\zeta d\zeta'\exp\left[\Delta\alpha(\zeta - \zeta') - \eta\int_{\zeta'}^\zeta d\zeta''(\omega - \omega'_f\zeta'')^2\right]$$

$$= \frac{1}{\zeta}\int_0^\zeta d\zeta'\exp\left\{\Delta\alpha(\zeta - \zeta') + \frac{\eta}{3\omega'_f}[(\omega - \omega'_f\zeta)^3 - (\omega - \omega'_f\zeta')^3]\right\} \qquad (5.236)$$

By performing the substitution $\zeta - \zeta' = x$, we get [20]

$$N(\omega, \zeta) = \frac{1}{\zeta} \int_0^\zeta dx \exp \left\{ \Delta \alpha x - \eta \left[\frac{\omega_f'^2 x^3}{3} + (\omega - \omega_f'\zeta)\omega_f x^2 + (\omega - \omega_f')^2 x \right] \right\}$$

(5.237)

The spectrum of the ASE at ζ is then proportional to $\zeta N(\omega, \zeta)$. In a reference frame that follows the sliding in the frequency domain, $\omega_z = \omega - \omega_f'\zeta$, the function inside the integral becomes independent of ζ. For values of ζ such that $\eta \omega_f'^2 \zeta^3 / 3 \gg 0$, $\zeta N(\omega, \zeta)$ saturates because of the presence of the negative exponential $\exp(-\eta \omega_f'^2 x^3 / 3)$ inside the integral.

The total ASE power is then

$$\langle |q(\tau, \zeta)|^2 \rangle = \int \frac{d\omega}{2\pi} \exp(i\omega\tau) \int \frac{d\omega'}{2\pi} \exp(-i\omega'\tau) \langle \tilde{q}^*(\omega, \zeta) \tilde{q}(\omega', \zeta) \rangle$$

$$= \langle n \rangle \int \frac{d\omega'}{2\pi} \zeta N(\omega, \zeta)$$

(5.238)

Inserting equation (5.237) into (5.238), we get

$$\langle |q(\tau, \zeta)|^2 \rangle = \frac{\langle n \rangle}{2\sqrt{\pi}} \int_0^\zeta dx \, (\eta x)^{-1/2} \exp \left(\Delta \alpha x - \frac{\eta}{12} \omega_f'^2 x^3 \right)$$

(5.239)

According to (5.239), the total ASE power saturates with sliding, as expected. The point where saturation begins may be estimated by the flex of the curve that gives the total ASE power against distance. By differentiating (5.239) twice with respect to ζ, we find that this point is the solution of

$$-\frac{1}{2\zeta_f} + \Delta \alpha - \frac{\eta}{4} \omega_f'^2 \zeta_f^2 = 0$$

(5.240)

We must at this point anticipate some of the results that we will derive in Section 5.3.4. As we will see below, the sliding cannot exceed, for stability, the critical value

$$\omega_{f,\mathrm{cr}}' = \sqrt{\frac{2}{27}} \eta$$

(5.241)

so it will be very useful to have the definition of a sliding rate normalized to the critical sliding rate,

$$\omega_r = \frac{\omega_f'}{\omega_{f,\mathrm{cr}}'} = \frac{\omega_f'}{\eta} \sqrt{\frac{27}{2}}$$

(5.242)

Since the critical value of sliding is usually, in soliton units, much less than one, we have $\zeta_f \gg 1$, and an excellent approximation of ζ_f is

$$\zeta_f = \frac{2}{\omega_f'} \sqrt{\frac{\Delta\alpha}{\eta}} \tag{5.243}$$

From equations (5.227), (5.228), and (5.242) we get

$$\Delta\alpha = \frac{\eta}{3}\left(1 + \frac{\omega_r^2}{2}\right) \tag{5.244}$$

which, inserted into (5.243), gives

$$\zeta_f = \frac{3\sqrt{2}}{\eta\omega_r}\left(1 + \frac{\omega_r^2}{2}\right)^{1/2} \tag{5.245}$$

In Fig. 5.6 we show the normalized total ASE power $\langle|q(\tau,\zeta)|^2\rangle/\langle n\rangle$ given by a numerical integration of (5.238) plotted against ζ, with $\eta = 0.4$, and a sliding rate one-half of the critical sliding rate, $\omega_r = 0.5$. For comparison, the ASE growth with fixed ($\omega_r = 0$) parabolic filters is reported for the same value of η. The calculated value of the position of the flex by (5.243) is $\zeta_f \approx 22.5$, against

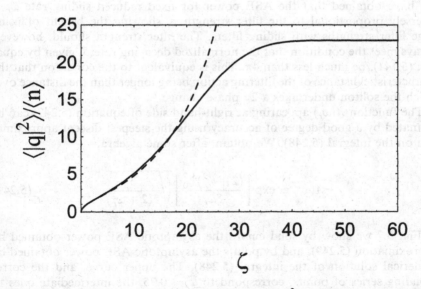

FIGURE 5.6 Solid line: normalized ASE power $|q(\tau,\zeta)|^2/\langle n\rangle$ obtained by numerical integration of equation (5.238) versus ζ. The parameters are $\omega_f' = \omega_{f,\text{cr}}'/2$. Dashed line: ASE growth with fixed filters and the same value of η.

the exact value $\zeta_f = 24.0$ obtained by solving (5.240). This value give a good qualitative indication of the distance from the line input where ASE begins to saturate because of filter sliding.

Equation (5.239) permits the estimation of the dependence of the asymptotic value of the ASE noise on the parameters of the line (filter strength, sliding rate, etc.). Equation (5.239), after (5.242) and (5.244) are used to eliminate ω_f' and $\Delta\alpha$ and substitution $y = \sqrt{\eta x}$ is performed in the integral, becomes

$$\langle |q(\tau,\zeta)|^2 \rangle = \frac{\langle n \rangle}{\sqrt{\pi \eta}} \int_0^{\sqrt{\eta \zeta}} dy \exp\left[\frac{y^2}{3}\left(1 + \frac{\omega_r^2}{2}\right) - \frac{1}{162} \omega_r^2 y^6 \right] \tag{5.246}$$

Performing the limit for $\zeta \to \infty$, we obtain

$$\frac{\langle |q_\infty|^2 \rangle}{\langle n \rangle} = \lim_{\zeta \to \infty} \frac{\langle |q(\tau,\zeta)|^2 \rangle}{\langle n \rangle} = \frac{1}{\eta} A(\omega_r) \tag{5.247}$$

where

$$A(\omega_r) = \frac{1}{\sqrt{\pi}} \int_0^\infty dy \exp\left[\frac{y^2}{3}\left(1 + \frac{\omega_r^2}{2}\right) - \frac{1}{162} \omega_r^2 y^6 \right] \tag{5.248}$$

We have obtained that the ASE power for fixed reduced sliding rate ω_r is inversely proportional to the filter strength η, showing the benefit of using large filter strengths with sliding filters. The filter strength should, however, always meet the condition that the normalized decaying rate β, given by equation (5.141), be much less than 4π. This is equivalent to the condition that the characteristic distance of the filtering action being longer than the distance over which the soliton undertakes a 2π phase change.

The function $A(\omega_r)$ appearing at right-hand side of equation (5.247) can be estimated by a good degree of accuracy using the steepest discent approximation on the integral (5.248). We obtain, after some algebra,

$$A(\omega_r) \simeq \exp\left[\frac{(2 + \omega_r^2)^{3/2}}{3\omega_r} \right] \sqrt{\frac{3}{2(2 + \omega_r^2)}} \tag{5.249}$$

In Fig. 5.7 we show by solid curves the asymptotic ASE power obtained by approximation (5.249), and by points the asymptotic ASE power obtained by numerical solution of the integral (5.248). The upper curve, and the corresponding series of points, correspond to $\omega_r = 0.75$, the intermediate ones to $\omega_r = 0.5$, and the lowest to $\omega_r = 0.25$. Figure 5.8 provides a comparison between the exact expression of $A(\omega_r)$ (solid line) and the approximate one (dashed line) given by equation (5.249).

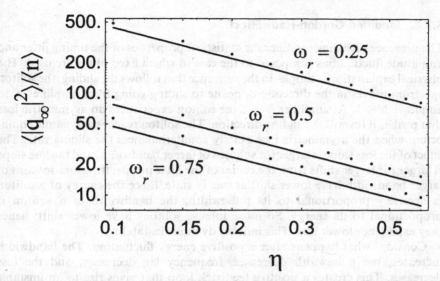

FIGURE 5.7 Asymptotic ASE power with sliding. Solid lines were obtained by approximation (5.249). Points were obtained by numerical solution of the integral at the right side of equation (5.239). Upper series $\omega_r = 0.75$, intermediate $\omega_r = 0.5$, lowest $\omega_r = 0.25$.

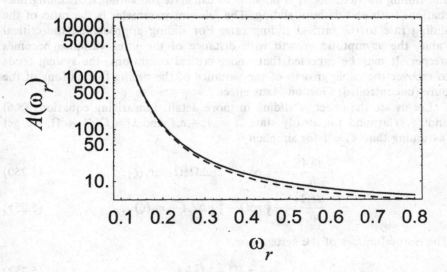

FIGURE 5.8 Exact expression of $A(\omega_r)$ (solid line) and the approximate one (dashed line) given by equation (5.249) versus the normalized sliding rate.

5.3.4 Modified Gordon-Haus Effect

The presence of sliding modifies the statistical properties of the timing jitter and amplitude fluctuations compared to the case in which fixed filters are used. The physical explanation is simple. In the reference that follows the sliding, the soliton spectrum moves in the direction opposite to sliding going from amplifier N to amplifier $N + 1$. At amplifier $N + 1$ the soliton experiences an asymmetric loss that pushes it toward the sliding direction. The soliton reaches a dynamic equilibrium when the asymmetric loss exactly counterbalances the sliding rate. The effect of the loss slope is larger for solitons of larger bandwidth, and the loss slope is larger for larger shifts from the center of the filter profile; therefore solitons of larger bandwidth have lower shift at steady state. Since the energy of a soliton is inversely proportional to its pulsewidth, the bandwidth of a soliton is proportional to its energy. So more intense solitons have lower shift; hence they experience lower loss. This induces dynamic instability.

Consider what happens after a positive energy fluctuation. The bandwidth increases, the pulsewidth decreases, frequency lag decreases, and the loss decreases. This creates a positive feedback loop that gives rise to an unstable system behavior. At the same time, though, there is a stabilizing effect of the *fixed* filters. The wing of the spetrum rises, the loss increases, and this stabilizes the system through a negative feedback loop. It is then quite intuitive that there is a maximum sliding rate above which the positive feedback loop takes over and transmission becomes unstable. This was predicted in the early paper from Mollenauer, Gordon, and Evangelides [20]. It makes also good sense that sliding increases the timing jitter because of the extra coupling between energy and timing fluctuations. It is possible to calculate the variance of timing fluctuations with and without sliding. The relevant parameter is the ratio of the sliding rate to the critical sliding rate. For sliding approaching the critical value, the asymptotic growth with distance of the jitter variance becomes steeper. It may be expected that under critical conditions, the system tends to recover the cubic growth of the variance of the timing fluctuations of the pure (uncontrolled) Gordon-Haus effect.

Let us see the effect of sliding in more detail. Linearizing equations (8.6) and (8.8) around the steady state $A = A_s + \delta A$ and $\Omega = \Omega(\zeta) + \delta\Omega$, we get (assuming that $A_s = 1$ for simplicity)

$$\frac{d\delta A}{d\zeta} = -\tfrac{2}{3}\eta\delta A - 2\eta\Delta\Omega\delta\Omega + n_A(\zeta) \qquad (5.250)$$

$$\frac{d\delta\Omega}{d\zeta} = -\tfrac{2}{3}\eta\delta\Omega - \tfrac{4}{3}\eta\Delta\delta A + n_\Omega(\zeta) \qquad (5.251)$$

The normal modes of the system are

$$x_1 = \delta\Omega - \sqrt{\tfrac{2}{3}}\delta A \qquad (5.252)$$

$$x_2 = \delta\Omega + \sqrt{\tfrac{2}{3}}\delta A \qquad (5.253)$$

and the damping strengths

$$\beta_1 = \tfrac{2}{3}(1 - \sqrt{6}\Delta\Omega)\eta \tag{5.254}$$

$$\beta_2 = \tfrac{2}{3}(1 + \sqrt{6}\Delta\Omega)\eta \tag{5.255}$$

The dynamic equations for $x_{1,2}$ are

$$\frac{dx_1}{d\zeta} = -\beta_1 x_1 + n_{x_1}(\zeta) \tag{5.256}$$

$$\frac{dx_2}{d\zeta} = -\beta_2 x_2 + n_{x_2}(\zeta) \tag{5.257}$$

where

$$n_{x_1}(\zeta) = n_\Omega(\zeta) - \sqrt{\tfrac{2}{3}}n_A(\zeta) \tag{5.258}$$

$$n_{x_2}(\zeta) = n_\Omega(\zeta) + \sqrt{\tfrac{2}{3}}n_A(\zeta) \tag{5.259}$$

Although the deterministic dynamics of the normal modes are uncoupled, they are correlated by the noise terms (5.258) and (5.259). The correlation function of the noise terms is

$$\langle n_{x_i}(\zeta)n_{x_j}(\zeta')\rangle = n_{i,j}\delta(\zeta - \zeta') \tag{5.260}$$

where, from equations (5.100)–(5.105),

$$n_{1,1} = n_{2,2} = n_\Omega + \tfrac{2}{3}n_A \tag{5.261}$$

$$n_{1,2} = n_\Omega - \tfrac{2}{3}n_A \tag{5.262}$$

From the dynamic equations (5.256) and (5.257) it can be seen that there is a critical value of the frequency lag $|\Delta\Omega|$, that is, $\delta\Omega_{0,\mathrm{cr}} = 1/\sqrt{6}$, beyond which the system becomes unstable. From (5.227) with $A_s = 1$, this value can be seen to correspond to a normalized sliding rate of

$$\omega'_{f,\mathrm{cr}} = \sqrt{\tfrac{2}{27}}\eta \tag{5.263}$$

For sliding rates larger in modulo than this value, one of the two damping coefficients becomes positive, the fluctuations instead of being damped exponentially grow, and the system becomes unstable. This is the manifestation of the instabilty described, in physical terms above, and discovered by Mollenauer et al. [20]. For the following derivation, it is convenient to rewrite (5.254) and (5.255) in the following way:

$$b_1 = \tfrac{2}{3}(1 - \omega_r)\eta \tag{5.264}$$

$$b_2 = \tfrac{2}{3}(1 + \omega_r)\eta \tag{5.265}$$

where

$$\omega_r = \sqrt{6}\Delta\Omega = \frac{\omega_f'}{\omega_{f,\mathrm{cr}}'} \tag{5.266}$$

is the ratio of the sliding rate to the critical sliding rate.

The amplitude and frequency fluctuations are, from (5.252) and (5.253),

$$\delta A(\zeta) = \tfrac{1}{2}\sqrt{\tfrac{3}{2}}[x_2(\zeta) - x_1(\zeta)] \tag{5.267}$$

$$\delta\Omega(\zeta) = \tfrac{1}{2}[x_1(\zeta) + x_2(\zeta)] \tag{5.268}$$

The average of the amplitude and frequency fluctuations is zero, and the correlation functions are

$$\langle \delta A(\zeta)\delta A(\zeta')\rangle = \tfrac{3}{8}\,[\langle x_1(\zeta)x_1(\zeta')\rangle - \langle x_1(\zeta)x_2(\zeta')\rangle$$
$$- \langle x_2(\zeta)x_1(\zeta')\rangle + \langle x_2(\zeta)x_2(\zeta')\rangle] \tag{5.269}$$

$$\langle \delta\Omega(\zeta)\delta\Omega(\zeta')\rangle = \tfrac{1}{4}\,[\langle x_1(\zeta)x_1(\zeta')\rangle + \langle x_1(\zeta)x_2(\zeta')\rangle$$
$$+ \langle x_2(\zeta)x_1(\zeta')\rangle + \langle x_2(\zeta)x_2(\zeta')\rangle] \tag{5.270}$$

To find the expressions for the correlation functions, we give the solution of (5.252) and (5.253) the initial condition $x_i(0) = 0$:

$$x_i(\zeta) = \int_0^\zeta d\zeta' \exp[-b_i(\zeta - \zeta')]n_{x_i}(\zeta') \tag{5.271}$$

The correlation functions are

$$\langle \delta x_i(\zeta)\delta x_j(\zeta')\rangle = \frac{n_{i,j}}{b_i + b_j}\,\{\exp[-b_i\zeta - b_j\zeta' + (b_i + b_j)\min(\zeta, \zeta')]$$
$$- \exp(-b_i\zeta - b_j\zeta')\} \tag{5.272}$$

The correlation functions (5.269) and (5.270) are symmetric, so we might assume without, loss of generality, that $\zeta > \zeta'$. By this assumption, equation (5.272) becomes

$$\langle \delta x_i(\zeta) \delta x_j(\zeta') \rangle = \frac{n_{i,j}}{b_i + b_j} \{\exp[-b_i(\zeta - \zeta')] - \exp(-b_i\zeta - b_j\zeta')\} \qquad (5.273)$$

We are interested to the variance of timing fluctuations and amplitude fluctuations. Defining $\delta T(\zeta) = T(\zeta) - T_a(\zeta)$ where $T_a(\zeta)$ is given by equation (5.229), the equation for the timing fluctuations is, from equation (5.225),

$$\frac{d\delta T(\zeta)}{d\zeta} = -\delta\Omega(\zeta) + n_T(\zeta) \qquad (5.274)$$

The variance of the timing fluctuations is then

$$\langle \delta T(\zeta)^2 \rangle = \int_0^\zeta d\zeta' \int_0^\zeta d\zeta'' \langle \delta\Omega(\zeta')\delta\Omega(\zeta'') \rangle + \int_0^\zeta d\zeta' \int_0^\zeta d\zeta'' \langle n_T(\zeta')n_T(\zeta'') \rangle$$

$$= 2 \int_0^\zeta d\zeta' \int_0^{\zeta'} d\zeta'' \langle \delta\Omega(\zeta')\delta\Omega(\zeta'') \rangle + n_T\zeta \qquad (5.275)$$

The integration of (5.275) is a trivial but tedious task. The result is [44]

$$\langle \delta T(\zeta)^2 \rangle = \frac{3\langle n \rangle}{4\eta^2} \{a_0 + a_1\zeta + a_2 \exp(-b_1\zeta) + a_3 \exp(-b_2\zeta)$$

$$+ a_4 \exp(-2b_1\zeta) + a_5 \exp(-2b_2\zeta)$$

$$+ a_6 \exp[-(b_1 + b_2)\zeta]\} + \frac{\langle n \rangle \pi^2}{12} \zeta \qquad (5.276)$$

where

$$a_0 = -\frac{3(6 + 29\omega_r^2 + \omega_r^4)}{8\eta(1 - \omega_r^2)^3} \qquad (5.277)$$

$$a_1 = \frac{1 + 2\omega_r^2}{(1 - \omega_r^2)^2} \qquad (5.278)$$

$$a_2 = \frac{3(1 + 2\omega_r)}{2\eta(1 + \omega_r)(1 - \omega_r)^3} \qquad (5.279)$$

$$a_3 = \frac{3(1 - 2\omega_r)}{2\eta(1 - \omega_r)(1 + \omega_r)^3} \qquad (5.280)$$

$$a_4 = -\frac{9}{16\eta(1 - \omega_r)^3} \tag{5.281}$$

$$a_5 = -\frac{9}{16\eta(1 + \omega_r)^3} \tag{5.282}$$

$$a_6 = \frac{3}{8\eta(1 - \omega_r^2)} \tag{5.283}$$

To derive equation (5.276), we have used equations (5.102), (5.104), and (5.105) with $A = 1$.

The variance of the amplitude fluctuations is easily found by using equation (5.269). We get

$$\langle \delta A(\zeta)^2 \rangle = \frac{9\langle n \rangle}{32\eta} \left\{ \frac{2}{3} [1 - \exp(-(b_1 + b_2)\zeta)] \right.$$
$$\left. + \frac{1 - \exp(-2b_1\zeta)}{1 - \omega_r} + \frac{1 - \exp(-2b_2\zeta)}{1 + \omega_r} \right\} \tag{5.284}$$

The limit of (5.276) for $\omega_r \to 0$ gives the expressions for timing jitter and amplitude fluctuations for fixed filters, equations (5.148) and (5.149). The effect of sliding is to enhance both amplitude and timing fluctuations. Consider timing jitter, for example. The asymptotic growth with distance of the variance of the timing fluctuations is, for large ζ and neglecting the direct contribution of the ASE noise on timing jitter,

$$\langle \delta T(\zeta)^2 \rangle \sim \frac{3\langle n \rangle}{4\eta^2} a_1 \zeta \tag{5.285}$$

Without sliding, $a_1 = 1$. So a_1 can be considered as the enhancement factor of timing jitter caused by sliding, a measure of the effect of sliding on timing jitter. For instance, for sliding rate one-half of the critical sliding, $\omega_r = 0.5$, the enhancement factor is $a_1 = 2.667$ for $\omega_r = 0.75$; that is to say, for sliding 75% of the critical sliding rate, the enhancement factor is $a_1 = 11.102$. Amplitude fluctuations are affected by the amplitude-frequency coupling mechanism as well, and they experience a similar enhancement.

It is particularly instructive to see what happens at the critical sliding rate. Although this condition cannot be reached because the system is unstable, one expects that at the critical point the presence of a zero eigenvalue in the dynamic equaitons makes the system recovering the behavior without filtering control. This is indeed the case. The asymptotic expansion for large distances of equation (5.276) and (5.284) is

$$\langle \delta T(\zeta)^2 \rangle \sim \frac{\langle n \rangle}{12} \zeta^3 \tag{5.286}$$

$$\langle \delta A(\zeta)^2 \rangle \sim \frac{3\langle n \rangle}{8} \zeta \tag{5.287}$$

This asymptotic growth should be compared with that without filtering:

$$\langle \delta T(\zeta)^2 \rangle \sim \frac{\langle n \rangle}{9} \zeta^3 \tag{5.288}$$

$$\langle \delta A(\zeta)^2 \rangle \sim \langle n \rangle \zeta \tag{5.289}$$

Although the rate of growth of fluctuations is smaller at the critical sliding than without filtering control, the asymptotic trend is the same.

In Fig. 5.9 we show the normalized timing jitter $\langle \delta T(\zeta)^2 \rangle / \langle n \rangle$ against ζ for $\eta = 0.4$. The solid curve refers to fixed filters, the short-dashed one to $\omega_r = 0.25$, the long-dashed one to $\omega_r = 0.5$, and the dot-dashed one to $\omega_r = 0.75$. The increase of the timing jitter caused by sliding shows up clearly. Figure 5.10 shows the normalized amplitude noise $\langle \delta A(\zeta)^2 \rangle / \langle n \rangle$ against ζ for $\eta = 0.4$, with the same values of the sliding rate of the corresponding curves as in Fig. 5.9.

To have an idea of the sliding rates used in experiments, assume the typical parameters of Bell-Laboratories experiments, $D = -\omega / \lambda \beta_2 = 0.45 \, \text{ps/(nm km)}$, 1 mm air-gap étalon (filter delay $t_c = 10 \, \text{ps}$), with reflectivity with $R = 9\%$ and filter spacing $L_f = 33 \, \text{km}$. With these numbers we have $\eta = 0.43$. The critical sliding rate in soliton units is $\omega_{f,\text{cr}} = 0.1172$. Assuming pulses of $T_F = 16 \, \text{ps}$, we get $T_s = 16/1.763 = 9.077 \, \text{ps}$, soliton length $L_D = T_s^2 / |\beta_2| = 107.7 \, \text{km}$, and the critical sliding rate in dimensional units $\omega'_{0,\text{cr}} / (2\pi) = \omega_{f,\text{cr}} / (z_0 T_s 2\pi) =$

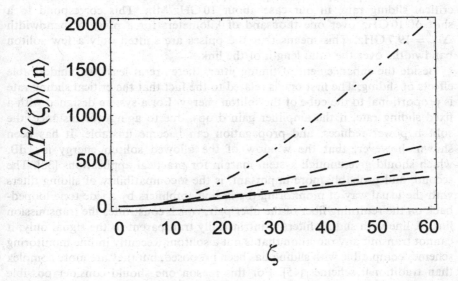

FIGURE 5.9 Normalized time jitter $\langle \delta T(\zeta)^2 \rangle / \langle n \rangle$ versus ζ for $\eta = 0.4$. Solid line corresponds to fixed filters, short-dashed line to $\omega_r = 0.25$, long-dashed line to $\omega_r = 0.5$, and dot-dashed line to $\omega_r = 0.75$.

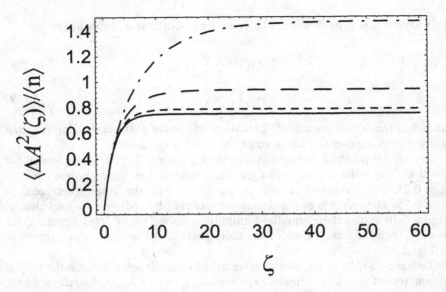

FIGURE 5.10 Normalized amplitude noise $\langle \delta A(\zeta)^2 \rangle / \langle n \rangle$ versus ζ for $\eta = 0.4$. Solid line corresponds to fixed filters, short-dashed line to $\omega_r = 0.25$, long-dashed line to $\omega_r = 0.5$, and dot-dashed line to $\omega_r = 0.57$.

19.077 GHz/Mm. Typical values of the sliding rates are about half of the critical sliding rate, in our case about 10 GHz/Mm. This correspond to a shift of 10 GHz over one thousand of kilometers for a pulse of bandwidth $\Delta \nu_F = 19.7$ GHz. This means that the pulses are shifted only a few soliton bandwidths over the total length of the link.

Beside the enhancement of timing jitter, there are at least two undesirable effects of sliding. The first one is related to the fact that the critical sliding rate is proportional to the cube of the soliton energy. For a system designed with a fixed sliding rate, if the amplifier gain drops, due to aging, for instance, the soliton power reduces, and propagation can become unstable. It has been shown, however, that the window of the allowed soliton energy is 3 dB, which should give enough system margin for practical applications [38]. The second, and probably more important, is the incompatibility of sliding filters with the usual way of monitoring the in-line amplifiers by a side tone looped-back on the returning fiber of the fiber pair, which constitutes the transmission line. A line with sliding filters is intrinsically transparent to the signal only; it cannot transmit any radiation that is not a soliton. Recently in-line monitoring schemes compatible with sliding has been proposed, but they are more complex than traditional schemes [45]. For this reason one should consider possible alternatives to the use of sliding filters. One could be the optimization of the filter profile to reduce the filter excess gain. The discussion of filter shape optimization will be the subject of the following section.

5.3.5 Butterworth Filters

Let us generalize the parabolic filter shape that we used as an approximation of étalon filters, in normalized units

$$\frac{\eta}{2}\left(i\,\frac{\partial}{\partial t}\right)^2 \tag{5.290}$$

into the following:

$$\frac{\eta}{2}\left(i\,\frac{\partial}{\partial t}\right)^{2n} \tag{5.291}$$

For $n > 1$ the curvature of the filter at the band-pass center is zero. Filters with zero curvature at the band-pass center are called *Butterworth filters*. In a Butterworth filter of order n, the lowest derivative at the band-pass center has order $2n$. Inserting the above filter shape in the normalized nonlinear Schrödinger equation, one gets (see equations 5.127–5.129)

$$\frac{\partial q}{\partial \zeta} = \frac{i}{2}\frac{\partial^2 q}{\partial \tau^2} + i|q|^2 q + \frac{\Delta\alpha}{2}\,q + \frac{\eta^{(n)}}{2}\frac{\partial^2 q}{\partial \tau^2} + n(\tau, \zeta) \tag{5.292}$$

where $\Delta\alpha$ remains related to the excess gain in the usual way,

$$\Delta\alpha = \delta g z_0 \tag{5.293}$$

and $\eta^{(n)}$ is $1/(2n)!$ times the $2n$th derivative with respect to the angular frequency of the filter spectral profile at the band-pass center, averaged over the filter spacing, in normalized units. By applying the perturbation approach used earlier for parabolic filters, we get [46]

$$\frac{dA}{d\zeta} = \Delta\alpha A - \eta^{(n)}\sum_{j=0}^{n}\binom{2n}{2j}M_{n-j}\Omega^{2j}A^{2(n-j)+1} + n_A(\zeta), \tag{5.294}$$

$$\frac{d\phi}{d\zeta} = \frac{1}{2}(A^2 - \Omega^2) + T\frac{d\Omega}{d\zeta} + n_\phi(\zeta) \tag{5.295}$$

$$\frac{d\Omega}{d\zeta} = -\eta^{(n)}\sum_{j=0}^{n-1}\binom{2n}{2j+1}M_{n-j}\Omega^{2j+1}A^{2(n-j)} + n_\Omega(\zeta) \tag{5.296}$$

$$\frac{dT}{d\zeta} = -\Omega + n_T(\zeta) \tag{5.297}$$

In the above equations the numbers M_k are defined as the $2k$th moments of the soliton spectrum

$$M_k = \frac{\pi}{4} \int_{-\infty}^{\infty} \omega^{2k} \operatorname{sech}^2\left(\frac{\pi}{2}\omega\right) d\omega \tag{5.298}$$

The values of M_k can be calculated as the $2k$th derivatives of the generating function $f(s) = s/\sin(s)$ calculated at $s = 0$. The values of M_k for the lowest values of k are $M_0 = 1$, $M_1 = 1/3$, $M_2 = 7/15$, and so on.

Equations (5.294)–(5.297) can be derived by the usual projection operation, where the integrals are more conveniently carried out in the Fourier domain. At steady state the soliton sits at the fiber center frequency $\Omega = 0$. The excess gain required to compensate for the filter loss at $\Omega = 0$ is

$$\Delta\alpha_n = \eta^{(n)} A^{2n} M_n \tag{5.299}$$

Linearizing equations (5.294) and (5.296) around the steady state, we get

$$\frac{d\delta A(\zeta)}{d\zeta} = -b_n \delta A(\zeta) \tag{5.300}$$

$$\frac{d\delta\Omega(\zeta)}{d\zeta} = -b_n \delta\Omega(\zeta) \tag{5.301}$$

The restoring force for amplitude and frequency fluctuations b_n is

$$b_n = 2n\eta^{(n)} A^{2n} M_n = 2n\Delta\alpha_n \tag{5.302}$$

If we define a "factor of merit" $\mathcal{F}$ of control filters as the ratio of the restoring force for amplitude and frequency fluctuations and the excess gain, we find that

$$\mathcal{F} = \frac{b_n}{\Delta\alpha_n} = 2n \tag{5.303}$$

We recall that $n = 1$ corresponds to the gain of the parabolic filters. For the same restoring force, Butterworth filters of order n have an excess gain that is n times smaller than that of étalon filters. The ASE power spectral density, normalized to the value that one would obtain without filters, is

$$N_n(\omega, \zeta) = \frac{\exp[(\Delta\alpha_n - \eta^{(n)}\omega^{2n})\zeta] - 1}{(\Delta\alpha_n - \eta^{(n)}\omega^{2n})\zeta} \tag{5.304}$$

The ASE power at the center of the filter profile $\omega = 0$ is approximately proportional to $\exp(\Delta\alpha_n\zeta)$. A soliton system that uses fixed filters for transmission control is limited by the accumulation of ASE power on zeros. The use of

Butterworth filters of order n permits transmission to approximately n times the distance of the case in which simple parabolic (étalon) filters are employed. It is possible to show that Butterworth filters are compatible also with wavelength division multiplexing if the Butterworth filters are designed to be periodic [46, 47]. Butterworth filters were used in a transmission experiment by Suzuki et al. at KDD, and they permitted transmission of single-channel transmission up to more than 11,000 km at 20 Gbit/s [48].

5.3.6 Summary of Soliton Transmission Control Schemes

In the previous sections we gave a yet incomplete account of different methods that have been proposed and successfully experimentally implemented for controlling soliton transmission. The purpose of this section is to summarize the main characteristics of each of them, and give some brief details of other methods that have been proposed for soliton control and have not been described at length.

Frequency filtering controls the soliton frequency and hence the soliton velocity. The direct timing fluctuations are not controlled. The residual frequency fluctuations, present in filtered transmission systems because of the ASE noise and other disturbances that directly couple to the soliton timing, still produce a linear growth with distance of the timing fluctuations. This is not a problem if solitons are used for transmission, even across a long distance. It can be a problem if solitons are used within a storage device. Another problem arising with filters is the ASE background. To compensate for the loss in the soliton wings, the use of filters requires excess gain at the center of the filter profile. The ASE noise added by the in-line amplifier then experiences an exponential growth within a bandwidth around the center of the filter passband. The growth of the ASE noise is more significant with strongest filters, and this is the main reason preventing the use of very narrowband fixed filters for soliton transmission control.

Timing fluctuations may be made bounded only if a device is inserted along the line which is locked to a precise time reference. Amplitude modulators are a possible choice. If amplitude modulators are inserted along the line together with frequency filters, both timing fluctuations are bounded and the ASE noise is kept below the threshold of exponential growth. This control scheme is stable, however, only if the modulator depth and the filter bandwidth are within given ranges. Indeed, amplitude modulators along makes the soliton amplitude unstable, and this instability is compensated by the stabilizing effect of the in-line frequency filters.

Another possibility for synchronous retiming is the use of phase modulators [49]. A harmonic phase modulation produces a harmonic frequency modulation in quadrature with it. Frequency modulations in the presence of a frequency filter converts into amplitude modulation, which in turn controls the soliton timing as discussed above for the scheme employing amplitude modulators.

The complete stability against timing and amplitude fluctuations with modulators and filtering is obtained at the expenses of a significant system complication. Synchronous retiming requires the clock to be regenerated locally by a clock recovery device. Also compatibility with wavelength division multiplexing is not straightforward, since the channels must be demultiplexed, retimed separately, then multiplexed again at each retiming stage. For these reasons the use of active retiming in transmission lines will probably be confined to very high bit-rate transmission with a single channel or a limited number of channels. On the other hand, active control is the only choice for soliton control if solitons must be stored for an indefinite time within a fiber loop, as required in optical memory devices.

The problem of the ASE noise growth can be efficiently solved by the use of sliding filters. Sliding filters obviously cannot solve the problem of the long-term linear divergence of the timing fluctuations. As already mentioned, however, long-term stability is not required if solitons are used in point-to-point transmission. Sliding filters have the benefit of preserving the full compatibility of filtering control with wavelength division multiplexing.

Another option for controlling the ASE growth is the use of in-line saturable absorbers [50]. The principle is similar to that used to stabilize pulses in passively mode-locked lasers. A saturable absorber is a device whose loss depends on the intensity, and it is large for low-intensity fields. The intense soliton experiences less loss in the saturable absorber than the weak ASE power generated close to the center of the filter profile. Saturable absorbers therefore keep the ASE power below threshold. A saturable absorber, however, tends to increase the amplitude fluctuations of the solitons. The reason can be readily understood. Weaker solitons experience larger loss than stronger solitons, and this tends to generate instability. Also they can be made compatible with wavelength division multiplexing only by demultiplexing before the saturable absorber, employing a different saturable absorber for each channel and then re-multiplexing as required with active modulators.

Finally, the rate of ASE noise generation might be reduced by a suitable profiling of the filter pass-band and, in particular, the use of flat-topped filters like the Butterworth filter. Although a careful filter design can keep the ASE power under control, it does not completely eliminate its exponential growth.

5.4   SOLITON INTERACTION AND COLLISION

5.4.1   Soliton Interaction

Solitons at the same frequency spaced a timing T apart interact by a coherent interaction that depends on the relative phase between the two and by an incoherent interaction. The coherent interaction, experimentally observed by Mitschke and Mollenauer in 1987 [51], is due to the overlap of soliton tails. Since two isolated fundamental solitons at infinite distance are a higher-order soliton solution of the propagation equation, this soliton-soliton interaction is the result of the intrinsic dynamics of a two-soliton solution [52]. Two solitons spaced so far apart that the coherent interaction is not effective yet experience a force that is independent of their relative phase. This incoherent interaction, discovered by Smith and Mollenauer [53], has been explained with the excitation by the solitons, through electrostriction, of an acoustic wave in the fiber core [54]. In this section we will examine in some detail only the short range coherent interaction.

To analyze the effect of the perturbation induced on a soliton by another soliton separated by a time interval T, we can again follow the perturbative approach [55].

Assume that two solitons are generated at the fiber input with timing T_1 and T_2,

$$q(\tau, 0) = q_1(\tau, 0) + q_2(\tau, 0) \tag{5.305}$$

where

$$q_i(\tau, 0) = A_i \operatorname{sech}[A_1(T - T_i)] \exp(i\phi_i) \tag{5.306}$$

If each soliton were isolated, it would propagate along the fiber undistorted, with unchanged amplitudes and frequency, and the phase and timing evolving according to equations (5.92) and (5.94) with $R = 0$. The presence of the other soliton instead perturbs the soliton evolution. Assume that the soliton separation is much larger than each soliton pulsewidth, that is, $|T_2 - T_1| \gg \max(1/A_1, 1/A_2)$. The effect of each soliton on the other can then be studied by perturbation theory. Assume that no other perturbation is experienced by soliton 1 but the presence of soliton 2. Since we consider only weak perturbations, we can assume that there is no interference between them, so the linear superposition principle applies. Other perturbations can be included by adding the corresponding terms to the right-hand side of the dynamic equation for the soliton observables. Let us make the ansatz

$$q(\tau, \zeta) = q_1(\tau, \zeta) + q_2(\tau, \zeta) \tag{5.307}$$

where

$$q_i(\tau, \zeta) = A_i \operatorname{sech}\{A_1[\tau - T_i(\zeta)]\} \exp[-i\Omega_i(\zeta) + i\phi_i(\zeta)] \tag{5.308}$$

Consider again (5.49). For $\tau \sim T_1$, roughly the average timing of soliton 1, the term at right-hand side of (5.49) arising from the nonlinear phase shift, which for the isolated soliton 1 would be $i|q_1|^2 q_1$, becomes

$$i|q_1 + q_2|^2(q_1 + q_2) \sim i|q_1|^2 q_1 + 2i|q_1|^2 q_2 + iq_1^2 q_2^*. \tag{5.309}$$

Only the first-order terms in $q_2 = q_2(\tau, \zeta)$ have been retained because they are the dominant terms when $\tau \sim T_1$. Therefore the perturbation due to soliton 2 on soliton 1 is

$$R = 2i|q_1|^2 q_2 + iq_1^2 q_2^* \tag{5.310}$$

Since only the tails of soliton 2 interact with 1, we can use approximation

$$q_2(\zeta, \tau) \sim 2A_2(\zeta) \exp\{-A_2(\zeta)[T - T_2(\zeta)] - i\Omega_2(\zeta) + i\phi_2(\zeta)\} \tag{5.311}$$

Assume, without loss of generality, that $T_1 > T_2$. Then the perturbative equations for soliton 1 are

$$\frac{dA_1}{d\zeta} = 4A^3 \exp[-(T_1 - T_2)\zeta] \sin(\phi_1 - \phi_2) \tag{5.312}$$

$$\frac{d\phi_1}{d\zeta} = \frac{1}{2}(A_1^2 - \Omega_1^2) + T_1 \frac{d\Omega_1}{d\zeta} + 6A^2 \exp[-(T_1 - T_2)\zeta] \cos(\phi_1 - \phi_2) \tag{5.313}$$

$$\frac{d\Omega_1}{d\zeta} = 4A^3 \exp[-(T_1 - T_2)\zeta] \cos(\phi_1 - \phi_2) \tag{5.314}$$

$$\frac{dT_1}{d\zeta} = -\Omega_1 - 2A \exp[-(T_1 - T_2)\zeta] \sin(\phi_1 - \phi_2) \tag{5.315}$$

To obtain the previous equations, we have approximated across the soliton 1, $\exp[\pm i(\Omega_1 - \Omega_2)T] \approx 1$, which corresponds to assuming that the frequency separation of the two pulses is much less than their bandwidth. This is the relevant case for our pruposes, since we are considering the interaction of consecutive pulses belonging to the same soliton stream being initially of the same frequency. We have also used approximation $A_1 \approx A_2$ at the right-hand sides of equations (5.312)–(5.315), and we have defined the average amplitude $A = (A_1 + A_2)/2$. Finally we make use of the following integrals:

$$\int_{-\infty}^{\infty} \operatorname{sech}^3(x)[2\exp(-x)]\,dx = 4 \tag{5.316}$$

$$\int_{-\infty}^{\infty} [1 - x\tanh(x)]\operatorname{sech}^3(x)[2\exp(-x)]\,dx = 2 \tag{5.317}$$

$$\int_{-\infty}^{\infty} x\operatorname{sech}^3(x)[2\exp(-x)]\,dx = 2, \tag{5.318}$$

$$\int_{-\infty}^{\infty} \tanh(x)\operatorname{sech}^3(x)[2\exp(-x)]\,dx = -\frac{4}{3} \tag{5.319}$$

The perturbation equation for the second solitons can be analogously evaluated, obtaining

$$\frac{dA_2}{d\zeta} = -4A^3 \exp[-(T_1 - T_2)\zeta]\sin(\phi_1 - \phi_2) \tag{5.320}$$

$$\frac{d\phi_2}{d\zeta} = \frac{1}{2}(A_2^2 - \Omega_2^2) + T_2\frac{d\Omega_2}{d\zeta} + 6A^2 \exp[-(T_1 - T_2)\zeta]\cos(\phi_1 - \phi_2) \tag{5.321}$$

$$\frac{d\Omega_2}{d\zeta} = -4A^3 \exp[-(T_1 - T_2)\zeta]\cos(\phi_1 - \phi_2) \tag{5.322}$$

$$\frac{dT_2}{d\zeta} = -\Omega_2 - 2A\exp[-(T_1 - T_2)\zeta]\sin(\phi_1 - \phi_2) \tag{5.323}$$

With the average amplitude already defined, $A = (A_1 + A_2)/2$, we now define the average phase $\phi = (\phi_1 + \phi_2)/2$, average frequency $\Omega = (\Omega_1 + \Omega_2)/2$, and average timing $T = (T_1 + T_2)/2$. We also define the differences $\Delta A_{1,2} = A_1 - A_2$, $\Delta\phi_{1,2} = \phi_1 - \phi_2$, $\Delta\Omega_{1,2} = \Omega_1 - \Omega_2$, and $\Delta T_{1,2} = T_1 - T_2$. We can assume, without loss of generality, that $\Omega \ll 1$. We can also assume that $\Delta\Omega_{1,2} \ll 1$, condition that adds to $\Delta A_{1,2} \ll A$, which we have already used to derive equations (5.312)–(5.323). The equations for the average amplitude and frequency are

$$\frac{dA}{d\zeta} = 0 \tag{5.324}$$

$$\frac{d\Omega}{d\zeta} = 0 \tag{5.325}$$

The average amplitude and frequency are constant. The equations for the amplitude, phase, frequency, and timing differences are

$$\frac{d\Delta A_{1,2}}{d\zeta} = 8A^3 \exp[-A\Delta T_{1,2}]\sin(\Delta\phi_{1,2}) \tag{5.326}$$

$$\frac{d\Delta\Omega_{1,2}}{d\zeta} = 8A^3 \exp[-A\Delta T_{1,2}]\cos(\Delta\phi_{1,.2}) \tag{5.327}$$

$$\frac{d\Delta\phi_{1,2}}{d\zeta} = A\Delta A_{1,2} \tag{5.328}$$

$$\frac{d\Delta T_{1,2}}{d\zeta} = -\Delta\Omega_{1,2}, \tag{5.329}$$

If we define two complex variables

$$P = A\Delta\Omega_{1,2} + iA\Delta A_{1,2} \qquad (5.330)$$

$$Q = -A\Delta T_{1,2} + i\Delta\phi_{1,2} \qquad (5.331)$$

equations (5.326) and (5.327) become

$$\frac{dP}{d\zeta} = 8A^4 \exp(Q) = -\frac{\partial V(Q)}{\partial S} \qquad (5.332)$$

$$\frac{dQ}{d\zeta} = P \qquad (5.333)$$

where

$$V(Q) = -8A^4 \exp(Q) \qquad (5.334)$$

These are the dynamical equations for a particle with a complex "position" Q and "momentum" P, in the complex "potential" $V(Q) = -8A^4 \exp(Q)$. One may easily find the general solutions (5.333) and (5.334) by applying the invariant of motion (a complex "energy" $-E$)

$$-E = \tfrac{1}{2} P^2 + V(Q) \qquad (5.335)$$

Inserting (5.333) into (5.335), we obtain

$$\frac{dQ}{d\zeta} = 2[V(Q) - E]^{1/2} \qquad (5.336)$$

The constant E can be evaluated using (5.335) for $\zeta = 0$. We obtain

$$E = 8A^4 \exp[Q(0)] - \tfrac{1}{2} P(0)^2 \qquad (5.337)$$

Equation (5.336) can be solved by separation of variables. The result is

$$-Q(\zeta) = -\ln \frac{E}{8A^4} + \ln \cos^2\left(-\sqrt{\frac{E}{2}}\zeta + c\right) \qquad (5.338)$$

where

$$c = \arccos \alpha \qquad (5.339)$$

$$\alpha = \sqrt{\frac{E}{8A^4}} \exp\left[-\frac{Q(0)}{2}\right] \qquad (5.340)$$

and the complex arccos is defined as $\arccos(\alpha) = -i \ln(\alpha \pm \sqrt{\alpha^2 - 1})$. Let us consider now some special cases.

Case $P(0) = 0$

For two solitons with the same amplitude and the same frequency, equation (5.338) gives $c = 0$ and (5.338) becomes

$$-Q(\zeta) = -Q(0) + \ln \cos^2 \left\{ 2A^2 \exp \left[\frac{Q(0)}{2} \right] \zeta \right\} \tag{5.341}$$

Two subcases are relevant

$\Delta\phi_{i,2}(0) = 0$. This case corresponds to initially in-phase solitons. From (5.331), $Q(\zeta) = -A\Delta T_{1,2}(\zeta) + i\Delta\phi_{1,2}(\zeta)$, and we get $\Delta\phi_{1,2}(\zeta) = \Delta\phi_{1,2}(0) = 0$ and

$$A\Delta T_{1,2}(\zeta) = A\Delta T_{1,2}(0) + \ln \cos^2 \left\{ 2A^2 \exp \left[-\frac{A\Delta T_{1,2}(0)}{2} \right] \zeta \right\} \tag{5.342}$$

The two solitons attract, as can be seen from their spacing $\Delta T_{1,2}(\zeta)$ which is decreasing with distance. An approximate estimation of the distance at which the two solitons will collide (approximate because the validity of the perturbation theory requires that the two solitons overlap only by their tails) is obtained by setting $A\Delta T(\zeta_c) = 0$. We get

$$\zeta_c \simeq \frac{\pi}{4A^2} \exp \left[\frac{A\Delta T_{1,2}(0)}{2} \right] \tag{5.343}$$

Inverse scattering theory shows that the solitons form in this case a bound state in which the two solitons periodically collide at distances that are multiple of ζ_c.

$\Delta\phi_{1,2}(0) = \pi$. This case corresponds to solitons initially out-of-phase. We obtain in this case $\Delta\phi_{1,2}(\zeta) = \Delta\phi_{1,2}(0) = \pi$ and

$$A\Delta T_{1,2}(\zeta) = A\Delta T_{1,2}(0) + \ln \cosh^2 \left\{ 2A^2 \exp \left[-\frac{A\Delta T_{1,2}(0)}{2} \right] \zeta \right\} \tag{5.344}$$

The two solitons repel. Their spacing, after an initial transience, asymptotically increases as

$$A\Delta T_{1,2}(\zeta) = A\Delta T_{1,2}(0) + 4A^2 \exp \left[-\frac{A\Delta T_{1,2}(0)}{2} \right] \zeta \tag{5.345}$$

Case $\Delta A_{1,2}(0) \neq 0$, $\Delta\phi_{1,2}(0) = \Delta\Omega_{1,2}(0) = 0$

For two solitons with initially the same frequency and in phase, but with unequal amplitudes, we have

$$E = 8 \exp[-\Delta T_{1,2}(0)] + \frac{\Delta A_{1,2}^2(0)}{2} \tag{5.346}$$

$$\alpha = \sqrt{\frac{E}{8}} \exp\left[-\frac{\Delta T_{1,2}(0)}{2}\right]$$

$$= \left\{1 + \frac{\Delta A_{1,2}^2(0)}{16} \exp[\Delta T_{1,2}(0)]\right\}^{1/2} \tag{5.347}$$

$$c = -i \ln(1 \pm \sqrt{\alpha^2 - 1}) = -i|c| \tag{5.348}$$

Assume $A = 1$. Combining the above equations with equation (5.338), we obtain, after some algebra,

$$\Delta T_{1,2}(\zeta) = \Delta T_{1,2}(0) + \ln\left[1 - \frac{\sin^2(\beta\zeta)}{\alpha^2}\right] \tag{5.349}$$

$$\beta = 2\alpha \exp\left[-\frac{\Delta T_{1,2}(0)}{2}\right]$$

$$= \left\{4 \exp[-\Delta T_{1,2}(0)] + \frac{\Delta A_{1,2}^2(0)}{4}\right\}^{1/2} \tag{5.350}$$

The two solitons oscillate around an equilibrium position. The minimum distance that they reach during propagation is obtained by setting $\sin(\beta\zeta) = 1$,

$$\Delta T_{\min} = \Delta T_{1,2}(0) + \ln\left(1 - \frac{1}{\alpha^2}\right)$$

$$= \Delta T_{1,2}(0) - \ln\left\{1 + \frac{16}{\Delta A_{1,2}^2(0)} \exp[-\Delta T_{1,2}(0)]\right\} \tag{5.351}$$

If $\Delta A_{1,2}(0)$ is large enough, the two solitons form a bound state and never collide [56]. The amplitude difference required so that $\Delta T_{\min}$ is a fraction f of the initial separation $\Delta T_{1,2}(0)$, that is, $\Delta T_{\min} = f\Delta T_{1,2}(0)$, is

$$\Delta A_{1,2}^2(0) = \frac{16 \exp[-\Delta T_{1,2}(0)]}{\exp[(1-f)\Delta T_{1,2}(0)] - 1} \tag{5.352}$$

In this numerical example, we assume that the separation between the two solitons changes by 10% maximum, therefore we set $f = 0.9$. If $\Delta T_{1,2}(0) = 4 \times 1.763 = 7.052$ (the initial separation is four times the full width at half power), equation (5.352) gives $\Delta A_{1,2}(0) = 0.116$; in other words, the initial soliton amplitude must be different by less than 12%. Suppression of soliton collision between two solitons of different amplitudes has a simple physical

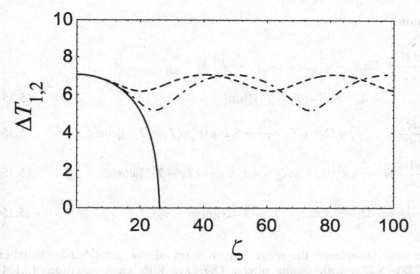

FIGURE 5.11 Normalized timing separation $\Delta T_{1,2}$ versus ζ for initially in-phase solitons. The initial separation is $\Delta T_{1,2}(0) = 7.052$. Dotted line corresponds to $\Delta A_{1,2} = 0.1$, dot-dashed line to $\Delta A_{1,2} = 0.05$, and solid line to equal amplitude solitons.

explanation. Two solitons of different amplitudes have different wavevectors; hence their interaction evolves, during propagation, from attractive to repulsive, and vice versa, averaging to zero over one period.

In Fig. 5.11 we show $\Delta T_{1,2}$ plotted against normalized distance ζ for initially in-phase solitons. The initial separation is four times the full width at half power, corresponding to $\Delta T_{1,2}(0) = 7.052$. The dotted line corresponds to $\Delta A_{1,2} = 0.1$, and the dot-dashed line to $\Delta A_{1,2} = 0.05$. The solid line is the first half-period of $\Delta T_{1,2}$ for two colliding equal-amplitude solitons. We see that an amplitude difference as low as 5% is enough to efficiently suppress the soliton collision.

5.4.2 Soliton Interaction with Filters

The presence of filters changes the soliton dynamics in soliton interaction as well. In the presence of filters, there is a fixed point in the (A, Ω) phase space, and only one value of the amplitude is admitted at steady state. For this reason we restrict ourselves to the case of initially equal amplitudes, $\Delta A_{1,2}(0) = 0$. On the single soliton, the perturbation due to filtering and that due to the presence of the other soliton add linearly. By combining equations (5.312)–(5.315) and (5.222)–(5.225), we obtain the perturbation equations for the dynamic evolution of soliton 1 in the presence of filtering and interaction with another soliton

$$\frac{dA_1}{d\zeta} = \Delta\alpha A_1 - \eta\left[(\Omega_1 - \omega_f'\zeta)^2 + \frac{A_1^2}{3}\right]A_1$$
$$- 4A^3 \exp[-(T_1 - T_2)\zeta]\sin(\phi_1 - \phi_2) \tag{5.353}$$

$$\frac{d\phi_1}{d\zeta} = \frac{1}{2}(A_1^2 - \Omega_1^2) + T_1\frac{d\Omega_1}{d\zeta} + 6A^2 \exp[-(T_1 - T_2)\zeta]\cos(\phi_1 - \phi_2) \tag{5.354}$$

$$\frac{d\Omega_1}{d\zeta} = -\frac{2}{3}\eta A_1^2(\Omega_1 - \omega_f'\zeta) - 4A^3 \exp[-(T_1 - T_2)\zeta]\cos(\phi_1 - \phi_2) \tag{5.355}$$

$$\frac{dT_1}{d\zeta} = -\Omega_1 - 2A \exp[-(T_1 - T_2)\zeta]\sin(\phi_1 - \phi_2) \tag{5.356}$$

We have considered the most general form of the perturbation equations (5.222)–(5.225) with sliding filters. The fixed filter case, equations (5.131)–(5.134), can be obtained by setting $\omega_f' = 0$. Analogously, for the second soliton, from equations (5.320)–(5.323), we get

$$\frac{dA_2}{d\zeta} = \Delta\alpha A_2 - \eta\left[(\Omega_2 - \omega_f'\zeta)^2 + \frac{A_2^2}{3}\right]A_2$$
$$+ 4A^3 \exp[-(T_1 - T_2)\zeta]\sin(\phi_1 - \phi_2) \tag{5.357}$$

$$\frac{d\phi_2}{d\zeta} = \frac{1}{2}(A_2^2 - \Omega_2^2) + T_2\frac{d\Omega_2}{d\zeta} + 6A^2 \exp[-(T_1 - T_2)\zeta]\cos(\phi_1 - \phi_2) \tag{5.358}$$

$$\frac{d\Omega_2}{d\zeta} = -\frac{2}{3}\eta A_2^2(\Omega_2 - \omega_f'\zeta) + 4A^3 \exp[-(T_1 - T_2)\zeta]\cos(\phi_1 - \phi_2) \tag{5.359}$$

$$\frac{dT_2}{d\zeta} = -\Omega_2 - 2A \exp[-(T_1 - T_2)\zeta]\sin(\phi_1 - \phi_2) \tag{5.360}$$

With sliding filters, the perturbation due to filters is usually stronger than that due to the soliton interaction. So we can assume that the deviations of the frequency and amplitude of the two solitons from the values $\Omega_j(\zeta)$ and A_j given by

$$\Omega_J(\zeta) = \omega_f'\zeta + \Delta\Omega \tag{5.361}$$

$$\Delta\Omega = -\frac{3\omega_f'}{2\eta A^2} \tag{5.362}$$

$$\Delta\alpha = \eta(\Delta\Omega^2 + \tfrac{1}{3}A_j^2) \tag{5.363}$$

are small. By this assumption the equations for the ΔA, $\Delta \Omega$, $\Delta \phi$, and ΔT become

$$\frac{d\Delta A_{1,2}}{d\zeta} = -\tfrac{2}{3}, \eta A^2 \Delta A_{1,2} - 2\eta A \Delta \Omega \Delta \Omega_{1,2}$$

$$+ 8A^3 \exp[-A\Delta T_{1,2}]\sin(\Delta \phi_{1,2}) \qquad (5.364)$$

$$\frac{d\Delta \Omega_{1,2}}{d\zeta} = -\tfrac{2}{3} \eta A^2 \Delta \Omega_{1,2} - \tfrac{4}{3} \eta A \Delta \Omega \Delta A_{1,2}$$

$$+ 8A^3 \exp[-A\Delta T_{1,2}]\cos(\Delta \phi_{1,2}) \qquad (5.365)$$

$$\frac{d\Delta \phi_{1,2}}{d\zeta} = \omega'_f \Delta T_{1,2} + A\Delta A_{1,2} \qquad (5.366)$$

$$\frac{d\Delta T_{1,2}}{d\zeta} = -\Delta \Omega_{1,2} \qquad (5.367)$$

Let us consider first the case without sliding, $\omega'_f = 0$. In this case equations (5.364)–(5.367) can be conveniently written in terms of the complex variables P and Q defined in equations (5.330) and (5.331). We obtain

$$\frac{dP}{d\zeta} = -\frac{2}{3} \eta A^2 P - \frac{\partial V(Q)}{\partial S} \qquad (5.368)$$

$$\frac{dQ}{d\zeta} = P \qquad (5.369)$$

where $V(Q)$ is still given by equation (5.334). The presence of filtering adds a friction force to the dynamic evolution of Q. The friction force produces a slowdown in the soliton dynamics. Unfortunately, however, one cannot use this slowdown to significantly reduce the timing between transmitted solitons, therefore increasing the bit-rate. For a given collision distance the soliton spacing can be reduced by no more than a few percent for practical filter bandwidths because of the exponential dependence of the potential on the soliton spacing $\Delta T_{1,2}$ [37].

If sliding instead of fixed filters are used, the friction force does not play a noticeable role in reducing the soliton interaction. In this case the relative phase of the two solitons evolves approximately as $\Delta T\omega'_f\zeta$ (see equation 5.366). The soliton interaction evolves from attractive to repulsive, and vice versa, averaging to zero over a distance where the relative phase goes through a 2π change. Soliton interaction is suppressed in the sliding filters by a mechanism similar to that underlying the suppression of interaction between solitons with alternating amplitudes [57].

For large initial soliton spacings, the transition of the relative phase from attractive to repulsive, and vice versa, is fast enough that solitons can form a

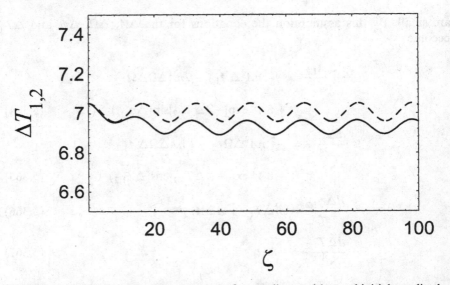

FIGURE 5.12 Timing separation versus ζ of two solitons with equal initial amplitude, frequency, and phase and initial timing separation $\Delta T_{1,2}(0) = 7.052$. The filter strength is $h = 0.4$, and the sliding rate is half of the critical sliding rate. Solid line: numerical integration of equations (5.364)–(5.367); dashed line: equation (5.370).

bound state. If we assume equal initial amplitudes, frequencies, and phases of the two solitons, we can approximately solve equations (5.364)–(5.367) by setting $\delta T_{1,2}(\zeta) \simeq \Delta T_{1,2}(0)$, $\Delta\phi(\zeta) = \omega_f' \Delta T_{1,2}(0)\zeta$, and $\Delta A_{1,2}(\zeta) = \Delta\Omega_{1,2}(\zeta) = \Delta A_{1,2}(0) = \Delta\Omega_{1,2}(0) = 0$. Solving by quadrature equations (5.365) and (5.367), we get

$$\Delta T_{1,2}(\zeta) \simeq -16 \exp[-\Delta T_{1,2}(0)] \frac{\sin^2[\omega_f' \Delta T_{1,2}(0)\zeta/2]}{[\omega_f' \Delta T_{1,2}(0)]^2} \tag{5.370}$$

In Fig. 5.12 we show by a solid line the numerical integration of equations (5.364)–(5.367) with $\eta = 0.4$ and sliding rate one-half of the critical sliding rate, $\omega_r = 0.5$. The initial amplitudes, frequencies, and phases of the two solitons are equal and the initial separation is $\Delta T_{1,2}(0) = 7.052$ (four times their full width at half power). With a dashed line is also plotted the timing separation given by equation (5.370). We see a good qualitative agreement between the two.

If the initial soliton spacing is decreased, the soliton interaction will be strong enough to prevent formation of a bound state, and the two solitons will eventually drift away. This is shown in Fig. 5.13, where we report the timing separation of two solitons as opposed to ζ obtained by numerical integration of equations (5.364)–(5.367). The solid line refers to

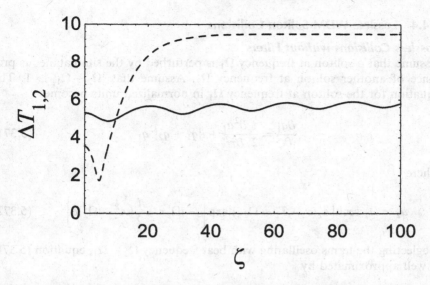

FIGURE 5.13 Timing separation versus ζ obtained by numerical integration of equations (5.364)–(5.367) under the conditions of Fig. 5.12 but with $\Delta T_{1,2}(0) = 5.289$ (solid line) and $\Delta T_{1,2}(0) = 3.526$ (dashed line).

$\Delta T_{1,2}(0) = 5.289$, and the dashed line to $\Delta T_{1,2}(0) = 3.526$. The other parameters are the same as those of Fig. 5.12.

5.4.3 Wavelength Division Multiplexing

One of the key advantages of soliton systems is that initially well separated solitons with different center frequencies pass through each other without changing their shape. The result of the collision with a soliton at a different frequency is only a time shift of magnitude inversely proportional to the square of the initial frequency separation. Wavelength division multiplexing with solitons is easily achieved. A detailed study of wavelength division multiplexed soliton systems is reported in ref. [62]. In that paper it is shown that the presence of lumped amplifiers does not change the qualitative picture of soliton collision compared to the lossless soliton propagation provided that the ratio of the collision length to the amplifier spacing is two or more. In the next section we will analyze a multiplexed soliton system by using the perturbative approach on the average soliton model; therefore we implicitly assume that the amplifier spacing meets this condition. We will release this condition in Section 5.4.5 where, following the analysis of ref. [62], we will give the conditions under which the average soliton model applies.

5.4.4 Lossless WDM Soliton Collisions

Lossless Collisions without Filters
Assume that a soliton at frequency Ω_1 is perturbed by the simulataneous presence of another soliton at frequency Ω_2. Assume that $|\Omega_2 - \Omega_1| \gg 1$. The equation for the soliton at frequency Ω_1 in normalized units becomes

$$\frac{\partial q_1}{\partial \zeta} = \frac{i}{2} \frac{\partial^2 q_1}{\partial \tau^2} + i|q_1 + q_2|^2 q_1 \tag{5.371}$$

where

$$q_2 = A_2 \operatorname{sech}[A_2(\tau - T_2 + \Omega_2\zeta)]\exp\left[-i\Omega_2\tau + \frac{i}{2}(A_2^2 - \Omega_2^2)\zeta\right] \tag{5.372}$$

Neglecting the terms oscillating with beat frequency $\Omega_2 - \Omega_1$, equation (5.371) is well approximated by

$$\frac{\partial q_1}{\partial \zeta} = \frac{i}{2} \frac{\partial^2 q_1}{\partial \tau^2} + i|q_1|^2 q_1 + 2i|q_2|^2 q_1 \tag{5.373}$$

We use the perturbation approach with the perturbation term

$$R = 2i|q_2|^2 q_1 \tag{5.374}$$

Although R formally is of order one, a perturbative approach is justified for large frequency separation because, when the two solitons overlap, their relative phase changes many times over each soliton pulsewidth, averaging the interaction to zero. We will assume as a working hypothesis that the effect of the perturbation is small. The expressions found at the end of our derivation will give a self-consistent check to support our approach.

Above we assumed that the two colliding solitons are copolarized, so that we could use the scalar propagation equation (with nonlinear coefficient n_2 scaled to accommodate the 8/9 factor accounting for averaging over the polarization). This case corresponds to the maximum collision-induced frequency shift. For orthogonally polarized solitons, the term $2i|q_2|^2 q_1$ is substituted by $i|q_2|^2 q_1$, and therefore the effects of the soliton collision can be scaled down by a factor 2. The rigorous analysis of the soliton collision between solitons of arbitrary polarization is due to Mollenauer, Gordon, and Heismann and presented in ref. [63]. In that paper it is also shown that collision changes the soliton polarization, preventing the simultaneous use of polarization division multiplexing and WDM.

Assume solitons of equal amplitudes, $A_1 = A_2 = 1$, and no filtering or other control schemes. The effect of soliton control may easily be added at a later stage because of the superposition principle that holds in our linear approximation. The equation for the center frequency of soliton j is (see equation 5.89)

$$\frac{d\delta\Omega_j}{d\zeta} = \text{Re} \int d\tau \, f_{\Omega_j}^* R \tag{5.375}$$

where

$$f_{\Omega_j} = \tanh(\tau - T_j + \Omega_j\zeta)q_j \tag{5.376}$$

Inserting equations (5.375) and (5.374) into equation (5.375), we are left with the integrations

$$2\int d\tau \, \tanh[A(\tau - T_1 + \Omega_1\zeta)]|q_1|^2|q_2|^2 = \int d\tau \, \frac{\partial|q_1|^2}{\partial\tau}|q_2|^2 \tag{5.377}$$

$$2\int d\tau \, \tanh[A(\tau - T_2 + \Omega_2\zeta)]|q_1|^2|q_2|^2 = \int d\tau \, \frac{\partial|q_2|^2}{\partial\tau}|q_1|^2 \tag{5.378}$$

Assume for simplicity that $T_1 = T_2 = 0$. This means that the reference frame along the fiber is chosen such that the point $\zeta = 0$ corresponds to overlapping solitons. Expand $\Omega_2 = \overline{\Omega}_2 + \delta\Omega_2$ and $\Omega_1 = \overline{\Omega}_1 + \delta\Omega_1$, where $\overline{\Omega}_j$ are the unperturbed soliton center frequencies, and $\delta\Omega_j$ are the perturbations caused by the soliton collision. Assume also, without loss of generality, that $\overline{\Omega}_2 = -\overline{\Omega}_1 = \Omega$. For symmetry, $\delta\Omega_1 = -\delta\Omega_2 = \delta\Omega$. Subtracting equation (5.375) for $j = 2$ and 1, we get

$$\frac{d\delta\Omega}{d\zeta} = \frac{1}{2\Omega}\frac{\partial}{\partial\zeta}\int d\tau \, \text{sech}^2(\tau + \Omega\zeta)\text{sech}^2(\tau - \Omega\zeta) \tag{5.379}$$

where we have used that

$$\frac{\partial|q_j|^2}{\partial\tau} = \frac{1}{\Omega_j}\frac{\partial|q_j|^2}{\partial\zeta} \tag{5.380}$$

and we have neglected the dependence of Ω on ζ at the right-hand side of equation (5.379). Performing the integral at right-hand side of equation (5.379), we get

$$\frac{d\delta\Omega}{d\zeta} = \frac{\partial}{\partial\zeta}\frac{g(2\Omega\zeta)}{2\Omega} = f(2\Omega\zeta) \tag{5.381}$$

where

$$g(x) = \frac{4[x\cosh(x) - \sinh(x)]}{\sinh^3(x)} \tag{5.382}$$

$$f(x) = -4\frac{x[1 + 2\cosh^2(x)] - 3\sinh(x)\cosh(x)}{\sinh^4(x)} \tag{5.383}$$

Integration of equation (5.381) from $\zeta = \zeta_0$ to ζ gives

$$\delta\Omega(\zeta) = \frac{1}{2\Omega} [g(2\Omega\zeta) - g(2\Omega\zeta_0)] \tag{5.384}$$

where ζ_0 corresponds to the fiber input. Recall that $\zeta = 0$ corresponds to the two solitons overlapping. In equation (5.384) we also used the fact that $\delta\Omega(\zeta_0) = 0$. Note that $g(x)$ has a symmetric shape with a maximum at $x = 0$ and that $g(0) = 4/3$; hence the maximum frequency deviation is $\delta\Omega_{max} = 2/(3\Omega)$. The perturbation caused by the collision is proportional to the inverse of the frequency separation; hence for large Ω the perturbative approach is well founded.

The perturbation R caused by the presence of the colliding soliton is in quadrature with the soliton; therefore it does not directly contribute to the soliton displacement. So the equation for the timing displacement is

$$\frac{d\delta T}{d\zeta} = -\delta\Omega \tag{5.385}$$

Without loss of generality, we can assume that $\delta T(\zeta_0) = 0$. The effect of the collision on the soliton at larger frequency $\overline{\Omega}_1 = \Omega$, the faster soliton, is

$$\delta T(\zeta) = -\int_{-\infty}^{\zeta} d\zeta' \delta\Omega(\zeta') \tag{5.386}$$

that is,

$$\delta T(\zeta) = -\frac{1}{2\Omega} \left\{ \frac{1}{2\Omega} [h(2\Omega\zeta) - h(2\Omega\zeta_0)] - g(2\Omega\zeta_0)(\zeta - \zeta_0) \right\} \tag{5.387}$$

where

$$h(x) = 2 \frac{\sinh(x)\cosh(x) - x}{\sinh^2(x)} \tag{5.388}$$

Note that $h(x)$ is an odd function with $h(x) \to 2$ for $x \to \infty$. After a complete collision of two initially nonoverlapping solitons $\zeta_0 \to \infty$, the one at Ω experiences the timing shift

$$\delta T_\infty = -\frac{1}{\Omega^2} \tag{5.389}$$

The soliton at larger frequency, the faster, is advanced. Conversely, the soliton at lower frequency, the slower, is delayed.

Assume that the two solitons are initially overlapping, that is, $\zeta_0 = 0$. Asymptotically the faster soliton acquires a constant frequency lag equal to

$$\delta\Omega(\zeta)|_{\zeta\to\infty,\zeta_0=0} = -\frac{2}{3\Omega} \tag{5.390}$$

The persistent frequency lag of the slower soliton has the opposite sign. The frequency lag is such that the two solitons pull each other. The effect of the constant frequency lag is an asymptotic drift of the two solitons from the timing they had if there were no collision. The time drift of the faster soliton is

$$\delta T(\zeta)|_{\zeta_0=0} \sim \frac{1}{2\Omega}\left(\frac{4}{3}\zeta - \frac{1}{\Omega}\right), \qquad \zeta\to\infty \tag{5.391}$$

that of the slower has the opposite sign. By the effect of the initial collision, the faster soliton (i.e., the soliton with the larger frequency) slows down, the slower soliton (i.e., the soliton with lower frequency) speeds up [64]. This effect is detrimental in optical communication systems that use solitons. If one does not take particular care in avoiding overlapping of solitons belonging to different WDM channels at line input, the persistent frequency shift produces unbearable errors in the transmission. We will see below that this problem is alleviated if in-line filters are used.

If the soliton are initially nonoverlapping, that is, $|\Omega|\zeta_0 \ll -1$, we can approximate the functions calculated at ζ_0 with their limit for $\zeta_0 \to -\infty$. There is no persistent frequency shift, only a persistent timing shift. When the collision is over, the soliton of larger frequency acquires a constant timing shift of

$$\delta T(\zeta)_{\zeta\to\infty,\zeta_0\to\infty} = \Omega^{-2} \tag{5.392}$$

The asymptotic timing shift of the soliton with larger frequency (the faster soliton) is negative; that is, the soliton is advanced by the collision. Conversely, the slower soliton, which has lower frequency, acquires a net delay.

Lossless Collisions with Filters

Assume now that the propagation of the two solitons is controlled by a multiple band-pass filter such as a periodic Fabry-Perot filter. Assume that two passbands of the filter are centered at $\overline{\Omega}_1 = \Omega$ and $\overline{\Omega}_2 = -\Omega$ and that their curvature at the band-pass center is the same. The coupled dynamic equations for the frequency and timing displacements are

$$\frac{d\delta\Omega}{d\zeta} = -\frac{2}{3}\eta\delta\Omega + f(2\Omega\zeta) \tag{5.393}$$

$$\frac{d\delta T}{d\zeta} = -\delta\Omega \tag{5.394}$$

The asymptotic displacement after collision can also be easily calculated. By integration of both sides of equation (5.394), we get

$$\delta T(\zeta) - \delta T(\zeta_0) = -\int_{\zeta_0}^{\zeta} d\zeta \delta\Omega(\zeta') \tag{5.395}$$

and solving for $\delta\Omega$ and integrating both sides of equation (5.393), we get

$$\int_{\zeta_0}^{\zeta} \delta\Omega(\zeta) = \frac{3}{2\eta} \int_{\zeta_0}^{\zeta} d\zeta' \left[-\frac{d\delta\Omega}{d\zeta'} + f(2\Omega\zeta) \right]$$

$$= \frac{3}{2\eta} \left[-\delta\Omega(\zeta) + \delta\Omega(\zeta_0) + \int_{\zeta_0}^{\zeta} d\zeta' f(2\Omega\zeta) \right] \tag{5.396}$$

Assuming that at the fiber input $\delta\Omega(\zeta_0) = 0$ and $\delta T(\zeta_0) = 0$, by combining the equations (5.395) and (5.396), we get [65]

$$\delta T(\zeta)|_{\zeta \to \infty} = \frac{3}{2\eta} \int_{\zeta_0}^{\zeta} d\zeta' f(2\Omega\zeta) = -\frac{3}{4\eta\Omega} g(2\Omega\zeta_0) \tag{5.397}$$

In equation (5.397) we have used the asymptotic property, which is the consequence of the restoring force for the frequency fluctuations induced by the filters, $\delta\Omega(\zeta) \to 0$ for $\zeta \to \infty$. If the solitions on the two channels are initially exactly overlapping, $\zeta_0 = 0$, the asymptotic timing shift is

$$\delta T(\zeta)|_{\zeta \to \infty, \zeta_0 = 0} = -\frac{1}{\eta\Omega} \tag{5.398}$$

if the solitons are initially well separated, we have instead $\zeta_0 \to -\infty$; hence

$$\delta T(\zeta)|_{\zeta \to \infty, \zeta \to -\infty} = 0 \tag{5.399}$$

Note the benefit of filters. Without filters, solitons initially overlapping suffer for an asymptotic *velocity* shift that produces a uniform drift of the timing of the two solitons while they propagate along the fiber. With filters, the two solitons initially overlapping acquire a constant *timing* shift. Solitons intially nonoverlapping acquire without filters a constant timing shift; with filters the timing shift reduces ideally to zero, although computer simulations have showed that a residual asymptotic timing shift persists also with filters [22].

In Fig. 5.14 we show the normalized frequency displacement, obtained by integration of equations (5.393) and (5.394), of the faster soliton (that at positive frequency) of two colliding solitons with $\Omega = 4$ plotted against ζ. Solid, dashed, and dot-dashed curves are for $\eta = 0.6$ and 0.2 and 0, respectively. In Fig. 5.15 we show the normalized timing for the same conditions. Although,

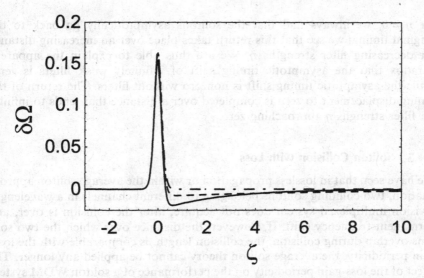

FIGURE 5.14 Normalized frequency displacement, obtained by numerical integration of equations (5.393) and (5.394), of the faster of two colliding solitons with $\Omega = 4$, versus ζ. Solid, dashed, and dot-dashed curves are for $\eta = 0.6$, 0.2, and 0, respectively.

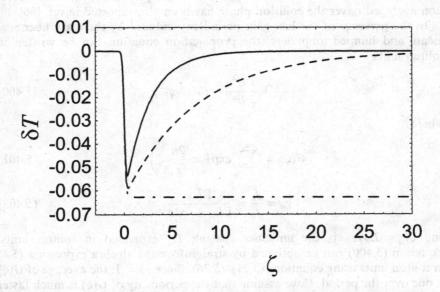

FIGURE 5.15 Normalized timing displacement, obtained by numerical integration of equations (5.393) and (5.394), of the faster of two colliding solitons with $\Omega = 4$, versus ζ. Solid, dashed, and dot-dashed curves are for $\eta = 0.6$, 0.2, and 0, respectively.

for $\eta \neq 0$, we always find that the solitons asymptotically go back to the original timing; we see that this return takes place over an increasing distance for decreasing filter strengths η. We are thus able to explain the apparent paradox that the asymptotic timing shift of infinitely weak filters is zero, while the asymptotic timing shift is nonzero without filters. The return of the timing displacement to zero is completed over a distance that tends to infinity for filter strenghs η approaching zero.

5.4.5 Soliton Collision with Loss

We have seen that in lossless propagation or within the average soliton approximation, two colliding solitons belonging to different channels in a wavelength division multiplexed system does not acquire, after the collision is over, any permanent frequency shift. If, however, the distance over which the two solitons overlap during collision, the collision length, is comparable with the loss-gain periodicity, the average soliton theory cannot be applied any longer. The effect of the loss-gain periodicity on the performance of a soliton WDM system has been analyzed by Mollenauer, Evangelides, and Gordon in ref. [62]. We will follow in this section the analysis of that paper. The same procedure allows the study of the filtered case. Here we follow the analysis of ref. [67] by exactly including the effect of lumped filtering in the expression for the collision-induced timing displacement. An analysis of wavelength division multiplexing collisions with filters arriving an expression of the permanent shift after collision averaged oaver the collision phase has been also reported in ref. [66].

In the presence of the loss-gain periodicity induced by the lossy fiber segments and lumped amplifiers, the propagation equation can be written in soliton units

$$\frac{\partial q}{\partial \zeta} = \frac{i}{2} \frac{\partial^2 q}{\partial \tau^2} + iG(\zeta)|q|^2 q \qquad (5.400)$$

where

$$G(\zeta) = \sum_{n=-\infty}^{\infty} \exp\left(in\frac{2\pi}{\zeta_{L_A}}\zeta\right)g_n \qquad (5.401)$$

$$g_n = \frac{f_n}{f_0} = \frac{\Gamma L_A}{\Gamma L_A + in2\pi} \qquad (5.402)$$

and $\zeta_{L_A} = L_A/z_0$ is the amplifier spacing L_A expressed in soliton units. Equation (5.400) can be obtianed by straightforward algebra expressing (5.6) in soliton units using equations (5.21)–(5.26). Since $g_0 = 1$, the average of $G(\zeta)$ is one over the period. If we assume that the periodicity of $G(\zeta)$ is much faster than the soliton length, that is, $\zeta_{L_A} \ll 1$, it might be shown that the shape of the soliton and its spectrum are not affected, to first order, by the periodic spatial modulation of $G(\zeta)$. The effect of $G(\zeta)$ is to produce a small-phase modulation on the soliton, which averages to zero over one period, thus sheding of radiation from the soliton with an energy-loss coefficient proportional to the spectral content of the soliton at frequencies $\pm(4\pi n/\zeta_{L_A} - 1)^{1/2}$, where n is an integer $n \geq 1$ that exponentially vanishes for $\zeta_{L_A} \gg 1$ [62, 33]. The analysis of soliton collisions in the presence of loss-gain periodicity follows the lines of the lossless case. Before going into the analysis, however, it might be worthwhile to generalize the analysis a little, including the case in which dispersion is not constant along the line. We will consider only the case in which the dispersion variations have the periodicity of the amplifier spacing, although generalizations are straightforward. If we assume that the soliton units are defined for the average dispersion, equation (5.400) can be generalized to include a nonuniform dispersion, which we assume with the periodicity of the amplifier spacing, and in-line filtering [62]

$$\frac{\partial q}{\partial \zeta} = \frac{\eta(\zeta)}{2}\frac{\partial^2 q}{\partial \tau^2} + i\frac{\Delta(\zeta)}{2}\frac{\partial^2 q}{\partial \tau^2} + iG(\zeta)|q|^2 q \tag{5.403}$$

where $\Delta(\zeta)$ is the ratio of the dispersion at ζ and the average dispersion. One can easily reduce equation (5.403) to a form similar to (5.400) by definition of a new normalized propagation distance by the transformation

$$\zeta_\Delta = \int_0^\zeta \Delta(\zeta)d\zeta \tag{5.404}$$

In the new variable equation (5.403) becomes

$$\frac{\partial q}{\partial \zeta_\Delta} = \frac{\eta_\Delta(\zeta_\Delta)}{2}\frac{\partial^2 q}{\partial \tau^2} + \frac{i}{2}\frac{\partial^2 q}{\partial \tau^2} + iG_\Delta(\zeta_\Delta)|q|^2 q \tag{5.405}$$

where

$$G_\Delta(\zeta_\Delta) = \frac{G(\zeta)}{\Delta(\zeta)} \tag{5.406}$$

$$\eta_\Delta(\zeta_\Delta) = \frac{\eta(\zeta)}{\Delta(\zeta)} \tag{5.407}$$

and the ζ at right-hand side of (5.406) and (5.407) are functions of ζ_Δ by equation (5.404). From (5.406) we can immediately see that if the dipersion profile exactly matches the power variations along the line consequence of linear attenuation, $G_\Delta = 1$ and the soliton collision behaves as if the fiber were lossless. However, we will see below that lumped filters, if present, make the system deviate a little from the ideal loss-less case discussed in the preceding section.

The following analysis applies to the case of constant dispersion. We will also give the results for the case of variable dispersion; for more detail, see [67].

The loss-gain periodicity affects only the functional form of the perturbation. To first order, the perturbation added by the soliton in the other channel is given by

$$R = 2iG(\zeta)|q_2|^2 q_1 \tag{5.408}$$

which replaces (5.374), which is valid in the lossless case. Since the derivation arriving at (5.379) involves only integration over time, it can be easily repeated in the more general case, obtaining without filters

$$\frac{d\delta\Omega}{d\zeta} = C(\zeta) \tag{5.409}$$

and with in-line filtering

$$\frac{d\delta\Omega}{d\zeta} = -\frac{2}{3}\eta(\zeta)\delta\Omega + C(\zeta) \tag{5.410}$$

where

$$C(\zeta) = \frac{G(\zeta)}{2\Omega}\frac{\partial}{\partial\zeta}\int d\tau \, \text{sech}^2(\tau + \Omega\zeta)\text{sech}^2(\tau - \Omega\zeta) \tag{5.411}$$

Assume that the collision begins with two well-separated solitons and, as in the previous section, that the two solitons overlap at $\zeta = 0$. Without filters the asymptotic frequency shift after a complete collision, $\delta\Omega_\infty = [\delta\Omega(\zeta) - \delta\Omega(-\zeta)]_{\zeta\to\infty}$, is given by

$$\delta\Omega_\infty = \mathcal{I}_1 \tag{5.412}$$

where

$$\mathcal{I}_1 = \int_{-\infty}^{\infty} d\zeta C(\zeta) \tag{5.413}$$

With in-line filtering, (5.410) can be readily integrated, obtaining

$$\delta\Omega(\zeta) = \int_{-\infty}^{\zeta} d\zeta' \exp\left[-\frac{2}{3}\int_{\zeta'}^{\zeta}\eta_\Delta(\zeta'')d\zeta''\right]C(\zeta') \tag{5.414}$$

The asymptotic timing shift $\delta T_\infty = [\delta T(\zeta) - \delta T(-\zeta)]_{\zeta\to\infty}$ is

$$\delta T_\infty = -\int_{-\infty}^{\infty} d\zeta\delta\Omega(\zeta) \tag{5.415}$$

Inserting (5.414) into (5.415), using

$$\int_{-\infty}^{\infty} d\zeta \int_{-\infty}^{\zeta} d\zeta' = \int_{-\infty}^{\infty} d\zeta' \int_{\zeta'}^{\infty} d\zeta \qquad (5.416)$$

changing ζ into $\zeta = x + \zeta'$, we get

$$\delta T_\infty = -\int_0^\infty dx \int_{-\infty}^{\infty} d\zeta' C(\zeta') \exp[-\tfrac{2}{3} \eta \zeta_{L_A} N(x,\zeta')] \qquad (5.417)$$

where

$$N(x,\zeta') = \frac{1}{\eta \zeta_{L_A}} \int_0^x \eta(\zeta'' + \zeta') d\zeta'' \qquad (5.418)$$

The lumped nature of the filters, which we assume are inserted after each amplifier, are taken into account in equaiton (5.410) by the ζ-dependent η (assume that $2\eta\zeta_{L_A}/3 \ll 1$)

$$\eta(\zeta) = \sum_{k=-\infty}^{\infty} \eta \zeta_{L_A} \delta(\zeta - k\zeta_{L_A}) \qquad (5.419)$$

Using equation (5.419) in (5.418), we obtain that $N(x,\zeta')$ is the number of amplifiers inserted within the fiber segment of length x having the lower end at $\zeta = \zeta'$. The function $N(x,\zeta')$ is periodic in ζ' with period ζ_{L_A} and assumes for $0 \le \zeta' < \zeta_{L_A}$ the integer values

$$N(x,\zeta') = n, \quad n\zeta_{L_A} - \zeta' \le x < (n+1)\zeta_{L_A} - \zeta' \qquad (5.420)$$

Because of equation (5.420), the exponential appearing in (5.417) is also a stepwise function of x. Therefore the integral over x in equation (5.417) can be readily evaluated by a series obtianing

$$\int_0^\infty dx \exp[-\tfrac{2}{3} \eta \zeta_{L_A} N(x,\zeta')] = -\zeta' + \zeta_{L_A} \sum_{n=0}^{\infty} \exp(-\tfrac{2}{3} \eta n \zeta_{L_A})$$

$$= -\zeta' + \zeta_{L_A}[1 - \exp(-\tfrac{2}{3} \eta_{L_A})]^{-1} \qquad (5.421)$$

Inserting this result into equation (5.417), we finally get

$$\delta T_\infty = -\zeta_{L_A}[1 - \exp(-\tfrac{2}{3} \eta \zeta_{L_A})]^{-1} \mathcal{I}_1 + \mathcal{I}_2 \qquad (5.422)$$

where $\mathcal{I}_1$ is given by equation (5.413),

$$\mathcal{I}_2 = \int_{-\infty}^{\infty} d\zeta S(\zeta) C(\zeta) \qquad (5.423)$$

and $S(\zeta')$ is a periodic sawtoothlike function assuming the values $S(\zeta') = \zeta'$ in its period $0 \leq \zeta' < \zeta_{L_A}$.

The asymptotic frequency shift without filters and the asymptotic timing shift with filters (the asymptotic frequency shift with filters is obviously zero) are related to the two integrals (5.413) and (5.423) which, by using definition (5.411) can both be written as

$$I_j = \int_{-\infty}^{\infty} d\zeta \, \frac{A_j(\zeta)}{2\Omega} \frac{\partial}{\partial \zeta} \int d\tau \, \mathrm{sech}^2(\tau + \Omega\zeta)\mathrm{sech}^2(\tau - \Omega\zeta) \qquad (5.424)$$

where $A_1 = G$ and $A_2 = SG$. To keep the notation simple, we will omit the index j in the following derivation, with the advise that it applies to both I_1 and I_2.

Expand $A(\zeta)$ as a spatial Fourier integral

$$A(\zeta) = \int \frac{dk}{2\pi} \exp(ik\zeta)A(k) \qquad (5.425)$$

and perform the ζ derivative. We get

$$I = -\frac{1}{2\Omega} \int_{-\infty}^{\infty} \frac{dk}{2\pi} \int_{-\infty}^{\infty} d\zeta \int_{-\infty}^{\infty} d\tau \, \exp(ik\zeta)ikA(k)\mathrm{sech}^2(\tau + \Omega\zeta)\mathrm{sech}^2(\tau - \Omega\zeta)$$

$$(5.426)$$

If we insert in equation (5.426) the expression

$$\exp(ik\zeta) = \exp\left[i\frac{k}{2\Omega}(-\tau + \Omega\zeta)\right]\exp\left[i\frac{k}{2\Omega}(\tau + \Omega\zeta)\right] \qquad (5.427)$$

and perform the integration over the entire (τ, ζ) plane with substitution $s_1 = -\tau + \Omega\zeta$ and $s_2 = \tau + \Omega\zeta$, which implies that $d\tau \, d\zeta = ds_1 ds_2/(2\Omega)$, we get

$$I = -\frac{1}{4\Omega^2} \int_{-\infty}^{\infty} \frac{dk}{2\pi} \, ikA(k)\left[\int_{-\infty}^{\infty} ds \, \exp\left(i\frac{k}{2\Omega}s\right)\mathrm{sech}^2(s)\right]^2 \qquad (5.428)$$

The integral over s is $2x/\sinh(x)$ where $x = \pi k/(4\Omega)$. Using the property that $A(z)$ is real and hence $A(-k) = A(k)$, we get

$$I = \frac{32}{\pi^2} \, \mathrm{Im} \int_0^{\infty} \frac{dk}{2\pi} \frac{A(k)}{k} \frac{x^4}{\sinh^2(x)} \qquad (5.429)$$

For periodic amplifier and filter spacing, $A(\zeta)$ can be expanded as in (5.401), obtaining

$$A(k) = 2\pi a_n \delta\left(k - n\frac{2\pi}{\zeta_{L_a}}\right) \qquad (5.430)$$

Substituting into equation (5.429), we get

$$\mathcal{I} = \frac{16\zeta_{L_A}}{\pi^3} \operatorname{Im} \sum_{n=0}^{\infty} a_n \frac{n^3 y^4}{\sinh^2(ny)} \tag{5.431}$$

where $y = \pi^2/(2\Omega\zeta_{L_A})$. The a_n are the coefficients of a Fourier series expansion of the periodic function $A(\zeta)$. In the more general case of variable dispersion, with G substituted by G_Δ, we obtain

$$a_{n,1} = \frac{1}{\zeta_{L_A}} \int_0^{\zeta_{L_A}} d\zeta_\Delta G_\Delta(\zeta_\Delta)\exp(-ik_n\zeta_\Delta) \tag{5.432}$$

$$a_{n,2} = \frac{1}{\zeta_{L_A}} \int_0^{\zeta_{L_A}} d\zeta_\Delta \zeta_\Delta G_\Delta(\zeta_\Delta)\exp(-ik_n\zeta_\Delta)$$

$$= \frac{i}{\zeta_{L_A}} \frac{\partial}{\partial k} \int_0^{\zeta_{L_A}} d\zeta_\Delta G_\Delta(\zeta_\Delta)\exp(-ik\zeta_\Delta)\Big|_{k=k_n} \tag{5.433}$$

$$k_n = \frac{2\pi n}{\zeta_{L_A}} \tag{5.434}$$

Combining (5.406) and $d\zeta_\Delta = \Delta(\zeta)d\zeta$ obtained differentiating both sides of (5.404) (note that the ζ and ζ_Δ axes coincide at the amplifier locations, so the integration extrema remain unchanged), we get

$$a_{n,1} = \frac{1}{\zeta_{L_A}} \int_0^{\zeta_{L_A}} d\zeta G(\zeta)\exp(-ik_n\zeta_\Delta) \tag{5.435}$$

where ζ_Δ is a function of ζ by equation (5.404).

Let us distinguish three cases. For constant dispersion, $\zeta_\Delta = \zeta$, and $G = G_\Delta$ is a decaying exponential that averages to unit over one period,

$$G(\zeta) = \frac{\alpha L_A}{1 - \exp(-\alpha L_A)} \exp(-\alpha\zeta z_0) \tag{5.436}$$

therefore we get (see also equation 5.401)

$$a_{n,1} = \frac{\alpha L_A}{\alpha L_A + in2\pi} \tag{5.437}$$

and

$$a_{n,2} = \frac{\zeta_{L_A}\alpha L_A[1 - (1 + \alpha L_A + in2\pi)\exp(-\alpha L_A)]}{[1 - \exp(-\alpha L_A)](\alpha L_A + in2\pi)^2} \tag{5.438}$$

Let us consider the case where dispersion is a stepwise function assuming the value D_1 for $0 = L_0 < z \leq L_1$, D_2 for $L_1 < z \leq L_2, \ldots, D_N$ for $L_{N-1} < z \leq L_N = L_A$, and periodically repeating over z. The constant normalized dispersion of each step is $\Delta_j = D_j / \overline{D}$, where $\overline{D}$ is the average dipersion. Using (5.435) and (5.433), we can show that

$$a_{n,1} = \frac{\alpha L_A}{1 - \exp(-\alpha L_A)} \sum_{j=0}^{N-1} \frac{e_j - e_{j+1}}{\alpha L_A + in2\pi\Delta_{j+1}} \tag{5.439}$$

$$a_{n,2} = \frac{\zeta_{L_A}\alpha L_A}{1 - \exp(-\alpha L_A)} \sum_{j=0}^{N-1} \frac{1}{(\alpha L_A + in2\pi\Delta_{j+1})^2}$$

$$\times \left\{ \left[\Delta_{j+1} + \frac{\phi_j}{2\pi} (\alpha L_A + in2\pi\Delta_{j+1}) \right] e_j \right.$$

$$\left. - \left[\Delta_{j+1} + \frac{\phi_{j+1}}{2\pi} (\alpha L_A + in2\pi\Delta_{j+1}) \right] e_{j+1} \right\} \tag{5.440}$$

$$e_j = \exp(-\alpha L_j - in\phi_j) \tag{5.441}$$

$$\phi_j = \begin{cases} 0, & j = 0 \\ 2\pi \sum_{k=1}^{j} \frac{\Delta_k(L_k - L_{k-1})}{L_A}, & j \geq 1 \end{cases} \tag{5.442}$$

Since Δ is normalized such that its average is unit, we have $\phi_N = 2\pi$.

Consider now the third and last case of dispersion following the loss profile. We have already shown that this case corresponds to the lossless case and can be studied performing the limits of (5.437) and (5.438) for $\alpha \rightarrow 0$. We obtain

$$a_{n,1} = 0 \tag{5.443}$$

$$a_{n,2} = \begin{cases} \dfrac{\zeta_{L_A}}{2}, & n = 0 \\ \dfrac{i\zeta_{L_A}}{(2\pi n)}, & n \geq 1 \end{cases} \tag{5.444}$$

which imply that $\mathcal{I}_1 = 0$ but, in general, that $\mathcal{I}_2 \neq 0$. As is obvious, if the dispersion along the line exactly compensates for linear loss or, equivalently, the power along the line is constant, in the unfiltered case soliton collisions are ideal, and there is no permanent frequency shift after collision. If distributed filtering is inserted along the line, there is no residual timing shift, as was shown in the previous section. The effect of lumped filtering is to produce a small residual timing shift for collision between solitons whose frequency spacing is so large and collision distance so short that no filters will be encountered during the collisions. In practice, instead of the difficult exponential tapering of dispersion, it is more practical to use a suitable stepwise profile approximating the ideal exponential one [22, 68]. The residual timing shift accounted for by $\mathcal{I}_2$ may be of some importance for multichannel systems of many Gbit/s each employing such a dispersion management scheme.

So far, we have assumed that the collision is centered at $\zeta = 0$ and that this corresponds to the position of one amplifier. The shifting of the amplifier chain with respect to the center of the soliton collision is easily accounted for by a shift of the functions $G(\zeta)$ and $S(\zeta)G(\zeta)$. In the Fourier space this adds to the Fourier coefficients the phase factor $\exp(ik\zeta_a/\zeta_{L_A})$ or $\exp[ik\int_0^{\zeta_a} d\zeta\Delta(\zeta)/\zeta_{L_A}]$ for nonconstant dispersion, where ζ_a is the position of the reference amplifier in the reference frame having the origin at the center of the collision. In the periodic case the phase factor to multiply the $a_{n,1}$ and $a_{n,2}$ by is

$$\exp[i\phi_a(n)] = \exp\left[i\,\frac{2\pi n}{\zeta_{L_A}}\int_0^{\zeta_a} d\zeta\Delta(\zeta)\right] \tag{5.445}$$

so that equation (5.431) becomes

$$\mathcal{I} = \frac{16\zeta_{L_A}}{\pi^3}\,\mathrm{Im}\sum_{n=0}^{\infty}\exp[i\phi_a(n)]a_n\,\frac{n^3 y^4}{\sinh^2(ny)} \tag{5.446}$$

The collision distance, defined as the distance across which the two solitons overlap more than their half power width, is in soliton units

$$\zeta_{\mathrm{coll}} = 1.7627\Omega^{-1} \tag{5.447}$$

In terms of the full width at half maximum of the soliton spectrum $\Delta\nu_F$ defined by (5.48) and of the frequency separation of the two colliding solitons $\Delta\nu_{2\Omega} = 4\pi\Omega T_0$ (we recall that the normalized frequency separation of the two solitons is 2Ω), the collision distance is

$$\zeta_{\mathrm{coll}} = \pi\,\frac{\Delta\nu_F}{\Delta\nu_{2\Omega}} \tag{5.448}$$

The function $y^4/\sinh^2(y)$ is very small beyond $y \approx 6$, so being $y = \pi^2/(2\Omega\zeta_{L_A}) \approx 2.8\zeta_{\mathrm{coll}}/\zeta_{L_A}$, for $\zeta_{\mathrm{coll}}/\zeta_{L_A} > 1$, the spatial harmonics other than the fundamental will give negligible contributions to the two integrals. As long as this condition is met, since the two series reduce to a single term, the phases of the periodic perturbation that correspond to a maximum frequency or timing shift are independent of $\zeta_{\mathrm{coll}}/\zeta_{L_A}$. Let us analyze in some detail the cases without filtering control and with filtering control.

No In-Line Control

In Fig. 5.16 we plot the residual frequency displacement $|\delta\Omega|$ in soliton units for two colliding solitons (we will omit the subscript ∞ for simplicity in both

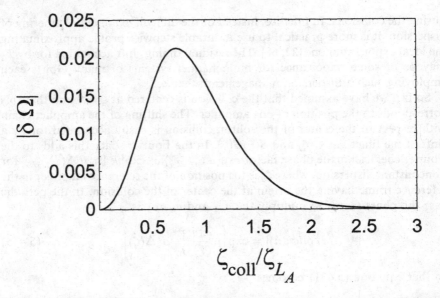

FIGURE 5.16   Residual frequency displacement, in soliton units, for two colliding solitons, versus the ratio of the collision distance to the amplifier spacing.

frequency and timing displacements from now on) against the ratio of the collision distance to the amplifier spacing. We have fixed the parameters of the soliton and of the line but varied the frequency separation of the two solitons. The values of the parameters are amplifier spacing $L_A = 33$ km, fiber loss 0.21 dB/km, soliton period $z_p = 353$ km (soliton length $z_0 = 224$ km), dispersion 0.45 ps/(nm km), soliton full width at half power $T_F = 20$ ps ($T_s = 11.34$ ps). The effect of the periodic power variation is negligible for $\zeta_{coll}/\zeta_{L_A} > 2$ [62].

We can easily convert the abscissa into a frequency separation in GHz in our numerical example by $\zeta_{coll}/\zeta_{L_A} = 336.35/\Delta\nu_{2\Omega}$. The ordinates can be converted into frequency shifts in GHz by $|\Delta\nu_{coll}| = 14.0295|\delta\Omega|$. The residual frequency shift added over many collisions causes an extra timing jitter that can be a problem for multichannels unfiltered soliton systems. Let us now see if it is possible to draw some general conclusions on the impact of collisions on multichannels systems. We have seen that $|\delta\Omega|$ is exponentially small for $\zeta_{coll}/\zeta_{L_A} > 2$. Therefore the frequency shift is negligible when this condition is met, and this converts into the practical design rule [62]

$$\Delta\nu_{max} = \frac{T_F}{2\pi L_A|\beta_2|} \tag{5.449}$$

Theoretically the timing shift caused by the collision of two solitons is the cause for the occurrence of the minimum allowed channel spacing. The residual timing shift after collision of two solitons spaced with frequency 2Ω, in soliton units, is given by equation (5.392). Each collision between solitons belonging to channels i and j and spaced $2\Omega_{i,j}$ then produces the asymptotic time shift, in soliton units,

$$\delta T_i = \frac{4}{(2\Omega_{i,j})^2} \tag{5.450}$$

If the average spacing between solitons in the same channel is T and their relative frequency spacing is $2\Omega_{i,j}$, the number of collisions occurring over the distance ζ_{tot} will range from none to $N_{i,j} = 2\Omega_{i,j}\zeta_{\text{tot}}/T$. The spread of the soliton arrival times, integrated over all channels, is

$$\delta T_{\text{tot},i} = \sum_{j \neq i} \frac{N_{i,j}}{2} \, \delta T_{i,j} = \frac{\zeta_{\text{tot}}}{T} \sum_{j \neq i} \frac{2}{2\Omega_{i,j}} \tag{5.451}$$

or in dimensional units,

$$\frac{\delta t_{\text{tot},i}}{T_B} = 0.5611 \left(\frac{T_F}{T_B}\right)^2 \frac{|\beta_2|}{T_F^3} \sum_{j \neq i} \frac{1}{\Delta\nu_{i,j}} \tag{5.452}$$

where $\Delta\nu_{i,j}$ is the frequency spacing between the ith and the jth channel and T_B, the bit-time, is T of equation (5.451) in dimensional units. Practical considerations dictate that the spread of the soliton arrival times does not exceed a significant fraction of the bit-time T_B, so $\delta t_{\text{tot},i}/T_B < J_{\text{max}}$. Also the ratio T_F/T_B is usually independent of the soliton pulsewidth and therefore of the bit-rate. Consider a comb of N equally spaced channels, and assume N to be odd for simplicity. The collision-induced time shift is more significant for the central channel, the worst case for a soliton belonging to this channel being that all channels of one side of the spectrum give no collisions while the channels on the other side give all possible collisions. If we number the channels from $-(N-1)/2$ to $(N-1)/2$, we get for the central channel

$$\frac{\delta t_{\text{tot},0}}{T_B} = 0.5611 \left(\frac{T_F}{T_B}\right)^2 L_{\text{tot}} \frac{|\beta_2|}{T_F^3} \frac{2}{\Delta\nu_c} \sum_{j=1}^{(N-1)/2} \frac{1}{j} \tag{5.453}$$

where $\Delta\nu_c$ is the channel spacing and L_{tot} is the total link length. The minimum channel spacing permitted by the collision-induced time jitter is then

$$\Delta\nu_{c,\text{min}} = 0.5611 \left(\frac{T_F}{T_B}\right)^2 L_{\text{tot}} \frac{|\beta_2|}{T_F^3} \frac{2}{J_{\text{max}}} \sum_{j=1}^{(N-1)/2} \frac{1}{j} \tag{5.454}$$

The maximum number of allowed channels is given by the ratio of the maximum channel spacing to avoid significant collision-induced frequency shift to the minimum channel spacing required to avoid significant collision-induced timing jitter. Using equations (5.449) and (5.454), we get

$$
\frac{\Delta\nu_{max}}{\Delta\nu_{c,min}} = 0.1418 \frac{J_{max}}{L_{tot}L_A} \left(\frac{T_B}{T_F}\right)^2 \frac{T_F^4}{|\beta_2|^2}
$$

$$
\times \left[0.577 + \ln\left(\frac{N-1}{2}\right) + \frac{1}{N-1}\right]^{-1} \tag{5.455}
$$

In the expression above we used the approximation, valid within 1.3% for $N > 2$,

$$
\sum_{j=1}^{(N-1)/2} j^{-1} \simeq 0.577 + \ln\left(\frac{N-1}{2}\right) + \frac{1}{N-1} \tag{5.456}
$$

The maximum number of channels N_{max} is $N_{max} = 1 + \Delta\nu_{max}/\Delta\nu_{c,min}$. From equation (5.455) we get

$$
N_{max} = 1 + 0.1418 \frac{J_{max}}{L_{tot}L_A} \left(\frac{T_B}{T_F}\right)^2 \frac{T_F^4}{|\beta_2|^2}
$$

$$
\times \left[0.577 + \ln\left(\frac{N_{max}-1}{2}\right) + \frac{1}{N_{max}-1}\right]^{-1} \tag{5.457}
$$

The right-hand side of (5.457) depends only weakly on N_{max}; therefore the maximum number of chanels is roughly proprotional to the fourth power of the pulsewidth. Since the information carried by each channel is inversely proportional to the pulsewidth, the total bit-rate is proportional to the cube of the pulsewidth. It is then convenient to use many channels with low bit-rates than a few channels with a high bit-rate each. To give an example, consider a disperison of 0.5 ps/(nm km) that corresponds to $|\beta_2| = 0.637\,ps^2/km$ at a 1.55 μm wavelength, $L_A = 50$ km, $L_{tot} = 9000$ km, $J_{max} = 0.1$ (maximum collision-induced timing spread 10% of the bit time), and $T_B/T_F = 5$. By equation (5.457) we see that for 7 channels the soliton pulsewidth is $T_F = 48.3$ ps and the bit-rate $R = 1/(5T_F) = 4.1\,4.1$ GBit/s. The channel spacing is 42 GHz. For 3 channels we have $T_F = 32.1$ ps, and a bit-rate 6.2 GHz. The overall channel capacity of a system with 7 channels at 4 GBit/s is 28 GBit/s, and the channel capacity of a system with 3 channels at 6 Gbit/s is 18 Gbit/s. The example above shows that if wavelength division multiplexing is used, it is convenient for the two considered sources of error that as many channels as possible be used. The optimum number of channels is, however, determined by other factors as well. When the number of channels increases, the soliton pulsewidth increases, and the soliton energy decreases. This would eventually lower the signal-to-noise ratio at the receiver down to intolerable values. Such lowering of the soliton energy could be counterbalanced by an increase of the dispersion, having also beneficial effects on the overall channel capacity, since, by equation (5.457), the overall capacity is roughly proportional to the inverse of the dispersion squared. However, this possibility is precluded by the simultaneous increase of single-channel Gordon-Haus jitter with dispersion. This last point will be discussed in some detail in the Section 5.5.

In-Line Filters

Consider now the filtered case. In Fig. 5.17 we plot the residual timing displacement $|\delta T|$, in soliton units, for two colliding solitons against the ratio of the collision distance to the amplifier spacing. The solid line corresponds to a normalized filter strength $\eta = 0.2$, and the dashed line to $\eta = 0.6$. The values of the line parameters are the same as those of Fig. 5.16, and we repeat them here for clarity: amplifier spacing $L_A = 33$ km, fiber loss 0.21 dB/km, soliton length $z_0 = 224$ km, dispersion 0.45 ps/(nm km), and soliton full width at half power $T_F = 20$ ps. Again the timing shifts can be converted to picoseconds by $|\delta t| = 11.34|\delta T|$ in our numerical example. By the dot-dashed line is provided a comparison to the normalized timing shift $|\delta T| = \Omega^{-2}$ in the lossless case with

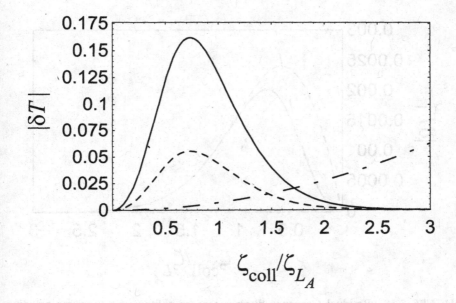

FIGURE 5.17 Residual frequency displacement, in soliton units, for two colliding solitons, versus the ratio of the collision distance to the amplifier spacing when in-line control filters are used. Solid line corresponds to a normalized filter strength $\eta = 0.2$; dashed line to $\eta = 0.6$. With a dot-dashed line is reported, for comparison, the normalized timing shift $|\Delta T| = \Omega^{-2}$ for the lossless case without filters.

no in-line filters. Note that the periodic variations, without filters, produce asymptotic timing displacements that monotonically increase with distance because of the residual frequency displacement that persists after the collision is over.

The contribution to the residual timing displacement of the term due to lumped filtering, $\mathcal{I}_2$ in (5.402) is usually negligible with respect to that caused by the periodic power variation, proportional to $\mathcal{I}_1$. We have seen, however, that $\mathcal{I}_2$ is the only term left when we use a dispersion profile proportional to the loss profile, which perfectly balances the effect of the periodic power variation [62, 22]. In Fig. 5.18 we show the residual timing jitter with relation to $\zeta_{\text{coll}}/\zeta_{L_A}$ for the filtered case, under the same conditions as in Fig. 5.17 but with exponential tapered dispersion. The solid line represents the residual timing shift when the center of the collision is placed in the middle between two consecutive amplifiers. For small $\zeta_{\text{coll}}/\zeta_{L_A}$ the collision distance is so short that no filters are encountered during collision, and the colliding solitons evolve as in the lossless case without filters. This is evident in Fig. 5.18 by the dot-dashed line which shows the asymptotic timing shift without filters. The worst case, the one with the largest timing shift for a small collision length and opposite in sign to that

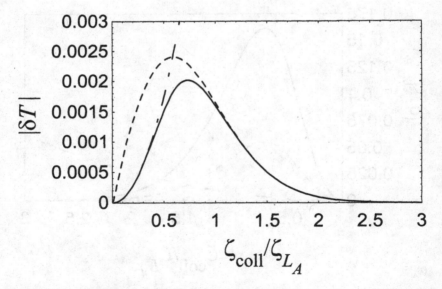

FIGURE 5.18 Residual frequency displacement, in soliton units, for two colliding solitons, versus the ratio of the collision distance to the amplifier spacing when in-line control filters are used and with exponential tapering of dispersion. Solid line is for the center of the collision in the middle of two consecutive amplifiers; dashed line is for the center of the collision at an amplifier position. With a dot-dashed line is reported, for comparison, the normalized timing shift $|\Delta T| = \Omega^{-2}$ for the lossless case without filters.

of the lossless case without filters, corresponds to the center of the collision at a filter position. This case is shown by a dashed line in Fig. 5.18.

The timing jitter is independent of the average filter strength η in this case but still depends on ζ_{L_A}, that is, on the filter periodicity. As mentioned earlier at the end of Section 5.4.4, though it might not seem so at first thought, this is not an unphysical result for filters with $\eta \to 0$. The asymptotic timing shift shown in the above figures is indeed reached for infinitely weak filters across a distance that tends to infinity.

Let us turn now to some issues arising from the above analysis of WDM soliton transmission with filters. We will use four numerical examples. Some system parameters are taken from the experiment of Mollenauer et al. [22]: filter strength $\eta = 0.4$, dispersion $D = 0.5\,\text{ps/(nm km)}$ corresponding to $|\beta_2| = 0.6373\,\text{ps}^2/\text{km}$, pulsewidth $T_F = 20\,\text{ps}$, bit-rate $10\,\text{GBit/s}$ (bit-time $T_B = 100\,\text{ps}$), loss $\alpha = 0.021\ln(10)\,\text{km}^{-1}$, and total length of the link $L_{\text{tot}} = 9000\,\text{km}$. We assume periodic channel spacing with multiples of $75\,\text{GHz}$. The amplifier and the filter spacings are chosen as $L_A = L_f = 50\,\text{km}$ for the first example. In the second and the third examples we assume perfect compensation in the power variation by the exponential tapering of dispersion, with path average dispersion $\overline{D} = 0.5\,\text{ps/(nm km)}$. The filter spacing is $50\,\text{km}$ in the second example, 33.3 and 66.6 periodically repeated in the third (only two amplifiers out of three are followed by a filter). The third case with unequal filter spacing can be analyzed as an extension of the preceding theory. If the filters are inserted periodically with spacing L_1 and L_2, the overall period being $L_{\text{per}} = L_1 + L_2$, it can be shown that $a_{n,1} = 0$ and that for $n > 1$ ($a_{n,0}$ does not contribute to the jitter because it is always real)

$$a_{n,2} = \frac{i\zeta_{L_{\text{per}}}}{2\pi n}\left\{1 - \frac{L_1}{L_{\text{per}}}\left[1 - \exp\left(-2i\pi n\frac{L_1}{L_{\text{per}}}\right)\right]\right\} \tag{5.458}$$

The fourth and last example is a line with $50\,\text{km}$ amplifier and filter spacing and a stepwise approximation of the ideal exponential dispersion profile that cancels the effect of loss. The length and the dispersion of the steps of constant dispersion is calculated by an algorithm suggested in ref. [68].

Our aim is to estimate the jitter for a generic number of channels of the system. Let us examine first the timing jitter between solitons in channel 1 and 8, namely on the soliton at lower frequency. The frequency separation between the solitons is $525\,\text{GHz}$. In Fig. 5.19 we show the timing shift in picoseconds induced by the collisions in the lossy case (with no exponential tapering of the dispersion) plotted against the phase of the collision L_a/L_A (L_a is the position of the center of the collision with respect to one amplifier). In Fig. 5.20 we show the timing shift per collision for exponential tapering of the dispersion, with relation to L_a/L_{per}, where L_{per} is the filter periodicity. The solid line refers to the case of 50-km periodic filter spacing for which $L_{\text{per}} = 50\,\text{km}$, and the dashed line to the case of 33.3 and 66.6 filter spacing for which the overall

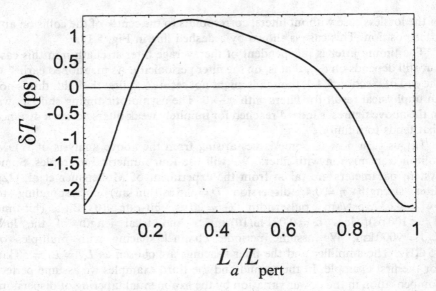

FIGURE 5.19 Timing shift per collision in picoseconds versus the phase of the collision, for constant dispersion.

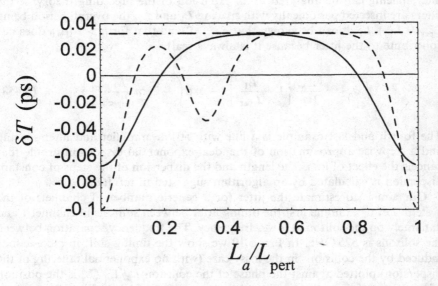

FIGURE 5.20 Timing shift per collision in picoseconds versus the phase of the collision, for exponential tapering of the dispersion. Solid line is for equal filter spacing, $L_{pert} = L_f = 50$ km; dashed line is for $L_{pert} = 100$ km, and filter spacing of 33.3 and 66.6 km periodically repeated. Dot-dashed line is the constant value of timing shift with exponential tapering and no filters.

periodicity is $L_{per} = 100$ km. As expected, the timing shift is much smaller than in the case with constant dispersion. When the center of the collision is far from the position of the filters, since the collision length is in our case only 12 km, the timing shift per collision is close to the value without filters, which is $\delta t = 0.032$ ps with our parameters. The value of timing shift with no filters is reported by the dot-dashed line in the figure.

In Fig. 5.21 we plot the timing shift in the lossy case, using 50-km amplifier and filter spacing and a stepwise approximation of the exponential profile of the dispersion that cancels the effect of loss, against the collision phase. The solid line refers to an approximation with 3 steps, the dashed with 4, and the dot-dashed with 100. The approximation with 100 steps is obviously indistinguishable from the solid line of Fig. 5.20 which refers to the ideal case of exponential tapering under the same conditions. Likely the algorithm chosen is not the best for the optimization of the dispersion that will reduce the effect of collision in filtered systems, as shown by the large timing shift for large values of the collision phase.

Timing jitter induced by collision in the filtered case is significantly different from that induced by collisions without filters. We have seen that without filters, the need to avoid a significant frequency shift after collision requires that the collision length be large enough that the effect of the amplifier

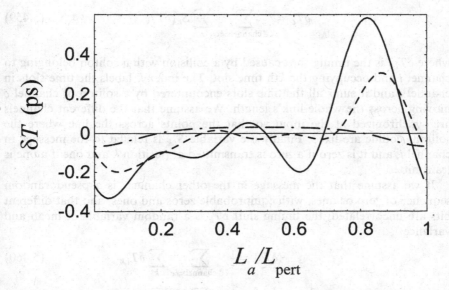

FIGURE 5.21 Timing shift per collision in picoseconds versus the phase of the collision, for stepwise approximation of the optimum exponential tapering of the dispersion. Solid line is a 3-step approximation, dashed a 4-step approximation, dot-dashed a 100-step approximation. The length of the steps and the values of the dispersion were chosen by the algorithm of ref. [68].

periodicity is averaged out. Even if this condition is met, however, what is left is the timing jitter induced by collisions in the lossless case, which depends only on channel spacing, and the effect on the two colliding solitons is always the slowing down of the slower and the advance of the faster soliton. With filters the amount of timing shift per collision and its effect depend on the phase of the collision; that is to say, on the position of the amplifier sequence with respect to the center of the collision. Without filters the timing jitter is proportional to the inverse of the frequency spacing squared; therefore it sets a limit to the minimum channel spacing of unfiltered systems. With filters, as we will see below, timing jitter sets instead a limit to the maximum number of channels that can be used. Finally in multichannel filtered systems the number of collisions is much larger than in unfiltered systems because of the larger number of channels and the longer transmission distance of filtered systems. Therefore our analysis of timing jitter in filtered systems follows a different line than that of unfiltered systems, where we could proceed with only an estimate of the maximum possible jitter induced by collisions. Here we will study the statistical properties of the timing jitter based on the assumption that the number of collisions is large.

In a WDM soliton system the total timing shift of a given soliton depends on the particular soliton sequence in the other channels. The total timing shift of a particular soliton belonging to channel c is

$$\delta T_c = \sum_{i \in \text{channels} \neq c} \sum_k S_{i,k} \delta T_{i,k} \tag{5.459}$$

where $\delta T_{i,k}$ is the timing shift caused by a collision with a soliton belonging to channel i and occupying the kth time slot. The index k labels the time slots in channel i and it sums all the time slots encountered by a soliton of channel c moving across the whole link's length. We assume that the different channels are synchronized at the input so that the points across the line where the solitons collide are fixed. Finally the variable $S_{i,k}$ is related to the message in channel i, and it is zero if a zero is transmitted at position k and one if a one is transmitted.

If we assume that the message in the other channels is a pseudorandom sequence of zero or ones, with equiprobable zeros and ones, and that different bits are uncorrelated; the timing shift δT_c is a random variable of mean and variance

$$\langle \delta T_c \rangle = \tfrac{1}{2} \sum_{i \in \text{channels} \neq c} \sum_k \delta T_{i,k} \tag{5.460}$$

$$\langle \delta T_c^2 \rangle - \langle \delta T_c \rangle^2 = \tfrac{1}{4} \sum_{i \in \text{channels} \neq c} \sum_k \delta T_{i,k}^2 \tag{5.461}$$

Because of the large number of collisions which, as we will see below, easily exceeds 100 in a multichannel system over transoceanic distances, the total timing shift is the sum of many independent random variable. Therefore, by the central limit theorem, its distribution is well approximated by a gaussian. This is the case if one assumes that the timing shift per collision is constant and equal to δT. Assuming a probability one-half that the time slot is empty and no collision occurs, after $N = 200$ potential collisions (100 collisions on average) the probability distribution of the timing of a soliton initially at $T = 0$ is

$$P(T) = \frac{1}{2^N} \sum_{i=0}^{N} \binom{N}{i} \delta(T - i\delta T) \tag{5.462}$$

This probability distribution is well approximated by a gaussian centered at $N\delta T/2$ and of variance $N\delta T^2/4$. For instance, the gaussian approximation gives 5.97×10^{-10} for the probability that $T > 143\delta T$, while the exact value obtained by equation (5.462) is 5.15×10^{-10}.

For a given system and a given channel, the variance given by (5.461) characterizes the collision-induced timing jitter. It is independent of the transmitted messages provided that the ones and zeros are equiprobable, and it depends just on the relative delays of the different channels at the input. The variance depends, in general, on the phases of the collisions, or rather on the location of the center of each collision with respect to the amplifier sequence. An estimate of the variance of the timing jitter induced by collisions can be written as

$$\sum_{k} \delta T_{i,k} = N_{c,i} \sum_{k} \frac{\delta T_{i,k}}{N_{c,i}} \tag{5.463}$$

$$\sum_{k} \delta T_{i,k}^2 = N_{c,i} \sum_{k} \frac{\delta T_{i,k}^2}{N_{c,i}} \tag{5.464}$$

where $N_{c,i}$ is the total number of potential collisions between channel c and i (the number of collisions if all ones are transmitted in channel i). In practice, it is very unlikely that the minimum distance between two collisions is a multiple of the amplifier spacing, which outside the laboratory is never exactly periodic. One can then assume that each collision occurs with a random phase with uniform distribution over 2π, and that the phase of each collision is independent of the others. Thus the sum appearing at right-hand side of equations (5.463) and (5.464) can be approximated by averages evaluated over a uniform distribution of the collision phase. The timing shift averaged over the amplifier or filter periodicity is zero, such as it might be obtained by averaging equations (5.422) and (5.446) over the phases $\phi_a(n)$,

$$\sum_{k} \frac{\delta T_{i,k}}{N_{c,i}} \approx \overline{\delta T(2\Omega_{c,i})} = 0 \tag{5.465}$$

where with the overline we have indicated the averaging over the collision phase. Analogously, the variance of the timing shift can be obtained by averaging (5.422),

$$\sum_k \frac{\delta T_{i,k}^2}{N_{c,i}} \approx \overline{\delta T (2\Omega_{c,i})^2} \tag{5.466}$$

$$= \frac{128 \zeta_{L_A}^2}{\pi^6} \sum_{n=1}^{\infty} \sum_{n=1}^{\infty} \left| \frac{\zeta_{L_A} a_{n,1}}{1 - \exp(-2\eta \zeta_{L_A}/3)} - a_{n,2} \right|^2$$

$$\times \frac{n^6 y (2\Omega_{c,i})^8}{\sinh^4 [ny(2\Omega_{c,i})]}, \tag{5.467}$$

where we have explicitly indicated the dependence of $y = \pi^2/(2\Omega_{c,i}\zeta_{L_A})$ on the dimensionless angular frequency separation between the two channels $2\Omega_{c,i}$. The number of potential collisions of a soliton propagating across the whole link length ζ_{tot}, with solitons belonging to a channel whose frequency is spaced apart $2\Omega_{c,i}$ is

$$N_{c,i}(2\Omega_{c,i}) = \frac{2\Omega_{c,i}\zeta_{\text{coll}}}{T} \tag{5.468}$$

where T is the bit length T_B in soliton units. By (5.464) we then get

$$\sum_k \delta T_{i,k}^2 = N_{c,i}(2\Omega_{c,i}) \overline{\delta T (2\Omega_{c,i})^2} \tag{5.469}$$

Therefore the variance of the timing fluctuations induced on solitons of a given channel c by all interchannel collisions is

$$\langle \delta T_{c,\text{tot}}^2 \rangle = \frac{1}{4} \sum_{i \in \text{channels} \neq c} N_{c,i}(2\Omega_{c,i}) \overline{\delta T (2\Omega_{c,i})^2} \tag{5.470}$$

An estimate of the root-mean-squared value of the timing shift for a soliton is then

$$\delta T_c = \langle \delta T_{c,\text{tot}}^2 \rangle^{1/2} \tag{5.471}$$

Note that by (5.471), (5.470), and (5.468), the timing jitter induced by collisions is proportional to $\sqrt{L_{\text{tot}}}$.

In Fig. 5.22 we show the average number of collisions for the channel at lower frequency in a WDM system with N channel plotted against N, calculated by

$$N_{\text{coll}} = \sum_{i=2}^{N} \frac{N_{1,i}}{2} = \frac{N(N-1)(2\Omega_{1,2})\zeta_{\text{tot}}}{4T} \tag{5.472}$$

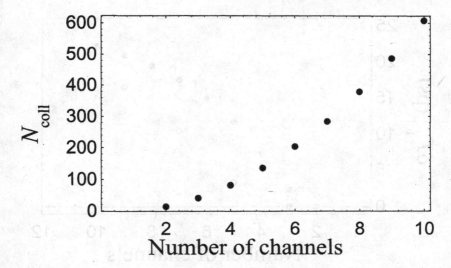

FIGURE 5.22 Average number of collisions for the channel at lower frequency of the WDM spectrum versus the total number of channels.

where $2\Omega_{1,2}$ is the normalized angular frequency separation of two adjacent channels. As we anticipated above, already with 5 channels the average number of collisions exceeds 100.

In Fig. 5.23 we show the timing jitter caused by collisions in picoseconds plotted against the number of channels for the case of constant dispersion. The points refer to the channels at the highest or lowest frequencies with all channels on one side of the spectrum and circles to the central channel of the WDM comb. The timing jitter is higher for the channels at the border of the WDM comb because solitons belonging to these channels experience on average the largest number of collisions. Notice that the jitter of the central channel of a comb of $2N + 1$ is $\sqrt{2}$ times larger than the jitter of the channel at the border in a WDM comb with $N + 1$ channels, and it is usually smaller than the jitter of the channels at the border of the WDM comb for the same number of channels.

Since it is independent from other sources of jitter, the timing jitter induced by collisions adds to the jitter coming, for instance, from the Gordon-Haus effect by

$$\delta T_{\text{tot}} = [\langle \delta T_{c,\text{tot}}^2 \rangle + \langle \delta T_{\text{GH}}^2 \rangle]^{1/2} \tag{5.473}$$

Anticipating what we will discuss in some detail in the Section 5.5, if we use a gaussian distribution for the jitter and if the only source of jitter are collisions, one error occurs when a soliton moves out of a detection window of amplitude $2t_w$. The error probability is then

$$E_{\text{coll}} = \text{erfc}\left[\frac{t_w}{\sqrt{2}\delta T_c}\right] \tag{5.474}$$

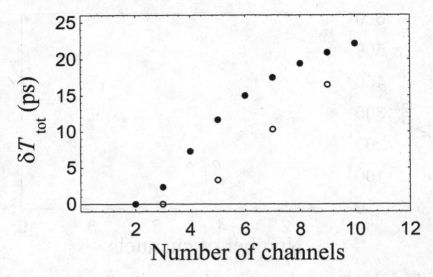

FIGURE 5.23 Dots are for the channels at the longer or shorter wavelength; circles for the channel at the central frequency of the WDM comb.

The acceptance window $2t_w$ usually ranges from 2/3 to 4/5 of the bit-time T_B. Since $\mathrm{erfc}[x/\sqrt{2}] > 10^{-9}$ for $x < 6.10941$, if we assume that $2t_w = (2/3)T_B = 66.6\,\mathrm{ps}$, we have E_{coll} becoming larger than 10^{-9} for $\delta T_c > 5.45\,\mathrm{ps}$, or if we assume that $2t_w = (4/5)T_B = 80\,\mathrm{ps}$, we have $\delta T_c > 6.55\,\mathrm{ps}$. For the channels at the extremity, the collision-induced jitter is 0.015, 2.3, and 7.3 ps for 2, 3, and 4 channels, respectively. It is then impossible, with a constant value of dispersion, to obtain an error probability less than 10^{-9} with more than 3 channels, and even with 3 channels the simultaneous presence of other sources of jitter can make it difficult to achieve this goal.

In Fig. 5.24 we show the timing jitter caused by collisions in picoseconds against the number of channels for the channel at the border of the WDM comb and with dispersion exponentially tapered. The points refer to the case of one filter every 50 km, and the circles to filters every 33.3 and 66.6 km (two out of three amplifiers, with 33 km of amplifier spacing). In both cases, timing jitter is considerably smaller than for constant dispersion. In Fig. 5.25 we show by circles the timing jitter estimated for a three-step approximation of the optimum exponential profile of dipersion, and by dots the timing jitter estimated for a four-step approximation. The length of the steps and the values of the dispersion have been chosen by the algorithm of ref. [68]. Although the dispersion map suggested, for unfiltered systems, in ref. [68] is probably not the optimal for minimizing timing jitter in filtered systems, one might note that with a three-step approximation the timing jitter is significantly larger than that obtained with the exponential profile. A careful optimization of the dispersion

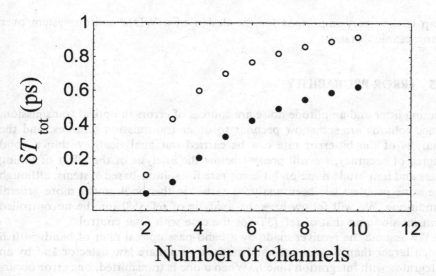

FIGURE 5.24 Root mean square timing jitter in picoseconds versus the number of channels, for exponential tapering of dispersion and for the channels at shorter or longer wavelength. Dots are for equal filters spacing, $L_{pert} = L_f = 50\,km$. Circles are for $L_{pert} = 100\,km$, and filters spacing of 33.3 and 66.6 km periodically repeated.

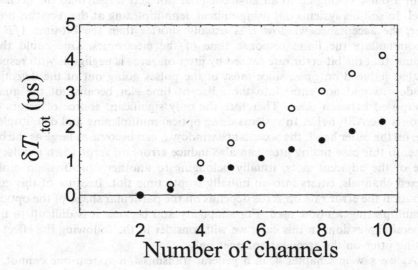

FIGURE 5.25 Root means square timing jitter in picoseconds versus the number of channels, for a stepwise approximation of the optimum exponential dispersion tapering and for the channels at shorter or longer wavelength. Circles are for a 3-step approximation; dots for a 4-step approximation. The length of the steps and the values of the dispersion have been chosen by the algorithm of reference [68].

map is then very important for the design of a WDM soliton system over transoceanic distances.

5.5 ERROR PROBABILITY

Timing jitter and amplitude noise are sources of errors in optical transmission. Since solitons are somehow peculiar to other transmission schemes, and the analysis of the bit-error rate can be carried out analytically within a good degree of accuracy, we will present below the analysis of the effect of timing jitter and amplitude noise on bit-error rate for soliton-based systems, although the same problem has been analyzed earlier in this book under more general conditions. We will follow here the analysis of ref. [58] for the uncontrolled transmission and that of ref. [37] for the case with filter control.

We assume the receiver made by a band-pass optical filter of bandwidth B much larger than the soliton bandwidth, by a square law detector and by an integrator with integration time t_i. When a one is transmitted, one error occurs when the detected power gets below the decision level. This may occur because of energy fluctuations of the transmitted pulse or because the pulse moves out of the acceptance window, which we assume has amplitude $t_i = 2t_w$. If a zero is transmitted, one error occurs when the detected ASE noise power integrated over the detection window overcomes the decision level, or the power leaking from a pulse belonging to an adjacent time slot gets larger than the decision level. In soliton systems not using optical demultiplexing at the receiver, however, the acceptance window t_i is usually shorter than the bit-time $1/B$ to accommodate the finite response time of the electronics. One could then assume that the bit-error rate caused by jitter on zeros is negligible with respect to that induced on ones, since most of the pulses going out of the detection window would not enter into the adjacent time slot because of the guard interposed between them. Therefore the only significant source of errors on zeros is the ASE noise. In systems using optical multiplexing and demultiplexing, on the other hand, the acceptance window t_i can become as large as the bit-time. In this case timing jitter can also induce errors on zeros when a pulse of one of the adjacent slots, usually belonging to another time-division multiplexed channels, enters into an initially empty time slot. Because of this contribution the error rate on zeros depends on the particular shape of the optical-demultiplexing window used. For simplicity and because it is difficult to find general expressions in this case, we will consider in the following the effect of timing jitter on the error-rate of ones only.

As we saw in Chapter 4, in a general transmission system one cannot, in general, decouple timing jitter and energy fluctuations in calculating the bit-error rate. Both sources must be considered simultaneously. At a fixed level the timing jitter depends on the amplitude of the pulse, is larger (smaller) for pulses undertaking a positive (negative) amplitude fluctuation. This couples the effect on the error probability of timing and amplitude fluctuations.

If we define the normalized window time $\tau_w = t_w/T_s$, the power of a pulse of amplitude A and timing T integrated over the detection window will be proportional to

$$E_{\text{det}} = \int_{-\tau_w}^{\tau_w} A^2 \, \text{sech}^2[A(t - T)]$$

$$= A \tanh[A(\tau_w - T)] + A \tanh[A(\tau_w + T)] \qquad (5.475)$$

Without loss of generality, we can always assume that the normalized amplitude of the pulse is $A = A_0 + \delta A$ (the average normalized amplitude is A_0) and that the timing is $T = \delta T$ (the average timing is zero). We assume, as above, that δA and δT have zero mean. Since both variables are obtained by the solution of a linear stochastic differential equation, their statistics is gaussian. The energy noise, however, is also affected by the ASE power background. With filtered systems, we will see below that by a good degree of approximation the ASE noise of the background gives a negligible contribution; therefore the amplitude fluctuations of a soliton are representative of the energy fluctuations of the detected energy being transmitted. This is intuitive because for sliding filters the ASE noise is kept small and with fixed filters this approximation fails only very close to the critical distance above which the ASE power becomes untolerably high. The inclusion of a more accurate probability distribution for the amplitude fluctuations would not, however, change the general lines of the discussion below; it would only make it computationally heavier.

Neglecting the ASE power, the average detected energy for large detection window is $\langle E_{\text{det}} \rangle = 2A_0$. An error occurs when the detected energy is less than the decision level $E_{\text{dec}} = 2A_{\text{dec}}$:

$$(A_0 + \delta A)\tanh[(A_0 + \delta A)(\tau_w - \delta T)]$$

$$+ (A_0 + \delta A)\tanh[(A_0 + \delta A)(\tau_w + \delta T)] < 2A_{\text{dec}} \qquad (5.476)$$

Equation (5.476) can be inverted, obtaining

$$|\delta T| > \mathscr{F}(A_0 + \delta A) \qquad (5.477)$$

where

$$\mathscr{F}(A) = \frac{1}{A} \text{arccosh}\left\{ \left[\frac{A}{2A_{\text{dec}}} \sinh(2A\tau_w) - \sinh^2(A\tau_w) \right]^{1/2} \right\} \qquad (5.478)$$

for $A \geq A_{\text{low}}$, where A_{low} is the lowest value of the amplitude that still gives a faithful detection of a one for $\delta T = 0$, which is for a soliton at the center of the time slot. The lowest value of the amplitude is the solution of the transcendental equation $\tanh(A_{\text{low}}\tau_w) = A_{\text{low}}/A_{\text{dec}}$. The region where conditions

(5.477)–(5.478) are met is the region outside the "safe" region labeled with S in Fig. 5.26. The values of the parameters used for the figure are $A_0 = 1$, $A_{\text{dec}} = \frac{1}{2}$, $\tau_w = 5.877$. The value of τ_w corresponds to a pulsewidth (full width at half power) 10 times the bit-time and to a detection window $2t_w$ of $\frac{2}{3}$ the bit-time. These values correspond to those used in the original Gordon-Haus paper [8]. The detection window is not extended to the total bit-period to accommodate for the finite response time of the electronics. In today's experiments, however, there is the tendency to transmit solitons more closely spaced, down to four times their full width at half maximum. Furthermore most of them uses optical time-domain multiplexing and demultiplexing, which allows the widening of the decision window and therefore a significant reduction of the system's sensitivity to the timing jitter. Likely the most practical method for optical demultiplexing employs electroabsorption modulators, as first proposed and demonstrated by the KDD group in Japan [59, 60]. The use of electroabsorption modulators permits the widening of the decision window to about $\frac{4}{5}$ of the bit-time at 10 Gbit/s [22]. Although electroabsorption modulators do not have the square response that we assume [59], the discussion below qualitatively applies also to this case.

As we mentioned earlier, we assume gaussian distributions for δA and δT. The error probability is then

$$P_{\text{err}} = 1 - \int_S d\delta A d\delta T \, \frac{1}{2\pi\sqrt{\langle \delta A^2 \rangle \langle \delta T^2 \rangle}}$$

$$\times \exp\left[-\frac{\delta A^2}{2\langle \delta A^2 \rangle} - \frac{\delta T^2}{2\langle \delta T^2 \rangle}\right] \tag{5.479}$$

$$= \text{erfc}\left[-\frac{\delta A_{\text{low}}}{\sqrt{2\langle \delta A^2 \rangle}}\right] + \int_{\delta A_{\text{low}}}^{\infty} d\delta A \, \frac{1}{\sqrt{2\pi\langle \delta A^2 \rangle}}$$

$$\times \exp\left[-\frac{\delta A^2}{2\langle \delta A^2 \rangle}\right] \text{erfc}\left[\frac{\mathcal{F}(\delta A)}{\sqrt{2\langle \delta T^2 \rangle}}\right] \tag{5.480}$$

where erfc is the complementary error function

$$\text{erfc}(x) = 1 - \frac{2}{\sqrt{\pi}} \int_0^x \exp(-x^2) \, dt \tag{5.481}$$

and $\delta A_{\text{low}} = A_{\text{low}} - A_0$. With $A_{\text{dec}} = \frac{1}{2}$ and $\tau_w = 5.877$, we have $\delta A_{\text{low}} = -0.4973$. To giver a numerical example, assume that the variance of the amplitude fluctuations and the variance of the timing fluctuations are such that each one alone would have given 10^{-9} error probability. This corresponds, in our numerical example, to $\langle \delta A^2 \rangle = 6.87403 \, 10^{-3}$ (note that the probability density function of the δA should be integrated down to $-\delta A_{\text{low}}$) and

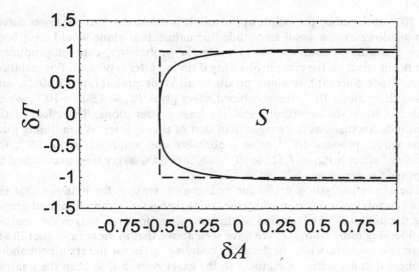

FIGURE 5.26 Region S in the δA–δT plane where a mark is transmitted faithfully, for $A_0 = 1$, $A_{\text{dec}} = \frac{1}{2}$ and $\tau_\omega = 5.877$.

$\langle \delta T^2 \rangle = 0.925259$. Applying equation (5.480), we find that the total error probability is $P_{\text{err}} = 2.542 \ 10^{-9}$. This value is slightly higher than that we would have obtained by the simple approach, assuming that one error occurs when either the timing jitter or the amplitude noise gives errors. With this approach the error probability is the sum of the error probabilities caused by the two processes, from which we subtract the product of the two error probablities to account for those processes, which are counted twice, in which both amplitude and timing jitter give error. We obtain $P_{\text{err}} = 2 \ 10^{-9}$. The simple approach and the exact one gives different results, though not too different. Why this occurs is once again understood with the help of Fig. 5.26. A simple analysis evaluates the error probability by integrating the joint distribution of energy and timing fluctuations outside the dashed contour of the figure. For positive amplitude fluctuations, the dashed contour includes narrow regions that do not belong to S. Conversely, for negative amplitude fluctuations, there are wider regions not included in the dashed contour that instead contribute to the error probability. Those wider regions, however, are at the tails of both energy and timing fluctuations, this explaining the small difference between the exact and approximate analysis.

If the solitons are more closely spaced the difference between the simple analysis and the exact one becomes more significant. Consider the same case as above but with $\tau_w = 2.938$, which corresponds to pulses spaced 5 pulse-widths at half power and detection window still $\frac{2}{3}$ of the bit-time. Now the total error probability is $P_{\text{err}} = 5.175 \ 10^{-9}$, which should be compared with

2×10^{-9} obtained by the simple approach. It is noticeable that for lower detection windows even a small amplitude fluctuation that alone would have been given a negligible contribution to the error probability, can yet produce a significant effect on the error probability if timing jitter is present. For instance, if amplitude fluctuations alone produces an error probability of 10^{-12}, and timing jitter alone 10^{-9}, the combined effect gives $P_{err} = 1.908 \times 10^{-9}$, which is almost twice the contribution of the timing jitter alone. The effect of the amplitude fluctuations is stronger than that of timing jitter: When timing jitter alone would produce 10^{-12} error probability and amplitude noise 10^{-9}, the combined effect is $P_{err} = 1.522 \times 10^{-9}$, which is 50% larger than that caused by amplitude noise alone.

The above analysis is useful for pedagogical reasons for it shows that the error probability is determined by the coupled effect of timing jitter and amplitude fluctuations. The inclusion of this coupling, however, makes the analysis considerably more complex. We have seen above that in most cases (not in all) the simple analysis, which neglects the coupling, gives for the error probability numbers that are within a factor 2 to the exact ones. So, to keep the analysis simple, we will consider below energy fluctuations and timing jitter separately as it is usually done. The error probability is simply given by the sum of the two error probabilities (less their product if that is not negligible). This approach greatly helps in extracting the error probability from computer simulations, because it permits to clearly distinguish between two different source of errors. The error probabilities due energy fluctuations and timing jitter is estimated by suitable statistical methods, and the total error probabilities is calculated by simply adding at the end the error probability of the two processes. This procedure gives for the error probability numbers that should be, at least, of the same order of the exact ones. One is, however, advised that a more precise calculation of the error probabilities should follow the more accurate derivation outlined above.

We will therefore study below energy fluctuations and timing jitter separately. Let us analyze timing jitter first.

5.5.1   Timing Jitter and the Gordon-Haus Limit

Timing jitter, within the approximated analysis described above, produces errors only when one is transmitted. If we assume that the decision level is one-half the average energy of the pulse, an error occurs when the transmitted pulse moves out of the detection window of amplitude $t_i = 2t_w$. Without filters, and with fixed and sliding filters, the equations describing the evolution of the observables of an isolated soliton are well approximated by a linear set of stochastic differential equations. If the initial values for the variables are deterministic, it is well known that the probability distribution of quantities described by a linear set of stochastic differential equations is gaussian. For interacting solitons of the same phase, the presence of an unstable attracting potential of exponential shape makes the system of equation nonlinear, and the probability distribution of the variables is not, in general, gaussian [61]. In the following, we will neglect the small nongaussian corrections to the gaussian statistics.

The error probability induced by timing jitter is, assuming a gaussian distribution,

$$P_{\text{err}} = \text{erfc}\left[\frac{\tau_w}{\sqrt{2\langle\delta T^2\rangle}}\right] = \text{erfc}\left[\frac{t_w}{\sqrt{2\langle\delta t^2\rangle}}\right] \tag{5.482}$$

This equation holds for uncontrolled and controlled transmission, provided that to the variance of the timing fluctuations is given the pertinent expression, and permits to evaluate the error probability caused by timing jitter alone. For uncontrolled transmission, for instance, one gets form (5.117), neglecting the direct coupling with the timing fluctuations (the second term at right side)

$$\frac{t_w^2}{\langle\delta t^2\rangle} = \frac{9}{\hbar\omega_0\gamma F\alpha f|\beta_2|}\frac{T_s t_w^2}{z^3} \tag{5.483}$$

If we substitute $\gamma = n_2\omega_0/(cA_{\text{eff}})$, $|\beta_2| = \lambda/\omega_0 D$ and $T_s = T_F/1.76275$, we obtain

$$\frac{t_w^2}{\langle\delta t^2\rangle} = \frac{5.10567 A_{\text{eff}} T_F t_w^2}{h n_2 DF\alpha f z^3} \tag{5.484}$$

For an error probability less than 10^{-9}, for instance, one should have $t_w^2/\langle\delta t^2\rangle \geq 37.3249$. Multiplying (5.484) by the bit-rate R, we find that the maximum distance that can be reached is then

$$(Rz_{GH})^3 = \frac{0.13679 A_{\text{eff}} t_w^2 T_F R^3}{Dn_2 hF\alpha f} \tag{5.485}$$

where $\hbar$ is the Planck constant, $2\pi\hbar$. The full width at half power of the pulse is usually a fraction of the bit-time, set by the requirement that the soliton-soliton interaction be weak. The product $T_F R$ is then constant (ranging between 4 and 10 in practical use depending on the type of control used). Also we have already mentioned (see the discussion after equation 5.478) that the detection window $2t_w$ is a fraction, ranging from 2/3 to 4/5, of the bit-time. Therefore also $t_w R$ is constant. For the above reasons one can immediately see that the Gordon-Haus effect sets a limit to Rz_{GH} which is roughly independent of the bit-rate. The Gordon-Haus effect therefore does not set a limit to the transmission distance z_{GH} or to the bit-rate R separately but to their product. To have an idea of the magnitude of the Gordon-Haus limit, consider an amplifier spacing of $26\,\text{km}$, $n_2 = 2.8 \times 10^{-20}\,\text{W/m}^2$, $A_{\text{eff}} = 50\,\mu\text{m}^2$, fiber loss $0.21\,\text{dB/km}$ corresponding to $\alpha = 0.0484\,\text{km}^{-1}$, $F = 1.6$, fiber dispersion $D = 0.6 \, \text{ps}/(\text{nm km})$, $t_w R = 1/3$, $T_F R = 1/10$. We obtain $R z_{GH} = 42.63 \, \text{GHz Mm}$. This means that for a bit-rate of 10 Gbit/s, the maximum distance before the error rate caused by the Gordon-Haus effect exceeds 10^{-9} is 4263 km.

One can immediately see from (5.484) how the Gordon-Haus limit can be extended. The fiber loss α and the fiber nonlinear coefficient n_2 cannot be modified. The added ASE noise per unit distance can be reduced in two ways. One is by reducing the amplifier spacing L_A. This brings f close to the ideal value for distributed amplification $f = 1$. The second is by inverting the amplifier active medium as much as possible, such that F gets as close as possible to the minimum value, $F = 1$, obtained for perfect inversion. The third is by reducing the group-velocity dispersion of the line. The Gordon-Haus limit is infact inversely proportional to dispersion. This makes very good sense because the fiber dispersion introduces the coupling between the frequency noise and the soliton velocity at the origin of the Gordon-Haus effect. Dispersion cannot, however, be reduced at will, since, if the dispersion is reduced, the pulse energy also proportionally decreases, equation (5.46), and this reduces the optical signal-to-noise ratio and ultimately produces the increase of the bit-error rate from the energy fluctuations side. The product $R \times z_{GH}$ can also be increased by using fibers with a large effective area [69]. A large effective area means a low nonlinear coefficient γ. For the lowest soliton energy compatible with a good signal-to-noise ratio at the receiver, a low nonlinearity permits the use of fibers with lower values of dispersion. Finally the most promising method to increase the Gordon-Haus limit, which will be that most likely to be used in future applications, is to reduce D while keeping the intensity of the solitary pulse high enough to have a good signal-to-noise ratio. This goal can be accomplished by using suitable dispersion management of the transmission line [70].

With filter control the modified Gordon-Haus can be readily found by using equation (5.151):

$$\frac{t_w^2}{\langle \delta t^2 \rangle} = \frac{4|\beta_2|t_w^2}{3T_0^3 \hbar \omega_0 F \alpha \gamma f z} \left(\frac{F_r t_d^2}{4L_f |\beta_2|} \right)^2 \tag{5.486}$$

which in practical units becomes

$$\frac{t_w^2}{\langle \delta t^2 \rangle} = \frac{7.30313 A_{\text{eff}} t_w^2}{a_1 T_F^2 h n_2 D F \alpha f z} \left(\frac{F_r t_d^2}{4L_f} \right)^2 \tag{5.487}$$

We have added in (5.487) a_1, defined by (5.278), to account for possible filter sliding. Once again, for error probability less than 10^{-9}, we should have $t_w^2/\langle \delta t^2 \rangle \geq 37.3249$. The modified Gordon-Haus limit is then

$$R z_{GH_f} = \left(\frac{F_r t_d^2}{4L_f T_F^2} \right)^2 \frac{0.195664 R^3 t_w^2 T_F A_{\text{eff}}}{R^2 a_1 n_2 D h F \alpha f} \tag{5.488}$$

With filtering control it is more difficult to find the optimal conditions for transmission. Without sliding, the excess gain produces an increase of the ASE noise proportional to $\exp(\delta g z)$, with δg given by (5.128) which, translated into practical units, gives

$$\delta g = 1.03576 \frac{F_r t_d^2}{4 L_f T_F^2} \tag{5.489}$$

for étalon filters. The excess gain is scaled by n for Butterworth-like filters of order n. Therefore the factor between brackets in (5.488) is set by the ASE noise growth. It is evident then that fixed filters are well suited to relatively low single-channel bit-rates, possibly combined with WDM, due to the factor R^2 in the denominator of (5.488). How low the bit-rate is, is determined by the type of filter used; it is higher for higher-order filters that are able to keep the ASE growth to a moderate level.

If sliding filters are adopted, since the ASE growth is eliminated by sliding, it might seem that increasing the filter strength $F_r t_d^2 / (4 L_f)$ would indefinitely increase the Gordon-Haus limit. Increasing the filter strength, however, translates into an increase of ASE power, which must be balanced by an increase of the sliding rate. The raised sliding rate produces a decrease of the Gordon-Haus limit through the increase of a_1, which also appears in denominator (5.488).

5.5.2 Energy Fluctuations

The fluctuations of the detected energy at the receiver depends on whether or not control filters are used along the line. Let us consider first the case of uncontrolled transmission.

Assume that there is no significant interaction between the optical field within the bandwidth B and time interval t_i and the field outside. The field contained in the bandwidth B and within time interval t_i can be described by a finite number $m = 2Bt_i$ of parameters, as dictated by the sampling theorem. We have restricted our phase space, which in principle has an infinite number of dimensions, to a finite-dimensional space. The frequency resolution is, by our assumption, $1/t_i$. Assume now that m is much larger than unity. Let us distinguish the two cases in which a zero or a one are transmitted.

When there is no signal, that is, a zero is transmitted, the only detected field is the ASE noise. The ASE noise belonging to the different frequency intervals of amplitude equal to our frequency resolutions $1/t_i$ is uncorrelated, and its intensity is low enough that propagation in the presence of Kerr nonlinearity does not induce significant correlation among them. The statistical properties of the detected signal x at the output of the integrator are those of the sum of $2m$ independent variables of gaussian statistics. Each of the $2m$ random variables has variance equal to the intensity of the ASE radiation, in normalized units, per unit bandwidth per unit time, $n_{sp}/2$, where

$$n_{sp} = \langle n \rangle \zeta \tag{5.490}$$

In the expression above the $\langle n \rangle$ is given by equation (5.35). If we normalize the photon number at the output of the integrator to n_{sp}, by defining

$$X_0 = \frac{n}{n_{sp}} \tag{5.491}$$

the variable at the output of the integrator is written as

$$X_0 = \sum_{j=1}^{2m} x_i^2 \tag{5.492}$$

with the x_i gaussian variable of variance $\frac{1}{2}$. Let us define the generating function of a random variable y with probability density function $P(y)$ as

$$G_y(s) = \int_{-\infty}^{\infty} dy \, \exp(-sy) P(y) \tag{5.493}$$

The generating function of the square of a gaussian variable of variance $\frac{1}{2}$ is

$$G_{x_j} = \frac{1}{(1+s)^{1/2}}, \tag{5.494}$$

The generating function of the sum of independent random variables is the product of the generating functions, so the generating function of X_0 is

$$G_{X_0}(s) = \frac{1}{(1+s)^m} \tag{5.495}$$

By inverse Laplace transformation, we get the probability density function of X_0 as

$$P_0(X_0) = \frac{X_0^{m-1}}{\gamma(m)} \exp(-X_0) \tag{5.496}$$

where $\Gamma(m)$ is the Euler gamma function.

Let us assume now that a one is transmitted. Then, within the linearization approximation in which the noise of the continuum is orthogonal to the soliton, the soliton is simply added to the noise. The noise for a zero can be interpreted as a multidimensional gaussian noise with $2m$ degrees of freedoms. Its statistical distribution is centered at the origin of a $2m$-dimensional space. When a soliton is added, the center of mass of the distribution is simply shifted by an amount equal to the signal which, in units normalized to n_{sp}, is

$$S_1^{1/2} = \left[\frac{E}{n_{sp}} \right]^{1/2} \tag{5.497}$$

where E is the energy of the soliton in the particular units that we consider (in normalized units, $E = 2A$). The reference frame of the $2m$-dimentional phase space can be rotated at will. It is convenient to choose one of the axes, say the one labeled with $2m$, parallel to the direction of the field representing the soliton so that of the $2m$ variables only that one labeled with $2m$ has an average different from zero. The field representing a one can then be written as

$$X_1 = \sum_{j=1}^{2m-1} x_j^2 + (x_{2m} + S_1^{1/2})^2 \tag{5.498}$$

Again the generating function is the product of the generating functions of the terms of the sum. Since the generating function of the square of a gaussian variable centered in $S_1^{1/2}$ is

$$G_{x_{2m}} = \frac{1}{(1+s)^{1/2}} \exp\left(-S_1 + \frac{S_1}{1+s}\right) \tag{5.499}$$

we get

$$G_{X_1}(s) = \frac{1}{(1+s)^m} \exp\left(-S_1 + \frac{S_1}{1+s}\right) \tag{5.500}$$

The inverse Laplace transform is performed by expanding $\exp[S_1/(1+s)]$ into a series. The result is

$$P_1(X_1) = \exp(-S_1 - X_1)\left(\frac{X_1}{S_1}\right)^{(m-1)/2} I_{m-1}(2\sqrt{S_1 X_1}) \tag{5.501}$$

where I_m is the mth-order Bessel function of imaginary argument. Since the noise is usually much lower than the signal, we can assume that $S_1 \gg 1$. Also we are interested into values of X_1 of the same order of S_1. For these reasons the asymptotic expansion of the Bessel function for high values of the argument is justified,

$$I_{m-1}(z) \sim \frac{1}{\sqrt{2\pi z}} \exp(z) \tag{5.502}$$

and (5.501) becomes

$$P_1(X_1) = \exp[-\sqrt{S_1} - \sqrt{X_1})^2]\left(\frac{\sqrt{X_1}}{\sqrt{S_1}}\right)^{m-1} \frac{1}{2\sqrt{\pi}(\sqrt{S_1}\sqrt{X_1})^{1/2}} \tag{5.503}$$

The above equation can be conveniently written in terms of the probability distribution of $P_1(\sqrt{X_1})$ by transformation

$$P_1(\sqrt{X_1})d\sqrt{X_1} = P_1(X_1)dX_1 = 2\sqrt{X_1}P_1(X_1)d\sqrt{X_1} \tag{5.504}$$

We finally obtain [58]

$$P_1(\sqrt{X_1}) = \frac{1}{\sqrt{\pi}} \exp[-(\sqrt{S_1} - \sqrt{X_1})^2] \left(\frac{\sqrt{X_1}}{\sqrt{S_1}}\right)^{m-1/2} \qquad (5.505)$$

Note the weak dependence of $P_1(\sqrt{X_1})$ on m and hence, being t_i determined by the bit-rate, on the filter bandwidth. Reduced by a factor of 2, the filter bandwidth changes the value of $P_1(\sqrt{X_1})$ at $X_1 = S_1/2$ only by the factor $2^{1/4} \approx 1.189$, which is less than 19%. This is a consequence of the fact that most of the noise comes from the beating of the ASE with the signal, and this contribution is independent of the filter bandwidth. It is only the noise caused by the ASE–ASE beat that gets smaller when the filter bandwidth is reduced, and this causes the reduction of $P_1(\sqrt{X_1})$. These results for zeros and ones, although obtained for integer values of m only, hold also for noninteger values of m by analytical continuation.

Consider now the case where filters are used. In this case the bandwidth of the linear radiation is smaller than the inverse of the bit-rate for filter bandwidth effective in reducing the Gordon-Haus effect. The bandwidth of the ASE radiation is such that $B_{ASE} t_i \ll 1$. This property holds with sliding filters as well, that is, for a moderate sliding rate. If a zero is transmitted, then the probability density function of the detected ASE power is a negative exponential. The ASE power with fixed filters is given by equation (5.168) and with sliding by equation (5.239). The ASE power integrated over t_i has then only two degrees of freedoms, its modulus and phase, and its probability density function is a negative exponential

$$P_0(E_0) = \frac{1}{\langle E_0 \rangle} \exp\left(-\frac{E_0}{\langle E_0 \rangle}\right) \qquad (5.506)$$

where, in normalized units,

$$\langle E_0 \rangle = \langle |q(\tau, \varsigma)|^2 \rangle \tau_i \qquad (5.507)$$

and τ_i is the integration time t_i in normalized units, $\tau_i = t_i/T_0$. The error probability, the probability the integrated ASE power overcomes the decision level, is then

$$P_{err,0} = \exp\left(-\frac{E_{dec}}{\langle E_0 \rangle}\right) \qquad (5.508)$$

The probability density function of the energy fluctuations of ones within the integration time with in-line filtering are given by the convolution of the probability distribution of the energy fluctuations of the background ASE and the energy fluctuations of the soliton. Again this property holds because of the linearization approximation. The energy fluctuations of the solitons are gaussian, at least if we linearize the dynamical equations describing the soliton propagation and if we neglect the soliton-soliton interaction [61]. So the probability density function of the energy fluctuations of one is the convolution of a gaussian with a negative exponential. The analytical expression is given in ref. [37]. The negative exponential, however, just gives a tail of the probability distribution at large energies. So the error probability of ones, which are the probability that the energy dectected when a zero is transmitted is below a given threshold, are found within a good degree of approximation taking into account just the energy fluctuations due to the soliton, with a modified average value given by the sum of the energy of the soliton and the ASE energy integrated over the detection window. The approximation gives better results for sliding filters, where the integrated ASE power on zeros is much lower than the soliton energy.

To give just one example of the use of the expressions found in Section 5.5, let us show a simplified calculation of the error probability in the case where sliding filters are used. Assume that the main contribution of the error rate on ones is timing jitter, and on zeros is amplitude noise. To simplify the analysis further, we do not optimize the decision level at the receiver. Limited to energy fluctuations, this can be done by evaluating the total error probability given by

$$P_{\text{tot}} = \pi_0 \int_{E_{\text{dec}}}^{\infty} P_0(X)dX + \pi_1 \int_{0}^{E_{\text{dec}}} P_1(X)dX \qquad (5.509)$$

where π_0 and π_1 are the probabilities that a zero and a one occur in the message, one-half each for equiprobable zeros and ones. The decision energy that gives the minimum error probability may be found by differentiating (5.509) with respect to E_{opt} and setting to zero the result. The resulting equation is

$$\pi_0 P_0(E_{\text{opt}}) = \pi_1 P_1(E_{\text{opt}}) \qquad (5.510)$$

Since the amplitude fluctuations of ones are strongly suppressed by sliding filters (5.510) usually gives values for the threshold energy larger than one-half the soliton energy. Compared, however, with the dependence of the error probability on the soliton pulsewidth, or equivalently on the soliton energy, the dependence of the error probability on the threshold energy is not very strong. This is because the limiting factor for a low error probability is mainly the amplitude fluctuations of zeros caused by the ASE background. The ASE background critically depends on the normalized sliding rate ω_r, which for a fixed sliding rate dpeends on the cube of the pulsewidth. For this reason we will assume below a threshold energy equal $\frac{1}{2}$ the energy of the soliton. In normalized units, in which the soliton energy is 2 and the soliton amplitude is unit, this corresponds to $E_{\text{dec}} = 1$.

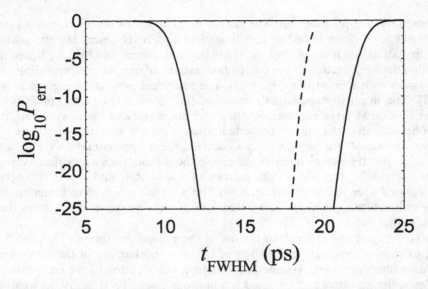

FIGURE 5.27 Window of allowed pulsewidths for a soliton transmission with sliding filters. Dashed line: timing jitter limit; solid line: limit induced by energy fluctuations of spaces.

The error probability induced by timing jitter and that due to amplitude noise on zeros are easily calculated with the help of (5.482) and (5.508), respectively. In Fig. 5.27 we show by a solid line the error probability induced by amplitude fluctuations on zeros and by a dashed line the error probability due to timing jitter, plotted against the full width at half power of the soliton T_F. The energy of the ASE background is evaluated with the help of (5.246), although the use of the asymptotic expression (5.247) and (5.249) gives almost undistinguishable results in the stable region of sliding rate which is lower than the critical sliding rate. The values of the parameters are taken from ref. [38]; these are dispersion $D = -\omega/\lambda\beta_2 = 0.6$ ps/(nm km), sliding rate 13 GHz/Mm, amplifier and filter spacing $L_A = L_f = 26$ km, nonlinear coefficient $n_2 = 2.8\,10^{-20}$ m^2/W, effective area $A_{eff} = 50\,\mu$m^2, $F = 1.6$, fiber loss 0.21 dB/km, filter reflectivity $R_f = 9\%$, $t_d = 10$ ps, corresponding to 100 GHz free spectral range of the étalon, and $R = 10$ Gbit/s, and the detection window is taken $\frac{2}{3}$ of the bit-time, $t_i = 2t_w = 2/(3R)$. The propagation distance is 10 Mm. For a given error rate, the allowed region is between the solid and dashed curves. For smaller soliton widths, the sliding rate in normalized units becomes too small for the sliding to sweep out the ASE noise generated by the large excess gain, which is inversely proportional to the square of the soliton pulsewidth. For larger pulsewidth, the normalized sliding rate, proportional to the cube of the pulsewidth, approaches the critical sliding rate reached for $T_F = 19.6859$ ps. Before the pulsewidth reaches this value, however, the timing jitter will become too large for a faithful transmission. For example, a bit-error rate of 10^{-10} will be reached for $T_F = 18.5579$ ps. The solid curve in the unstable region is unphysical. If the system were stable, this curve would correspond to the case where the threshold energy (which is one-half of the soliton energy, inversely proportional to the pulsewidth) is so low that the error probability is dominated by the energy fluctuations at zero. Experimental evidence of the existence of the window of allowed soliton energies shown in Fig. 5.27 has been reported in ref. [38].

REFERENCES

1. A. Hasegawa and F. D. Tappert, Transmission of stationary non-linear optical pulses in dispersive dielectric fibers. I. Anomalous dispersion, *Applied Physics Letters* 23: 142–144, 1973.

2. G. P. Agrawal, Fiber-optic communication systems, Wiley, New York, 1992, ch. 1.

3. V. E. Zakharov and A. B. Shabat, Exact theory of two dimensional self focusing and one-dimensional self-modulation of waves in nonlinear media, *Zh. Eksp. Teor. Fiz* 61: 118–134, 1971 [*Sov. Phys. JETP* 34: 62–69, 1972].

4. C. S. Gardner, J. M. Green, M. D. Kruskal, and R. M. Miura, Method for solving the Korteweg-de Vries equation, *Phys. Rev. Lett.* 19: 1095–1097, 1967.

5. L. F. Mollenauer, R. H. Stolen, and J. P. Gordon, Experimental observation of picosecond pulse narrowing and solitons in optical fibers, *Phys. Rev. Lett.* 45: 1095–1098, 1980.

6. A. Hasegawa, Amplification and reshaping of optical solitons in a glass fiber–IV: Use of the stimulated Raman process, *Opt. Lett.* 8: 650–652, 1983.

7. L. F. Mollenauer and K. Smith, Demonstration of soliton transmission over more than 4,000 km in fiber with loss periodically compensated by Raman gain, *Opt. Lett.* 13: 675–677, 1988.

8. J. P. Gordon and H. A. Haus, Random walk of coherently amplified solitons in optical fiber transmission, *Opt. Lett.* 11: 665–667, 1986.

9. H. A. Haus and J. A. Mullen, Quantum noise in linear amplifiers, *Phys. Rev.* 128: 2407–2413, 1962.

10. R. J. Mears, L. Reekie, I. M. Jauncey, and D. N. Payne, Low noise erbium-doped fibre amplifier operating at 1.54 μm, *Electron. Lett.* 23: 1026–1028, 1987.
    E. Desurvire, J. R. Simpson, and P. C. Beeker, High-gain erbium-doped traveling-wave fiber amplifier, *Opt. Lett.* 12: 880–890, 1987.

11. M. Nakazawa, Y. Kimura, and K. Suzuki, Soliton amplification and transmission with Er^{3+}-doped fiber repeater pumped by GaInAsP laser diode. *Electron. Lett.* 25: 199–200, 1989.
    M. Nakazawa, K. Suzuki, and Y. Kimura, 20-GHz soliton amplification and transmission with Er^{3+}-doped fiber, *Opt. Lett.* 14: 1065–1067, 1989.

12. L. F. Mollenauer, S. G. Evangelides, and H. A. Haus, Long-distance soliton propagation using lumped amplifiers and dispersion shifted fibers, *J. Lightwave Technol.* 9: 194–196, 1991.

13. A. Hasegawa and Y. Kodama, Guiding-center soliton in optical fibers, *Opt. Lett.* 16: 1385–1387, 1991.

14. K. J. Blow and N. J. Doran, Average soliton dynamics and the operation of soliton systems with lumped amplifiers, *IEEE Photon. Technol. Lett.* 3: 369–371, 1991.

15. L. F. Mollenauer, M. J. Neubelt, M. Haner, E. Lichtman, S. Evangelides, and B. M. Nyman, Demonstration of error-free soliton transmission at 2.5 Gbit/s over more than 14,000 km, *Electron. Lett.* 27: 2055–2056, 1991.

16. A. Mecozzi, J. D. Moores, H. A. Haus, and Y. Lai, Soliton transmission control, *Opt. Lett.* 16: 1841–1843, 1991.

17. Y. Kodama and A. Hasegawa, Generation of asymptotically stable optical solitons and suppression of the Gordon-Haus effect, *Opt. Lett.* 17: 31–33, 1992.

18. L. F. Mollenauer, E. Lichman, M. J. Neubelt, and B. M. Nyman, Demonstration of error-free soliton trnasmission over more than 15000 km at 5 Gbit/s, single channel, and more than 11000 km at 10 Gbit/s in two channel WDM, *Electron. Lett.* 29: 910–911, 1993.

19. L. F. Mollenauer, E. Lichtman, G. T. Harvey, M. J. Neubelt, and B. M. Nyman, Demonstration of error-free soliton transmission over more than 15,000 km at 5 Gbit/s, single channel, and over more than 11,000 km at 10 Gbit/s in two channel WDM, *Electron Lett.* 28: 792–794, 1992.

20. L. F. Mollenauer, J. P. Gordon, and S. G. Evangelides, The sliding frequency-guiding filters: An improved form of soliton jitter control, *Opt. Lett.* 17: 1575–1577, 1992.

21. L. F. Mollenauer, E. Lichtman, M. J. Neubelt, and G. T. Harvey, Demonstration, using sliding-frequency guiding filters, of error-free soliton transmission over more than 20 Mm at 10 Gbit/s, single channel, and over more than 13 Mm at 20 Gbit/s in a two-channel WDM, *Electron. Lett.* 29: 910–912, 1993.

22. L. F. Mollenauer, P. V. Mamyshev, and M. J. Neubelt, Demonstration of soliton WDM transmission at 6 and 7 × 10 Gbit/s, error free over transoceanic distances, *Electron. Lett.* 32:471–473, 1996.

23. M. Nakazawa, K. Yamada, H. Kubota, and E. Suzuki, 10 Gbit/s soliton data transmission over one million of kilometers, *Electron. Lett.* 27: 1270–1272, 1991.

24. M. Nakazawa, K. Suzuki, H. Kubota, and E. Yamada, 60 Gbit/s WDM (20 Gbit/s × 3 unequally spaced channels) soliton transmission over 10,000 km, *Proceedings of Optical Amplifiers and Their Application OSA Topical Meeting*, Monterey, CA July 11–13, 1996, paper PDP 7.

25. M. Sukuki, I. Morita, N. Edagawa, S. Yamamoto, H. Taga, and S. Akiba, Reduction of Gordon-Haus timing jitter by periodic dispersion compensation in soliton transmission, *Electron. Lett.* 31: 2027–2029, 1995.

26. M. Suzuki, I. Morita, N. Edagawa, S. Yamamoto, and S. Akiba, 20 Gbit/s-based soliton WDM transmission over transoceanic distances using periodic compensation of dispersion and its slope, *Proceedings of the 22nd European Conference on Optical Communication*, pp. 15–18, September 15–19, 1996, Oslo, Norway, Volume 5, postdeadline papers.

27. A. Hasegawa and Y. Kodama, *Solitons in Optical Communications*, Oxford Series in Optical and Imaging Sciences, Clarendon Press, Oxford, 1995.

28. H. A. Haus and W. S. Wong, Solitons in Optical Communications, *Rev. Mod. Phys.* 68: 432–444, 1996.

29. R.-J. Essiambre and G. P. Agrawal, Soliton communication systems, in *Progress in Optics*, vol. 37, E. Wolf (ed), North-Holland Physics and Elsevier Science, Amsterdam, 1997.

30. S. G. Evangelides, L. F. Mollenauer, J. P. Gordon, and N. S. Bergano, Polarization multiplexing with solitons, *J. Lightwave Technol.* 10: 28–35, 1992.

31. E. Desurvire, *Erbium-Doped Fiber Amplifiers: Principles and Applications*, Wiley, 1994.

32. H. A. Haus and Y. Lai, Quantum theory of soliton squeezing: a linearized approach, *J. Opt. Soc. Am.* B7: 386–392, 1990.

33. J. P. Gordon, Dispersive perturbations of solitons of the nonlinear Schödinger equation, *J. Opt. Soc. Am.* B9: 91–97, 1992.

34. D. J. Kaup, Perturbation theory for solitons in optical fibers, *Phys. Rev.* A9: 5689–5694, 1990.

35. H. A. Haus, Quantum noise in a soliton-like repeater system, *J. Opt. Soc. Am.* B8: 1122–1126, 1991.

36. E. A. Golovchenko, A. N. Pilipetskii, C. R. Menyuk, J. P. Gordon, and L. F. Mollenauer, Soliton propagation with up- and down-sliding-frequency guiding filters, *Opt. Lett.* 20: 539–541, 1995.

37. A. Mecozzi, Long-distance soliton transmission with filtering, *J. Opt. Soc. Am.* B10: 2321–2330, 1993.

38. P. V. Mamyshev and L. F. Mollenauer, Stability of soliton propagation with sliding-frequency guiding filters, *Opt. Lett.* 19: 2083–2085, 1994.

39. A. Mecozzi, J. D. Moores, H. A. Haus, and Y. Lai, Modulation and filtering control of soliton transmission, *J. Opt. Soc. Am.* B9: 1350–1357, 1992.

40. H. A. Haus, Waves and fields in optoelectronics, Prentice-Hall Series in *Solid State Physical Electronics*, Nick Holonyak, Jr., (ed), Prentice-Hall, Englewood Cliffs, New Jersey, 1984.

41. H. A. Haus and A. Mecozzi, Long-term storage of a bit-stream of solitons, *Opt. Lett.* 17: 1500–1502, 1992.

42. K. Iwatsuki, K. Suzuki, and S. Kawai, 40 Gbit/s adiabatic and phase-stationary soliton transmission with sliding-frequency filter over 4000 km reciprocating dispersion-managed fiber, *Proceedings of Optical Amplifiers and Their Application OSA Topical Meeting*, Monterey, CA July 11–13, 1996, paper PDP 6.

43. Y. Kodama, M. Romagnoli, and S. Wabnitz, Stabilization of optical solitons by an acousto-optic modulator and filter, *Electron. Lett.* 30: 261–262, 1994.
F. Fontama, L. Bossalini, P. Franco, M. Midrio, M. Romagnoli, and S. Wabnitz, Self starting sliding frequency fiber soliton laser, *Electron. Lett.* 30: 321–322, 1994.

44. A. Mecozzi, M. Midrio, and M. Romagnoli, Timing jitter in soliton transmission with sliding filters, *Opt. Lett.* 21: 402–404, 1996.

45. F. I. Khatri, S. G. Evangelides, P. V. Mamyshev, B. N. Nyman, and H. A. Haus, A line monitoring system for undersea soliton transmission systems with sliding-frequency guiding filters, *IEEE Photon. Technol. Lett.* 8: 730–732, 1996.

46. A. Mecozzi, Soliton transmission control by Butterworth filters, *Opt. Lett.* 20: 1859–1861, 1995.

47. A. Mecozzi, Soliton dynamics with fixed and sliding filters, in *Nonlinear Guided Waves and their Applications*, vol. 15, 1996 OSA Technical Digest Series, Optical Society of America, Washington DC, 1996, pp. 132–134.

48. M. Suzuki, N. Edagawa, H. Taga, H. Tanaka, S. Yamamoto, and S. Akiba, Feasibility demonstration of 20 Gbit/s single channel soliton transmission over 11,500 km using alternating amplitude solitons, *Electron. Lett.* 30: 1083–1085, 1994.

49. S. Wabnitz, Suppression of soliton interactions by phase modulations, *Electron. Lett.* 29: 1711–1713, 1993.
    J. N. Smith, K. J. Blow, W. J. Firth, and K. Smith, Soliton dynamics in the presence of phase modulators, *Opt. Commun.* 102: 324–328, 1993.
    J. N. Smith, K. J. Blow, W. J. Firth, and K. Smith, Suppression of soliton interaction by periodic phase modulators, *Opt. Lett.* 19: 16–18, 1994.

50. Y. Kodama, M. Romagnoli, and S. Wabnitz, Soliton stability and interactions in fiber lasers, *Electron. Lett.* 28: 1981–1983, 1992.
    M. Matsumoto, H. Ikeda, and A. Hasegawa, Suppression of noise accumulation in bandwidth limited soliton transmission by means of nonlinear loop mirrors, *Opt. Lett.* 19: 183–185, 1994.

51. F. N. Mitschke and L. F. Mollenauer, Experimental observation of interaction forces between solitons in optical fibers, *Opt. Lett.* 12: 355–357, 1987.

52. J. P. Gordon, Interaction forces among solitons in optical fibers, *Opt. Lett.* 8: 596–598, 1983.

53. K. Smith and L. F. Mollenauer, Experimental observation of soliton interaction over long fiber paths: Discovery of a long-range interaction, *Opt. Lett.* 14: 1284–1286, 1989.

54. E. M. Dianov, A. V. Luchnikov, A. N. Pilipetskii, and A. N. Starodumov, Electrostriction mechanism of soliton interaction in optical fibers, *Opt. Lett.* 15: 314–316, 1990.

55. V. I. Karpman and V. V. Solov'ev, A perturbational approach to the two-soliton system, *Physica D* 3: 487–502, 1981.

56. P. L. Chu and C. Desem, Mutual interaction between solitons of unequal amplitudes in optical fiber, *Electron. Lett.* 21: 1133–1134, 1985.

57. Y. Kodama and S. Wabnitz, Analysis of soliton stability and interactions with sliding filters, *Opt. Lett.* 19: 162–164, 1994.

58. J. P. Gordon and L. F. Mollenauer, Effects of fiber nonlinearities and amplifier spacing on ultra-long distance transmission, *J. Lightwave Technol.* 9: 170–173, 1991.

59. M. Suzuki, H. Tanaka, N. Edagawa, and Y. Matsushima, New applications of sinusoidally driven InGaAsP electroabsorption modulator to in-line optical gates with ASE noise reduction effect, *J. Lightwave Technol.* 10: 1912–1918, 1992.

60. M. Suzuki, N. Edagawa, H. Taga, H. Tanka, S. Yamamoto, Y. Takahashi, and S. Akiba, Long-distance soliton transmission up to 20 Gbit/s using alternating-amplitude solitons and optical TDM, *IEICE Trans. Electron.* E78-C: 12–21, 1995.

61. C. R. Menyuk, Non-gaussian corrections to the Gordon-Haus distribution resulting from soliton interactions, *Opt. Lett.* 20: 285–287, 1995.

62. L. F. Mollenauer, S. G. Evangelides, and J. P. Gordon, Wavelength division multiplexing with solitons in ultra-long distance transmission using lumped amplifiers, *J. Lightwave Technol.* 9: 362–369, 1991.

63. L. F. Mollenauer, J. P. Gordon, and F. Heismann, Polarization scattering by soliton-soliton collision, *Opt. Lett.* 20: 2060–2062, 1995.

64. Y. Kodama and A. Hasegawa, Effects of the initial overlap on the propagation of optical solitons at different wavelengths, *Opt. Lett.* 16: 208–210, 1991.

65. A. Mecozzi and H. A. Haus, Effect of filters on soliton interactions in wavelength division multiplexing systems, *Opt. Lett.* 17: 988–990, 1992.

66. M. Midrio, P. Franco, F. Matera, M. Romagnoli, and M. Settembre, Wavelength division multiplexed soliton transmission with filtering, *Opt. Comm.* 112: 283–288, 1994.

67. A. Mecozzi, Timing jitter in wavelength division multiplexed filtered soliton transmission, *J. Opt. Soc. Am. B.*, December 1997.

68. A. Hasegawa, S. Kumar, and Yuji Kodama, Reduction of collision induced time jitters in dispersion managed soliton transmission systems, *Opt. Lett.* 21: 39–41, 1996.

69. B. Biotteau, J.-P. Hamaide, F. Pitel, O. Audouin, P. Noichi, and P. Sansonetti, Enhancement of soliton system performance by use of new large effective area fibers, *Electron. Lett.* 31: 2026–2027, 1995.

70. N. J. Smith, F. M. Knox, N. J. Doran, K. J. Blow, and I. Bennion, Enhanced power solitons in optical fibers with periodic dispersion management, *Electron. Lett.* 32: 54–55, 1996.

CHAPTER SIX

Repeaterless Systems

So far electronic switching is adopted in the network nodes, the optical technologies introduced in the transport and access network has been limited to point-to-point transmission between nodes.

In a large mesh network, a great number of fiber links are present spanning a few hundreds of kilometer and this has stimulated the investigation of transmission systems without in-line optical amplifiers working at such distances. This solution seems attractive in terms of cost reduction and simplicity.

The substitution of coaxial cables with optical fibers in the transport network has been often carried out by installing step-index fibers operating in the second transmission window. This is the case of Germany among the European countries and of large areas of the transport network in the United States. Replacing this large quantity of already installed step-index fibers with DS fibers in order to operate in the third transmission window at a low dispersion has a huge cost. Thus the optimization of systems operating on step-index fibers is a key issue for a first upgrading of transport and access networks in which systems adopting electrical repeaters are substituted by repeaterless optical systems. For this reason the main part of this chapter will be devoted to the case of step-index fiber links.

All of the transmission systems considered will be at $1.55\,\mu$m although, in principle, the analysis can apply as well to the 1.3-μm window. Optical transmission at $1.3\,\mu$m is characterized by a lower chromatic dispersion ($\sim 1\,$ps/nm/km), but the higher attenuation ($\sim 0.4\,$dB/km) makes this choice more attractive for amplified links than repeaterless systems.

Finally soliton transmission will be not treated because fiber loss makes repeaterless soliton transmission not particularly convenient. Soliton systems perform better in very long systems where the optical amplification restores the energy lost during the propagation. However, nonlinear assisted transmission will be considered in the case of NRZ signals.

This chapter is divided into seven sections. The main impairments for optical repeaterless systems are dealt with in Section 6.1, while Sections 6.2

to 6.6 are dedicated to techniques that maximize the span length by compensating for critical performance degrading effects. In particular, Section 6.2 deals with prechirping techniques, consisting in the introduction of a chirp in the transmitted intensity modulated signal to compensate the chirp introduced by fiber dispersion. In Section 6.3 the so-called dispersion supported transmission is analyzed. It consists in transmitting a frequency-modulated signal that is transformed during propagation into an intensity-modulated signal by fiber dispersion. Transmission systems exploiting the negative chirp introduced by Kerr effect to partially compensate for fiber dispersion are analyzed in Section 6.4, while the possibility of using optical phase conjugation in the middle of the optical link to obtain dispersion and Kerr effect compensation is considered in Section 6.5. Section 6.6 turns to systems adopting passive dispersion compensating devices. In this case passive devices as a piece of suitably designed fiber, or more compact devices as grating filter or interferometers, are inserted at the receiver or at the transmitter in order to introduce a frequency-dependent phase shift that compensates for fiber dispersion. The use WDM in optical repeaterless links is finally analyzed in Section 6.7.

6.1 MAIN PERFORMANCE LIMITATIONS OF REPEATERLESS TRANSMISSION SYSTEMS

To extend the span length in repeaterless transmission systems, it seems opportune to increase the fiber-launched power and improve the receiver sensitivity.

Receiver sensitivity may gain advantage from coherent detection schemes, but their economical feasibility and practical convenience are questionable, especially in the case of TDM transmission. On the other hand, the adoption of optical EDFA preamplifiers or postamplifiers cab [1] brings the sensitivity of direct detection receivers quite close to the theoretical limit. The receiver sensitivity can be improved also by means, of Forward Error Correction (FEC) [2]. It consists of properly encoding the transmitted bit pattern in a way that a suitable algorithm can be applied in the receiver to correct possible errors. Accepting a negligible bandwidth expansion, the receiver sensitivity can be increased by approximately 4 dB and the overall system reliability results also optimized.

When a high-sensitivity receiver has been adopted, it is important to avoid transmission impairments causing a floor in the error probability curve. In fact, if no error probability floor is present, transmission on a longer span may be achieved by increasing the fiber launched power.

In Section 4.2.1 it was shown that pulse broadening caused by fiber dispersion induces an error probability floor. This means that in the presence of fiber dispersion, the maximum link span is determined by the dispersion coefficient and by the bit-rate independently of the transmitted power. Intuitively this derives from the fact that dispersion induces intersymbol interference: The increase in transmitted power causes both the power of the considered bit and the power of the interferring bit to increase at the same rate; thus the signal-to-interference ratio remains the same.

If no spurious chirp is added to the signal during the modulation process, as when using an external modulator, the link length L for which the dispersion-induced penalty is equal to 1 dB can be evaluated with the method described in Section 4.2.1. Within a first approximation this dispersion limit must satisfy the following relation [3]:

$$R^2DL = \frac{c}{2\lambda^2} \tag{6.1}$$

where R is the bit-rate, D is the chromatic dispersion coefficient, c is the vacuum light velocity, and λ is the signal wavelength.

When the semiconductor laser used at the transmitter is directly modulated, a spurious chirp is added to the transmitted signal, generally enhancing the impact of dispersion. For example, for $\lambda = 1532$ nm, $D = 17$ ps/nm/km, and $R = 10$ Gbit/s, the dispersion limited link length is about 50 km, while it is reduced at approximately 10 km for direct laser modulation [4]. Hence the primary challenge for high bit-rate transmission is to reduce the impact of dispersion to an acceptable penalty.

Several solutions to this problem have been suggested in the literature: Techniques that apply to the transmitter, to the receiver, or affect the pulse while propagating. Some of these techniques involve time-dependent manipulation of input signals, some rely on time-independent processing; some need a complete *retuning* when link parameters as lengths or bit-rate are changed, while other techniques are less critical with respect to modifications of link parameters. In Fig. 6.1 is illustrated a *road map* of the main different strategies proposed to combat the chromatic dispersion and that will be analyzed in the following sections of this chapter.

When the dispersion impairment is attenuated by adopting some compensation method, the link length might be increased by increasing the transmitted power. Erbium-doped fiber amplifiers, with their high-output power, low-noise contribution, and simple implementation are suitable for this purpose. But if, on one hand, the injection of high powers at the transmitter permits fiber losses to be overcome, on the other hand, it enhances fiber nonlinearities [5] such as stimulated Brillouin scattering (SBS) [6], and self-phase modulation (SPM) [7] that can cause a strong degradation of the system performances.

As described in Section 2.2.3, SBS introduces a limitation on the maximum injected optical power. The Brillouin threshold P_{th} of an IM signal with bit-rate R whose optical carrier has a Lorentzian lineshape with a linewidth $\Delta\nu_s$, can be evaluated by applying equation (2.51). If R is much greater than the Brillouin gain linewidth $\Delta\nu_B$ (i.e., generally smaller than 30 MHz), the result can be written as

$$P_{th} = 2P_{th0}\frac{\Delta\nu_B + \Delta\nu_s}{\Delta\nu_B} \tag{6.2}$$

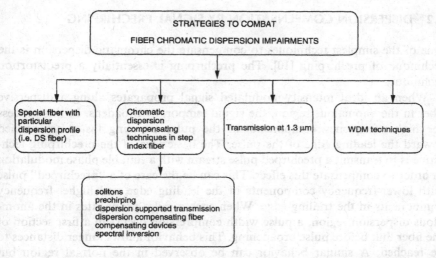

FIGURE 6.1 "Road map" of the main strategies adopted by the scientific community to combat fiber chromatic dispersion impairment.

where P_{th0} is the SBS threshold for a monochromatic signal. From equation (6.2) it is clear that P_{th} increases by increasing the carrier linewidth. However, a large carrier linewidth also degrades the system performances due to fiber dispersion. In order to avoid this undesirable effect, the laser can be modulated with a low-frequency dither outside the receiver bandwidth. An experimental confirmation of the results obtained by equation (6.2) has been presented in [8] where an externally modulated DFB laser was used to investigate the SBS threshold on a 347 km step-index fiber at a bit-rate of 2.5 Gbit/s. It has been demonstrated that the application of a 3%, 10-KHz dither joined to the modulation at 2.5 Gbit/s suppresses the SBS completely.

As far as SPM is considered, it introduces additional frequency components, inducing spectral broadening. In the absence of dispersion, the spectral broadening will not result in a performance degradation in the case of direct detection, since the phase terms disappear when the optical power is detected. In the presence of chromatic dispersion SPM can generate pulse broadening and hence system performance degradation. This effect is particularly severe with step-index fibers, for which the chromatic dispersion is higher. In the case of step-index fiber the problem is further aggravated if the transmitter is a directly modulated laser diode because of the large chirp at the signal wavelength. Typically at a bit-rate of 2.5 Gbit/s, degradation appears for optical powers above +17 dBm in step-index fibers. Besides the chromatic dispersion, relevance of the effect depends on the propagation distance, the bit-rate, the modulation format, and the pulse profile, among other parameters [9].

6.2 DISPERSION COMPENSATION BY SIGNAL PRECHIRPING

One of the simplest techniques to compensate the chromatic dispersion is the technique of prechirping [10]. The prechirping is essentially a predistortion technique.

When an ideal intensity-modulated signal propagates along a dispersive fiber in the anomalous region, the signal temporally broadens. In this process the higher-frequency components of the pulse traveling faster are pushed toward the leading edge of the pulse. The basic idea of the prechirping technique is to transmit a prechirped pulse stream with a suitable phase modulation in order to compensate this effect. This can be the case of a "prechirped" pulse with lower-frequency components in the leading edge and higher-frequency components in the trailing edge. When such a pulse propagates in the anomalous dispersion region, a pulse width compression occurs in a first section of the fiber link before pulse broadening. This behavior allows longer distances to be reached. A similar behavior can be observed in the normal region but with opposite prechirp in the pulse frequency. A visual description of the prechirping technique is shown in Fig. 6.2, where the signal waveform is shown at different positions along the link.

Commonly signal prechirping is realized by combining an external IM modulator with a frequency modulation of the transmitting laser bias current. The frequency modulation of the bias current (generating the signal chirp) and the driving current of the external modulator are synchronized by the same clock in order to correctly generate the prechirped signal. In order to maximize the allowable transmission length, the frequency modulation parameters have to be optimized for each individual link. Moreover a spurious direct IM modulation of the laser or imperfections in the chirp linearity may limit the system performance.

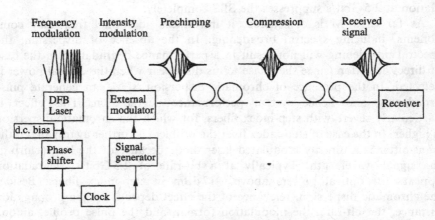

FIGURE 6.2 Schematic representation of prechirping technique.

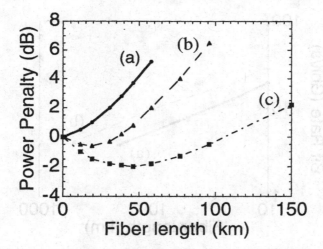

FIGURE 6.3 Eye penalty versus transmission distance calculated in the absence of prechirping (a), with prechirping (b), and with modified prechirping (c).

A greater dispersion compensation capability can be obtained for NRZ transmission by means of a modified version of the prechirping technique. It consists in adopting optical polarization multiplexing to generate a NRZ signal starting by two orthogonal RZ prechirped signals at a half the bit-rate [11]. The increase of the complexity of the system is the price to be paid.

The effectiveness of the prechirping technique is shown in Fig. 6.3, where the eye opening penalty (EOP) has been reported as a function of the transmission length for a 10 Gbit/s NRZ signal propagating along a step-index fibers ($D = 18$ ps/nm/km) without prechirping technique, with prechirping technique and with the modified prechirping technique. Dispersion compensation capabilities are also shown in Fig. 6.4 in terms of the maximum bit-rate at which the dispersion-induced eye penalty is equal to 1 dB versus the fiber length. The data have been extracted from ref. [10], and they refer to the case where a nonideal external modulator is adopted with a chirping parameter, α equal to 0.6 (for an ideal modulator $\alpha = 0$) [12].

Prechirping technique can be useful not only in a linear propagation regime but also in the presence of SPM. In this case controlled interaction among prechip, SPM, and chromatic dispersion allows the maximum transmission distance to be almost doubled compared to the case where prechiping is absent.

The feasibility of the prechirp technique by external modulators have been experimentally confirmed. Repeaterless transmission at 10 Gbit/s over the record length of 204-km standard fibers with an average dispersion of 16.8 ps/nm/km has been demonstrated, as a result of a suitable balance of dispersion, SPM, and prechirping [13]. Analogously, if three in-line amplifiers are also included in the link, the dispersion-limited transmission distance of a 5-Gbit/s

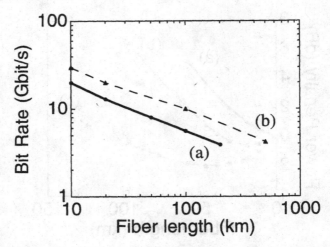

FIGURE 6.4 Bit-rate versus the fiber link length in the absence of any compensation (a) and with prechirping technique (b), assuming an acceptable power penalty of 1 dB.

NRZ system has be extended from 300 to 450 km demonstrating the applicability of step-index fibers in the framework of inland networks [14].

The main advantage of prechirping is the conceptual simplicity, while the disadvantages are essentially the presence of an external modulator, the requirement of a narrow linewidth laser, and the necessity of a complete *retuning* when the link parameters are changed, for example, to upgrade the bit-rate.

6.3 DISPERSION-SUPPORTED TRANSMISSION

The method of dispersion-supported transmission (DST) can be considered a particular case of the prechirping technique. In this scheme the optical transmitter generates a frequency-modulated signal that is converted into an amplitude-modulated signal at the receiver by propagating along a dispersive fiber. The most evident advantage of this method is its simplicity, since it does not require additional components such as external modulators, interferometers, or optical dispersion compensation elements: conventional directly modulated DFB–laser diodes can be employed. The main limitation is that the frequency modulation depth must be tuned to match the span length.

The principle of operation is well described in a paper by Wedding et al. [15] and it is summarized in Fig. 6.5 where the signal waveforms in the key points of the link are sketched: The transmitter generates frequency modulated signal with a constant optical power. The signal frequency can assume two different values: ν_1 and $\nu_2 = \nu_1 + \Delta\nu$. The dispersive fiber differently delays the signal frequency components: As a result interference occurs, and at the link output, the frequency modulation is converted into amplitude modulation.

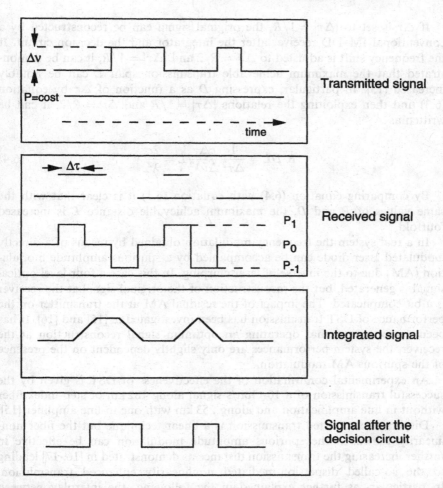

FIGURE 6.5 Scheme of the principle of operation of dispersion-supported transmission technique. At the transmiter the signal is purely frequency modulated, and the optical power is constant. At the receiver amplitude modulation is present, and after the integrator and the decision circuit the original signal can be reobtained.

In particular, the received power can assume three different levels: an intermediate one corresponding to the input level, a positive pulse due to constructive interference, and a negative pulse due to the absence of the signal.

The delay $\Delta\tau$ between ν_1 and ν_2 depends on $\Delta\nu$, on the link dispersion D, and on the propagating distance L, according to equation (2.17), which in this case can be rewritten as

$$\Delta\tau = \frac{c\Delta\nu}{\lambda^2}DL = \Delta\lambda DL \qquad (6.3)$$

If $\Delta\tau$ is set to $|\Delta\tau| = 1/R$, the original signal can be reconstructed by a conventional IM–DD receiver after the integrator and the decision circuit. If the frequency shift is adjusted to $\Delta\nu = R/2$ and $|\Delta\tau| = 1/R$, it can be demonstrated that the maximum achievable transmission span L can be notably increased [15]. In particular, expressing D as a function of $\Delta\tau$ by equation (6.3) and then exploiting the relations $|\Delta\tau| = 1/R$ and $\Delta\nu = R/2$, it can be written as

$$R^2 DL = \frac{1}{\Delta\tau^2} \frac{c\Delta\tau}{\Delta\nu\lambda^2} L = \frac{2c}{\lambda^2} \tag{6.4}$$

By comparing equation (6.4) with equation (6.1) it is clear that with the same values of R and D, the maximum achievable distance L is increased fourfold.

In a real system the frequency modulation obtained by means of a directly modulated laser diode can be accompanied by a spurious amplitude modulation (AM) due to the inverse of laser chirping. In this case a four-level optical signal is generated, but the reconstruction of the original signal at the receiver is a bit complicated. The impact of the residual AM at the transmitter on the performance of DST transmission has been investigated in [15] and [16]. It has been demonstrated that operating an optimum signal reconstruction at the receiver, the system performances are only slightly dependent on the presence of the spurious AM modulation.

An experimental confirmation of the effectiveness of DST is given by the successful transmission of a 10-Gbit/s signal along 182 km of step-index fiber without in line amplification and along 253 km with one in-line amplifier [15].

Dispersion-supported transmission is a linear technique but the fiber nonlinearity joined to the spurious amplitude modulation can be effective in further increasing the transmission distance as demonstrated in [16, 17] leading to the so-called dispersion-mediated nonlinearity enhanced transmission. In particular, as further explained in the following, the interplay between fiber dispersion and nonlinearity on amplitude modulated signal helps in the stabilization of the power distribution of the signal and improves the transmission performance.

In Fig. 6.6 a comparison among the performances of externally modulated IM–DD systems, linear DST systems, and dispersion-mediated nonlinearity enhanced systems is shown for a bit-rate $R = 10$ Gbit/s and $D = 17$ ps/nm/ km. The figure shows that for a link length below 50 km when the dispersive effect is not so critical, externally modulated IM–DD system show better performances; when the chromatic dispersion affects significantly the performance of IM–DD system and induces frequency to amplitude modulation conversion, the DST technique allows longer distances to be reached, and even better results can be obtained when the dispersion-supported transmission technique is enhanced by fiber nonlinearity. In the latest case the maximum achievable distance depends on a suitable combination of input signal power and on the

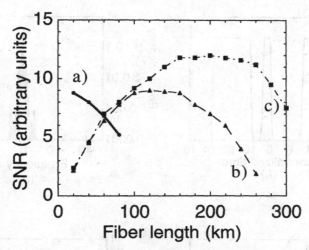

FIGURE 6.6 System performance evaluated in terms of the signal-to-noise ratio (SNR) versus the fiber link length. The bit-rate is 10 Gbit/s and $D = 17$ ps/nm/km. Case (*a*) refers to IM-DD system, case (*b*) to linear DST, case (*c*) to DST enhanced by nonlinearity, ($P_{ave} = 10$ mW). The amplitude modulation index is 0.25. The data have been extracted from ref. [17].

residual amplitude modulation. In particular, curve *c* refers to an input power of 10 mW and to a residual AM with a modulation depth equal to 25%, and in-line amplifiers have been introduced every 100 km to ensure a solitonlike behavior. The nonlinear coefficient is $\gamma = 2.2$ (W km)$^{-1}$.

6.4   NONLINEAR ASSISTED TRANSMISSION

As shown in Section 2.3, in the anomalous dispersion region the linear chirp induced by fiber dispersion and the nonlinear one due to SPM have opposite sign; thus partial compensation occurs. This phenomenon is well known when the transmitted pulse presents the particular shape of a hyperbolic secant. In this condition the nonlinear compensation is particularly efficient, and the pulse, called soliton, experiences dispersion-free propagation. However, partial dispersion compensation can be attained also for other pulse shapes: The method of nonlinear-assisted transmission is based on this effect.

In order to better understand the role of nonlinearity, the output power spectra and the bit patterns of a 10-Gbit/s NRZ intensity-modulated signal propagating along 120 km of step-index fiber for different launched powers are shown in Fig. 6.7. As predicted by the theory (see Section 2.2), in the linear regime corresponding to a peak power, $P_{peak} = 0.1$ mW (Fig. 6.7c, 6.7d), the optical spectrum remains almost the same as the input one, but the pulses are broadened by chromatic dispersion; after increasing the input power up to

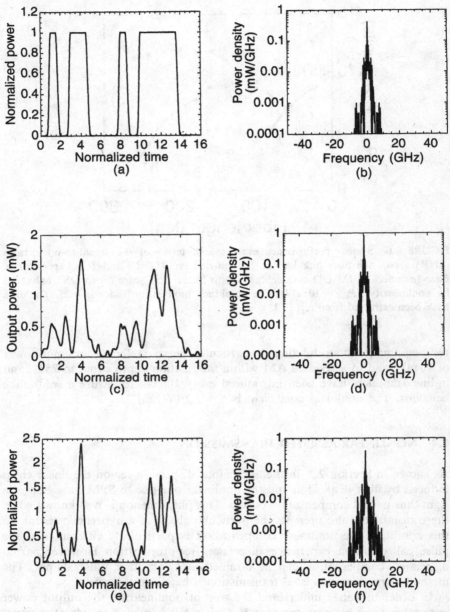

FIGURE 6.7 Input and output spectra and relative bit-patterns of an intensity-modulated NRZ signal at 10 Gbit/s after propagating along 120 km of a step-index fiber with $D = 15.6\,\text{ps/nm/km}$, $\gamma = 2.7\,(\text{Wkm})^{-1}$. (a) and (b) refer to the input bit-pattern and spectrum; (c) and (d) to the output signal after propagation in the linear regime ($P_{\text{peak}} = 0.1\,\text{mW}$); (e) and (f) to the output signal after propagation in the nonlinear regime ($P_{\text{peak}} = 20\,\text{mW}$).

$P_{peak} = 20\,mW$ (Fig. 6.7e, 6.7f), the spectrum becomes wider because of the effect of SPM, and the pulse energy remains more confined within the bit-time because of pulse compression, and noiselike behavior appears in the bit-pattern [18].

A quantitative analysis of the capability of Kerr nonlinearity to partially compensate dispersion on step-index fiber can be obtained in terms of Q factor as a function of the transmitted power. The Q factor, estimated by simulations, is shown in Fig. 6.8 for a 100-km-long system operating at a bit-rate of 10 Gbit/s. The fiber parameters adopted in the simulations are $D = 16\,ps/nm/km$ and $\gamma = 2.7\,(W\,km)^{-1}$, and $\alpha = 0.25\,dB/km$. The optical and electrical filters at the receiver have, respectively, a bandwidth of $B_{FWHM} = 40\,GHz$ and $B_{FWHM} = 8\,GHz$. In the linear regime the system performance are strongly degraded by intersymbol interference due to chromatic dispersion, and as predicted by formula (6.1), at this distance the transmission is not allowed (Q factor < 9). As the power increases, both chromatic dispersion and nonlinear Kerr effect generate a chirp on the pulse. In the anomalous dispersion region of the fiber, the net chirp on the pulse is smaller than the chirp in the linear case. As a result the system performance improves by means of nonlinear compensation.

A limitation on the power increase is essentially determined by the stimulated Brillouin scattering threshold and the occurrence of overcompensation of the chromatic dispersion by the nonlinear Kerr effect. In fact for intensity-modulated systems with a bit-rate far larger than the SBS linewidth the threshold is almost independent on the bit-rate, and it is in the range of 10–20 mW. So if a SBS suppression technique is not adopted, the beneficial effect of nonlinear compensation will not be observed [6].

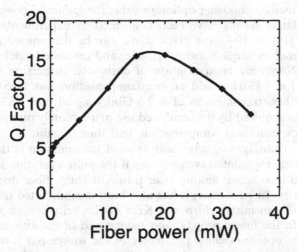

FIGURE 6.8 Q factor as a function of input peak powers for a 10-Gbit/s IM-DD NRZ signal propagating along 100 km of step-index fiber.

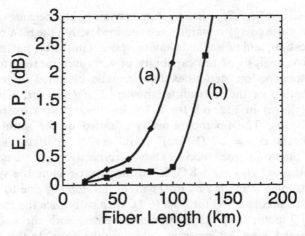

FIGURE 6.9 Eye opening penalty as a function of propagating distance for a 10-Gbit/s IM-DD NRZ for linear (a) ($P_{peak} = 1\,mW$) and nonlinear (b) regime ($P_{peak} = 20\,mW$).

If the power is increased up to the state where nonlinearity dominates the linear dispersion effect ($L_{NL} < L_D$), the Kerr effect overcompensates the chromatic dispersion and pulse splitting and pulse interactions occur as a result of modulation instability effect, producing signal degradation.

In Fig. 6.9 the eye opening is evaluated versus the propagating distance in the linear (a) and nonlinear (b) regime. The results show that the dispersion-limited distance (50 km for a bit-rate of 10 Gbit/s and without dispersion compensation) can be extended almost of 50%, achieving transmission over 100 km of step-index fiber by choosing optimum value for launched power.

These simulative results have been confirmed by experiments, as demonstrated in ref. [19]. A 10-Gbit/s NRZ signal can be transmitted over 107 km of step-index fiber by single booster amplifier and generate launched power up to +15 dBm. Nonlinear compensation of chromatic dispersion impairments was obtained for +13 dBm and an overcompensation was also observed at +15 dBm. Further, transmission of a 2.5-Gbit/s signal over 357 km of step-index fibers was achieved by the combined use of a remotely pumped amplifier, SBS suppression nonlinear compensation, and Raman gain.

One of. the advantages of nonlinear assisted transmission is that it can be applied to directly modulated systems even if the pulse evolution is more complicated due to interaction among laser transient chirp, fiber dispersion, and Kerr nonlinearity. In particular, it has also been demonstrated that a suitable combination of transmitter chirp and Kerr nonlinearity can improve system performance. In the linear propagation regime and in the anomalous dispersion region, the optimum chirp parameter of the source has a negative sign; then, as the powers is increased, the chirp parameter shifts toward more positive values [20].

6.5 DISPERSION COMPENSATION BY OPTICAL PHASE CONJUGATION

Optical phase conjugation as a distortion compensator of a signal propagating in a dispersive channel in optical communication systems was first proposed in [21]. An optical phase conjugator (OPC) is a device that provides the spectral inversion of the signal, preserving causality. This means that the pulses retain their temporal order, since the group delay is not affected by the OPC.

In this section we will discuss this technique first assuming the existence of an ideal OPC and analyzing the system performances, then discussing the feasibility of applying the OPC in a telecommunication environment.

6.5.1 Signal Propagation in Systems Adopting Phase Conjugators

An optical pulse distorted by the propagation along a dispersive fiber can be reshaped by an OPC followed by a fiber with the same overall dispersion of the link before the OPC.

Signal propagation along the first fiber link can be described by the non-linear Schrödinger equation (2.57). Defining carefully a suitable functional space for the normalized field envelope $U(\xi, \tau)$, equation (2.57) can be rewritten as

$$\left[\frac{\partial}{\partial \zeta} + i\hat{D}_{\beta 2} + \hat{D}_{a} + i\hat{D}_{\gamma} + \hat{D}_{\beta 3}\right] U(\zeta, \tau) = 0 \tag{6.5}$$

where the *real* operators appearing in equation (6.5) are defined as

$$\hat{D}_{\beta 2} = \frac{1}{2}\text{sign}(\beta_2)\frac{\partial^2}{\partial \tau^2}, \quad \hat{D}_{a} = \frac{1}{2}\frac{L_D}{L_a}$$

$$\hat{D}_{\gamma} = \frac{L_D}{L_{NL}}|U|^2, \quad \hat{D}_{\beta 3} = \frac{1}{6}\frac{L_D}{L_D'}\frac{\partial^3}{\partial \tau^3} \tag{6.6}$$

The formal solution of equation (6.5) is

$$U(\zeta, \tau) = U(0, \tau)\exp(i\hat{D}_{\beta 2} + \hat{D}_{a} + i\hat{D}_{\gamma} + \hat{D}_{\beta 3})L_1 \tag{6.7}$$

where L_1 is the length of fiber before the OPC.

The field $U(\zeta, \tau)$ is conjugated by the OPC and propagated in a second fiber link. The propagation in this second link is governed by equation (6.5) with new operators evaluated by considering the parameters of the considered fiber link. Indicating the operators relative to the second fiber link with the apex, the field in front of the receiver can be written as

$$U(\zeta, \tau) = U^*(0, \tau)\exp([-i\hat{D}_{\beta 2} + \hat{D}_{a} + -i\hat{D}_{\gamma} + \hat{D}_{\beta 3}]L_1$$

$$+ [i\hat{D}'_{\beta 2} + \hat{D}'_{a} + i\hat{D}'_{\gamma} + \hat{D}'_{\beta 3}])L_2 \tag{6.8}$$

where L_2 is the length of the link after the OPC.

If the second-order dispersion and the fiber loss are equal to zero, $\hat{D}_a = 0$ and $\hat{D}_{\beta 3} = 0$. In this condition, if the nonlinear coefficients of the two fiber links are equal and the dispersion parameters β_2 and β_2' satisfy the equation $\beta_2 L_1 = \beta_2' L_2$, equation (6.8) gives $U(\zeta, \tau) = U^*(0, \tau)$. This means that in the absence of third-order dispersion and losses, suitably placing an OPC along a fiber link chromatic dispersion and Kerr effect can be compensated.

If the fiber loss is not equal to zero, total compensation is possible, in repeaterless systems provided that the optical power is evolving symmetrically along the fiber with respect to the OPC. This means that the loss in the first fiber must be compensated by an appropriately distributed gain in the second fiber or the fiber length must be short enough to assume that the system behaves almost as a lossless systems. This second condition can be approximately satisfied if the fiber effective length $L_e = (1 - e^{-\alpha L})/\alpha$ is quite shorter than the nonlinear length L_{NL} [22].

In the case of systems adopting in-line optical amplification, the effective length is substituted by the amplifier spacing, that must be considerably shorter than the nonlinear length. In a periodically amplified soliton system compensation is achievable if the amplifier spacing is small compared to the soliton period. Within this range optical phase conjugation can *invert* also soliton interactions and reduce the Gordon-Haus jitter due to the interaction of the soliton stream with the optical noise generated by in-line amplifiers [23].

6.5.2 Phase Conjugators for Telecommunication Systems

The phase conjugation is commonly implemented by the process of FWM in a nonlinear medium. The principle is sketched in Fig. 6.10. The signal is

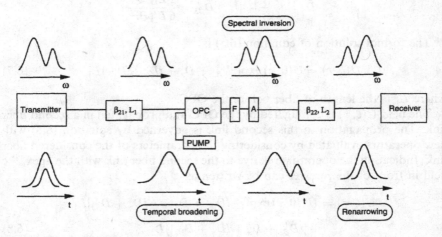

FIGURE 6.10 Principle of operation of spectral inversion by optical phase conjugation in communication systems. OPC: optical phase conjugator, F: filter, A: amplifier.

transmitted over a first section of fiber of length L_1. In the anomalous dispersion region blue components travel faster than red components. This leads to temporal pulse broadening. At the optical phase conjugator a pump at the frequency ν_p and the signal beat together generating a conjugate signal at the frequency $\nu_c = 2\nu_p - \nu_s$ which has the spectrum reversed with respect to the input signal. By filtering and amplifying the output field, the conjugate signal can be extracted and propagated along a second span of fiber of length L_2. As a result the fiber chromatic dispersion, acting on the inverted spectrum signal, can restore the original shape of the pulse, and a temporal renarrowing can be observed.

The two most promising candidates as nonlinear media for optical phase conjugation are optical fibers or semiconductor amplifiers. In the case of optical fiber, the phase-matching requirement suggests to use FWM in a dispersion shifted fiber. The weak fiber nonlinearity imposes the use of a long fiber (typically 20–25 km) and of high pump powers. Typical conversion efficiencies in the range of $-20\,\mathrm{dB}$ and $-25\,\mathrm{dB}$ can be obtained considering the limitation on the pump power induced by stimulated Brillouin scattering [24]. Moreover the DS fiber dispersion can induce further distortions on the conjugate signal that can alter the phase relationship between the input and the conjugate signal so reducing the achievable amount of nonlinearity compensation.

Conversion efficiencies at least one order of magnitude higher can be obtained by using semiconductor amplifiers [25]. In this case the effect due to the device dispersion is considerably reduced, since the devices length is of the order of millimeters and the effective gain bandwidth is of the order of tens of nanometers. Disadvantages of semiconductor amplifiers are the presence of ASE noise at the amplifier output and the rising of pattern effects due to the dynamic behavior of the carrier density in the active region.

Due to the nature of FWM, both considered devices are polarization dependent: The conversion efficiency varies from a maximum value when the polarization of the pump are copolarized to zero in the case where the pump and signal are orthogonal. The polarization dependency can be compensated by locking the pump polarization to the input signal by an active polarization controller in single-channel transmission, but this inhibits the practical realization of WDM systems.

Polarization-independent optical phase conjugation can be obtained by adopting the scheme of nondegenerate FWM: If two orthogonally polarized pumps are adopted, the FWM efficiency is independent of the state of polarization of the input signal because of the symmetry of the tensor $\chi^{(3)}$ [26].

Polarization-independent optical phase conjugation in an optical semiconductor amplifier has been demonstrated with a 10-Gbit/s directly modulated DFB laser propagating over 160 km of a standard single-mode fiber.

Another method to obtain a polarization-independent OPC is to use a single pump linearly polarized at $\pi/4$ with respect to the x axis. The x and y linear polarization components are divided and decorrelated by delaying one of them a time interval greater than the coherence time of the laser. After injecting the two uncorrelated pumps into the nonlinear medium, the polarization-independent FWM is obtained [26].

An example of the improvement that can be achieved with spectral inversion by OPC in uncompensated transmission is demonstrated in ref. [25]. Bit-error rate curves are measured as a function of received optical power in the case of back-to-back transmission, uncompensated transmission, and compensated transmission considering a sequence of $2^{15} - 1$ NRZ pseudorandom sequence. By comparing the performance achieved in the three cases, it is clear how this technique, known as the midpoint spectral inversion technique (MPSI), can remove all the dispersion-induced intersymbol interference that is strongly evident in the uncompensated curve, and performances very close to the back-to-back curve are obtained. The effectiveness of midpoint spectral inversion technique has been experimentally demonstrated for both single channel and WDM systems [27].

Transmission of 10 Gbit/s directly modulated DFB (high-chirped source) over 200 km of dispersive fiber has been achieved by adopting an OPC based on FWM in a semiconductor amplifier or through 360 km of a step index fiber by using an external cavity diode laser with a Mach Zehnder modulator and four-wave mixing in 21 km of a dispersion-shifted fiber to get spectral inversion. As far as the WDM experimental systems are concerned, two 10-Gbit/s wavelength division multiplexed channels have been transmission over 560 km of single-mode fiber by adopting a fiber-based OPC.

6.6   LINEAR DEVICES FOR DISPERSION COMPENSATION

Fiber dispersion can be compensated not only by using highly nonlinear devices as in phase conjugation but also by linear devices introducing a linear chirp opposite to those introduced by fiber propagation.

Linear dispersion-compensating devices can be utilized at the transmitter end or at the receiver end of a repeaterless system, or they can be joined with in-line amplifiers in repeatered systems. Since these devices introduce a linear chirp on the signal phase, they can effectively compensate fiber dispersion in the absence on nonlinear effects. In the presence of SPM the interplay between this effect and fiber dispersion makes linear compensation less effective. Therefore linear compensation is not suitable for long-distance systems where nonlinear effects cannot be neglected.

The two key parameters for an ideal dispersion-compensating device are the overall dispersion introduced by the device and its bandwidth. In fact it should have a dispersion opposite to that of the fiber and nearly constant over the transmitted signal bandwidth.

A figure of merit M for such devices has been introduced in [28] to quantify their equalizing capability. A significant figure of merit must increase with the device dispersion $\phi_{\nu\nu}$, which is defined as

$$\phi_{\nu\nu} = \left. \frac{\partial^2 \phi(\nu)}{\partial \nu^2} \right|_{\nu_0} \tag{6.9}$$

where $\phi(\nu)$ is the phase of the complex frequency response of the device, and also increase with the bandwidth B_0 over which $\phi_{\nu\nu}$ is constant. To obtain an adimensional figure of merit, it is reasonable to define M as proportional to $\phi_{\nu\nu}B_0^2$. The scale factor can be meaningfully determined by considering a real case: A gaussian pulse of 3 dB width equal to τ_0 is injected into a fiber link of length z and then is processed by a dispersion-compensating device with dispersion $\phi_{\nu\nu}$ to recover the fiber-induced broadening. In the absence of nonlinear effects, the pulses at the system output is a gaussian pulse whose standard deviation T can be evaluated by equation (2.60) in which the overall dispersion length is given by

$$L_D = \frac{T_0^2}{|\beta_2 + \phi_{\nu\nu}/z|} = \frac{4\ln(2)\tau_0^2}{|\beta_2 + \phi_{\nu\nu}/z|} = \left| \frac{L_D' L_D''}{L_D' - L_D''} \right| \tag{6.10}$$

where T_0 is the input pulse standard deviation, L_D' is the fiber dispersion length and L_D'' the compensation device dispersion length. Analogously to the fiber dispersion length, the compensator dispersion length is defined as $L_D'' = T_0^2 z/|\phi_{\nu\nu}|$. In order to recover the initial pulse width at the system output, it must be $\beta_2 = -\phi_{\nu\nu}/z$, that is $L_D' = L_D''$. Therefore the adimentional quality factor of the dispersion compensation device can be defined as $M = z/L_D''$. By noting that the optical bandwidth of the input pulse is given by $2\pi B_0 \tau_0 = 1$, we obtain

$$M = \frac{\pi^2}{\ln(2)} \phi_{\nu\nu} B_0^2 \tag{6.11}$$

As desired, the quality factor M defined in equation (6.11) is proportional to $\phi_{\nu\nu}B_0^2$ by a scale factor with a well-defined physical meaning. In the case of resonant devices, a dispersion increase corresponds to a bandwidth decrease so that devices with the same value of M can be characterized by different values of $\phi_{\nu\nu}$ and B_0.

Besides a high value of M, other requirements for dispersion-compensating devices are low insertion loss and insensitivity to the input field polarization. The main dispersion compensating devices are dispersion-compensating fibers (whose index profile is designed so that a few fiber kilometers provide a sufficient high dispersion at the transmission bandwidth), Gires-Tournois interferometers, chirped Bragg gratings, and cascaded Mach Zehnder filters.

In view of the large overall dispersion that is present in long links (for a chromatic dispersion $D = 17$ ps/nm/km, a fiber length $L = 100$ km, and bandwidth $B = 1\,\text{nm}$, the differential delay is $1700\,\text{ps}$ and the differential optical path length is $50\,\text{cm}$), it seems that good results can be obtained particularly with fiber devices: This technique has the advantage of being broadband and relatively simple. On the other hand, dispersion-compensating fibers present a large loss and a high nonlinear parameter. The large loss is due to the high level of germanium doping needed to obtain a high dispersion parameter inducing an attenuation of the order of 0.4–$0.5\,\text{dB/km}$ and to the length of the fiber (generally almost 25% of the overall link length). The high nonlinear coefficient (γ of the order of $5.2 \times 10^{-20}\,\text{m}^2/\text{W}$) is due to the small fiber core (generally the effective mode radius ρ_c is of the order of $4\,\mu\text{m}$).

Dispersion compensators based on optical filters are interesting, since they are compact (typically a few centimeters long), low loss, polarization insensitive, and offer high negative dispersion without exhibiting nonlinear behavior as dispersion compensating fiber devices.

In the following section we will mainly analyze dispersion-compensating fibers but include a brief discussion of optical filters for dispersion compensation.

6.6.1   Dispersion Compensation by Dispersion-Compensating Fibers

The basic principle of fiber-based dispersion-compensating devices is that by joining fibers with chromatic dispersion of opposite sign and suitable lengths, an average dispersion close to zero can be obtained [29]. The compensating fiber can be several kilometers long, but the reel is compact. The reel can be inserted at the transmitter, at the receiver, or at any point of the transmission link.

The length L' of dispersion-compensating fiber (DCF) necessary to compensate the chromatic dispersion β_2 accumulated along a link of length L is

$$L' = -\frac{\beta_2}{\beta_2'}L \tag{6.12}$$

where β_2' is the dispersion coefficient of the DCF.

The overall attenuation of the obtained fiber link (standard plus DCF fiber) is given by

$$L\alpha + L'\alpha' = L\left(\alpha + \frac{\beta_2}{|\beta_2'|}\alpha'\right) \tag{6.13}$$

where α' is the loss per unit of length of the DCF fiber. The insertion of the DCF fiber is thus equivalent to increase the fiber attenuation. In principle, this effect gets smaller as the DCF dispersion parameter is increased.

On the ground of the above discussion, a figure of merit for the DCF fiber can be defined starting from the fiber dispersion coefficient D' and from the fiber attenuation as

$$M = \frac{D'}{\alpha'} = \frac{2\pi c}{\lambda^2}\frac{|\beta_2'|}{\alpha'} \tag{6.14}$$

The value of M is a measure of the dispersion for unit of loss. Figures of merit up to 250 ps/nm/dB have been achieved by optimizing the refractive index profile of the core and cladding materials [30].

A first approach to increase the fiber dispersion coefficient is to increase the percentage of germanium oxide in the core. In this way two opposite effects result: a larger value of negative chromatic dispersion and an increasing of scattering loss. Thus the increase of the dispersion above a certain value causes the figure of merit M to decrease.

An alternative technique consists in doping the fiber cladding with fluorine ($F–SiO_2$); this increases the refractive index difference between the core and the cladding and results in reduced fiber loss and larger negative chromatic dispersion with respect to pure silica fibers.

Standard DCF have the same sign of β_3 of standard fibers; thus only second-order dispersion can be compensated. To obtain a zero average dispersion over a large bandwidth, even third-order dispersion must be compensated; thus DCF with negative β_3 must be obtained. By choosing exotic refractive index profile (e.g., a double-cladding structure), it is possible to satisfied the above condition. This property is particularly useful when WDM is adopted.

Another important DCF performance parameter is the nonlinear coefficient γ. The nonlinear coefficient of DCF tends to be larger than in standard fiber due to the small field diameter (4.5 μm) and a high concentration of germanium in the core. If the nonlinear behavior of the DCF is important, the nonlinear coefficient should be taken into account in the expression of the fiber quality parameter. In this case the definition for M can be modified as

$$M' = \frac{\rho_c'^2}{\rho_c^2} \frac{\alpha'}{\alpha} \frac{(e^{\alpha L} - 1)^2}{e^{\alpha L}} \frac{e^{\alpha' L'}}{(e^{\alpha' L'} - 1)^2} \qquad (6.15)$$

where ρ_c' is the mode effective radius of the DCF [31]. This new definition stresses that DCFs with the same figure of merit M can present different performances due to different values for the nonlinear figure of merit M'.

In designing repeaterless transmission systems using DCFs, besides DCFs performances, DCFs location is also a key issue because of SPM effect. They can be used in a pre- or postcompensation scheme depending on whether the DCF is placed just after the transmitter or before the receiver. Typically the postcompensation scheme seems to be more advantageous because, as in this case, the power at the input of the DCF is reduced with respect to the precompensation scheme, and the degrading effect due to SPM in DCF is reduced. Then, if a soliton system is considered, the postcompensation scheme is strongly recommended for avoiding undesirable chirping of the pulse before the transmission in the anomalous dispersion region of the SMF. Conversely, when the modulation instability effect in the transmission fiber dominates the other nonlinear effects, it is more convenient to adopt the precompensation scheme.

Finally DCF can present a high value of the polarization mode dispersion due to the large refractive index change between the core and the cladding and the small core diameter. Design optimization of DCF that incorporates this knowledge is another important issue [32].

The potentialities of systems adopting DCFs have been proved experimentally: A 2.4-Gbit/s signal has been transmitted over 306 km of standard single-mode fiber with the direct modulation scheme and up to 410 km with the help of remote amplification. A 10-Gbit/s signal transmission has been achieved along 150 km of the step-index fiber using DCF and a directly modulated source. By a combination of techniques such as prechiping, DCF, remotely pumped post- and preamplifiers, a 10-Gbit/s repeaterless system has been achieved, with a transmission distance of 411 km. The use of DCFs in WDM systems has permitted the operation at capacity as high as 4×10 Gbit/s on an 80-km step-index fiber link [33]. Already DCFs have been introduced in commercial Japanese telecommunication systems.

6.6.2 Optical Filters for Compensation of Fiber Dispersion

The operation principle is common to all filter-based dispersion compensators. In the following we will briefly describe the characteristics of the most common devices.

In conventional Bragg gratings and Gires-Turnois or Fabry-Perot interferometers, the basic idea is that each frequency component of the signal remains trapped in the interferometer structure for a time that is longer as the frequency approaches the interferometer resonance. Negative or positive delays are obtained depending on the position of the signal spectrum with respect to the resonance peak. The operation principle is illustrated in Fig. 6.11; the signal frequency component ν_2 is closer to the cavity resonance respect to ν_1 and as a result gains a greater delay.

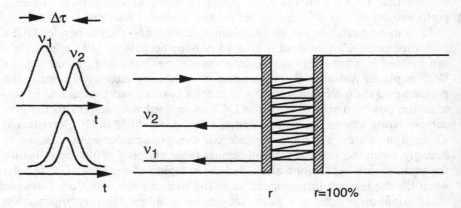

FIGURE 6.11 Principle of operation of Gires-Tournois interferometer.

The main problem of interferometric devices is that while the dispersion increases with the value of the cavity factor of the interferometer (i.e., with the mirror reflectivity), the resonance narrows, and hence the figure of merit decreases. This characteristic is partially counterbalanced by the periodic frequency characteristic of some interferometers such as the Gires-Tournois devices. In fact these compensators are effective in WDM systems because the periodical dispersion allows an equal dispersion compensation in each channel if the channel spacing is equal to the free spectral range of the interferometer.

The effectiveness of dispersion compensation is greatly enhanced if *linearly chirped* Bragg gratings are adopted instead of conventional Bragg gratings [34]. In this case the resonance frequency linearly depends on the position along the grating. It means that different frequency components of the broadened pulse can be reflected at different points, accumulating a delay that varies linearly with the frequency, as seen in Fig. 6.12. For this kind of device the delay is a key factor that fixes the length of the device (typically few centimeters). The sign of the dispersion is determined by the direction of propagation along the device. The capability of such device has been experimentally demonstrated in repeaterless systems up to a bit-rate × length product $RL = 1.6$ Tb/s × km [35].

In linearly chirped intermodal couplers the delay is introduced when the different frequency components of the signal are coupled to a waveguide mode with a smaller or greater group velocity by means of a refractive index grating in photosensitive optical fiber [36]. This produces a behavior analogous to chirped Bragg gratings.

A device that does not require index grating is a two-core fiber structure. In this case the group velocity in the two cores is different, and the diameter is linearly tapered along the fiber length. As a result different frequency components are coupled to one of the two cores in different positions, experiencing a different delay [37].

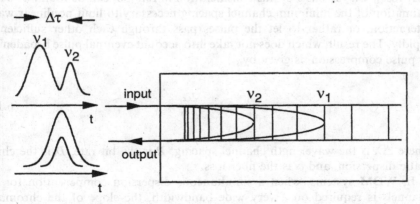

FIGURE 6.12 Principle of operation of linearly chirped fiber Bragg grating.

6.7 WAVELENGTH DIVISION MULTIPLEXING TECHNIQUES

The repeaterless WDM transmission technique seems to be a powerful way to upgrade the capacity of single-channel systems. In this case the bit-rate, and hence the bandwidth, per channel can be lowered at the expense of the number of channels. Thus degradation due to dispersive effect are reduced, but new contributions from fiber nonlinearities must be considered as the cross-phase modulation (XPM) effect and the crosstalk induced by four-wave mixing (FWM) and stimulated Raman scattering (SRS) among channels (see also Chapter 2).

In the absence of chromatic dispersion and loss, the XPM effects result in a pure phase term proportional to the sum of the power transmitted on each channel according to equation (2.37). From that equation it is clear that the contribution due to XPM is doubled compared to that of SPM. In a nondispersive medium the spectral broadening, at least in principle, will not result in a pulse degradation, particularly in the case of direct detection scheme where the phase terms disappear revealing the square of the optical field. In general, in the presence of chromatic dispersion the behavior is more complicated and the nonlinear Schrödinger equation cannot be solved analytically. The chromatic dispersion is responsible for collisions among channels at different wavelengths, since they travel with different group velocity. When the pulses spatially overlap, nonlinear FWM generates new spectral components. The power evolution of such components and the efficiency of the FWM process depend on the chromatic dispersion of the fiber according to equations (2.48) and (2.49), which reveal a dramatic buildup of spurious waves approaching the zero chromatic wavelength. If the time of collision is short enough, so that the loss of power due to the fiber is negligible, the spectral distortion due to the wave interaction disappears after the collision. As a consequence a moderate amount of chromatic dispersion can be helpful in letting sufficiently spaced channels to cross each other completely on an absorption length ($L_{loss} = 1/\alpha$, where α is the fiber loss). On the basis of this consideration in ref. [38] can be found a rough estimation of the minimum channel spacing necessary to limit nonlinear wave interaction, or rather to let the pulses pass through each other sufficiently rapidly. The result, which does not take into account eventual pulse broadening or pulse compression, is given by

$$\Delta\lambda = \frac{2\alpha}{RD} \tag{6.16}$$

where $\Delta\lambda$ is the wavelength channel spacing, R is the bit rate, D is the chromatic dispersion, and α is the fiber loss.

In WDM systems, when a simultaneous dispersion compensation for all channels is required on a very wide bandwidth, the slope of the chromatic dispersion of the fiber can play a key role [39]. This aspect will be investigated in Chapter 8, where this effect is more evident because it accumulates along the distance.

Another important nonlinear effect responsible for system performance degradation is the stimulated Raman scattering [40]. It leads to a transfer of power from short wavelength to longer wavelength channels within the Raman gain bandwidth (about 10 THz) with an efficiency that depends on the wavelength spacing (see Fig. 2.8). Thus the dependence of SRS on the channel spacing is in the opposite direction with respect to the behavior observed in the case of XPM. An important feature of SRS is that it exhibits a threshold-like behavior; namely a significant conversion of pump energy occurs only when the pump intensity exceeds a certain threshold level. It has been demonstrated that assuming a triangular Raman gain profile, if the peak power in each channel does not exceed the value estimated by equation (2.52), system degradation due to SRS may be neglected.

When standard SMF are adopted, the degradation due to FWM are limited by the high value of the chromatic dispersion of the fiber. Numerical simulations can indicate the dependence of XPM on input power and wavelength spacing among channels and its effect on system performance.

A comparison of the performance in terms of eye penalty as a function of input powers for one-channel and eight-channel transmission propagating in the regime of zero average chromatic dispersion is provided in Fig. 6.13. A 10-Gbit/s signal is transmitted over 150 km, and the chromatic dispersion of standard SMF ($\beta^2 = -20$ ps^2/km) is postcompensated by a DCF ($\beta^2 = 80$ ps^2/km). This configuration both prevents chromatic dispersion

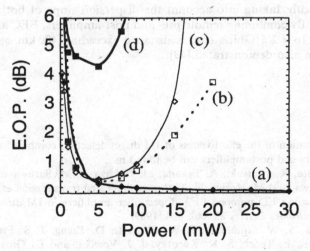

FIGURE 6.13 Eye opening penalty versus input power for a single channel (a) and 8-channel systems with a channel frequency spacing Δf equal to 150 GHz (b), 75 GHz (c), 35 GHz (d).

impairments and limits FWM efficiency, since the average chromatic dispersion is equal to zero, while its local value is different from zero. The channel spacing is posed at 35 GHz, 75 GHz, and 150 GHz. The system degradation in WDM curves appears as the nonlinearities become significant, increasing the power. The figure also shows that as the signal power is increased, more channel separation is required to match the same value of system performance. As the channel spacing is increased the WDM systems approach the single-channel curve.

By combining the technique for compensating dispersion on the single-channel system with the WDM technique, very high capacity can be achieved. But new requirements and tolerances have to be observed due to the enhanced contribution from the nonlinearities. These effects are particularly evident when DCFs are adopted, since the nonlinear coefficient of a DCF is typically four times larger than in the case of a standard SMF.

By prechirping a signal with an external modulator to suppress the SPM effect and by postcompensating the dispersion with a DCF, 10 Gbit/s, 4-channel (40 Gbit/s) over 200 km, and 16-channel over 150 km, repeaterless transmissions have been obtained over the standard SMF (24 Tbit/s × km) [41]. In this case the input power was the result of a compromise between the necessity to reduce the contribution from the nonlinearities and to maximize the signal-to-noise ratio (SNR), further degraded by the DCF loss. Moreover an optimum value for the residual chromatic dispersion was found as a result of a balance between the pulse compression due to the SPM and the XPM and pulse broadening in the DCF. The tolerance band of the residual chromatic dispersion is a relevant factor in simultaneous dispersion compensation for all channels. In this case it should be wide enough to match the WDM signal bandwidth, taking into account the dispersion slope of both the SMF and the DCF. By combining remote pre- and post-amplifiers, FEC and Raman amplification 16 × 2.5 Gbit/s over a distance exceeding 400 km on standard SMF has been also demonstrated [42].

REFERENCES

1. A demonstration of the effectiveness of IM/direct detection combined with EDFA preamplifiers and postamplifiers can be found in
K. Hagimoto, K. Iwatsuki, A. Takada, M. Nakazawa, M. Saruwatari, K. Aida, K. Nakagawa, and M. Horiguchi, 250 km nonrepeated transmission experiment at 1.8 Gbit/s using LD pumped Er^{3+} doped fiber amplifiers in IM/direct detection system, *Electronics Letters* 25: 662–664, 1989.
Y. K. Park, S. W. Granlund, T. W. Cline, L. D. Tzeng, J. S. French, J.-M. Delavaux, R. E. Tench, S. K. Korotky, J. J. Veselka, and Di Giovanni, 2.488 Gbit/s-318 km repeaterless transmission using Erbium doped amplifiers in a direct detection system, *IEEE-Photonics Technology Letters* 4: 179–182, 1992.
An exemplum of remotely pumped EDFA is reported in

O. Gautheron, S. S. Sian, G. Grandpierre, M. S. Chaudhry, J. L. Pamart, T. Barbier, E. Bertin, P. Bonno, E. Brandon, M. Genot, P. Marmier, M. Mesic, P. M. Gabla, and P. Bousselet, 481 km, 2.5 Gbit/s, and 501 km, 622 Mbit/s, unrepeatered transmission using forward error correction and remotely pumped postamplifiers and preamplifiers, *Electronic Letters* 31: 378–379, 1995.

2. An analysis of error correction coding is extensively discussed in
   G. C. Clark and J. B. Cain, Error correction coding for digital communication, Plenum, New York, 1981.
   A result of FEC implementation can be found in
   P. M. Gabla, J. L. Pamart, R. Uhel, E. Leclerc, J. O. Frorud, F. X. Ollivier, and S. Borderieux, 401 km, 622 Mb/s and 357 km, 2.488 Gb/s IM/DD repeaterless transmission experiments using erbium-doped fiber amplifiers and error correcting code, *IEEE-Photonics Technology Letters* 4: 1148–1151, 1992.

3. Computations of chromatic dispersion impairments on coherent detection systems using different modulation formats can be found in
   A. F. Elrefeie, R. E. Wagner, D. A. Atlas, and D. G. Daut, Chromatic dispersion limitations in coherent lightwave transmission systems, *Journal of Lightwave Technology* 6: 704–709, 1994.

4. Chromatic dispersion limitations for directly modulated systems can be found in
   S. Fujita, M. Kitamura, T. Torikai, N. Henmi, H. Yamada, T. Suzaki, I. Takano, and M. Shikada, 10 Gbit/s, 100 km optical fibre transmission experiment using high speed MQW DFB-LD and back illuminated GaInAs APD, *Electronics Letters* 25: 702–703, 1989.

5. An overview of the limitations on lightwave systems due to fiber nonlinearities can be found in
   A. R. Chraplyvy, Limitations on lightwave communications imposed by optical fiber nonlinearities, *Journal Lightwave Technology* 8: 1548–1557, 1990.

6. A general treatment of stimulated Brillouin scattering can be found in
   E. P. Ippen and R. H. Stolen, Stimulated Brillouin scattering in optical fiber, *Applied Physics Letters* 21: 539–541, 1972.
   For an analysis more focused on the limitations due to SBS on to lightwave systems, on suppressing SBS, see
   D. A. Fishman and J. A. Nagel, Degradations due to stimulated Brillouin scattering in multigigabit intensity-modulated fiber-optics systems, *Journal of Lightwave Technology* 11: 1721–1728, 1993.
   Y. Aoki, K. Tajima, and I. Mito, Input power limits of single mode optical fibers due to stimulated Brillouin scattering in optical communication systems, *Journal of Lightwave Technology* 6: 710–799, 1988.

7. A general treatment of self-phase modulation is found, for example, in
   R. H. Stolen and C. Lin, Self-phase modulation in silica optical fiber, *Physical Review* A 17: 1448–1453, 1978.

8. L. D. Pedersen, B. Velschow, C. F. Pedersen, and F. Ebskamp, Uncoded NRZ 2.488 Gbit/s transmission over 347 km standard fiber, 67 dB span loss, using remotely pumped amplifier, SBS suppression, SPM optimisation and Raman gain, *Technical Digest of OFC'94*, paper PD28.
   An example of BER performance under SBS suppression by broadening the laser linewidth can be found also in

Y. K. Park, O. Mizuhara, L. D. Tzeng, J. M. P. Delavaux, T. V. Nguyen, M. L. Kao, P. D. Yeates, and J. Stone, A 5 Gbit/s repeaterless transmission system using erbium-doped fiber amplifiers, *IEEE-Photonics and Technology Letters* 5: 78–82, 1993.

9. The influence on self-phase modulation of the pulse shape and the sequence length have been investigated in:
   O. Gautheron, J.-L. Beylat, and P. Bousselet, Experimental investigation of stimulated Brillouin scattering and self phase modulation effects on long distance 2.5 Gbit/s repeaterless transmission, *Proceedings of ECOC'93*, Montreux, Switzerland, September 1993, paper TUC4.5.
   O. Gautheron, SPM effects in 622-Mbit/s and 2.5-Gbit/s repeaterless transmission with launched powers as high as +24.5 dBm. *Technical Digest of OFC'94*, paper FC2, p. 289, 1994.

10. The idea of predistorted pulse is already present in the following paper, where the prechirp is electronically implemented
    T. L. Koch and R. C. Alferness, Dispersion compensation by active predistorted signal synthesis, *IEEE-Journal of Lightwave Technology*, 800–805, 1985.
    A detailed theoretical and experimental investigation can be found in
    N. Henmi, T. Saito, and T. Ishida, Prechirp technique as a linear dispersion compensation for ultrahigh speed long span intensity modulation directed detection optical communications systems, *IEEE-Journal of Lightwave Technology* 12: 1706–1719, 1994.

11. N. Henmi, T. Saito, M. Yamaguchi, and S. Fujita, 10 Gbit/s, 100 km normal fiber transmission experiment employing a modified prechirp technique, *Technical Digest of OFC'91*, San Diego, CA, 1991.

12. More details on the chirp parameter of external modulator can be found in
    F. Koyoma, and K. Iga, Frequency chirping in external modulators, *Journal of Lightwave Technology* 6: 87, 1988.

13. B. F. Jørgensen, R. J. S. Pedersen, and C. Rasmussen, Transmission of 10 Gbit/s beyond the dispersion limit of standard single mode fibers, *Proceedings of the 21st ECOC'95*, pp. 557–564, Brussels.

14. A. D. Ellis, S. J. Pycock, D. A. Cleland, and C. H. F. Sturrock, Dispersion compensation in 450 km transmission system employing standard fibre, *Electronics Letters* 28: 954–955, 1992.

15. The general principle of operation of DST and first experimental results are presented in
    B. Wedding, New method for optical transmission beyond dispersion limit, *Electronics Letters* 28: 1298–1300, 1992.
    A more extended analysis of the method and other experimental results can be found in
    B. Wedding, B. Franz, and B. Junginger, 10-Gb/s optical transmission up to 253 km via standard single mode fiber using the method of dispersion supported transmission, *IEEE-Journal of Lightwave Technology* 12: 1720–1727, 1994.
    B. Wedding, B. Franz, B. Junginger, B. Clesca, and P. Bousselet, Repeaterless optical transmission at 10 Gbit/s via 182 km of standard single mode fiber using a high power booster amplifier, *Electronics Letters* 29: 1498–1500, 1993.
    B. Wedding and B. Franz, Unregenerated optical transmission at 10 Gbit/s via

204 km of standard single mode fiber using a directly modulated laser diode, *Electronics Letters* 29: 402–404, 1993.

16. C. J. Rasmussen, B. F. Jørgensen, R. J. S. Pedersen, and F. Ebskamp, Optimum amplitude and frequency modulation in an optical communication system based on dispersion supported transmission, *Electronics Letters* 31: 746–747, 1995.

17. The role of nonlinearities in the scheme of DST has been investigated in
C. Kurtzke and A. Gnauck, Operating principle of in-line amplified dispersion-supported transmission, *Electronics Letters* 29: 1969–1970, 1993.
The impact of residual amplitude has been evaluated in
C. Kurtzke, S. Kindt, and K. Petermann, Impact of residual amplitude modulation on the performance of dispersion supported and dispersion mediated nonlinearity enhanced transmission, *Electronics Letters* 30: 988–990, 1994.

18. An analysis of the propagation of NRZ signal along optical fiber links, taking into account the effects of chromatic dispersion and Kerr nonlinearity, for different modulation formats can be found in
E. Iannone, F. S. Locati, F. Matera, M. Romagnoli, and M. Settembre, Performance evaluation of single channel coherent systems in presence of nonlinear effects, *Electronics Letters* 28: 645–646, 1992.
E. Iannone, F. S. Locati, F. Matera, M. Romagnoli, and M. Settembre, Nonlinear evolution of ASK and PSK signals in repeaterless fibre links, *Electronics Letters* 28: 1902–1903, 1992.
J. P. Hamaide and P. Emplit, Limitations in long haul IM/DD optical fibre systems caused by chromatic dispersion and nonlinear Kerr effect, *Electronics Letters* 26: 1451–1452, 1990.
N. Suzuki and T. Ozeki, Simultaneous compensation of laser chirp, Kerr effect, and dispersion in 10 Gbit/s long haul transmission systems, *IEEE-Journal of Lightwave Technology* 11: 1486–1494, 1994.

19. The effectiveness of nonlinear assisted transmission has been verified experimentally in
P. I. Kuindersma, P. P. G. Mols, G. L. A. Hofstad, G. Cuypers, M. Tomesen, T. Dongen, and J. J. M. Binsma, Non-linear dispersion compensation: Repeaterless transmission of 10 Gbit/s NRZ over 107 km standard fibre with an EA-MOD/DFB module. *Technical Digest of ECOC 93*, paper ThP12.10, 1993.
L. D. Pedersen, B. Velschow, C. F. Pedersen, and F. Ebskamp, Uncoded NRZ 2.488 Gbit/s transmission over 347 km standard fiber, 67 dB span loss, using a remotely pumped amplifier, SBS suppression, SPM optimisation, and Raman gain, *Technical Digest of OFC'94*, paper PD28, 1994.
O. Gautheron, SPM effects in 622 Mbit/s, and 2.5 Gbit/s repeaterless transmission with launched powers as high as +24.5 dBm, *Technical Digest of OFC'94*, Paper FC2, 1994.

20. The interplay between the source chirp and chromatic dispersion has been investigated in a purely linear regime in
A. H. Gnauck, S. K. Korotky, J. J. Veselka, J. Nagel, C. T. Kemmerer, W. J. Minford, and D. T. Moser, Dispersion penalty reduction using an optical modulator with adjustable chirp, *IEEE Photonics Technology Letters* 3: 916–918, 1991.
Also for interplay in the presence of fiber Kerr nonlinearity, see
A. H. Gnauck, R. W. Tkach, and M. Mazurczyk, Interplay of chirp and self phase

modulation in dispersion limited optical transmission systems, *Proceedings of ECOC'93*, Montreux, Switzerland, paper TuC4.4, 1993.

21. A. Yariv, D. Fekete, and D. M. Petter, Compensation for channel dispersion by nonlinear optical phase conjugation, *Optics Letters* 4: 52–54, 1979.

22. An investigation on the role of nonlinearity in a step-index fiber system based on optical phase conjugation in the presence of lumped optical amplification can be found in
W. Pieper, C. Kurtzke, R. Schanbel, D. Breuer, R. Ludwig, H. G. Weber, and K. Petermann, Nonlinearity-insensitivity standard fiber transmission based on optical phase conjugation in a semiconductor laser amplifier, *Proceedings of ECOC'94*, 2, pp. 729–732, Florence, 1994.

23. For a theoretical analysis of utilization of OPC in periodically amplified soliton transmission systems, see
W. Forysiak and N. J. Doran, Conjugate solitons in amplified optical fibre transmission systems, *Electronics Letters* 30: 154–155, 1994.

24. An application of OPC by means of FWM in dispersion shifted fibers can be found in
S. Watanabe, T. Naito, and T. Chikama, Compensation of chromatic dispersion, in a single mode fiber by optical phase conjugation, *IEEE-Photonics Technology Letters* 5: 92–95, 1993.
See also
A. H. Gnauk, R. M. Jopson, and R. M. Derosier, 10 GBit/s 360 km transmission over dispersive fiber using midsystem spectral inversion, *IEEE-Photonics Technology Letters* 5: 663–666, 1993.

25. The advantages offered by OPC by means of semiconductor laser amplifiers are outlined in the paper.
M. C. Tatham, G. Sherlock, and L. D. Westbrook, Compensation of fibre chromatic dispersion by optical phase conjugation in a semiconductor laser amplifier, *Electronics Letters* 29: 1851–1852, 1993.

26. A technique for OPC with an efficiency independent of the signal polarization is demonstrated using either DS fibers or semiconductor optical amplifiers as the nonlinear medium in
R. M. Jopson and R. E. Tench, Polarisation independent phase conjugation of lightwave signals, *Electronics Letters* 29: 2216–2217, 1993.
A demonstration of polarization independent OPA in SOA by two orthogonally polarized pumps at two different wavelengths is reported for a 10 Gbit/s signal propagating over 160 step-index fiber in
P.-Y. Cortès, M. Chbat, S. Artigaud, J.-L. Beylat, and J. Chesnoy, Below 0.3 dB polarization penalty in 10 Gbit/s directly modulated DFB signal over 160 km using mid-span spectral inversion in a semiconductor optical amplifiers, *Proceedings of ECOC'95*, pp 271–274, Brussels, paper Tu.B.2.3, 1995.

27. Some significative experimental results on mid-point spectral inversion technique can be found in
M. C. Tatham, X. Gu, L. D. Westbrook, G. Sherlock, and D. M. Spirit, 200 km transmission of 10 Gbit/s directly modulated DFB signals using mid-span spectral inversion in a semiconductor optical amplifier, *Proceedings of ECOC'94*, vol. 2, pp. 733–736, Florence, 1994.

A. Røyset, S. Y. Set, A. Goncharenjo, and R. I. Laming, Transmission of <10 ps pulses over 318 km standard fiber using midspan spectral inversion, *Proceedings of ECOC'95*, Brussels.

A. D. Ellis, M. C. Tatham, D. A. O. Davies, D. Nesset, D. G. Moodie, and G. Sherlock, 40 Gbit/s transmission over 202 km of standard fiber using midspan spectral inversion, *Electronic Letters* 30: 299–301, 1994.

A. H. Gnauck, R. M. Jopson, P. P. Iannone, and R. M. Derosier, Transmission of two wavelength multiplexed 10 Gbit/s channels over 560 km of dispersive fibre, *Electronic Letters* 30: 727–728, 1994.

28. The definition of a figure of merit for dispersion compensating devices can be found in
R. G. Priest and T. G. Giallorenzi, Dispersion compensation in coherent fiber optic communications, *Optics Letters* 12: 622–624, 1987.

29. C. Lin, H. Kogelnick, and L. G. Cohen, Optical pulse equalization of low dispersion transmission in single mode fibers in 1.3–1.7 μm spectral region, *Optics Letters* 5: 476–478, 1980.

30. The optimum range of the core's refractive index in achieving high figure of merit and the dependence of cladding materials on fiber loss are investigated in
M. Onishi, Y. Koyano, M. Shigematsu, H. Kanamori, and M. Nishimura, Dispersion compensating fibre with a high figure of merit of 250 ps/nm/dB, *Electronics Letters* 30: 161–163, 1993.

31. The new figure of merit M' was introduced in the following paper:
F. Forghieri, R. W. Tkach, A. R. Chraplyvy, and A. M. Vengsarkar, Dispersion compensating fiber: Is there merit in the figure of merit? *Proceedings of OFC'96*, Technical Digest, Paper ThM5, pp. 255–257, 1996.

32. Some indications about dispersion-compensating fiber characteristics and their dependence on the percentage of doping may be found in
M. Onishi, H. Kanamori, and T. Kato, Optimization of dispersion compensating fibers considering self phase modulation suppression, *OFC'96 Technical Digest*, paper ThA2, pp. 200–201, 1996.

33. Some significative experimental results obtained by using DCFs in unrepeatered systems can be found in
M. Kakui, T. Kato, T. Kashiwada, K. Nakazato, C. Fukuda, M. Onishi, and M. Nishimura, 2.4 Gbit/s repeaterless transmission over 306 km non-dispersion–shifted fiber using directly modulated DFB–LD and dispersion compensating fiber, *Electronics Letters* 31: 51–52, 1995.

M. S. Chaudhry, S. S. Sian, K. Guild, P. R. Morkel, and C. D. Stark, Single span transmission of 2.5 Gbit/s over 410 km with remote amplification and dispersion compensating, *Proceedings of ECOC'94*, vol. 4, pp. 19–22, Postdeadline papers, 1994.

P. B. Hansen, L. Eskildsen, S. G. Grubb, A. M. Vengsarkar, S. K. Korotky, T. A. Strasser, J. E. J. Alphonsus, J. J. Veselka, D. J. Di Giovanni, D. W. Peckham, D. Truxal, W. Y. Cheung, S. G. Kosinski, and P. F. Wysocki, 10 Gbit/s, 411 km repeaterless transmission experiment employing dispersion compensation and remote post- and pre-amplifiers, *Proceedings of ECOC'95*, Brussels, paper We.B.1.2, 1995.

G. Ishigawa, M. Sekiya, H. Onaka, H. Nishimoto, and T. Chikama, Optimization of pre-chirping and dispersion compensation for 10 Gbit/s, repeaterless transmission using standard single mode fiber, *Proceedings of ECOC'94*, vol. 2, pp. 693–696, 1994.

A. D. Ellis and D. M. Spirit, Unrepeatered transmission over 80 km standard fibre at 40 Gbit/s, *Electronics Letters* 30: 72–73, 1994.

34. For a review of all-fibers dispersion compensating devices based on distributed resonant coupling, see
F. Oullette, J. F. Cliche, and S. Gagnon, All fiber devices for chromatic dispersion compensation based on chirped distributed resonant coupling, *IEEE-Journal of Lightwave Technology* 12: 1728–1738, 1994.

35. The utilization of linearly chirped Bragg gratings for dispersion compensation has been proposed in
F. Ouellette, Dispersion cancellation using linearly chirped Bragg grating filters in optical waveguide, *Optics Letters* 12: 847–849, 1987.
Experimental demonstrations can be found in
D. Garthe, W. S. Lee, R. E. Epworth, T. Bircheno, and C. P. Chew, Practical dispersion compensation based on fibre gratings with a bitrate length product of 1.6 Tbs-km, *Proceedings of ECOC '94*, Postdeadline Papers, vol. 4, pp. 11–14, 1994.
S. V. Chernikov, J. R. Taylor, and R. Kashyap, Dispersion compensation of 100 Gbit/s optical fiber transmission using a chirped fiber grating, *Proceedings of OFC '95*, Paper WB5, 1995.
P. A. Krug et al., *Proceedings of OFC '95*, Postdeadline paper PD27, 1995 (compensazione su 270 km a 10 Gb/s).
R. I. Laming, N. Robinson, P. L. Scrivener, S. Barcelos, L. Reekie, J. A. Tucknott, and M. N. Zervas, A dispersion tunable grating in a 10 Gbit/s 100–220 km step index fibre links, *Proceedings of ECOC '95*, pp. 585–587, Paper We.B1.7, 1995.

36. C. D. Poole, J. M. Wiesenfeld, and A. R. McCormick, Broadband dispersion compensation by using the higher-order spatial mode in a two mode fiber, *Optics Letters* 17: 985–987, 1994.

37. F. Ouellette and Y. Duval, Optical equalisation with linearly tapered two dissimilar core fibre, *Electronics Letters* 27: 1668–1670, 1991.

38. D. Marcuse, A. R. Chraplyvy, and R. W. Tkach, Dependence of cross phase modulation on channel number in fiber WDM systems, *IEEE-Journal of Lightwave Technology* 12: 885–890, 1994.

39. K. Oda, M. Fukutoku, M. Fukui, T. Kitoh, and H. Toba, 16 × 10 Gbit/s optical FDM over a 1000 km conventional single mode fiber employing dispersion compensating fiber and gain equalization, *Proceedings of OFC '95*, San Diego, February 26, March 3, 1995, Postdeadline paper PD-22.

40. A. R. Chraplyvy, Optical power limits in multi-channel wavelength division multiplexed systems due to stimulated Raman scattering, *Electronics Letters* 20: 58–59, 1984.

41. A theoretical and experimental investigation on the power penalty induced by XPM in WDM systems can be found in
H. Onaka, H. Miyata, K. Otsuka, and T. Chikama, 10 Gbit/s, 4-wave 200 km and 16-Wave 150 km repeaterless transmission experiments over standard single mode fiber, *Proceedings of ECOC'94*, pp. 49–52, Postdeadline papers, 1994.

An extended analysis to the link in which two EDFA amplifiers with DCF in the middle are included can be found in

H. Miyata, H. Onaka, K. Otsuka, and T. Chikama, Dispersion compensation design for 10 Gbit/s, 16-wave WDM transmission system over standard single mode fiber, *Proceedings of ECOC'95*, Brussels, paper Mo.A.4.3, pp. 63–66, 1995.

42. S. S. Sian, S. M. Webb, and K. M. Guild, 16×2.5 Gbit/s WDM unrepeated transmission over 427 km (402 km without forward error correction), *Proceedings of ECOC'95*, Brussels, paper Th.A.3.3, pp. 975–978, 1995.

CHAPTER SEVEN

Long Distance TDM Transmission

Very long, high-capacity transmission links are present both in the transport area and in the submarine area of the telecommunication network. Point-to-point links with length of the order of 5000 km are present in the terrestrial telephone network in the United States and in Europe; submarine cables connecting Europe and North America are about 6000 km long, and cables connecting North America and Japan are as long as 9000 km.

Such long links are traditionally realized adopting optical transmission with electronic regeneration: The signal is periodically detected and decoded so as to recover the transmitted message; then this message is encoded again onto a newly generated signal. Due to electronic regeneration, signal attenuation and distortion do not accumulate along the link. However, the electronic regenerator is a complex and expensive device, especially when high-speed signals (at a bit-rate of 2.5 or 10 Gbit/s) have to be processed or when the device is designed for submarine use.

In the last ten years, high-gain, low-noise optical amplifiers have been developed. These devices can be used to compensate the fiber attenuation in long communication links, substituting electronic regenerators; in this way the optical link becomes *transparent*, since no optoelectronic conversion occurs.

Optical amplifiers are much cheaper than regenerators; moreover a transparent link can be upgraded by increasing the bit-rate or the transmission format without changing in-line devices while upgrading a transmission link and adopting electronic regenerator needs to upgrade all these devices. This last issue is particularly important in the case of submerged systems, where the operations needed to recover, upgrade, and reinstall an in-line device are particularly expensive. The price to be paid for the substitution of regenerators with amplifiers is the accumulation of optical noise and signal distortions along the link, thus limiting the transmission capacity.

Numerous experiments have shown that a huge information capacity (up to some tens of Gbit/s) can be transmitted along optically amplified systems a few thousand kilometers long. Point-to-point links up to 10,000 km long have been realized, maybe the maximum distance that can be useful in the telecommunication network. A list of important experiments is reported in [1]. Some of the experiments listed in [1] are also analyzed in more detail in Tables 7.1 and 7.2, where soliton-based and NRZ-based systems are considered, respectively.

In very long optical links, fiber dispersion and nonlinearities become important effects, since they cumulate along the link length. For this reason the analysis of propagation of a transmitted signal is generally complicated. One important exception is the soliton system for which complete analysis can be carried out, as shown in Chapter 5, obtaining accurate results. Even in soliton systems, however, the theoretical approach can be complicated by particular circumstances. For example, the highest cost of a long link is due to optical amplifiers, so increasing amplifier spacing is a main issue in system design. If the amplifier spacing is increased comparably with the soliton period, the average Schrödinger equation, on which the analytical theory of solitons in amplified systems is based, does not hold any more.

In this chapter, high-capacity, very long optical links using time division multiplexing (TDM) are analyzed. A similar analysis is carried out in Chapter 8 for WDM systems. The performance analysis is mainly based on the simulation technique detailed in Appendix A1; thus the main considered performance evaluation parameter is the Q factor (see Section 4.2). This choice is a consequence of the fact that when strong nonlinear interaction arises between the transmitted signal and the optical noise, the received signal probability density function cannot be evaluated theoretically, and to be accurate, even its statistical estimation requires a huge number of simulation runs. When it is the case, the estimated jitter variance is also reported. As transmitters, we assume the use of lasers as being extremely modulated, so that the source chirp contribution can be disregarded.

Both soliton and nonsoliton IM-DD systems are considered in this chapter. A separate analysis of these systems is not justified because transmitted pulses tend to change into solitons during propagation over very long links in the anomalous dispersion region, so that the difference among soliton and nonsoliton systems tends to vanish. Moreover, from a design point of view, it is more meaningful to compare difference solutions that can satisfy the same requirements.

This chapter is divided into three sections. Section 7.1 gives a detailed investigation of optical systems operating with a low PMD in the third-transmission window ($\lambda \sim 1.55\,\mu m$). In this case EDFAs are adopted.

Depending on the GVD distribution along the link, the various signal propagation regimes can be distinguished and Section 7.1 is divided into subsections in which systems operating in the same propagation regime are analyzed and compared. The categories used here will reappear in our studies of WDM systems (in Chapter 8) and transmission through all-optical networks (in

TABLE 7.1 Some important characteristics of long distance transmission experiments adopting soliton signaling.

λ (μm)	R (Gbit/s)	L (km)	β_2 (ps/(km·nm))	Amplifier spacing (km)	Soliton width (ps)	R × L (Gbit/s*km)	Notes	Lab.	Ref. [1]
1.555	2.5	14,000	−1.74	27	50	3.5×10^4	Recirculating loop, fix filtering after EDFA (4 nm filters)	AT&T	e6
1.554	20	11,500	−0.27	30	11.5	2.3×10^5	Recirculating loop, fix filtering (2.1 nm filters) unequal amplitude solitons	KDD	e7
1.556	10	20,000	−0.576	26	18	2×10^5	Recirculating loop, sliding filters	AT&T	e8
1.543	10	27,000	−0.512	35	22	2.7×10^5	Recirculating loop, sliding signal, fix filters	CNET	e9
1.552	10	10^6	−1.92	50	36/42	10^7	Recirculating loop, in-line filtering and pulse reshaping (Ti:LiNbO$_3$ modulator)	NTT	e10
1.543	10	10^6	−0.512	70	22	10^7	Recirculating loop, in-line filtering and pulse reshaping (Ti:LiNbO$_3$ modulator)	CNET	e11

Note: In the last column of the table the reference in which the experiment is reported in detail is indicated. All the papers cited in this table are listed in note [1] of this chapter's reference list.

TABLE 7.2 Some important characteristics of long distance transmission experiments adopting NRZ signaling.

λ (μm)	R (Gbit/s)	L (km)	Average β_c (ps/(km·nm))	In-line amplifier	R×L (Gbit/s*km)	Notes	Lab.	Ref.
1.552	10	6000	0.126	119	6×10^4	3-nm receiver optical filter, GVD optical compensation	NTT	[e12]
1.558	10	9000	0	274	9×10^4	Sawtooth dispersion link, 1-nm receiver optical filter	KDD AT&T	[e13] [e14]
≈1.55	2.5	9720	0	108	2.4×10^4	Actual submarine cable	NTT	[e14]
	10	6480	0	72	6.4×10^4	Actual submarine cable	KDD	[e15]
1.559	5.3	11,300	0	189	6×10^4	Actual submarine cable	Alcatel	[e16]
1.559	5	8100	≈0	181	40,500	Polarization scrambling	NTT	[e17]
≈1.55	20	5520	≈0	138	1.1×10^5	RZ transmission		

Note: In the last column of the table the reference in which the experiment is reported in detail is indicated. All the papers cited in this table are listed in note [1] of this chapter's reference list.

Chapter 9). In particular, in Section 7.1.1 the different propagation regimes are defined and analyzed; in Section 7.1.2 the performance of systems operating in the regime of low constant chromatic dispersion are analyzed and extended; in Section 7.1.3 to systems in the regime of high constant chromatic dispersion. The propagation condition of fluctuating GVD is analyzed in Section 7.1.4, where particular attention is devoted to the method of dispersion management both for limiting NRZ spectral broadening and for controlling soliton propagation.

It has to be emphasized that the terms *low dispersion* and *high dispersion* are used in this chapter in different ways from their usual meanings (adopted in other parts of this book). The dispersion of the fiber link is considered low when a wide spectral broadening occurs during transmission due to the nonlinear interaction between the transmitted signal and the optical noise. When the dephasing introduced by the GVD limits the spectral broadening, the GVD is considered high. Thus a GVD of $2\,\mathrm{ps^2/km}$ is considered high by this point of view, while generally it is referred as a low GVD, since it is compared implicitly with the GVD of step-index fibers in the third-transmission window of the order of $17\,\mathrm{ps^2/km}$. A GVD of $0.05\,\mathrm{ps^2/km}$ is considered low by the point of view adopted in this chapter in almost every practical condition. The concept of high and low GVD will be more accurately analyzed in Section 7.1.1 where we consider the different propagation regimes.

Transmission systems in which polarization evolution can impair transmission or increase system capacity are dealt with in Section 7.2. Finally, in Section 7.3, very long communication systems using step-index fibers and operating in the second-transmission window are analyzed. This is an important problem since many step-index fibers have been installed in Europe and the United States.

The possible use of doped fiber amplifiers in the second transmission window ($\lambda \sim 1.3\,\mu\mathrm{m}$) is not considered in this book, since such amplifiers are not yet completely developed for system applications. Thus in-line amplification is obtained by means of semiconductor amplifiers.

7.1 SYSTEMS IN THE THIRD-TRANSMISSION WINDOW WITH LOW PMD

In this section the performances of single-channel systems operating in the third-transmission window ($\lambda = 1.55\,\mu\mathrm{m}$) are analyzed, neglecting the effect of the PMD. In-line amplification is provided by EDFAs; these amplifiers are characterized by a long lifetime of the excited state (as shown in Section 3.4); thus gain saturation is not influenced by signal amplitude modulation and the gain can be assumed to be constant over time.

The most important nonlinear effect in these systems is the Kerr nonlinearity: Its effects accumulates along the link up to be the main factor influencing the system design. On one side, the Kerr effect introduces signal distortion and nonlinear coupling between signal and ASE; on the other side, the phase modulation introduced by the Kerr effect can be used to compensate the GVD-induced chirping. As shown in Chapters 2 and 5, this compensation is better exploited in transmitting optical solitons. Raman and Brillouin effects are threshold effects, and they can be neglected if signal power is kept below a certain value, which is the case of the systems analyzed in this chapter.

By these assumptions, optical transmission systems can be classified by the regime in which the signal transmission occurs. In particular, three different transmission regimes can be individuated: low constant GVD, high constant GVD, and fluctuating GVD.

The regime of low constant GVD is characterized by a high FWM efficiency. Due to this effect, ASE noise and the transmitted signal experience nonlinear coupling along the link, and the signal spectrum broadens up to occupy the entire EDFA bandwidth.

A small amount of GVD limits the nonlinear interaction between signal and noise. In this case different behaviors can be observed depending on the sign of β_2.

In the normal region the chirping induced by the GVD and by the Kerr effect has the same sign, generating a large pulse broadening and preventing the system from working. In the anomalous region the Kerr-induced chirp tends to compensate the GVD effect; thus long distance transmission is possible. Solitons can exist in this region.

If the GVD fluctuates along the link, the propagation regime critically depends on the fluctuation characteristic length. Under certain conditions, the presence of a high local GVD prevents spectral broadening by decreasing the FWM efficiency, while the low average GVD limits pulse broadening at the link output. Under this condition high-performance transmission systems can be designed.

The values of the physical parameters for the different types of optical fibers discussed in this chapter are reported in Table 2.1. The inversion factor F of the EDFA is assumed to be equal to 2.

7.1.1 Propagation Regimes with Constant Dispersion along the Link

Constant GVD with a Value Near Zero
When the value of β_2 is very low (roughly $|\beta_2| < 0.05 \, \text{ps}^2/\text{km}$ in practice) the propagation regime can be classified as a constant GVD with a *value near zero*. An example of signal propagation in this regime is shown in Fig. 7.1. The optical power and the optical spectrum have been numerically evaluated at the output of a 9000-km-long link with $\beta_2 = -0.01 \, \text{ps}^2/\text{km}$ for a 5-Gbit/s-NRZ signal feed. The amplifier spacing is 60 km.

In Fig. 7.1a and b the power profile and the spectrum are given for propagation in the absence of ASE, and the same quantities are given in the presence of ASE in Fig. 7.1c and d.

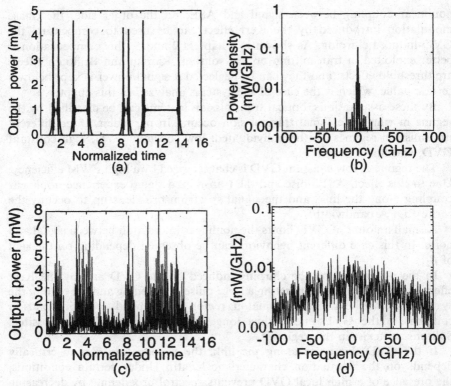

FIGURE 7.1   Temporal (a, c) and spectral (b, d) behavior of a 5-Gbit/s NRZ signal with an input peak power of 1 mW at the output of a link 9000 km long with $\beta_2 = -0.01\,\text{ps}^2/\text{km}$. The amplifier spacing is 60 km. (a) and (c) are evaluated without ASE, while (c) and (d) are in the presence of the ASE noise generated on a bandwidth $B_{\text{ASE}} = 320\,\text{GHz}$.

Figure 7.1*a* shows that in the absence of ASE, the signal has a negligible distortion that is almost completely located at the edges of the transmitted pulses. Such distortion is mainly due to the presence of SPM which introduces a spurious phase modulation in the signal that can be limited by low-pass filtering at the receiver.

Figure 7.1*b* shows a limited spectral broadening: The 3-dB width of the output spectrum is the same as at input; new frequency components arise up to about 60 GHz apart from the carrier, but their power is quite low. The spectral broadening is due to FWM and MI. In particular, the FWM efficiency is high on a large bandwidth (see Chapter 2, Fig. 2.5); thus new frequencies are generated starting from the signal spectrum. The power of each new frequency component is proportional to $P_1P_2P_3$, where P_i ($i = 1, 2, 3$) is the power of the ith interacting wave with frequency ν_i. Since the signal power is mainly limited to the frequency interval $[-R, R]$ (see Chapter 4, Fig. 4.2), the newly generated FWM component has negligible power if $|\nu_1 - \nu_3| > 2R$, and the FWM-induced spectral broadening is considerably reduced.

In the anomalous dispersion region, MI can induce amplification of frequency components far from the carrier; however, this effect is not so effective in the absence of ASE because the signal components far from the carrier are very small. The signal evolution is completely different in the presence of ASE noise [2]. As shown in Fig. 7.1d, the high FWM efficiency produces a large spectral broadening due to nonlinear interactions between the signal and the ASE frequency components. This effect is enhanced by the presence of MI, whose gain bandwidth is large when β_2 is close to zero.

Considering the power profile of the signal, it is evident from Fig. 7.1c that the spectral broadening induces a large ripple in the pulse profile; however, the energy of a single pulse remains almost constant.

The large spectral broadening occurring in the low GVD regime can be theoretically understood by analyzing a simple situation. This involves signal propagation in the absence of all dispersive effects (without PMD and where $\beta_2 = 0$) where the average Schrödinger equation, as derived in Section 5.1.1, can be applied. In this case MI is not present since $\beta_2 = 0$.

Starting from equation (5.12), the average propagation equation for the amplified link in the absence of dispersive effects can be written as [3]

$$\frac{\partial A(t,z)}{\partial z} = -i\gamma\rho|A(t,z)|^2 A(t,z) + A_{\text{ASE}}(t,z) \tag{7.1}$$

where

$$\rho = \frac{1 - e^{2\alpha L_A}}{2\alpha L_A} \tag{7.2}$$

L_A is the amplifier spacing, A the field complex envelope, and A_{ASE} is a complex zero average noise term accounting for the ASE noise averaged over the amplifier spacing. The noise power is given by

$$\langle |A_{\text{ASE}}(t,z')|^2 \rangle = \frac{Fh\nu(G-1)B}{L_A} \tag{7.3}$$

To analytically determine the statistical properties of the field at the fiber output from equation (7.1), some assumptions have to be made. First of all, ideal NRZ transmission is assumed, with perfectly squared input pulses. Moreover ideal square filters of bandwidth B are assumed after each optical amplifier; thus the system can support only bandlimited signals. For the sampling theorem, a pulse of bandwidth B and width T has $2m = 2BT$ degrees of freedom; that is, the pulse can be determined by its complex samples at m points, spaced $\Delta t = 1/B$.

Since no GVD is present and ASE noise samples in different time slots are uncorrelated, each pulse evolves independently from the others. Since the Kerr effect preserves the total number of photons in each time slot, the statistical properties of the intensity of the field are not affected by propagation. In fact, from equation (7.1), the following equation can be deduced for the average output signal power:

$$\langle |A(t,z)|^2 \rangle = |A(t,0)|^2 + Fh\nu(G-1)Bm_a \qquad (7.4)$$

where m_a is the amplifier number and $\langle \rangle$ indicates the ensemble average. In the absence of dispersive effects, the average signal power at the link output is the sum of the input signal power plus the accumulated ASE noise: The same average power as would be detected in a linear system. If the receiver has an optical bandwidth equal to B, the system performances can be evaluated as in a linear system in presence of the only ASE noise, as shown in Section 4.3.2.

If the Kerr effect does not influence the average signal power, it changes completely the signal average, which can be written as

$$\langle A(t,z) \rangle = \sqrt{\Psi(z)} A(t,0) \qquad (7.5)$$

where the function $\Psi(z)$ has a complex expression, reported in [3]. For large values of z (a few thousands kilometers) the asymptotic approximation of $\Psi(z)$ can be carried out as

$$\Psi(z) = \exp - 2[i\gamma\rho P_o z + \sqrt{2\gamma\rho\Xi}\, z + \tfrac{1}{3}(\gamma\rho)^2 \Xi P_o z^3] \qquad (7.6)$$

where P_o is the peak power of the input signal and the noise dependent parameter Ξ is defined as

$$\Xi = \frac{h\nu F(G-1)B}{L_A} \qquad (7.7)$$

From equations (7.5) and (7.6) can be inferred that the absolute value of the average field envelope decreases exponentially as the link length increases, since the average signal power remains instead constant, the exponential decay of the average field is due to phase fluctuations. Two different mechanisms can be identified that add phase noise to the propagating field.

The first mechanism is due to the beating between the ASE noise and the field. This beating produces amplitude fluctuations of the order of $2P_o\sqrt{\Xi z}$ that, due to the Kerr effect, are converted into phase fluctuations $\Delta\phi = 2\gamma\rho P_o z\sqrt{\Xi z}$ (see equation 2.35). The characteristic decay length of the cubic term in the exponent of (7.6) is

$$z_1 = \sqrt[3]{\frac{3}{(\gamma\rho)^2 \Xi P_o}} \qquad (7.8)$$

thus this term represents the decreasing of the average field due to the signal-ASE nonlinear beating at z, $\Delta\phi \approx 1$. The second phenomenon introducing phase noise is the ASE–ASE beating. In this case the Kerr phase shift can be written as $\Delta\phi = \gamma\rho\Xi z^2$ and the characteristic length z_2, is given by

$$z_2 = \sqrt{2\gamma\rho\Xi} \tag{7.9}$$

The distance z_2 is the distance at which the second term at the exponent of equation (7.6) is equal to one; thus this term represents the contribution of the ASE–ASE nonlinear beating. If we assume the parameters of a DS fiber, an input power $P_0 = 2.5\,\mathrm{mW}$, and a bandwidth B = 270 GHz, z_1 is of the order of 3500 km and z_2 is of the order of 4500 km.

In the absence of dispersive effects, the evolution along the link of the signal spectrum $S(\nu)$ can be analytically estimated. In particular, indicating by $S_0(\nu)$ the input spectrum, we obtain

$$S(\nu) = \Psi(z)S_0(\nu) + \left\{ \frac{P_0}{B}\left[1 - \Psi(z)\right] + Fh\nu(G-1)N_{\mathrm{amp}} \right\} \tag{7.10}$$

This expression shows that due to the nonlinear interaction between the signal and ASE and between ASE and ASE, the output spectrum is composed of two contributions: An attenuated signal spectrum and, between the curly brackets in (7.10), a flat noiselike spectrum that grows at a much faster rate than in the linear case. This means that during the propagation, information is transferred all along the bandwidth in which the ASE noise is generated. Such an effect induces large spectral broadening. In particular, when the link length is much larger than z_1, the output signal spectrum gets flat on the bandwidth B. This is clearly seen in Fig. 7.2 where the output spectrum of a 5-Gbit/s-NRZ signal having a peak power of 1 mW is shown after the propagation in a link with the same characteristics as in Fig. 7.1 but in the absence of dispersive effects and for three different propagating distances. In this case B = 320 GHz and consequently z_1 = 6000 km.

In the normal dispersion region of the fiber a different behavior can be observed. In Fig. 7.3 the example of Fig. 7.1 is shown in a normal dispersion region ($\beta_2 = 0.01\,\mathrm{ps}^2/\mathrm{km}$). In comparing Figs. 7.1 and 7.3 the different roles played by the MI in the two dispersion regions can be seen. In the anomalous dispersion region the MI induces the growth of components far from the carrier that contribute to the spectral broadening. Conversely, in the normal dispersion region MI may be responsible for an increase of the power of the central frequencies with respect to the lateral ones [4]. This effect limits the spectral broadening due to FWM; as a consequence in Fig. 7.3 the signal spectrum shows a reduced broadening with respect to Fig. 7.1. For the same reason the presence of ASE is less detrimental in the normal region than in the anomalous one.

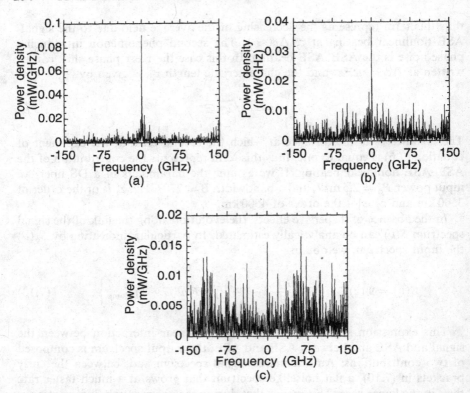

FIGURE 7.2 Output spectrum of a 5-Gbit/s NRZ signal having a peak power of 1 mW after the propagation in a link 9000 long with an amplifier spacing of 60 km in absence of dispersive effects and in presence of ASE noise with $B_{ASE} = 320$ GHz. (a) $L = 3800$ km, (b) $L = 5640$ km, (c) $L = 7200$ km.

Constant GVD with a High Value

When $|\beta_2|$ is greater than the limit value (≈ 0.05 ps^2/km), the propagation regime will be classified as a regime at a *high* constant GVD. This is the case of DS fibers not operating at the zero dispersion wavelength or step-index fibers. Figure 7.4 was obtained with the same parameters as Fig. 7.1, but in this case the value of β_2 is set to -1.28 ps^2/km. Figure 7.4a shows that after propagating, the input train of squared pulses experiences a considerable change in the shape, and solitonlike behavior appears. As explained in Section 2.3.2, this is due to the interplay between the Kerr effect and chromatic dispersion which in the anomalous dispersion region produces a solitonlike pulse independently of the input signal shape [5]. In Fig. 7.4b the signal shows a smaller spectral broadening with respect to Fig. 7.1b (see the spectrum width at the power density level of 0.001 mW/GHz). This is due to a decrease of the gain bandwidth of FWM and MI as the GVD increases.

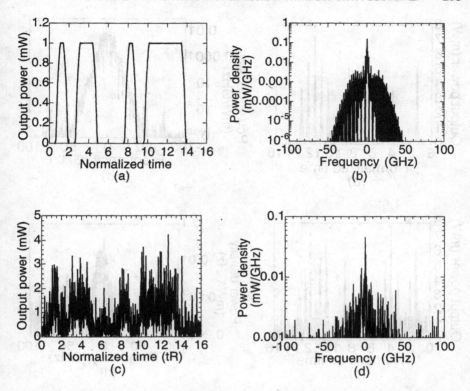

FIGURE 7.3 Temporal (a, c) and spectral (b, d) behavior of a 5-Gbit/s NRZ signal with an input peak power of 1 mW at the output of a link 9000 long with $\beta_2 = 0.01$ ps^2/km. The amplifier spacing is 60 km. (a) and (b) are without ASE, while (c) and (d) with the ASE noise emitted by the optical amplifiers with $B_{ASE} = 320$ GHz.

The peaks, located at approximately ±50 GHz and ±75 GHz with respect to the carrier, are due to the so called *sideband instability* induced by the periodical variation of the power along the link. As shown in [6], sideband instability peaks exist at a spacing $\Delta\nu(k)$ from the carrier, where k is a positive integer and

$$\Delta\nu(k) = \pm\sqrt{\frac{k}{2\pi L_A \beta_2}} \tag{7.11}$$

As shown in Fig. 7.4c, there is no perceptible observation of signal–ASE noise nonlinear coupling: The ASE noise is almost linearly added to the pulses shown in Fig. 7.4a. In fact the output spectrum, shown in Fig. 7.4d, does not broaden due to the ASE noise: This is due to the higher value of chromatic dispersion that limits the optical bandwidth of the efficiency curve of FWM and MI.

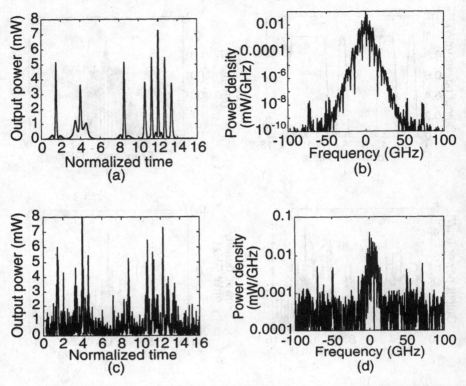

FIGURE 7.4 Temporal (a, c) and spectral (b, d) behavior of a 5-Gbit/s NRZ signal with an input peak power of 1 mW at the output of a link 9000 long with $\beta_2 = -1.28 \, \mathrm{ps^2/km}$. The amplifier spacing is 60 km. (a) and (b) are without ASE, while (c) and (d) with the ASE noise emitted by the optical amplifiers with $B_{ASE} = 320 \, \mathrm{GHz}$.

In the Fig. 7.5 the behavior of the same signal is shown in the normal dispersion region of the fiber ($\beta_2 = 1.28 \, \mathrm{ps^2/km}$); the parameters are the same as in the Fig. 7.1. The temporal evolution of the signal is quite different from the case of propagation in the anomalous region shown in Fig. 7.4. In the normal dispersion region the chirp due to the GVD adds to the chirp due to the Kerr effect, causing a large pulse broadening. In contrast, the behavior of the spectral evolution of the signal is similar to that of Fig. 7.4, even if the spectral broadening is reduced by the different kind of MI that affects the propagation in the normal region. In the particular case shown in Fig. 7.5, MI tends to transfer signal power from the sidebands to the central carrier so that the peaks of the sideband instability are almost suppressed. However, in other cases, sideband instability peaks can be evident even in the normal region.

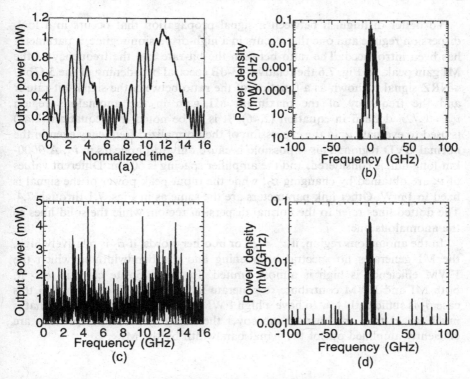

FIGURE 7.5 Temporal (a, c) and spectral (b, d) behavior of a 5 Gbit/s NRZ signal with an input peak power of 1 mW at the output of a link 9000 long with $\beta_2 = 1.28 \text{ ps}^2/\text{km}$. The amplifier spacing is 60 km. (a) and (b) are without ASE, while (c) and (d) with the ASE noise emitted by the optical amplifiers with $B_{\text{ASE}} = 320 \text{ GHz}$.

Limits of Applicability of the Different Propagation Regimes

In comparing Figs. 7.1, 7.3, 7.4, and 7.5, the following two conclusions can be drawn:

- In a regime of low chromatic dispersion, large spectral broadening occurs especially in the anomalous dispersion, while the output signal power consists of the input signal plus the ASE noise; this behavior is almost independent of the sign of the chromatic dispersion.

- In a regime of high chromatic dispersion, the signal develops limited spectral broadening, but its behavior significantly depends on the sign of the chromatic dispersion. Indeed, while in a regime of normal dispersion the signal shows a strong degradation, in an anomalous regime the Kerr effect can compensate the chirp induced by the GVD, producing soliton-like behavior.

To better distinguish between a signal propagation that occurs in a low-dispersion regime and one that occurs in a high-dispersion regime, a parameter has been introduced: The ratio between the bit-rate and the frequency of the MI gain peak. In Fig. 7.6 the relative -3-dB spectral broadening of the 5-Gbit/s-NRZ signal is shown as a function of the ratio between the signal bit-rate R and the frequency of the maximum MI gain in the anomalous region $\nu_c = \omega_c/2\pi$ defined in equation (2.46). It is to be noted that equation (2.46) is used to evaluate ν_c in the expression of the normalized frequency even in the normal GVD region: This is possible because only $|\beta_2|$ appears in ν_c. A 9000-km-long link is considered, and the amplifier spacing is 60 km. Different values of ν_c are obtained by changing β_2, while the input peak power of the signal is fixed to 1 mW. Other link parameters are the same as in Figs. 7.1 through 7.4. The dotted lines refer to the normal dispersion region, while the solid lines to the anomalous one.

In the anomalous region, if $\nu_c < R$, or in other words, if β_2 is relatively high, the MI generates no spectral broadening and the bandwidth in which the FWM efficiency is high is almost limited. In the opposite case, if $\nu_c > R$, both MI and FWM contribute to generate large spectral broadening. In this case β_2 is sufficiently low to have a high FWM efficiency on a bandwidth that is much larger than the bit-rate; moreover the newly generated frequencies are efficiently amplified out of the signal bandwidth by the MI.

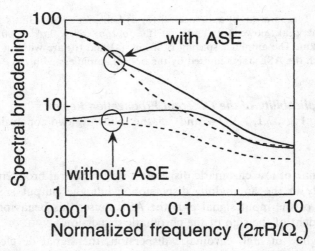

FIGURE 7.6 Relative -3 dB spectral broadening of the 5-Gbit/s NRZ signal versus the ratio between the signal bit-rate R and the frequency $\nu_c = \sqrt{2\gamma P/\beta_2}/2\pi$ defined in equation (2.46) considering a 9000-km-long link and an amplifier spacing of 60 km. Different values of ν_c are obtained by changing β_2, while the input peak power of the signal is fixed at 1 mW. Dotted lines refer to the normal dispersion region, while solid lines to the anomalous one.

In the normal region the spectral broadening is mainly due to the FWM process. In the presence of ASE, the signal shows a great spectral broadening when $R/\nu_c \ll 1$. Such a broadening is more evident in the anomalous dispersion region for the presence of modulation instability. Conversely, when $R/\nu_c > 1$, both in the normal and in the anomalous region, the ASE does not significantly affect the signal evolution, which is mainly determined by the interplay between the Kerr effect and chromatic dispersion.

On the basis of the above considerations, two propagation regimes can be distinguished by the value of R/ν_c: The condition $R/\nu_c \ll 1$ corresponds to the regime of low chromatic dispersion, while the condition $R/\nu_c \geq 1$ to the regime of high chromatic dispersion.

So far a constant GVD has been assumed along the link. In practice it is not easy to maintain the GVD constant. Unavoidable GVD fluctuations are always present due to the tolerance in the GVD value of the fiber pieces composing the link. Furthermore very long links are often composed by optical fibers of different kinds that have been installed at different moments in various parts of the networks.

When the fluctuations of the chromatic dispersion are very small and occur on very short scales, the signal propagation behaves as in the case of a link with constant chromatic dispersion. In the case of a link with a randomly fluctuating chromatic dispersion with a high average GVD (as a rule-of-thumb above $0.1\,\mathrm{ps}^2/\mathrm{nm}$), a constant GVD propagation regime can occur if the fluctuations are sufficiently small. The characteristic parameters of the GVD fluctuation are the characteristic length L_f, which is the length at which the correlation function of the GVD falls below 0.1, the GVD average value $\bar{\beta}_2$, and the GVD standard deviation $\sigma_{\beta2}$. The signal propagation occurs as in the case of constant GVD equal to $\bar{\beta}_2$ if the fluctuation is sufficiently small (in practice, $\bar{\beta}_2 > 10\sigma_{\beta2}$) and sufficiently rapid in space that the signal propagation is sensible only to the average GVD. This last condition is verified in practice if $L_f < 0.1 \min(L_{NL}, L_D)$, where L_{NL} and L_D are the nonlinear and the dispersion lengths, respectively, as defined in Section 2.3.1.

If the average GVD is very close to zero ($\bar{\beta}_2 \ll 0.1\,\mathrm{ps}^2/\mathrm{km}$), constant GVD propagation can again occur, but the requirement $\bar{\beta}_2 > 10\sigma_{\beta2}$ has to be substituted with a different condition. This condition must ensure a large spectral broadening, which is the characteristic of a low-dispersion propagation regime. To obtain large spectral broadening, the FWM efficiency has to be high in a bandwidth larger than the bandwidth B_{ASE} over which the ASE noise is generated. To put this requirement into an analytical form, the approximate expression of the FWM bandwidth reported in Section 2.2.2 must be used. In particular, if the fiber length over which the FWM efficiency is high is greater than the GVD characteristic length, with a GVD equal to its standard deviation, the following condition is obtained: $L_f\sigma_{\beta2}B_{ASE}^2 \ll 1$.

7.1.2 Performances of Systems in the Low Chromatic Dispersion Regime

The performances of transmission systems operating in a regime of constant, low GVD are strongly influenced by the spectral broadening experienced by the transmitted signal. In Fig. 7.7 the signal bandwidth of a 5-Gbit/s-NRZ signal is shown versus the propagation distance in the case of a link with an amplifier spacing equal to 60 km. The bandwidth is measured as the -3-dB width of the spectrum, and the input peak power is 1 mW. Curve a refers to the absence of dispersive effects, curve b to the propagation in the anomalous chromatic dispersion ($\beta_2 = -0.1\,\text{ps}^2/\text{km}$) and including the effect of third-order chromatic dispersion ($\beta_3 = +0.1\,\text{ps}^3/\text{km}$), and curve c is analogous to b but propagating in the normal dispersion region of the fiber ($\beta_2 = 0.1\,\text{ps}^2/\text{km}$). The ASE noise is generated on a bandwidth of 320 GHz. The values of GVD and input power used in this figure correspond to a MI frequency $\nu_c = 40$ GHz and a normalized frequency equal to 0.13. As shown by Fig. 7.6, this value of R/ν_c indicates a regime of low GVD, and hence large spectral broadening of the signal is expected.

The sharp knee for $L = 5500$ km present in curves a can be explained by equation (7.10), which shows that the signal spectrum presents an attenuation while the ASE noise uniformly grows. Thus the spectrum width is almost constant until reaching the condition at which the ASE level gets higher than

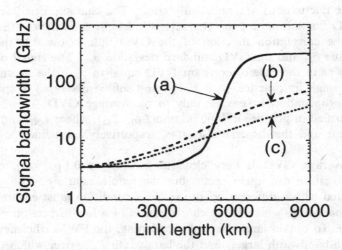

FIGURE 7.7 Spectral broadening versus link length of a 5-Gbit/s NRZ signal with an input peak power of 1 mW. The amplifier spacing is 60 km, and the bandwidth is measured as the -3-dB width of the spectrum. Curve (a) refers to the absence of dispersive effects, curve (b) to the presence of anomalous chromatic dispersion ($\beta_2 = -0.1\,\text{ps}^2/\text{km}$) and of third-order chromatic dispersion ($\beta_3 = +0.1\,\text{ps}^3/\text{km}$), and curve (c) is analogous to (b) but with normal dispersion ($\beta_2 = 0.1\,\text{ps}^2/\text{km}$). The ASE noise is generated on a bandwidth of 320 GHz.

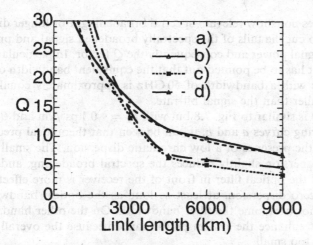

FIGURE 7.8 Q factor versus fiber link length is reported for a 5-Gbit/s IM-DD system considering the link parameters adopted in Fig. 7.7 and with $\beta_2 = 0$. The different curves refer to the following situations: (a) has been analytically obtained from equations (7.4) and (4.54), (b) is obtained by simulating the signal propagation along the link with the only bandwidth limitation provided by the ASE bandwidth of 320 GHz, (c) is obtained by adding a Fabry-Perot filter with a 3-dB bandwidth of 50 GHz in front of the receiver, (d) is obtained by inserting in-line Fabry-Perot filters with a 3-dB bandwidth of 50 GHz besides the filter in front of the receiver.

the signal level for $L \sim 5500$ km. In case b for long links the signal presents a smaller spectral broadening, since, for the presence of chromatic dispersion, the FWM bandwidth is narrower with respect to the case a. On the other hand, for $L < 5000$ curve b shows a larger broadening than curves a for the presence of the MI effect. Curve c is similar to curve b, even though the spectral broadening is smaller for the different role played by MI in the normal region.

In Fig. 7.8 the Q factor versus fiber link length is reported for a 5-Gb/s IM-DD system with the same link parameters as in Fig. 7.7 and with $\beta_2 = 0$. The different curves are defined as follows.

a is analytically obtained from equations (7.4) and (4.61).

b is obtained by simulating the signal propagation along the link with only the bandwidth limitation provided by the ASE bandwidth of 320 GHz.

c is obtained by adding a Fabry-Perot filter with a 3-dB bandwidth of 50 GHz in front of the receiver.

d is obtained by inserting in-line Fabry-Perot filters with 3-dB bandwidths of 50 GHz besides the filter in front of the receiver.

The figure shows a good agreement between curves a and b. The action of in-line optical filters is effective only for distances shorter than 3000 km, where the signal does not show relevant spectral broadening. For longer distances the filters start to cut the tails of the spectrally broadened signal and progressively reduce the signal power and consequently the Q factor. In particular, considering curve d, it has to be pointed out that the equivalent bandwidth of a cascade of 150 filters with a bandwidth of 50 GHz is approximately equal to 1 GHz, which is smaller than the signal bit-rate.

Figure 7.9 is similar to Fig. 7.8 but with $\beta_2 = -0.1 \, \mathrm{ps}^2/\mathrm{km}$ and $\beta_3 = 0.1 \, \mathrm{ps}^3/\mathrm{km}$. Comparing curves a and b, it can be seen that theoretical predictions are still valid in the presence of a low chromatic dispersion. The small amount of chromatic dispersion slightly reduces the spectral broadening, and as a result the action of the optical filter in front of the receiver is more effective than in the case of zero dispersion, at least as long as the optical bandwidth of the signal does not overcome the filter bandwidth. On the other hand, the in-line filters do not enhance the system performance because the overall link bandwidth is still too small.

The system performance in the normal dispersion region is shown in Fig. 7.10. In this regime and for distances longer than 6000 km, an increase in the Q factor can be observed in curve b with respect to curve a. This effect is due to the nonlinear interaction between the signal and ASE that induces a reduction in the standard deviation of accumulated noise along the link [7].

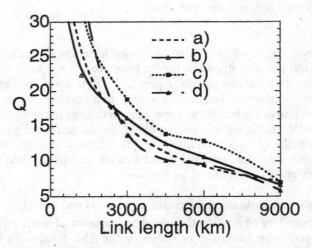

FIGURE 7.9 Q factor versus fiber link length is reported for a 5-Gb/s IM-DD system considering the link parameters adopted in Fig. 7.7 and with $\beta_2 = -0.1 \, \mathrm{ps}^2/\mathrm{km}$. The different curves refer to the following situations: (a) has been analytically obtained from equations (7.4) and (4.61), (b) is obtained by simulating the signal propagation along the link with the only bandwidth limitation provided by the ASE bandwidth of 320 GHz, (c) is obtained by adding a Fabry-Perot filter with a 3-dB bandwidth of 50 GHz in front of the receiver, (d) is obtained by inserting in-line Fabry-Perot filters with a 3-dB bandwidth of 50 GHz besides the filter in front of the receiver.

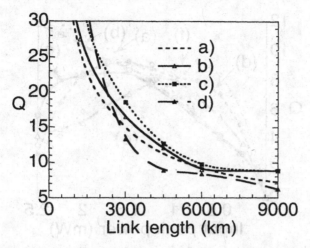

FIGURE 7.10 Q factor versus fiber link length is reported for a 5-Gbit/s IM-DD system considering the link parameters adopted in Fig. 7.7 and with $\beta_2 = 0.1\,\text{ps}^2/\text{km}$. The different curves refer to the following situations: (a) has been analytically obtained from equations (7.4) and (4.54), (b) is obtained by simulating the signal propagation along the link with the only bandwidth limitation provided by the ASE bandwidth of 320 GHz, (c) is obtained by adding a Fabry-Perot filter with a 3-dB bandwidth of 50 GHz in front of the receiver, (d) is obtained by inserting in-line Fabry-Perot filters with a 3-dB bandwidth of 50 GHz besides the filter in front of the receiver.

In Fig. 7.11 the systems analyzed in this section are compared in terms of their Q factor and input peak power for a link length of 9000 km.

Curve a is analytically evaluated in the absence of dispersive effects and without narrow optical filters, while curves b to f are obtained by simulations,

b is obtained the same as curve a.

c is obtained with $\beta_2 = -0.1\,\text{ps}^2/\text{km}$ and $\beta_3 = +0.1\,\text{ps}^3/\text{km}$.

d is obtained with $\beta_2 = -0.1\,\text{ps}^2/\text{km}$ and $\beta_3 = +0.1\,\text{ps}^3/\text{km}$ and using a Fabry-Perot filter with a bandwidth $B_o = 50\,\text{GHz}$ in front of the receiver.

e is obtained with $\beta_2 = 0.1\,\text{ps}^2/\text{km}$ and $\beta_3 = 0.1\,\text{ps}^3/\text{km}$.

f is obtained with $\beta_2 = 0.1\,\text{ps}^2/\text{km}$ and $\beta_3 = +0.1\,\text{ps}^3/\text{km}$ and using a Fabry-Perot filter with a bandwidth $B_o = 50\,\text{GHz}$ in front of the receiver.

The agreement between curves a and b is quite good, confirming the validity of equation (7.4). Equation (7.4) can be applied with a negligible error also in the presence of a low chromatic dispersion until a power threshold of about 1.5 mW for the assumed parameters values. For a higher transmitted power the

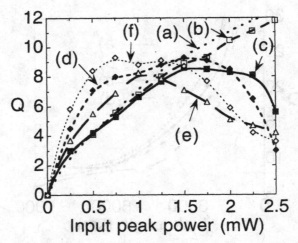

FIGURE 7.11 Q factor reported versus input peak power for a link length of 9000 km for the systems analyzed in the Figs. 7.8, 7.9, and 7.10. (a) is evaluated analytically in the absence of dispersive effects and without narrow optical filters, while curves (b) to (f) are obtained by simulations; in particular, (b) is obtained in the same situation of curve (a), (c) is obtained with $\beta_2 = -0.1\,\text{ps}^2/\text{km}$ and $\beta_3 = +0.1\,\text{ps}^3/\text{km}$, (d) is obtained with $\beta_2 = -0.1\,\text{ps}^2/\text{km}$ and $\beta_3 = +0.1\,\text{ps}^3/\text{km}$ and a Fabry-Perot filter with a bandwidth $B_o = 50\,\text{GHz}$ in front of the receiver, (e) is obtained with $\beta_2 = 0.1\,\text{ps}^2/\text{km}$ and $\beta_3 = +0.1\,\text{ps}^3/\text{km}$, (f) is obtained with $\beta_2 = 0.1\,\text{ps}^2/\text{km}$ and $\beta_3 = +0.1\,\text{ps}^3/\text{km}$ and a Fabry-Perot filter with a bandwidth $B_o = 50\,\text{GHz}$ in front of the receiver.

presence of chromatic dispersion cannot be neglected any more. The filter placed in front of the receiver will only weakly improve the system performance.

It is important to note that even though for high optical powers the case of zero GVD shows better performances, this is not easy to realize in practice. In long links some form of filtering along the line is always present, inducing both a residual dispersion and a cut of the signal spectrum tail and thus producing a huge performance degradation. Conversely, the presence of a small GVD along the link allows the spectral broadening to be limited avoiding spurious effects due to in-line filtering.

To conclude this section on systems operating in a regime of low chromatic dispersion, it can be stated that at almost zero GVD, very high-speed signals can, in principle, be transmitted over long distances (e.g., 5-Gbit/s transmission over 9000 km), but in practice, the unavoidable limitation of the system bandwidth due to filters, amplifiers, receiver, and so on, can affect the spectrally broadened signal, reducing system performance. An alternative more practical solution may be transmission in other regimes such as in a regime with a fluctuating GVD.

Nevertheless, the almost zero GVD regime may be suitable for high-speed systems over shorter distances (e.g., 10-Gbit/s transmission over 2500 to

3000 km). Then a regime of low GVD may be realized by a careful choice of the transmitter wavelength based on the measurement of the zero dispersion wavelength of the optical cable. In this way the complex optical cable design needed to realize dispersion management (see Section 7.1.4) can be avoided.

Distances of 2500 to 3000 km are typical of the highways networks spanning continental distances. Thus they are a logical area of application for the systems discussed in this section.

The potentialities of systems operating in a low chromatic dispersion regime have been variously tested. For example, an experimental system operating at 1.2 Gbit/s over a 904-km-long fiber link with an overall dispersion smaller than $0.128 \, \text{ps}^2/\text{km}$ is described in [1, e1], while a system operating at 2.5 Gbit/s over a fiber link 4520 km long with an overall dispersion of $0.128 \, \text{ps}^2/\text{km}$ is reported in [1, e2]. This last experiment is particularly important because it is very close to the theoretical limit for such systems.

7.1.3 Systems Operating in the High Chromatic Dispersion Regime

As the GVD is increased above the values that warrant the permanence in a regime of low chromatic dispersion ($R/\nu_c \ll 0.1$), the GVD-induced chirping becomes more and more important. In the normal dispersion region this induces a large pulse broadening and prevents the system from working correctly. In the anomalous region the GVD-induced chirp can be partially or even totally compensated by the Kerr effect, allowing transmission with good performance [5]. This behavior is shown in Fig. 7.12 where the Q factor of a 5-Gbit/s transmission system is given with relation to β_2 for a 9000-km-long link with an amplifier spacing of 60 km. Curve a refers to an NRZ signal with an input peak power of 1 mW, curve b to a RZ signal with a secant hyperbolic shape with a 3-dB width of 25 ps (time soliton: $T_s = 14.2 \, \text{ps}$) and with a peak power corresponding exactly to the soliton peak power P_k and depending on the value of β_2 according to the expression

$$P_k = \frac{\alpha L_A G |\beta_2|}{(G-1)\gamma T_s^2} = \frac{G \ln(G)}{G-1} \frac{|\beta_2|}{\gamma T_s^2} \tag{7.12}$$

of equation (5.26) [8], and curve c to an RZ signal with a secant hyperbolic shape with a 3-dB width of 25 ps and a peak power of 6.6 mW, corresponding to the same average power of case a. A Fabry-Perot filter is present in front of the receiver in all the cases; its bandwidth is 50 GHz for curve a and 20 GHz for curves b and c.

In the case of the NRZ signals better performances are attained for the values of β_2 within the interval $-0.2 \, \text{ps}^2/\text{km} < \beta_2 < -0.05 \, \text{ps}^2/\text{km}$ and $0.05 \, \text{ps}^2/\text{km} < \beta_2 < 0.25 \, \text{ps}^2/\text{km}$. The region strictly around $\beta_2 = 0$ must be avoided because there the signal exhibits a large spectral broadening, so it is significantly distorted by the filter at the receiver. The broadening is reduced by the presence of a small amount of GVD, and in particular, the signal band-

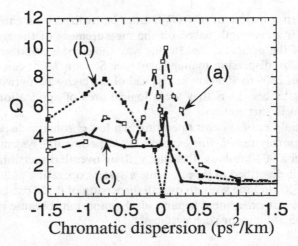

FIGURE 7.12 Q factor of a 5-Gbit/s transmission system versus β_2 considering a 9000-km-long link with an amplifier spacing of 60 km. Curve (a) refers to an NRZ signal with an input peak power of 1 mW, curve (b) to a soliton signal with a 3-dB width of 25 ps and exactly the soliton peak power (the time soliton $T_s = 14.2$ ps), P_k, that depends on the value of β_2 according to equation the expression $P_k = \alpha L_A G \beta_2 / (G-1) \gamma T_s^2$ [8], and curve (c) to an RZ signal with a secant hyperbolic shape a 3-dB width of 25 ps and a peak power of 6.6 mW. This peak power allows the same average power of case (a) to be attained. A Fabry-Perot filter is present in front of the receiver in all the cases; its bandwidth if 50 GHz for curve (a) and 20 GHz for curves (b) and (c).

width at the receiver is below 50 GHz when $|\beta_2| > 0.05$. For $|\beta_2| > 0.2 \, \text{ps}^2/\text{km}$ the GVD considerably degrades the system performance. The increase of the Q factor for β_2 values within the interval $-1 < \beta_2 < -0.05 \, \text{ps}^2/\text{km}$ is due to the partial compensation of the chirp induced by the GVD by means of the Kerr effect [5]; however, such compensation is not sufficient to give a value of Q higher than 6. In the case of soliton systems (curve b) the nonlinear compensation of chromatic dispersion is more effective, and higher Q factors can be achieved. As the GVD approaches zero, the input peak power given by equation (7.12) decreases, and hence the signal-to-noise ratio deteriorates, reducing the achievable Q factor. This effect is not present in curve c, where the power is kept constant as the β_2 varies.

In comparing curve a with curve c, clearly NRZ signals show better performances than RZ signals in the low-dispersion regime ($|\beta_2| < 0.3 \, \text{ps}^2/\text{km}$). This is because the peak power is lower and the nonlinear interaction between signal and ASE is weaker. Also it can be inferred from Fig. 7.12 that it is difficult to design long-distance, high-capacity NRZ systems in a high chromatic dispersion region, which is the region where soliton systems attain their best performances.

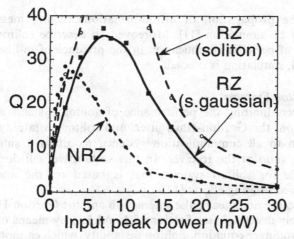

FIGURE 7.13 Q factor of 10-Gbit/s systems versus the transmitted peak power for an NRZ system, an RZ system using supergaussian pulses, and an RZ system using hyperbolic secant pulses (solitons). A 1000-km-long link with an amplifier spacing of 40 km and $\beta_2 = -1.28$ ps^2/km is considered. The hyperbolic secant pulses have a 3-dB width of 20 ps, while the supergaussian pulses have $m = 3$ (see Appendix A1).

As far as shorter distance systems are concerned, partial compensation of the GVD attained by the Kerr effect in the anomalous region can be exploited to design high-performance NRZ and nonsoliton RZ systems. This possibility is explored in Fig. 7.13 where the Q factor of a 10-Gbit/s system is shown with relation to the transmitted peak power of an NRZ system, an RZ system using supergaussian pulses, and an RZ system using hyperbolic secant pulses. A 1000-km-long link with an amplifier spacing of 40 km and $\beta_2 = -1.28$ ps^2/km is considered. The hyperbolic secant pulses have a 3-dB width of 20 ps, while supergaussian pulses have $m = 3$ (see Appendix A1). In all the cases the Q factor increases as the input power grows until reaching a maximum value for the optimum transmitted power which depends on the transmission format. In particular, the optimum power is equal to that which can be obtained by equation (7.12) if hyperbolic secant pulses are used. In this case perfect GVD compensation occurs for a peak power around 9 mW. Increasing the power above the optimum value, the Kerr-induced chirp prevails and the system performances get worst.

In the following section, the design of soliton systems will be discussed and their performances will be assessed by simulation. In Chapter 5 it is shown that a wide-ranging theory can be developed to describe the main effects arising in soliton systems. Despite the effectiveness of the soliton theory, simulative analysis is needed in designing real systems. In fact the theory can analyze the different effects separately, but their interplay cannot be described. For example, the interplay between the Gordon-Haus jitter [9] and the soliton interaction

[10] during the propagation of an information-carrying message can be described only by simulation [11]. Moreover, to describe soliton interaction and the effect of periodic amplification in the presence of in-line filtering for soliton control, simulation is needed.

Design of Soliton Systems

The main effects limiting the performance of soliton systems are nonlinear pulse interaction, the Gordon-Haus jitter, and soliton instability, besides the need, common to all communication systems, to attain a sufficiently high signal-to-noise ratio at the receiver. In this section we will derive a sort of operating table for soliton systems that is based on the analysis of the above-mentioned limiting effects [12].

In Chapter 5 the nonlinear pulse interaction and the Gordon-Haus jitter are more extensively discussed (see Sections 5.4 and 5.3) by means of the average nonlinear Schrödinger equation. Soliton instability, which cannot be described by the average nonlinear Schrödinger equation, for it needs a more complex theoretical approach, will be analyzed by means of simulative tools [13].

Soliton instability is strictly related to the periodical signal amplification along the link. Due to losses in the passive fiber, the soliton power decreases during propagation, and in order to maintain its soliton nature, the pulse width increases. In an amplified system, if the soliton period z_0 (see equation 2.67) is much greater than the amplifier spacing L_A and according to equation (7.12) the input peak power is given by $P_k = \alpha L_A G \beta_2 / (G - 1) \gamma T_s^2$ [8], a soliton can propagate with a fixed time duration T_s. If the condition $z_0 \gg L_A$ is not satisfied, the localized amplification restores the soliton input power but does not compensate the soliton broadening. As a consequence the broadened soliton looses more power during propagation through the next fiber piece before the next amplifier. After a few amplifier sections the solitons tends to stabilize its width, but the lost energy forms a radiative wave with a very large spectrum. Due to fiber dispersion, frequency components far from the soliton central frequency travel along the fiber with a group velocity different from those of the soliton so that the radiative wave tends to diffuse along the fiber. The radiative wave interacts with soliton pulses through Kerr effect inducing a sort of time jitter.

In practice, the impact of soliton instability on the system performances can be neglected if the following empirical condition is fulfilled [13]:

$$L_A \leq \tfrac{8}{10} z_0 \qquad (7.13)$$

The soliton interaction is due to the fact that two adjacent solitons show an interaction that tends to attract or to repulse according to their relative phase. Such an effect can be neglected if the time distance among adjacent pulses is higher than a quantity that depends on several system parameters.

Starting from equation (5.342) for in-phase solitons of the same amplitude and from equation (5.344) for out-of-phase solitons of the same amplitude, it

can be seen that soliton interaction can be neglected with a good approximation if the normalized soliton spacing ΔT satisfies the condition [13]

$$\Delta T = \frac{T}{T_F} \geq 1.2 \ln\left(\frac{4L}{z_0}\right) \tag{7.14}$$

where L is the link length, T the bit interval, and T_F the soliton 3-dB width.

As far as the Gordon-Haus jitter is concerned, in the absence of soliton in-
line control, the jitter variance is given by equation (5.113). Generally, the first
addendum in equation (5.116) is dominant, the jitter standard deviation can be
written as

$$\sigma_j = \sqrt{\frac{h\nu F \alpha \gamma |\beta_2| f(G) L^3}{9T_s}} \tag{7.15}$$

where the factor $f(G)$ represents the noise enhancement factor due to the use of
lumped amplifiers and is defined in equation (5.34),

$$f = \frac{(G-1)^2}{G[\ln(G)]^2} \tag{7.16}$$

As shown in Section 4.2.1, the presence of time jitter induces an error
probability floor. The exact value of the floor for a given jitter variance
depends on the transmitted pulse shape and on the PDF of the additive
noise and of the jitter. Assuming the additive noise and the jitter to have a
gaussian PDF, the performance evaluation model reported in Section 4.2.1 can
be applied. In this way it can be assumed that the error probability is lower
than 10^{-9} if the condition $\sigma_j R < 0.06$ is verified [13].

Unfortunately, if the Gordon-Haus induced jitter has really a gaussian
PDF, the additive noise at the output of a soliton system cannot be rigorously
assumed to be gaussian. However, a simulative analysis of the performances of
soliton systems allows one to state that the condition $\sigma_j R < 0.06$ provides a
correct system working [11]. This condition is assumed in this book in order to
fix the amount of acceptable jitter.

Finally the peak power of the soliton has to guarantee a sufficiently high
signal-to-noise ratio. The optical signal to noise ratio in the electrical bandwidth (see equation 4.59) in a soliton system is a function of the soliton peak
power P_o. Due to the nonlinear coupling between ASE and the signal, an exact
expression of the noise power at the receiver input is difficult to obtain.
However, it can be verified by simulations that in all the practical cases, the
received noise power can be evaluated with a negligible error by assuming a
linear noise accumulation along the link. In this condition equation (4.59) can
be rewritten in the case of a soliton system as

$$SNR_e = \frac{\bar{P}_o}{N_a h\nu F(G-1)R} = \frac{P_o T_F}{N_a h\nu F(G-1)} \tag{7.17}$$

where $\bar{P}_o$ is the average received optical power in the bit interval and $N_a = L/L_A$ the number of amplifiers.

Starting from equation (4.61), the Q factor can be evaluated as a function of the SNR$_e$; thus the condition $Q > 6$ induces a condition on the product $P_o T_F$. Equations (7.13) to (7.17) allow the main design constraints for soliton systems to be fulfilled with a suitable choice of the design parameters.

Generally, some link parameters are fixed by economical reasons as the amplifier spacing or the values of the GVD of already installed cables. In this case the maximum capacity of the system is attained by optimizing the transmitted solitons parameters.

In Fig. 7.14 the constraints of 9000-km-long soliton systems with an amplifier spacing of 40 km and a $\beta_2 = -1.28\,\mathrm{ps^2/km}$ ($D = 1\mathrm{ps/nm/km}$) are plotted in terms of bit-rate and the pulse width T_s. Other link parameters are reported in Table 2.1. Curve a shows the limitation due to the soliton instability, which can be avoided when the pulse width T_s is greater than the value fixed by equation (7.13) with the parameters of the link considered independently of the bit-rate. The areas under curve b and curve c are free from nonlinear pulse interactions and Gordon-Haus jitter, respectively. The area on the left side of curve d refers to condition $Q > 6$. The dashed area represents bit-rate/pulse-width combinations for which all the different constraints are fulfilled: In particular, Fig. 7.14

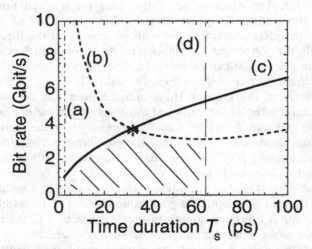

FIGURE 7.14 Constraints of 9000-km-long soliton systems with an amplifier spacing of 40 km, $\beta_2 = -1.28\,\mathrm{ps^2/km}$ and $\beta_3 = 0.066\,\mathrm{ps^3/km}$ ($D = 1\,\mathrm{ps/nm/km}$) reported in the plot. The various constrain are considered separately. In particular, for each different constraint the maximum attainable bit-rate is plotted versus the pulse width T_s. Curve (a) shows the limitation due to the soliton instability, curve (b) the limitation due to nonlinear pulse interaction, curve (c) the Gordon-Haus jitter, and curve (d) the condition $Q > 6$. The dashed area represents bit-rate/pulse-width combinations for which all the different constraints are fulfilled.

shows that in this case the maximum attainable capacity is 3.5 Gbit/s with $T_s = 30$ ps. Of course the bit-rate can be further increased by optimizing other link parameters (e.g., β_2 or the amplifier spacing), and it can be greatly increased by introducing in-line soliton control.

Performances of Soliton Systems without In-line Soliton Control

Plots similar to Fig. 7.14 allow the operating point of a soliton system to be individualized. However, simulation tools are necessary when real system performances have to be assessed.

The performances of soliton systems operating at different bit-rates with the same link parameters used in Fig. 7.14 are shown in Fig. 7.15. According to performance evaluation discussed in Section 4.2.1, in RZ pulses both the Q factor (Fig. 7.15a) and the time jitter standard deviation (Fig. 7.15b) have to be estimated. Acceptable system performances are obtained when the conditions $Q > 6$ and $\sigma_j R < 0.06$ are simultaneously satisfied. In Figs. 7.15a and 7.15b the Q factor and the time jitter standard deviation are plotted against the width T_s of the transmitted soliton. As a comparison, in Fig. 7.15b the theoretical behavior of the Gordon-Haus jitter has also been calculated (dashed lines).

At $R = 2.5$ Gbit/s the system shows good performance if 15 ps $< T_s < 60$ ps. For $T_s > 60$ ps the Q factor is smaller than 6, while for $T_s < 15$ ps the time jitter is unacceptable.

In the case of the 2.5-Gbit/s curve of Fig. 7.15b, soliton instability is responsible for the different behavior observed between the simulative estimation of the time jitter and the theoretical curve of Gordon-Haus jitter for $T_s < 25$ ps. In fact the radiative wave generated by soliton instability increases the noise background and hence the time jitter. Soliton interaction is not relevant at 2.5 Gbit/s in the considered range of the soliton width.

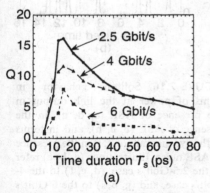
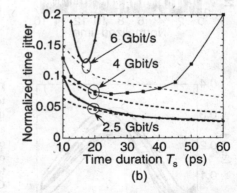

(a)   (b)

FIGURE 7.15 Q factor (a) and the normalized timing jitter, $\sigma_j R$, (b) versus with width, T_s, of three soliton systems operating in a link with the same characteristics as in Fig. 7.14. In (b) the theoretical behavior of the Gordon-Haus jitter is also reported (dashed lines).

At a bit-rate of 4 Gbit/s, the soliton interaction becomes important and generates almost entirely the extra jitter with respect to the Gordon-Haus limit reported in Fig. 7.15b. The random nature of the time shift is due to the random sequence of pulses constituting the message. The high value of the jitter inhibits the transmission at 4 Gbit/s, even if the Q factor were acceptable. This effect is more evident at $R = 6$ Gbit/s. In this case the Gordon-Haus effect, soliton instability, and soliton interaction contribute to increase the value of the time jitter above the acceptable level, so transmission is not possible. The bit patterns at the link output in the the presence of ASE (a, d, g) in the absence of ASE (b, e, h), and eye diagrams (c, f, i) in presence of ASE noise are reported in Fig. 7.16. Illustrations (a, b, c) refer to the 2.5-Gbit/s case, (d, e, f) to the 4-Gbit/s case, and (g, h, i) to the 6-Gbit/s case. All the cases were evaluated for a pulse time duration of 25 ps. At 2.5 Gbit/s the signal is transmitted with good performance and, in particular, the time jitter is very low. Conversely, in the case of the 4 Gbit/s, the solitons, even though they do not present strong modifications in their pulse shapes, are affected by a relevant jitter due to the presence of the ASE noise. In the case of a 6-Gbit/s system, the solitons show severe degradation in their pulse shape also in the absence of ASE noise: this is due to the strong interaction among the pulses.

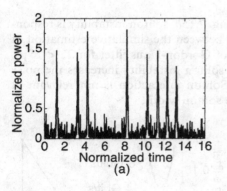

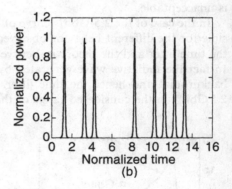

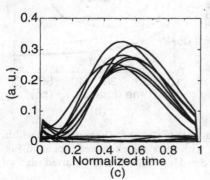

FIGURE 7.16 Soliton behavior in terms of power at the link output in the presence of ASE (a, d, g), in the absence of ASE (b, e, h), and in terms of the eye diagram (c, f, i) in the presence of ASE noise. Illustrations (a, b, c) refer to the 2.5-Gbit/s case, (d, e, f) to the 4-Gbit/s case, and (g, h, i) to the 6-Gbit/s case. All the cases are evaluated for a time duration of the pulses at $\tau_s = 30$ ps. The link parameters are the same as in Fig. 7.14.

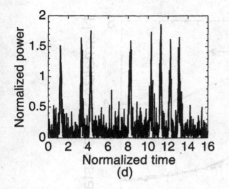

(d)

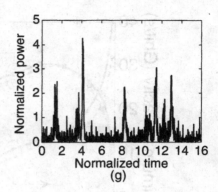

(g)

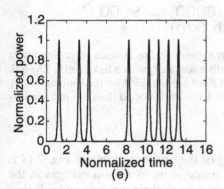

(e)

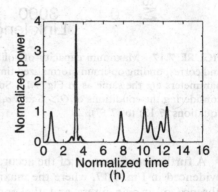

(h)

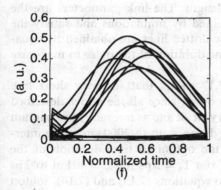

(f)

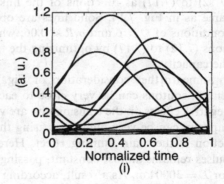

(i)

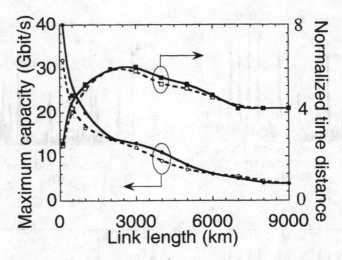

FIGURE 7.17   Maximum capacity of soliton systems in the absence of in-line filters and corresponding optimum normalized time distance, $T_F R$, versus link length. The link parameters are the same as in Fig. 7.14. Solid lines are obtained by the simulations by considering the conditions of $Q > 6$ and $\sigma_j R < 0.06$, while dotted lines are obtained by equations (7.12) to (7.17).

A further confirmation of the accuracy of the operating table of Fig. 7.14 is evidenced in Fig. 7.17, where the maximum capacity of soliton systems in the absence of in-line filters and the corresponding optimum normalized time separation are evaluated both by simulative tools and by the equations (7.12) to (7.17) as functions of the link length. The link parameters are the same as in Fig. 7.14. Solid lines are obtained by simulations and satisfy the conditions of $Q > 6$ and $\sigma_j R < 0.06$, while dotted lines are obtained by equations (7.12) to (7.17) by optimizing the time duration of the pulse to maximize the capacity.

Some further consideration of Fig. 7.17 shows that for very short link length, solitons can be very close to each other, since all the above described negative effects on the transmission are very weak and as a result the maximum capacity can be very high. Increasing the distance up to 3000 km, pulse interaction is the main limiting effect. Here the optimized time duration of the pulses remains almost constant, passing from $T_s = 5$ ps for $L = 40$ km to 7 ps for $L = 3000$ km. As a result, according to equations (7.13) and (7.14), soliton instability is still weak, while time distance between adjacent pulses increases with link length, inducing a decrease of maximum capacity. For distances longer than 3000 km, soliton interactions and the Gordon-Haus effect become dominant. To reduce the Gordon-Haus jitter, a longer time duration of the pulses must be adopted as the link length increases. For L between 3000 and 6000 km, the time distance between pulses can be narrower as shown by equation (7.14) and Fig. 7.17. For $L > 6000$ km the figure shows a small decrease of capacity due to the Gordon-Haus effect. The soliton interaction dependence on the link length is almost compensated by the increase of time duration and, hence of soliton period, due to Gordon-Haus effect.

The potentialities of soliton systems without in-line filters are outlined here according to experimental results. A 10-Gbit/s transmission over 4000 km has been successfully realized without in-line filters and is described in [1, e3]. In that paper good agreement between experimental results and theoretical predictions demonstrates that the performance of soliton systems can be accurately foreseen by theoretical or simulation methods.

Soliton systems without in-line control have been exploited to transmit very high bit-rate signals over medium distances. A first experiment of this type is reported in [1, e4] where a 40-Gbit/s signal was transmitted for 65 km using four in-line amplifiers. The chromatic dispersion of the link was $\beta_2 = -3.6\,\mathrm{ps}^2/$ km, so very high-power solitons were needed. A signal transmission at 160 Gbit/ s over a 200-km-long link has been reported in [1, e5]. In this case the link GVD was $\beta_2 = -0.2\,\mathrm{ps}^2/\mathrm{km}$, and hence the peak soliton power was limited to 109 mW, corresponding to an average optical power of 13.5 dBm.

Performances of Soliton Systems with In-line Soliton Control

The time jitter induced by the soliton stream interaction with ASE can be strongly reduced by in-line soliton control, as shown in Section 5.3. Soliton control based on in-line pulse reshaping by using amplitude modulators is complex to be implemented in practical systems, so it is not considered in this chapter. We will refer mainly to soliton control by in-line filtering because at the moment it is technologically the more mature method.

The effects of in-line filters is shown in Fig. 7.18. In Fig. 7.18a the Q factor is plotted against the filter bandwidth B_o for both sliding and fixed filters. In Fig. 7.18b the normalized time jitter, normalized to bit-time, versus the frequency shift $\delta\nu$, in GHz at each amplifier position, is plotted. The DS fiber link length is 9000 km long with a GVD of $\beta_2 = -1.28\,\mathrm{ps}^2/\mathrm{km}$. In the case of sliding filters, the curve is evaluated at a bit-rate $R = 18$ Gbit/s, while in the case of fixed filters at only $R = 13$ Gbit/s. These bit-rates are the maximum bit-rates for which the system fulfills the condition $Q > 6$ by optimizing the input power and the ratio $\Delta T = T/T_F$. The optimum values of ΔT are 5.5 in the case of sliding filters and 6 in the case of fixed filters. For each sliding rate the system presents an optimum filter bandwidth, which decreases with an increasing sliding rate. The sliding filters are characterized by the parameters ω'_f and η introduced in Chapter 5, equations (5.221) and (5.129), respectively. Assuming the expression of the 3-dB filter bandwidth B_o of a Fabry-Perot filter, as defined in the Appendix A1 in equation (A1.10), such parameters can be written as

$$\omega'_f = \frac{2\pi\delta\nu T_s^3}{L_A|\beta_2|} \quad \text{and} \quad \eta = \frac{2}{\pi^2 B_o^2 L_A|\beta_2|} \tag{7.18}$$

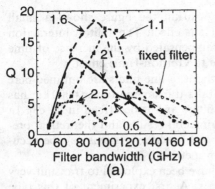
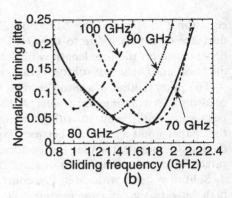

FIGURE 7.18 Effects of the in-line sliding filters on the performance of soliton systems in terms of Q factor versus filter bandwidth for different frequency shifts measured in GHz (a) and normalized timing jitter versus frequency shift for different filter bandwidths (b). Fabry-Perot filters are used, while the other link parameters are the same as in Fig. 7.14. The bit-rate is 18 Gbit/s, and the normalized time distance is 5.5. For comparison in (a) also the case of soliton system with fixed filters is reported with $R = 13$ Gbit/s and a normalized time duration ΔT equal to 6.

In the technical literature also another parameter, called *filter strength*, s, is used [15]. The relationship between s and η is simple: $s = \eta/2$. In Fig. 7.18 the best system performance can be achieved around the conditions $\delta\nu = 1.6$ GHz and $B_o = 80$ GHz which correspond to $\omega_f' = 0.036$ and $\eta = 0.3$.

The values of $\delta\nu$ and B_o corresponding to the best system performance are close to the critical sliding condition presented by equation (5.241), which indicates the maximum frequency shift that can be adopted without inducing instability effects. In particular, in Fig. 7.18, $\omega_f'/\eta = 0.12$ instead of 0.27. It has to be pointed out that a strong increase of time jitter is shown when ω_f'/η is higher than 0.15; this is clearly shown in Fig. 7.18 where for $B_o = 80$ GHz and $\delta\nu = 2$ GHz the Q factor is high but the time jitter unacceptable. Working too near to the critical sliding has two disadvantages: the increase of time jitter and system instability.

In general, simulations have shown that optimum values of the sliding parameters exist that gives a maximum value of Q and a minimum jitter variance. In the following, we will analyze soliton systems adopting sliding filters characterized by $\omega_f' = 0.036$ and $\eta = 0.3$. As a comparison, it can be noted that in the case of the fixed filters the optimum filter bandwidth is equal to 130 GHz, corresponding to $\eta = 0.114$.

In Fig. 7.19 the behavior of soliton systems with fixed filters is reported for different bit-rates in terms of the Q factor versus the normalized input peak power P/P_k (a) and of the normalized time jitter versus the normalized input peak power (b). The link parameters are the same as those of Fig. 7.18, and the normalized time distance between adjacent pulses is $\Delta T = 6$. From the figure it

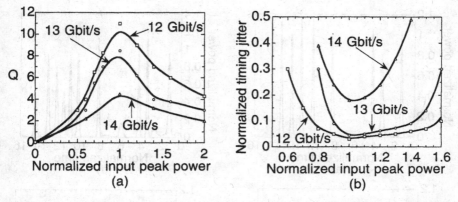

FIGURE 7.19 Behavior of soliton systems in the presence of fixed filters reported for different bit-rates in terms of the Q factor versus the normalized input peak power (a) and of the normalized time jitter versus the normalized input peak power (b). The link parameters are the same as in Fig. 7.18, and the normalized time distance between adjacent pulses is $\Delta T = 6$.

is clear that 13 Gbit/s is almost the maximum capacity of the link when fixed filters are used.

To better understand the system behavior, the corresponding time profile of the signal at the link output is reported in Fig. 7.20 for bit-rates of 13 Gbit/s (Fig. 7.20a) and 14 Gbit/s (Fig. 7.20d). As a comparison, the time profiles have been reported also in the absence of ASE in Fig. 7.20b and in Fig. 7.20e. Finally the eye diagram is reported for the two analyzed cases in Fig. 7.20c and in Fig. 7.20f, respectively.

The performances of the system with a bit-rate of 13 Gbit/s (case a) are mainly limited by the ASE noise whose frequencies close to the carrier are amplified by the excess gain that compensates the filters loss. Such an effect induces the growth of weak spurious pulses, absent in Fig. 7.20b. However, as reported in Fig. 7.19, the Q factor is higher than 6 and the normalized timing jitter lower than 0.06.

In the case of the system at 14 Gbit/s (Fig. 7.20d), soliton instability due to periodic amplification is not negligible, and the spurious pulses are not only due to the ASE noise but also to the presence of radiative waves generated at each amplifier position. In fact, as shown by the Fig. 7.20e, spurious waves are also present in the absence of ASE noise because the soliton period is comparable with the amplifier spacing, and they are amplified by the excess gain, inducing further growth of the spurious pulses. Obviously the presence of the ASE noise worsens the signal propagation.

The performance of soliton systems adopting sliding filters is shown in Fig. 7.21, where the Q factor (a) and the normalized time jitter (b) is plotted against the normalized input peak power (P/P_0) for different values of the bit-rate. The link parameters are the same as in Fig. 7.19. The sliding filters parameters

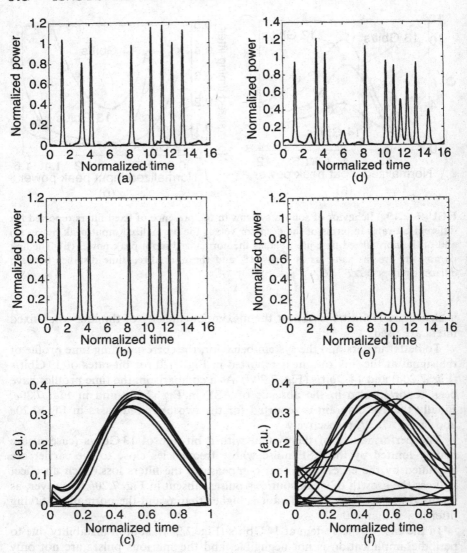

FIGURE 7.20 Time profile of the signal at the link output for the bit-rates of 13 Gbit/s (a) and 14 Gbit/s (d). The corresponding output signals after propagation in the absence of ASE is reported in (b) and (e). The eye diagram is reported for the two analyzed cases in (c) and in (f).

are $\omega'_f = 0.037$ and $\eta = 0.3$ and $\Delta T = 5.5$. In Fig. 7.21 the curves relative to the bit-rate $R = 20$ Gbit/s represent an exemplum where, even if the Q factor is greater than 6 for a certain range of input power, the normalized time jitter is always greater than 0.06. In this condition transmission at $R = 20$ Gbit/s is not possible.

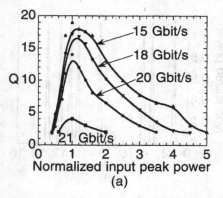
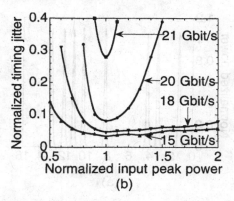

(a)   (b)

FIGURE 7.21 Performances of soliton systems adopting sliding in terms of the Q factor (a) and the normalized time jitter (b) versus the normalized input peak power (P/P_k) for different values of the bit-rate. The link parameters are those of Fig. 7.19. The sliding filters parameters are $\omega_f^i = 0.037$ and $\eta = 0.3$ and $\Delta T = 5.5 T_F$.

Figure 7.21 shows that when the soliton period is much larger than the amplifier spacing, a good tolerance with respect to input powers is obtained with sliding filters. For example, in the case $R = 15$ Gbit/s, the Q factor is higher than 6, and the normalized time duration lower than 0.06 for a quite large normalized power interval (0.7, 3.5).

In order to show the action of the sliding filters, the temporal bit-pattern of the received pulse train at $R = 20$ Gbit/s is reported with and without ASE in Figs. 7.22a and 7.22b, respectively. The eye diagram at the same bit-rate is shown in Fig. 7.22c. The same graphs, but at $R = 21$ Gbit/s, are plotted in Figs. 7.22d–f.

In comparing Fig. 7.22a with Fig. 7.20a, it can be observed how the use of the sliding filters greatly reduces the effect of the extra gain amplification of the ASE noise and suppresses the spurious pulses. In comparing Fig. 7.22b with Fig. 7.20e, the effect of sliding filters in reducing the soliton instability is evident by the fact that the spurious pulses present at 14 Gbit/s with fixed filters disappear at 20 Gbit/s when sliding filters are used. Moreover sliding filtering reduces the effect of nonlinear soliton interaction; in fact this figure has been obtained assuming a normalized time distance between the pulses equal to $\Delta T = 5.5$, smaller than in the case of fixed filters. It can be also observed how well sliding filters preserve the soliton shape. The main limitation for sliding soliton systems remains the Gordon-Haus effect.

At $R = 21$ Gbit/s the effects of the soliton instability and of the nonlinear pulse interaction suddenly reappear with the same sliding filter parameters because of the critical sliding condition. In fact, in the absence of ASE noise, the radiative waves are as shown in Fig. 7.22e, and the interaction among pulses and radiative waves induces a sort of time jitter as shown by the variation of the position of the pulses.

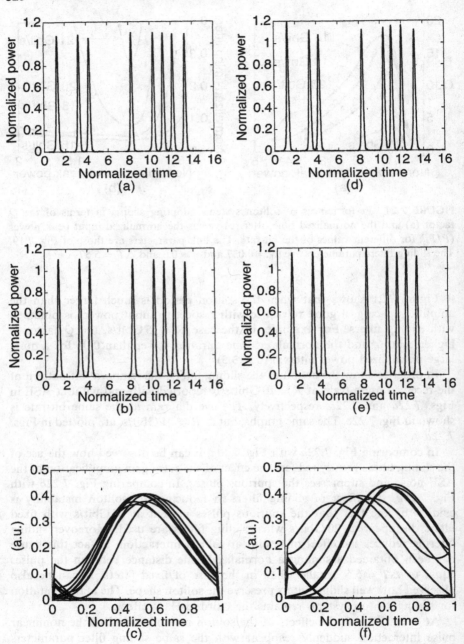

FIGURE 7.22 Time profile of the signal at the link output for the bit-rates of 20 Gbit/s (a) and 21 Gbit/s (d). The corresponding output signals after propagation in the absence of ASE is reported in (b) and in (e). The eye diagram is also reported for the two analyzed cases in (c) and in figure (f).

Until now the performances of very long soliton systems were analyzed by adopting the optimum time separation ΔT between adjacent solitons. In particular, the Q factor has been evaluated for different values of ΔT at the soliton power, and the value of ΔT corresponding to the highest Q has been selected.

The optimum time distance strongly depends on the link length L, as shown in Fig. 7.23 where the normalized time distance ΔT, estimated by optimizing system performance, is shown versus L. Links without in-line soliton control, links adopting fixed filters, and links adopting sliding filters are considered in the figure. The curve in absence of in-line filters is the same as that of Fig. 7.17. For $L < 3000$ km the benefits due to the presence of in-line filters in reducing the pulse interaction is quite evident. Such benefits are still present for $L > 3000$ km, even though they are not visible in Fig. 7.23, since one of the advantages of in-line filtering is that narrower pulses can propagate with respect to the case without filtering but according to equation (7.14) the time distance between pulses must be higher. If the two curves had been estimated with the same pulse duration, the benefits of in-line filtering would have been appeared also for $L > 3000$ km. The superiority of sliding filters with respect to fixed filters is quite evident.

A large amount of experimental work has been carried out to verify the performances of very long distance soliton systems adopting in-line soliton control. The main characteristic of some interesting experiments have been reported in Table 7.1. In this table four different control techniques are considered: fixed filters, sliding filters, sliding signal, and pulse reshaping by in-line modulators. The technique of sliding signal is quite similar to the technique of

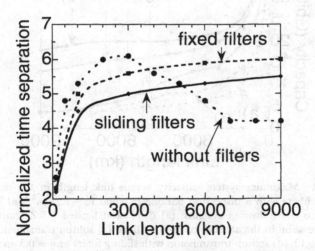

FIGURE 7.23 Optimum time separation, ΔT, versus the link length L for soliton systems in the absence of in-line filters and in the presence of fixed filters and sliding filters. The link parameters are those of Fig. 7.14.

sliding filters; the difference is that while in the latter the filters characteristic function slides along the link, in the former case the signal frequency is changed at the filters locations by placing a frequency shifter before each fixed filter.

The advantage of using the sliding filters with respect of fixed filters is clearly shown by the experimental results. The active pulse reshaping seems to be the most powerful technique. However, it is practically more complex, and it is questionable if the advantages offered by this technique really counterbalance its greater complexity.

Another interesting technique used to control soliton interactions and the Gordon-Haus effect is given by the transmission of solitons of unequal amplitude [1, e7]. This method, based on the observations carried out in Section 2.3.2 when dealing with soliton interactions, can be used, besides all the soliton controls method, to further improve the system's capacity.

Comparison among Systems in the High Chromatic Dispersion Regime

A first comparison among the performances of different systems operating in regime of high chromatic dispersion is presented in Fig. 7.24, where the maximum system capacity is reported as a function of the link length in the case of a link adopting DS fibers with a dispersion parameter equal to $-1.28 \, ps^2/km$ and $L_A = 40 \, km$. The maximum capacity has been obtained by optimizing the

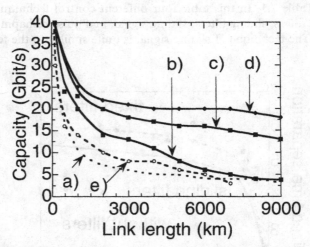

FIGURE 7.24 Maximum system capacity versus link length in the case of a link adopting DS fibers with a dispersion parameter equal to $-1.28 \, ps^2/km$. The different curves refer to the following systems: (a) dispersion limited NRZ transmission, (b) soliton transmission in the absence of in-line filters, (c) soliton transmission with fixed filters at $\eta = 0.12$, (d) soliton transmission with sliding filters at $\eta = 0.3$ and $\omega_f' = 0.036$, (e) NRZ transmission with nonlinear compensation of the GVD without in-line filters. All in-line filters are of the Fabry-Perot type. For soliton systems the time duration ΔT was chosen according to the results of Fig. 7.23.

system parameters (mainly the input power and the time duration of the pulses) and increasing the bit-rate until the conditions on Q and on the time jitter are both satisfied: This means that $Q = 6$ and $\sigma_j R < 0.06$ have been checked.

In Fig. 7.24 the different curves refer to the following systems:

a   for dispersion limited NRZ transmission.

b   for soliton transmission in absence of in-line filters.

c   for soliton transmission with fixed filters with $\eta = 0.12$.

d   for soliton transmission with sliding filters with $\eta = 0.3$ and $\omega_f' = 0.036$.

e   for NRZ transmission with nonlinear compensation of the GVD without in-line filters.

All in-line filters are of the Fabry-Perot type. For soliton systems the time duration ΔT was chosen according to the results of Fig. 7.23.

All the curves show a decrease of capacity as the link length increases. The behavior of curve b has already been explained by Fig. 7.17. In presence of in-line filters the system capacity greatly increases, especially in presence of sliding filters. The capacity of soliton systems is always higher with respect to the case of NRZ, also when the effect of the nonlinear compensation of GVD is used or if in-line filters are inserted along the line [15].

In Fig. 7.25 the curves have been obtained for step-index fiber links with a dispersion parameter of $-20\,\text{ps}^2/\text{km}$. In this case, due to high value of the GVD, the curves are reported up to a maximum length of 4000 km.

The different curves refer to the following systems:

a   for dispersion limited NRZ transmission.

b   for soliton transmission in absence of in-line filters.

c   for soliton transmission with fixed filters, $\eta = 0.12$.

d   for soliton transmission with sliding filters, $\eta = 0.3$ and $\omega_f' = 0.036$.

e   for NRZ transmission with nonlinear compensation of GVD and without in-line filters.

By comparing Figs. 7.23 and 7.24, a practical result can be derived. Calling R_{ds} and R_{st} the maximum capacity of a certain system over a DS link and step-index link, respectively, the following approximate relation holds:

$$\frac{R_{ds}}{R_{st}} \approx \sqrt{\frac{\beta_2(\text{st})}{\beta_2(\text{DS})}} \tag{7.19}$$

where $\beta_2(\text{st})$ and $\beta_2(\text{DS})$ are the dispersion parameters of the step-index and the DS fiber, respectively. This conclusion can be easily understood by looking at equation (2.58) which shows how the dispersion length depends on the square of the capacity and on the inverse of the GVD value.

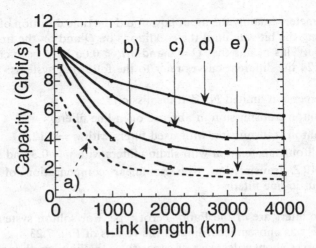

FIGURE 7.25 Maximum system capacity versus link length in the case of a link adopting step-index fibers with a dispersion parameter equal to $-20\,\text{ps}^2/\text{km}$. In this case, due to high value of the GVD, the curves are reported up to a maximum length of 4000 km. The different curves refer to the following systems: (a) dispersion limited NRZ transmission, (b) soliton transmission in the absence of in-line filters, (c) soliton transmission with fixed filters at $\eta = 0.12$, (d) soliton transmission with sliding filters at $\eta = 0.3$ and $\omega_f' = 0.036$, (e) NRZ transmission with nonlinear compensation of the GVD and without in-line filters.

So far we have considered an amplifier spacing equal to 40 km. Details on the capacity of optical point-to-point links with different amplifier spacing can be found elsewhere [11]. In particular, very high-capacity systems can be implemented also by considering amplifier spacing of the order of 100 km, especially if soliton systems with sliding filters are used. From numerical simulations it results that for two links that differ from each other just in amplifier spacing L_{A1} and L_{A2}, the maximum capacities R_{LA1} and R_{LA2} achievable in the regime of high GVD are approximately related by the relationship $R_{LA2} \approx R_{LA1}\sqrt{L_{A1}/L_{A2}}$ [15].

7.1.4 Systems Operating in Fluctuating Chromatic Dispersion

In Section 7.1.2 it is shown that the performances of optical systems in the regime of low chromatic dispersion are mainly limited by the nonlinear interaction between the signal and the ASE. In [16] it was shown for the first time that a particular map of GVD along the link, which results in a zero average dispersion, can strongly limit the nonlinear interaction between the signal and the ASE.

This can be explained by the fact that since the GVD is locally nonzero, the efficiency of FWM is low and the nonlinear interaction between the signal and

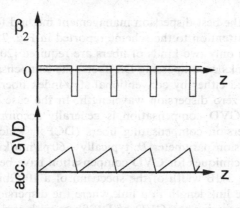

FIGURE 7.26 Scheme of dispersion management realized by two different fibers. The corresponding accumulated GVD is also reported.

the ASE is strongly reduced [16–17]. In the literature the construction of a particular dispersion map along the link is usually called *dispersion management*. In [16] the *dispersion management* was performed by a sawtooth distribution of GVD along the link, but efficient dispersion management can be achieved in many other ways.

In Fig. 7.26 a dispersion management scheme using two different fibers is illustrated by reporting both the GVD distribution along the link and the accumulated GVD in each link section [17]. The fiber link is characterized by a dispersion β_2. In the link the GVD β_2 is periodically compensated every L_L kilometers by a particular piece of fiber with length L_c km and a dispersion coefficient equal to β_2'. To obtain an overall average dispersion equal to $\bar{\beta}_2$, the parameters of the two fibers have to be selected according to the relation

$$\bar{\beta}_2 = \frac{L_L\beta_2 + L_c\beta_2'}{L_L + L_c} \tag{7.20}$$

The term $L_p = L_L + L_c$ indicates the periodic length of the dispersion management.

As shown in Sections 7.1.2 and 7.1.3, and in particular by Fig. 7.12, to obtain a satisfactory propagation of NRZ signals, the overall link dispersion has to be kept as low as possible, while to efficiently propagate solitons, a small amount of anomalous chromatic dispersion is useful [18]. Thus, for the sake of simplicity, in the following we will assume that NRZ signals propagate in the regime of $\bar{\beta}_2 = 0$, while solitons in the regime of $\bar{\beta}_2 = -1.28\,\text{ps}^2/\text{km}$.

For solitons a promising dispersion management scheme, in which the GVD uniformly decreases along the link, has been proposed. This configuration preserves the time duration of the soliton pulses as they loose energy propagating in a lossy fiber link, according to equation (7.12) [19]. At the moment it is not clear which is the best dispersion management method for soliton signals; here we limit our attention to the scheme reported in Fig. 7.26, since it seems more practical and only two kinds of fibers are required [20].

In the case of links encompassing DS fibers, the compensation of the GVD is generally attained either by conventional step-index fibers or by DS fibers having a different zero dispersion wavelength. In the case of links adopting step-index fibers, GVD compensation is generally accomplished by special fibers, called dispersion compensating fibers (DCFs), which at $\lambda = 1.55\,\mu$m have a high dispersion parameter D, typically $-60\,$ps/nm/km [21]. In Section 6.6.2 alternative techniques for GVD compensation have been summarized.

In Fig. 7.27 the 3-dB width of the spectrum of a 10-Gbit/s-NRZ signal is reported versus the link length in a link where the dispersion management is achieved by compensating the GVD of DS fibers with step-index fibers. The amplifier spacing is equal to 40 km for all the curves, while the length L_p is equal to 10, 80, and 640 km. This means that to attain dispersion management, a piece of step-index fiber is placed at each amplifier location in the first case, every two amplifiers in the second case, and every 16 amplifiers in the third case. The dispersion parameters of the link are $\beta_2 = 1.28\,$ps^2/km, $\beta_2' = -20\,$ps^2/km, and $\bar\beta_2 = 0$. The solid lines refer to the case where the

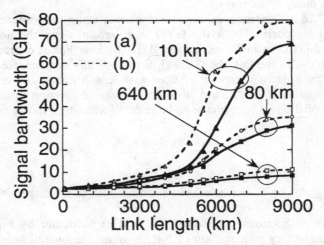

FIGURE 7.27 Spectral broadening of a 10-Gbit/s NRZ signal versus the link length in a link with dispersion management, (the GVD of DS fibers is compensated by step-index fibers) and with an amplifier spacing equal to 40 km. The length L_p is equal to 10, 80, and 640 km. In the first case, to attain dispersion management, a piece of step-index fiber is placed at each amplifier location; in the second case it is present every two amplifiers; and in the third case every 16 amplifiers. Other parameters are $\beta_2 = 1.28\,$ps^2/km, $\beta_2' = -20\,$ps^2/km, and $\bar\beta_2 = 0$. Solid lines refer to the case where the compensating fiber is placed at the input of the corresponding amplifier, while dashed lines to the case where it is placed at the amplifier output.

compensating fiber is placed at the input of the corresponding amplifier, while the dashed lines refer to the case where it is placed at the amplifier output.

The figure shows that the spectrum width is about 2 GHz (equal to the width of the transmitted spectrum) up to 500 km and, on increasing the link length, tends approximately to [22]

$$B_o = \frac{1}{\sqrt{4\pi^2 \beta_2' L_c}} \qquad (7.21)$$

This result reflects the fact that the signal spectrum tends to occupy all of the bandwidth in which the FWM efficiency is high according to equation (2.42).

Figure 7.27 shows also that the spectral broadening is higher when the compensating fiber is placed at the output of the corresponding amplifier. This is due to the fact that in the compensating fiber, propagation in the anomalous region occurs, and the MI cooperates to the spectral broadening. When the compensating fiber is placed at the amplifier output, the signal power during the transit in the compensating fiber is higher, and the MI is more effective.

The Q factor versus input peak power is reported in Fig. 7.28 for the cases of Fig. 7.27 for a 9000 km link length. The curves refer to the case where an optical Fabry-Perot filter with a 3-dB bandwidth of 50 GHz is placed in front of the photodiode. In all cases the Q factor increases with the transmitted power up to a threshold; above that threshold a further increase of the power results in a lowering of the Q factor.

In the approximation of the linear regime, valid for $P_o < 0.2$ mW, all the curves show the same behavior, since the zero average GVD produce no signal

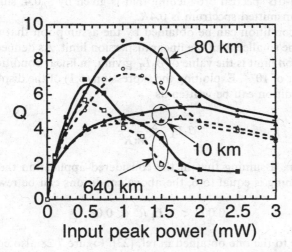

FIGURE 7.28 Q factor versus input peak power for the 10-Gbit/s NRZ systems reported in Fig. 7.27 in the case of a link 9000 km long.

distortion, and the ASE noise is the same, since it is assumed that both the DS fiber and the compensating fiber have the same loss.

When the Kerr effect manifests, in the case of $L_p = 10$ km a large spectral broadening occurs. As a consequence the signal spectrum does not fit the receiver filter, and the performances deteriorate.

In the case of $L_p = 640$ km, during the propagation in the longer transmission fiber segments with length L_L, the signal accumulates a large chirp due to GVD that combined with Kerr effect induces a nonnegligible degradation that cannot be compensated by the linear chirp induced by the compensating fiber.

Finally in the case of $L_p = 80$ km the best performances are obtained as a result of accompanying effects: A limited spectral broadening allows the use a narrow optical filter in front of the receiver along with a reduced chirp induced by the GVD because of the shorter L_L and negligible interplay between the Kerr effect and GVD.

Generally, good dispersion management is reached when the product $\beta_2 L_L$ is high enough to reduce the FWM efficiency but not so high as to cause interplay between the Kerr effect and GVD.

The first condition can be analytically formulated as follows: The 3-dB signal spectrum at fiber output can be estimated by the FWM bandwidth in the limit case of small attenuation and good phase matching (see equation 2.44) of the FWM efficiency. If the 3-dB maximum acceptable broadening at the fiber output is ξR, obtained is

$$\beta_2 L_L > \frac{0.45}{\xi^2 R^2} = \frac{7.2 \times 10^{-2}}{\eta^2 R^2} \tag{7.22}$$

where η is the 3-dB spectral broadening that is given by $\xi/0.4$, since the 3-dB width of the transmitted spectrum is $0.4R$.

The second condition can be obtained by the assumption that the product $\beta_2 L_L$ is by far the smaller than the linear dispersion limit. As defined in Section 6.1, the dispersion limit is the value of $\beta_2 L_L$ giving, in linear condition, an error probability floor of 10^{-9}. Exploiting the expression (6.1) of the dispersion limit, this second condition can be written as

$$\beta_2 L_L \ll \frac{1}{4\pi R^2} \tag{7.23}$$

In conclusion, assuming that in the considered application the acceptable spectral broadening is equal to 6, the above conditions can be rewritten as

$$0.002 < \beta_2 L_L R^2 \ll 0.08 \tag{7.24}$$

which is similar to the one obtained in ref. [23]. Figure 7.28 also confirms that the use of fibers with normal dispersion at the amplifier output is preferable for NRZ transmission, and this is the condition we will turn to next.

Dispersion management is very important for links encompassing step-index fibers in the third transmission window. As shown in Section 7.1.3, these links can also have severe limitations on distances of the order of 200 km, even when solitons are adopted. Using dispersion management, distances of the order of 1000 km can be reached by 10-Gbit/s signals. This is an important objective, considering that most of the already installed fibers are step-index fibers.

An example of the potentiality of the dispersion management method in the case of links with step-index fibers is reported in Fig. 7.29, where the Q factor is reported with relation to the input peak power for three 10-Gbit/s systems over a distance of 1000 km. In the figure the different systems are indicated as follows

a   refers to a system using NRZ format.

b   refers to a soliton without in-line filtering.

c   refers to a soliton system with sliding filters.

The amplifier spacing is 40 km, and the dispersion compensation is obtained by means of DCFs with a GVD equal to 80 ps²/km, a loss equal to 0.5 dB/km, and $\gamma = 3\,(\text{Wkm})^{-1}$. In case *a* the DCFs are located at the output of the optical amplifiers while in *b* and in *c* the DCFs are at the input of the optical

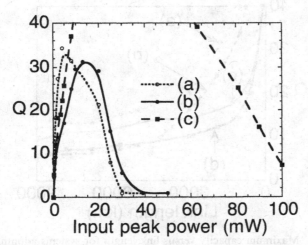

FIGURE 7.29 Q factor versus the input peak power for three 10-Gbit/s systems over a distance of 1000 km. (a) refers to a system using NRZ format, (b) refers to a soliton system without in line filtering, (c) refers to a soliton system with sliding filters. The amplifier spacing is 40 km, and the dispersion management consists of step-index fibers and DCFs with a GVD equal to 80 ps²/km and a loss equal to 0.5 dB/km. In case (a) the DCFs are located at the output of the optical amplifiers, while in (b) and in (c) the DCFs are at the input of the optical amplifiers. T_F for solitons was chosen equal to 20 ps.

amplifiers. Indeed, as demonstrated in ref. [18], in the case of soliton signals it is preferable to use the propagation in conditions of high power in the anomalous dispersion regime. The average GVD is chosen equal to 0 for NRZ signals and to $-1.28\,\mathrm{ps^2/km}$ for soliton signals. T_F for solitons was chosen equal to 20 ps.

In the case of NRZ signals, the Q factor grows as the input power increases due to the improvement of the signal-to-noise ratio. When a certain power level is reached, depending on the system, the Kerr effect predominates, lowering the Q factor.

The behavior of soliton systems is quite similar to the one reported in Section 7.1.3. In the case of dispersion management the power corresponding to the fundamental soliton can be found from equation (7.12), assuming for the value of β_2 the average GVD of the link. However, as observed in ref. [24], the best performance of the soliton system is obtained for a power higher than that of the fundamental soliton (9.4 mW). It has been checked that soliton systems reported in Fig. 7.29 show an acceptable time jitter every times the Q factor is higher than 10. Furthermore, by comparing the soliton systems of Fig. 7.13 and Fig. 7.29, the results show that in links with dispersion management the time jitter can be lower than in links with a constant GVD.

In Fig. 7.30 we report the maximum capacity versus the link length for systems adopting dispersion management with only two different fibers.

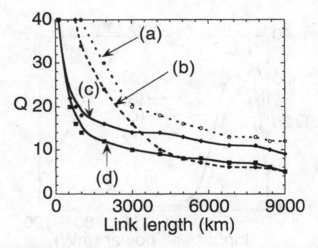

FIGURE 7.30 Maximum capacity versus link length for systems adopting dispersion management using only two different fibers. The four reported curves are evaluated by the following conditions: (a) refers to an NRZ system using DS fibers and step-index fibers for dispersion compensation, in this case $\bar{\beta}_2 = 0$; (b) refers to an NRZ system using step-index fibers and DCFs, even in this case $\bar{\beta}_2 = 0$; (c) refers to a soliton system without in-line filtering, in this case $\bar{\beta}_2 = -1.28\,\mathrm{ps^2/km}$; (d) refers to a solitons system with sliding filters, even in this case $\bar{\beta}_2 = -1.28\,\mathrm{ps^2/km}$. In this case the filter characteristics are the same as used in 7.13.

The four reported curves are evaluated by the following conditions:

a   refers to an NRZ system using DS fibers and step-index fibers for dispersion compensation; in this case $\bar{\beta}_2 = 0$.

b   refers to an NRZ system using step-index fibers and DCFs; likewise in this case $\bar{\beta}_2 = 0$.

c   refers to a soliton system using step-index and DCF without in-line filtering; in this case $\bar{\beta}_2 = -1.28\,\text{ps}^2/\text{km}$.

d   refers to a soliton system using step-index and DCF with sliding filters; likewise in this case $\bar{\beta}_2 = -1.28\,\text{ps}^2/\text{km}$. Also the filter characteristics are the same as used in Section 7.1.3 ($\eta = 0.3$ and $\omega_f' = 0.036$).

Curve a shows that when dispersion management on DS fibers is adopted, a 10-Gbit/s transmission of NRZ signals up to 9000 km is allowed. In comparing the curves b–d, obtained with dispersion management on step-index fibers, it is evident that for short distances NRZ systems outperform soliton systems, but for longer distances soliton systems, even if the average chromatic dispersion is higher than in the NRZ systems, show better performance, particularly when sliding filters are used.

The capacity of soliton systems can be improved by increasing the average GVD, for an instance close to $-0.2\,\text{ps}^2/\text{km}$. The use of dispersion management for soliton signals is also important in the case of links encompassing DS fibers. Although theory has not yet explained some behavior, dispersion management can limit the jitter effects. At the moment it seems that soliton signals during the propagation assume a shape more similar to gaussian than a hyperbolic secant [18]. Indeed, simulations have shown that 10-Gbit/s soliton systems can operate with a good power tolerance, without any in-line control, in links 9000 km long by using a link encompassing DS fibers and partially compensate the GVD with step-index fibers every 80 km, with an average GVD of $-0.2\,\text{ps}^2/$km and a duration of the pulses T_s equal to 5.7 ps.

The experimental activity on dispersion management–based systems has been pushed by the possibility of using dispersion management to design transoceanic systems operating at 10 Gbit/s without using soliton transmission. By this point of view the experiment of ref. [1, e13] is particularly significantly because it demonstrates the feasibility of designing transpacific NRZ systems using 10 Gbit/s.

7.2   SYSTEM PERFORMANCES IN PRESENCE OF PMD

In Section 7.1 signal evolution was studied without including polarization effects. Such an approximation was possible on the notion that the performances of IM-DD systems are determined only by the signal power received in the bit interval and not by the optical field polarization. However, the

presence of two polarization modes introduces effects that can become significant even for IM-DD systems.

The first effect is related to the fiber PMD [25]. Since it accumulates over the
whole system length, even if the fiber PMD is low, it can play an important role
in very long and high bit-rate systems. PMD manifests itself as a differential
group delay between the two polarization modes, and in presence of random
mode coupling a pulse spreading can be observed. Futhermore the presence of
Kerr nonlinearity and ASE noise can induce a strong depolarization of the
light in systems based on the polarization modulation [26].

Depolarization becomes more evident as the signal bandwidth exceeds the
bandwidth of the PSPs [27]. This situation can occur even when the input
optical bandwidth is not very large (e.g., for signals at 10 Gbit/s) but the
propagation is in a regime of low chromatic dispersion with a consequent
spectral broadening.

It should be reminded that when both polarization modes are considered,
the ASE noise power doubles with respect to the case of single polarization
[27]. However, in an IM-DD system the expression of the Q factor given in
equation (4.61) is still valid in the presence of two polarization modes because
the increase of the degree of freedom of the system compensates the increase of
the ASE noise.

Even if the PMD can be neglected, other polarization-related effects limiting
system performances can arise due to the in-line devices that are present especially in optically amplified systems.

The most important polarization related effects are the following [28]:

- *Polarization-dependent loss* (PDL). This effect arises when some in-line
  devices have a loss that depends on the field polarization. PDL can be
  present in imperfect splices, optical filter, and whenever the coupling
  between the fiber and an in-line device is not correct.

- *Polarization-dependent gain* (PDG). This effect arises when optical amplifiers are used whose gain depends on the signal polarization [29]. A certain
  amount of PDG is often present in semiconductor amplifiers, while
  EDFAs are almost immune by this problem.

- *Polarization hole burning* (PHB). Polarization hole burning is due to the
  different physical mechanisms in semiconductor amplifiers and EDFAs.
  In the quantum well semiconductor amplifier, the PHB is due to the fact
  that the valence band is split into two bands. The two-hole populations are
  coupled differently with the field polarizations, so when a polarized field
  saturates the amplifier, a gain asymmetry is created even if the linear
  amplifier gain is polarization independent. This asymmetry has a duration
  related to the characteristic time for the equilibrium between the two
  valence bands.

  In EDFAs, the polarization hole burning is generated by the spatial
  asymmetry of the emission cross section of the erbium ion. Since the

ions have random orientation inside the glass matrix, a polarized optical field introduces a selective saturation, causing a gain asymmetry. Differently from the case of semiconductor amplifiers, this asymmetry is not removed by the presence of the other population and remains until the presence of the saturating field is removed [30].

Due to the combined effect of PDL, PDG, and PHB, the Q factor of an IM-DD system fluctuates in time following the fluctuations of the received field polarization. Moreover, when the signal spectrum is wider than the PSP's bandwidth and depolarization occurs, the presence of PDL, PDG, and PHB further decreases the degree of polarization [27].

As discussed in more detail in Chapter 2, soliton propagation in the presence of two polarization modes can maintain the same characteristics of the propagation with only one mode even in the presence of birefringence and random mode coupling if the input peak power P_0 given by either equation (2.68) or equation (7.12) is multiplied by the factor 9/8 [31]. Consequently solitons seem suitable for polarization multiplexing systems, and in principle, the optical system capacity can be doubled without increasing the optical bandwidth. Polarization multiplexing consists in sending two different messages by means of fields with orthogonal polarizations [26]. The field polarization fluctuates along the link, but the orthogonality is maintained; thus the two signals can be separated at the receiver using a polarization controller. Different kinds of polarization controllers can be devised based on both optical components [32] and electronic processing [33], and almost ideal polarization evolution tracking can be obtained.

Even if soliton transmission is not adopted, transmission schemes based on polarization modulation can be designed that associate two polarization states with the transmitted binary symbols. These systems show the interesting properties of better sensitivity with respect to IM-DD systems [27], immunity to the effect of PHB, and very high efficiency in exploiting fiber capacity by multilevel transmission [34].

In this section we first describe the signal evolution in the presence of PMD, then the problem of light depolarization, and finally the capacity of soliton-based IM-DD systems enhanced by the method of *alternate polarization* [35].

In the following the condition of high birefringence will be assumed (i.e., the most common condition) considering a value of $\Delta\beta$ around $100\,\text{km}^{-1}$ and a random mode coupling with $L_h = 500\,\text{m}$.

7.2.1 IM-DD Systems in the Presence of PMD

In the presence of PMD, the pulses propagating along the fiber widen during propagation by causing intersymbol interference at the receiver.

At a first glance the transmission impairments induced by PMD seems similar to the one induced by GVD. Unlike GVD, which is a deterministic effect, PMD is random process due to thermal and mechanical oscillations. For this reason it cannot be easily equalized at the receiver. In order to better understand the random nature of the PMD, we first consider the case of linear signal propagation in the presence of PMD only. In this case it is possible to compare the error probability by assuming a deterministic PMD with a value equal to the average PMD and to evaluate the average error probability by taking into account the PMD random nature.

In the first case the error probability can be evaluated by the method described in Section 4.2.1, which holds for all cases where a deterministic pulse broadening occurs during propagation. Then the pulse broadening depends on the input field polarization and can be evaluated by decomposing the input polarization into the basis of the PSPs. In particular, the maximum pulse broadening occurs when the input polarization state has equal components along the PSPs. The pulse broadening is given by the differential group delay $\langle \Delta \tau \rangle$, as defined by equation (2.28). This condition represent a pessimistic estimation of the system performance because the link PSPs fluctuate. If a fluctuating PDM is considered, the system error probability calculated as in Section 4.2.1 depends on $\Delta \tau$. The average error probability P_e can be evaluated starting from the PMD-dependent error probability $P_e(\Delta \tau)$, the PDF of $\Delta \tau$ reported in equation (2.27), obtaining

$$P_e = \int_0^\infty P_e(\Delta \tau) p(\Delta \tau) d \, \Delta \tau \tag{7.25}$$

Assuming the system parameters reported in Table 4.1 and a link length of 800 km, in Fig. 7.31 the comparison between the error probability evaluated in

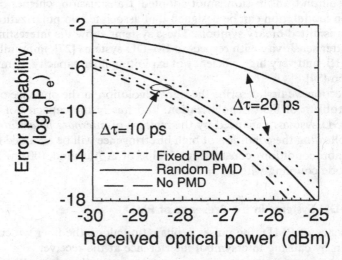

FIGURE 7.31 Error probability versus transmitted peak power for two values of $\langle \Delta \tau \rangle$: 10 ps and 20 ps. The link is 800 km long, and the system parameters are reported in Table 4.1.

the two cases versus the received optical power is reported for two values of $\langle \Delta\tau \rangle$: 10 ps and 20 ps. The corresponding values of $\Delta\beta'$ are 5.42×10^{-4} and 1.1×10^{-3} ps/km.

It is useful to observe that with the assumed parameters, the PMD standard deviation is $\sigma_{\Delta\tau} = 4.2$ ps and $\sigma_{\Delta\tau} = 8.4$ ps in the two considered cases. It is evident from the figure that the random nature of the PMD induces an additional penalty with respect to the case of constant PMD that must be carefully taken into account into the modeling.

If the use of the average error probability is opportune for a first understanding of the impact of a random PMD, this is perhaps not suitable to fix system requirements for designers. Since the PMD fluctuations are slow, when a high PMD occurs, the instantaneous error probability will be much higher than the average error for a long period (about a minute), causing a complete blinding of the receiver. A more suitable way to prescribe system requirements can be introduced by considering the percentiles of the error probability. A possible requirement is that the probability that $P_e(\Delta\tau) > 10^{-9}$ has to be lower than 10^{-5}. This is equivalent to requiring that the system be out of work no more than a 0.001% of the service time, which is an average of five minutes in a year.

In conclusion, it should be noted that the system requirements regarding the PMD can be formulated in different ways and no international standard exists at the moment, so the discussion introduced in this section must be considered as only an introduction to the problem. In the following section we will analyze the performances of transmission systems operating in a nonlinear regime of constant PMD with relation to different systems with PMD.

In Fig. 7.32 the Q factor is reported with relation to the peak power for a 10-Gbit/s soliton system operating in a link with $L_A = 40$ km, $L = 1000$ km, and $\beta_2 = -1.28$ ps^2/km. Three different values of DGD at the link output are (a) 10 ps, (b) 33 ps, and (c) 77 ps. The 3-dB width of the input pulses is 20 ps. The general behavior of the Q factor is similar to that observed in Figs. 7.13 and 7.29, even though the maximum value depends on the value of the PMD, and in particular in case c the value of 6 is reached only for a narrow power interval due to the strong PMD. By comparing Fig. 7.31, obtained in the linear regime, with the results shown in Fig. 7.32, it can be appreciated how the Kerr effect can reduce the impact of PMD by a self-trapping effect [36]. In fact the soliton system attains good performances for $\Delta\tau = 33$ ps, whereas in a linear system the performance would be quite degraded for $\Delta\tau = 20$ ps [37]. The behavior of the curves of Fig. 7.32 can be further understood by Fig. 2.14 in which the signal profile is reported at the link output under similar propagation conditions despite the absence of loss and optical amplification.

The behavior of a 10 Gbit/s NRZ system is reported in Fig. 7.33 under the same conditions as in Fig. 7.31 [38]. In this case the presence of the PMD also induces a degradation of system performance, compared with Fig. 7.13,

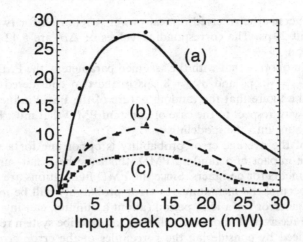

FIGURE 7.32 Q factor versus peak power for a 10-Gbit/s soliton systems operating in a link with $L_{amp} = 40\,km$, $L = 1000\,km$, and $\beta_2 = -1.28\,ps^2/km$. Three different values of DGD at the link output are considered: (a) 10 ps, (b) 33 ps, (c) 77 ps. The 3-dB width of the input pulses is 20 ps.

obtained in absence of PMD. The presence of the Kerr effect limits the impairment of the PMD as shown by the curve c, which further indicates how a Q factor higher than 6 can be obtained, though in a narrow power interval.

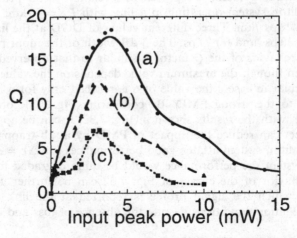

FIGURE 7.33 Q factor versus peak power for a 10-Gbit/s NRZ systems operating in a link with $L_{amp} = 40\,km$, $L = 1000\,km$, and $\beta_2 = -1.28\,ps^2/km$. Three different values of DGD at the link output are considered: (a) 10 ps, (b) 33 ps, (c) 77 ps.

For a soliton signal it can be demonstrated that self-trapping occurs if the following relationship holds [38]

$$\Delta\beta'\sqrt{L_c} < Y\left(\frac{L}{L_D}\right)\sqrt{\frac{\beta_2}{1.28}} \tag{7.26}$$

where β_2 is expressed in ps^2/km, $\Delta\beta'$ in ps/km, L_c in km, and Y is a function of the ratio L/L_D, where L is the link length. The function Y determines when the self-tapping effect shows up. In [36] it was found that Y is equal to 0.3 for $L/L_D = 57$.

Numerical results have shown that equation (7.26) holds also for NRZ signals; the value of Y versus L/L_D is reported in Fig. 7.34 for both NRZ and soliton signals. The simulation have been obtained for a 10-Gbit/s system, using a DS fiber with $L_A = 40$ km and $L_c = 500$ m. The values of Y are evaluated using the maximum value of the parameter $\Delta\beta'/\sqrt{L_c}$ which allows a minimum Q factor higher than 6. To isolate the effect of the PMD nonlinear compensation, the ASE noise is included only at the link output. Figure 7.34 highlights the different effectiveness of nonlinear compensation in the case of solitons and NRZ signals.

In Fig. 7.35 the maximum distance achievable in the presence of PMD by 10-Gbit/s systems is reported with relation to PMD value. The PMD is indicated in ps/$\sqrt{km}$, which is the usual measurement unit for PMD where the link length is not fixed. This unit derives from expression (2.28) for the average PMD of a link. Different systems are considered in the figure:

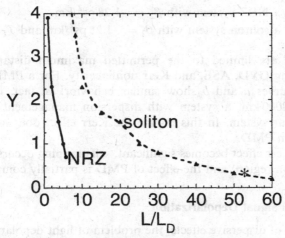

FIGURE 7.34 Behavior of function Y versus L/L_D for NRZ and soliton signals. The simulation are obtained for a 10-Gbit/s system, considering a DS fiber with $L_a = 40$ km and $L_c = 500$ m.

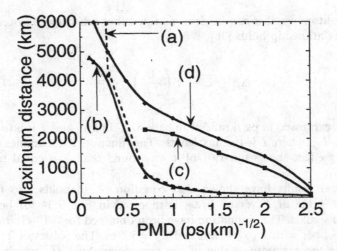

FIGURE 7.35 Maximum distance achievable in the presence of PMD by 10-Gbit/s systems versus PMD. Different systems are considered in the figure: (a) the NRZ system in the presence of only PMD, (b) the NRZ system under dispersion management with $L_c = 80$ km, $\beta_2 = -1.28$ ps^2/km, and with the step-index fiber to compensate dispersion located at the output of the optical amplifiers, (c) the NRZ system with $\beta_2 = -1.28$ ps^2/km, and (d) the soliton system with $\beta_2 = -1.28$ ps^2/km.

a   refers to a NRZ system in presence of only PMD.

b   refers to a NRZ system with dispersion management with $L_c = 80$ km, $\beta_2 = 1.28$ ps^2/km, and step-index fiber to compensate dispersion located at the input of each optical amplifiers.

c   refers to a NRZ system with $\beta_2 = -1.28$ ps^2/km.

d   refers to a soliton system with $\beta_2 = -1.28$ ps^2/km and $T_F = 20$ ps.

Curves *b–d* are limited to the permitted maximum distance given the interplay among GVD, ASE, and Kerr nonlinearity. For a PMD higher than 0.5 ps/$\sqrt{\text{km}}$, curves *a* and *b* show similar behavior: In fact, for a distance shorter than 2000 km, a system with dispersion management behaves very close to a linear system. In this regime the Kerr effect does not induce any improvement in PMD.

When the Kerr effect becomes significant, self-trapping occurs in the anomalous dispersion region, and the effect of PMD is partially compensated.

7.2.2 Optical Signal Depolarization

In the absence of dispersive effects, the problem of light depolarization can be analytically studied by the formalism of Stokes parameters [39]. The following analysis will be limited to the high-birefringence case with random mode coupling $L_c \ll L_{\text{NL}}$. Under this condition, which is generally satisfied in most optical communication systems, the study of signal evolution can be obtained by adding the terms of ASE noise to equations (2.78). In these equations the cross-polarization term has the same coefficient as the self-polarization term (Manakov equations) [40–41]. As a result the Kerr phase shift depends only on the total power of the field, and it is the same for both polarizations. The phase noise induced by the signal–ASE beating is the same for both polarizations, and it does not effect the phase difference between the two polarizations and thus the polarization state. If the system has enough bandwidth to accommodate the widened spectrum, the polarization evolves like in a linear system, and the depolarization coefficient is given by

$$\Delta = \frac{1}{1 + 2S_{sp}B_o} \qquad (7.27)$$

where S_{sp}, defined as in equation (4.49), is the ASE power spectral density at the receiver input and B_o is the optical bandwidth.

The validity of equation (7.26) is limited by the presence of PMD which introduces frequency-dependent coupling between the polarization modes. This effect becomes nonnegligible as the signal bandwidth approaches the PSP bandwidth, approximately given by $1/\Delta\tau$, where $\Delta\tau$ is the average PMD given by equation (2.28) [42]. Hence the evaluation of the spectral broadening is an important parameter that determines the validity of equation (7.24).

For example, a large spectral broadening can occur in the regime of low GVD as a result of nonlinear interaction between the signal and ASE.

In Fig. 7.36 the polarization degree Δ is a function of the propagation distance for the same link considered in Fig. 7.9, and in particular the amplifier spacing is 60 km, and the bit-rate is 5 Gbit/s.

The different curves refer to the following situations:

a   is obtained from equation (7.27).

b   is obtained by simulation in the absence of dispersive effects.

c   is obtained by simulation with $\beta_2 = -0.1\,\mathrm{ps^2/km}$ and without PMD.

d   is obtained by simulation with $\beta_2 = -0.1\,\mathrm{ps^2/km}$ and PMD of 0.14 $\mathrm{ps/\sqrt{km}}$.

e   is obtained by simulation with $\beta_2 = -0.1\,\mathrm{ps^2/km}$, a PMD of 0.14 $\mathrm{ps/\sqrt{km}}$, and a PDL = 0.1 dB located at each amplifier position.

The agreement between theory (curve *a*) and simulations (curves *b* and *c*) is good in the absence of PMD and PDL. The effect of the depolarization of the light is visible in cases *d* and *e*.

The light depolarization can be reduced if the signal at the link output is filtered by an optical filter having a bandwidth equal to the 3-dB bandwidth of the received signal. To show this effect, in Fig. 7.37 the polarization degree Δ is plotted against the propagation distance for a link with zero GVD and

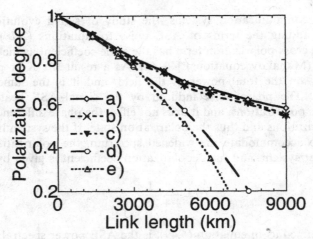

FIGURE 7.36 Polarization degree Δ versus the propagation distance for the link considered in Fig. 7.9. The different curves refer to the following situations: (a) is obtained from equation (7.27); (b) is obtained by simulation in the absence of dispersive effects; (c) is obtained by simulation with $\beta_2 = -0.1\,\mathrm{ps^2/km}$ and without PMD; (d) is obtained by simulation with $\beta_2 = -0.1\,\mathrm{ps^2/km}$ and a PMD of $0.14\,\mathrm{ps}/\sqrt{\mathrm{km}}$; (e) is obtained by simulation with $\beta_2 = -0.1\,\mathrm{ps^2/km}$, a PMD of $0.14\,\mathrm{ps}/\sqrt{\mathrm{km}}$, and a PDL $= 0.1$ dB located at each amplifier position.

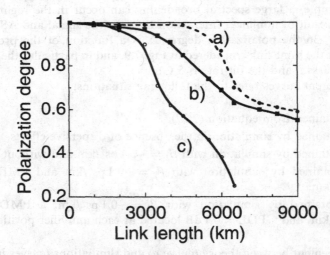

FIGURE 7.37 Polarization degree Δ versus the propagation distance for a link with zero GVD and standard birefringence, supposing that a Fabry-Perot filter is used in front of the receiver. The other parameters are the same as those of curve (a) of Fig. 7.36. The curves refer to the following situations: (a) is obtained from equation (7.27) assuming the optical bandwidth to be equal to the signal bandwidth, (b) is obtained by simulation in the absence of dispersive effects, (c) is obtained by simulation with a PMD of $0.14\,\mathrm{ps}/\sqrt{\mathrm{km}}$, a PDL $= 0.1$ dB located at each amplifier position and $\beta_2 = 0$.

standard birefringence, assuming that a Fabry-Perot filter is used in front of the receiver and operating at a bit-rate of 5 Gbit/s.

Different curves in Fig. 7.37 refer to the following situations:

a   is obtained from equation (7.27) assuming an optical bandwidth equal to the signal bandwidth.

b   is obtained by simulation in the absence of dispersive effects.

c   is obtained by simulation with $\beta_2 = 0$, a PMD of 0.14 ps/$\sqrt{\text{km}}$, and a PDL = 0.1 dB located at each amplifier position.

For $L < 3000$ km, the filter at the link output strongly limits the light depolarization. For $3000 < L < 5000$, the difference between curve *a* and curve *b* is due to the fact that though the signal spectrum has not broadened much, part of the signal information is transferred to the noise background, outside of the filter bandwidth. In case *c* a large nonlinear spectral broadening occurs, and the associated depolarization cannot be limited by the filtering process at the link output.

Since one of the main limiting effects inducing light depolarization is the signal spectral broadening, a strong reduction of the light depolarization can be obtained by adopting systems with dispersion management [41]. The effect of dispersion management on the polarization degree is visualized in Fig. 7.38 for a bit-rate of 5 Gbit/s. The polarization degree is treated as a function of the link length in both the case of a constant GVD ($\beta_2 = -0.1$ ps^2/km, dashed lines)

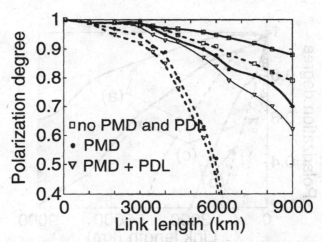

FIGURE 7.38 Polarization versus link length in the case of a constant GVD ($\beta_2 = -0.1$ ps^2/km, dashed lines) and in the case of the sawtooth GVD distribution (solid lines) for an input peak power of 1 mW. At the output of the optical link a Fabry-Perot filter is placed with a bandwidth equal to the 3-dB width of the received optical spectrum.

and a sawtooth GVD distribution (solid lines) for an input peak power of 1 mW. At the output of the optical link a Fabry-Perot filter is placed with a bandwidth equal to the 3-dB width of the received optical spectrum. The values of PMD and PDL are the same as in Fig. 7.37.

In the absence of PMD and PDL the adoption of dispersion management does not perceptibly improve the polarization degree of the output light. On the other hand, in the presence of PMD the use of a dispersion management technique, which can reduce spectral broadening, guarantees a high polarization degree. The PDL introduces a further, but small, polarization degradation.

The degree of polarization is different depending on the propagation regime. In Fig. 7.39, three different cases are summarized:

a refers to the propagation of a 10-Gbit/s soliton signal with $\beta_2 = -1.28\,\mathrm{ps}^2/\mathrm{km}$ and $T_F = 20\,\mathrm{ps}$.

b refers to a 5-Gbit/s-NRZ signal in a link with $\beta_2 = -1.28\,\mathrm{ps}^2/\mathrm{km}$.

c refers to a 5-Gbit/s-NRZ signal in a link with sawtooth distribution with $L_c = 80\,\mathrm{km}$.

The parameters are the same as in Fig. 7.35, and the amplifier spacing is 40 km. The solid lines refer to the case of a low PMD ($0.14\,\mathrm{ps}/\sqrt{\mathrm{km}}$), and the dashed lines to the case of high PMD ($1.4\,\mathrm{ps}/\sqrt{\mathrm{km}}$).

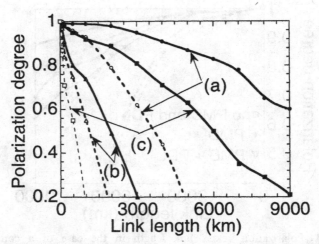

FIGURE 7.39 Polarization versus link length for signals of different shape: (a) refers to the propagation of a 10-Gbit/s soliton signal with $\beta_2 = -1.28\,\mathrm{ps}^2/\mathrm{km}$, (b) refers to a 5-Gbit/s NRZ signal in a link with $\beta_2 = -1.28\,\mathrm{ps}^2/\mathrm{km}$, (c) refers to a 5-Gbit/s NRZ signal in a link with sawtooth distribution and $L_c = 80\,\mathrm{km}$.

The figure clearly shows that solitons are more robust than NRZ signals with respect to light depolarization, even when the Gordon-Haus effect is quite strong. In the presence of a high PMD, good polarization degree is maintained for soliton systems with distances shorter than 2500 km because the pulse shape is strongly influenced by PMD. When NRZ signals are considered, the advantageous effect of dispersion management can again be seen if the PMD is low; conversely, in presence of high PMD, propagation in the anomalous region is preferred. Thus, polarization modulation schemes are feasible but only with solitons [43] or with NRZ signals with dispersion management [27].

7.2.3   Solitons with Alternate Polarizations

Recall from Chapter 5 and Section 7.2.2 that solitons exhibit three peculiar properties that make them interesting for polarization multiplexing schemes [27]. They are maintenance of a high degree of polarization, compensation of the polarization mode dispersion, and maintenance of the orthogonality between states of polarization [43].

One possible way to increase system capacity in exploiting the polarization properties of solitons and not increase too much the complexity of the system is to transmit adjacent pulses on orthogonal polarization states [35]. Since orthogonal solitons show lower nonlinear interaction, the method of alternate polarization permits the use of pulses with a longer duration and as a consequence with a longer soliton period, and detection is performed by a conventional direct-detection receiver.

In Fig. 7.40 the Q factor and the timing jitter are reported for a 4-Gbit/s soliton system with alternate polarizations as a function of the 3-dB time

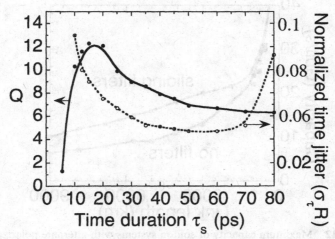

FIGURE 7.40 Q factor and timing jitter versus time duration of the pulses for a 4-Gbit/s soliton system with alternate polarizations as in Fig. 7.15.

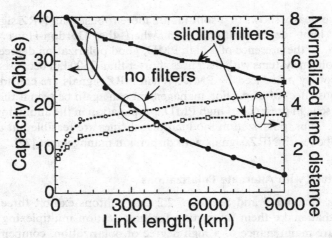

FIGURE 7.41 Maximum capacity and the optimum time spacing reported versus the link length for both absence and presence of in-line sliding filters.

duration of the pulses. The parameters are the same as in the Fig. 7.15. In comparing the two figures, it can be noted that while the improvement in the Q factor is not relevant, the time jitter has considerably decreased by the use of alternate polarization solitons, especially for a soliton duration higher than 20 ps.

The influence of sliding filters has been investigated in alternate polarization systems. The results are presented in Fig. 7.41 in terms of maximum capacity and optimum time spacing behavior versus the link length. From a comparison

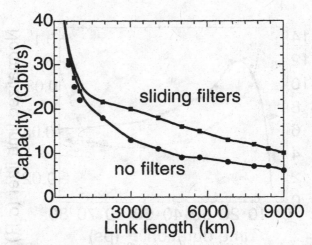

FIGURE 7.42 Maximum capacity of soliton systems with alternate polarizations versus link length in links with sawtooth GVD distribution. The link characteristics are those of Fig. 7.30.

with Fig. 7.40 and Fig. 7.24, it can be deduced that almost a 60% reduction occurs in the time separation among solitons with a consequent 60% improvement in the link capacity.

The advantage of alternate polarized soliton systems is evident also when a sawtooth GVD distribution that includes step-index fibers is considered, as the results show in Fig. 7.42. The link characteristics are the same as those of Fig. 7.30.

The feasibility of soliton systems using polarization-related techniques has been also experimentally verified. A 20-Gbit/s signal was transmitted successfully over a 19,000-km-long link simulated by a recirculating loop in [1, e18]. Sliding filters and polarization multiplexing were used, transmitting a 10-Gbit/s signal for each polarization mode of the fiber. The property of solitons of maintaining the polarization along the pulse and preserving orthogonality among states of polarization was thus exploited to increase system capacity. An interesting experimental study of the parameter range in which this kind of transmission system can operate is reported in [1, e19].

7.3 OPTICAL SYSTEMS ADOPTING SEMICONDUCTOR AMPLIFIERS

Since a large number of step-index fibers are already installed in the access and in the transport networks both in Europe and in United States, the issue of upgrading long-distance optical communication links using such fibers is important. One possible strategy is to use EDFAs by accepting signal propagation in the third-transmission window where the dispersion is quite high.

Another promising possibility is to exploit the 1.3-µm optical window, where the step-index fibers have the zero dispersion wavelength. The state of the art here it to use wide-bandwidth polarization-insensitive semiconductor optical amplifiers (SOAs) [44]. This method takes advantage of both the low dispersion of the step-index fibers at this carrier wavelength and the attractive features of SOAs. SOAs cover a wide wavelength range from 0.7 to 1.6 µm; they are pumped electrically and are substantially cheaper than fiber amplifiers. However, two negative factors in the utilization of SOAs in the communication systems have to be taken into account: saturation effects, which lead to unequal amplification of different pulses in the pattern in the case of high bit-rate transmission, and additional chirp, which a pulse acquires after passing the amplifier. Recent studies have demonstrated the feasibility of 10-Gbit/s transmission over a 500-km optical link based on step-index fibers and in-line SOAs [45].

In this section an investigation on the use of in-line SOAs to obtain high-capacity transmission links at 1.3 µm is presented [46] along with an optimization of the system parameters. In all of the simulations presented in this section, the fiber parameters have the values reported in Table 2.1; any other parameters have the values reported in Table 4.2.

To take into account the fast saturation mechanism of a SOA, the amplifier must be simulated by solving equations (3.18) and (3.20) and taking into account the relation (3.29) between the local gain and the local refraction index. Moreover the noise terms have to be added to the equation describing the field propagation and the carrier dynamics. If noise terms are directly added to equations (3.18) and (3.20), they become stochastic equations, and the standard Runge-Kutta method cannot be applied any more to solve them. Numerical solution of stochastic equations is possible by using the rules of Ito differential calculus [47], but a simulation program adopting such procedures can be more time-consuming than a standard simulation of systems adopting fiber amplifiers.

The ASE noise can also be included in the amplifier simulation in a more simple heuristic way. In particular, the amplifier is simulated neglecting the presence of ASE, and the optical noise is added at the amplifier output. In this model the instantaneous noise power is evaluated by equation (3.10) using the instantaneous value of the amplifier gain and of the carrier density.

This second simulation method is much less complex than the first, but its accuracy has to be verified [48]. Thus the second simulation model will be used in this section.

First we consider the signal propagation, assuming an amplifier spacing equal to 50 km, a fiber loss equal to 0.4 dB/km, and a nonlinear coefficient equal to $2 \, (\text{Wkm})^{-1}$. For the SOA we have an emission factor $F = 5$, a Henry constant equal to 5 [44], a saturation power of 30 mW, and a carrier lifetime equal to 200 ps.

Figure 7.43 shows the behavior of a soliton signal at the output of a link 500 km long both in the presence (a) and in absence (b) of ASE noise for a value of the GVD equal to $-1.5 \, \text{ps}^2/\text{km}$. The effect of saturation of the SOA is clearly shown by the different peak powers of the pulses. In particular, it can be

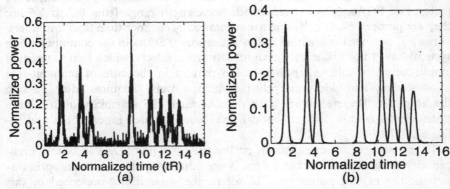

FIGURE 7.43 Behavior of a soliton signal at the output of a link 500 km long both in presence (a) and in absence (b) of ASE noise (b).

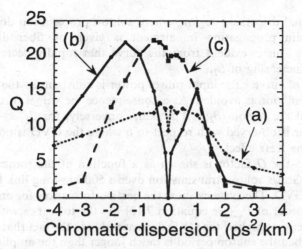

FIGURE 7.44 Q versus input GVD for 10-Gbit/s IM-DD systems operating in a link 500 km long. (a) NRZ signals with $P = 4$ mW, (b) soliton signal with P equal to nominal peak power of the fundamental solitons, and (c) soliton signal with $P = 4$ mW.

seen that the pulse experiences a lower gain when it is placed immediately after another pulse.

In Fig. 7.44 the Q factor is reported as a function of the link GVD, β_2, for a 10-Gbit/s IM-DD system operating in a 500-km-long link. The curves of Fig. 7.44 are evaluated for the following circumstances:

a   refers to NRZ signals with a transmitted peak power $P = 4$ mW.

b   refers to hyperbolic secant RZ signals having an input peak power equal to the nominal peak power (equation 7.12) of the fundamental solitons in the presence of periodical amplification (this means that the power value depends on the GVD value).

c   refers to hyperbolic secant RZ signals with a transmitted peak power $P = 4$ mW.

NRZ signal systems exhibit good performance for $|\beta_2| < 1$ ps^2/km, which means that when the GVD is negligible, in the anomalous region the Kerr effect partially compensates the GVD.

In case b the Q factor decreases, lowering the value of $|\beta_2|$ for $|\beta_2| < 1$ ps^2/km; this is because P_0 is proportional to $|\beta_2|$, as shown in equation (7.12). Thus, if $|\beta_2|$ decreases, the optical SNR and the Q factor decrease as well. Lowering the GVD, the soliton period z_0 decreases (see equation 2.67), and when the soliton period becomes comparable with the amplifier spacing, soliton instability arises with the degradation of the Q factor.

In the normal dispersion region, where soliton propagation does not take place, the main propagation impairment is given by fiber dispersion if $\beta_2 > 1\,\text{ps}^2/\text{km}$. This is evident from the curve that rapidly decreases in this region at the increasing of β_2.

In the case of curve c the input pulses power is constant, so the problem of the SNR degradation is avoided. As a consequence the values of the Q factor are high in all the region $|\beta_2| < 1\,\text{ps}^2/\text{km}$; conversely, for $\beta_2 < -2\,\text{ps}^2/\text{km}$ a worse behavior is observed with respect to b where the GVD is not well compensated by the Kerr effect.

In Fig. 7.45 the Q factor is shown as a function of the transmitted peak power for 10-Gbit/s soliton transmission over a 500-km-long link for different values of the GVD. The systems show the best performances for an input peak power of the order of P_k (see equation 7.12). As the tolerance with respect to the input power gets higher, the GVD decreases due to the fact that solitons are more stable when the soliton period is much longer than the amplifier spacing.

Similar curves have been evaluated in the case of NRZ signals in Fig. 7.46. The systems show their best performances for an input peak power in the range of 5 to 15 mW. It should be noted that for $\beta_2 = 0$ the Q factor assumes lower values with respect to the cases in which β_2 is small but different from zero: This is due to the fact that the nonlinear interaction between the signal and ASE is stronger in the first case. It has to be pointed out that in a link with SOAs, a better tolerance for the input power is shown with respect to the links encompassing the erbium amplifiers. This is due to the fact that the maximum power that propagates in the links depends on the saturation power of the

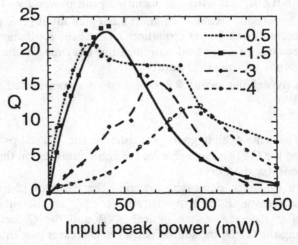

FIGURE 7.45 Q versus input peak power for 10-Gbit/s soliton IM-DD systems operating in a link 500 km long. The values of the GVD are reported in the figure.

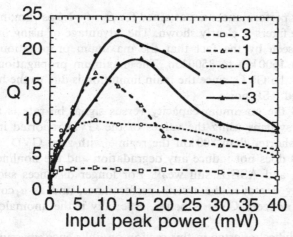

FIGURE 7.46 Q versus input peak power for 10-Gbit/s NRZ IM-DD systems operating in a link 500 km long. The values of the GVD are reported in the figure.

SOAs. In this way the SOA induces a form of control on the signal power, avoiding propagation with levels that are too high.

The maximum capacity versus signal bit-rate for soliton transmission is shown in Fig. 7.47, both in the presence of and in the absence of in-line sliding filters. Two values of the GVD (-1.5 and $-3\,ps^2/km$) are considered, and the sliding parameters η and ω_f' are 0.3 and 0.036, respectively. As already shown in

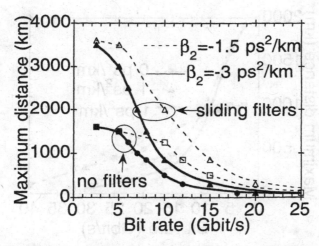

FIGURE 7.47 Maximum capacity versus signal bit-rate for IM-DD soliton systems operating in a link with a SOA amplifiers spaced 50 km apart. The values of the GVD are reported in the figure.

Fig. 7.42 better performances are attained for $\beta_2 = -1.5\,ps^2/km$; moreover the effect of in-line filters is clearly shown. The advantage of using in-line sliding filters can be seen by the fact that the maximum propagation distance is upgraded from 1600 km to 3500 km. The maximum propagation distance is independent of the GVD, since the main limitation is due to the high value of the accumulated ASE noise.

In Fig. 7.48 the maximum capacity versus signal bit-rate is reported for IM-DD NRZ systems using the values of the GVD reported in the figure. For distances shorter than 800 km the regime with zero GVD is preferable, since the GVD does not induce any degradation and the nonlinear coupling between signal and ASE is still weak. For longer distances such nonlinear coupling deeply degrades the system performance, and as a consequence a regime with a nonzero GVD is better, especially if the anomalous region is used.

From the studies reported in this section on links encompassing SOAs, the following conclusions can be made. For propagation over distances shorter than 500 km, a regime with a GVD close to zero is preferred both with RZ or NRZ signals. For longer distances, such a regime should be avoided because the nonlinear interaction between the signal and ASE dominate, and as a consequence a higher absolute value of the GVD is required to limit the interaction. Transmission in the anomalous region provides better performances especially if soliton signals are adopted. To obtain propagation over distances longer than 2000 km, it is necessary to use soliton systems with sliding filters.

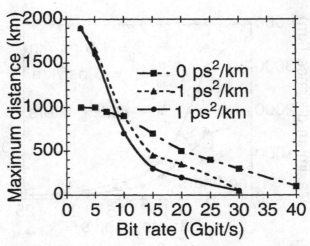

FIGURE 7.48 Maximum capacity versus signal bit-rate for IM-DD NRZ systems operating in a link with the SOA amplifiers spaced 50 km apart. The values of the GVD are reported in the figure.

REFERENCES

1. The main experimental results on the performance of the high-capacity single channel are found in the following references, while the main characteristics of the systems with higher capacties are shown in Tables 7.1:

   e1 N. Edagawa, Y. Yoshida, H. Taga, S. Yamamoto, K. Mochizuki, and H. Wakabayashi, 904 km, 1.2 Gbit/s non-regenerative optical fiber transmission experiment using 12 Er-doped fiber amplifier, *Electronics Letters*, 26: 66–67, 1990.

   e2 S. Saito, M. Murakami, A. Naka, Y. Fukada, T. Imai, M. Iaki, and T. Ito, In-line amplifiers transmission experiment over 4500 km at 2.5 Gbit/s, *IEEE-Journal of Lightwave Technology*, 10: 1117–1124, 1992.

   e3 The following paper reports two different experiments, one without in-line soliton control over 4000 km and one adopting synchronous intensity modulation over 5200 km. Besides the reported experimental data, the comparison between experimental data and theoretical predictions carried out in the paper is quite interesting.
   A. Naka, T. Matsuda, S. Saito, and K. Sato, 5200 km straight-line soliton transmission experiment at 10 Gbit/s, *Electronics Letters*, 31: 1679–1682, 1995.

   e4 K. Iwatsuki, K. Suzuki, S. Nishi, and M. Saruwatari, 40 Gbit/s optical soliton transmission over 65 km, *Electronics Letters*, 28: 1821–1822, 1992.

   e5 M. Nakazawa, K. Suzuki, E. Yoshida, E. Yamada, T. Kitoh, and M. Kawachi, 160 Gbit/s soliton data transmission over 200 km, *Electronics Letters*, 31: 565–566, 1995.

   e6 L. F. Mollenauer, M. J. Neubelt, M. Haner, E. Lichtman, S. G. Evangelides, and B. M. Nyman, Demonstration of error free soliton transmission at 2.5 Gbit/s for more than 14000 km, *Electronics Letters*, 27: 2055–2056, 1991.

   e7 K. Suzuki, N. Edagawa, H. Taga, H. Tanaka, S. Yamamoto, and S. Akiba, Feasibility demonstration of 20 Gbit/s single channel soliton transmission over 11500 km using alternating amplitude solitons, *Electronics Letters*, 30: 1083–1084, 1994.

   e8 L. F. Mollenauer, E. Lichtman, M. J. Neubelt, and G. T. Harvey, Demonstration, using sliding-frequency guiding filters, of error free soliton transmission over more than 20,000 km at 10 Gbit/s, single channel, and over more than 13000 km at 20 Gbit/s in a two channel WDM, *Proceedings of OFC/IOOC'93*, paper PD8-1, pp. 37-40, San Jose, CA, February 21–26, 1993.

   e9 G. Aubin, T. Montalant, J. Moulu, B. Nortier, F. Pirio, and J.-B. Thomine, Demonstration of soliton transmission at 10 Gbit/s up to 27 Mm using "signal frequency sliding" technique, *Electronics Letters*, 31: 52–54, 1995.

   e10 M. Nakazawa, E. Yamada, H. Kubota, and K. Suzuki, 10 Gbit/s soliton data transmission over one million kilometers, *Electronics Letters*, 27: 1270–1273, 1991.

   e11 G. Aubin, T. Montalant, J. Moulu, B. Nortier, F. Pirio, and J.-B. Thomine, Soliton transmission at 10 Gbit/s with a 70 km amplifier span over one million kilometers, *Electronics Letters*, 30: 1163–1165, 1994.

e12 M. Murakami, T. Kataoda, T. Imai, K. Hagimoto, and M. Aiki, 10 Gbit/s, 6000 km transmission experiment using erbium-doped fiber in-line amplifiers, *Electronics Letters*, 28: 2254–2255, 1992.

e13 H. Taga, N. Edagawa, H. Tanaka, M. Suzuki, S. Yamamoto, H. Watabayashi, N. S. Bergano, C. R. Davidson, G. M. Homsey, D. J. Kalmus, P. R. Trischitta, D. A. Gray, and R. L. Maybach, 10 Gbit/s, 9000 km IM-DD transmission experiment using 274 Er-Doped Fiber Amplifiers Repeaters, *Proceedings of OFC/IOOC'93*, PD1-1, pp. 9–12, San Jose, CA, February 21–26, 1993.

e14 M. Murakami, T. Takahashi, M. Aoyama, M. Amemiya, M. Sumida, N. Ohkawa, Y. Fukada, T. Imai, and M. Aiki, 2.5 Gbit/s-9720 km, 10 Gbit/s-6480 km transmission in the FSA commercial system with 90 km spaced optical amplifier repeaters and dispersion-managed cables, *Electronics Letters*, 31: 814–817, 1995.

e15 T. Otani, K. Goto, H. Abe, M. Tanaka, H. Yamamoto, and H. Wakabayashi, 5.3 Gbit/s 11300 km data transmission using actual submarine cables and repeaters, *Electronics Letters*, 31: 380–381, 1995.

e16 V. Letellier, G. Bassier, P. Marmier, R. Morin, R. Uhel, and J. Artur, Polarization scrambling in 5 Gbit/s 8100 km EDFA based system, *Electronics Letters*, 30: 589–590, 1994.

e17 A. Naka, T. Matsuda, and S. Saito, Optical RZ signal straight-line transmission experiment with dispersion compensation over 5520 km at 20 Gbit/s and 2160 km at 2 × 20 Gbit/s, *Electronics Letters*, 32: 1694–1695, 1996.

e18 F. Favre and D. Le Guen, 20 Gbit/s soliton transmission over 19 Mm using sliding-frequency guiding filters, *Electronics Letters*, 31: 991–992, 1995.

e19 F. Favre, D. Le Guen, and M. L. Moulinard, Robustness of 20 Gbit/s 63 km span, 6 Mm sliding filter controlled soliton transmission, *Electronics Letters*, 31: 1600–1601, 1995.

2. Studies on the spectral broadening in presence of ASE can be found in
D. Marcuse, Single-channel operation in very long nonlinear fibers with optical amplifiers at zero dispersion, *IEEE-Journal of Lightwave Technology*, 9: 356–361, 1991.
F. Matera and M. Settembre, Nonlinear compensation of chromatic dispersion for phase- and intensity-modulated signals in the presence of amplified spontaneous emission noise, *Optical Letters*, 19: 1198–1200, 1994.
F. Matera and M. Settembre, Nonlinear evaluation of amplitude and phase modulated signals and performance evaluation of single-channel systems in long haul optical fiber links, *Journal of Optical Communications*, 17: 13–24, 1996.

3. A. Mecozzi, Long-distance transmission at zero dispersion: Combined effect of Kerr nonlinearity and the noise of the in-line amplifiers, *Journal of the Optical Society of America B*, 11: 462–469, 1994.
A. Mecozzi, Error probability of amplified IMDD systems at zero dispersion, *Electronics Letters*, 29: 2136–2137, 1993.

4. A. Barthelemy and R. De La Fuente, Unusual modulation instability in fibers with normal and anomalous dispersions, *Optics Communications*, 73: 409–412, 1989.

5. The nonlinear compensation of chromatic dispersion for a train of squared pulses (an NRZ signal) was studied for the first time in
J. P. Hamaide and P. Emplit, Limitations in long haul IM/DD optical fiber systems

caused by chromatic dispersion and nonlinear Kerr effect, *Electronics Letters*, 26: 1451–1453, 1990.

A similar effect was found for a phase modulated signal in

E. Iannone, F. S. Locati, F. Matera, M. Romagnoli, and M. Settembre, Nonlinear evolution of ASK and PSK signals in repeaterless fiber links, *Electronics Letters*, 28: 1902–1903, 1992.

The compensation of GVD by means of the Kerr effect and in presence of source chirp is investigated in

N. Suzuki and T. Ozeki, Simultaneous compensation of laser chirp, Kerr effect, and dispersion in 10 Gbit/s long-haul transmission systems, *IEEE-Journal of Lightwave Technology*, 11: 1486–1494, 1993.

A comparison between the cases of intensity and phase modulated signals is reported in

F. Matera and M. Settembre, Nonlinear compensation of chromatic dispersion for phase- and intensity-modulated signals in the presence of amplified spontaneous emission noise, *Optics Letters*, 19: 1198–1200, 1994.

Nonlinear compensation of chromatic dispersion for nonsoliton signals is also experimentally demonstrated in

A. H. Gnauk, R. W. Tkach, and M. Mazurczyk, Interplay of chirp and self phase modulation in dispersion-limited optical transmission systems, *Proceedings of ECOC'93*, pp. 104–108, Montreux, Switzerland, September 12–16, 1993.

6. F. Matera, A. Mecozzi, M. Romagnoli, and M. Settembre, Sideband instability induced by periodic power variation in long-distance fiber links, *Optics Letters*, 18: 1499–1501, 1993.

7. M. Midrio, F. Matera, and M. Settembre, Reduction of the amplified spontaneous emission noise impact on optical communication systems, *Proceedings of OFC'97*, Dallas, February 16–22, 1997, paper WL40, 1997.

8. D. Marcuse, Simulations to demonstrate reduction of the Gordon-Haus effect, *Optics Letters*, 17: 34–36, 1992.

9. J. P. Gordon and H. A. Haus, Random walk of coherently amplified solitons in optical fiber links, *Optics Letters*, 11: 665– 667, 1986.

10. Soliton interaction has been widely studied, both for the theoretical interest in the behavior of this particle-like pulse and for its practical importance. See, for example, J. P. Gordon, Interaction forces among solitons in optical fibers, *Optics Letters*, 8: 596–598, 1983.

For the perturbative approach, see

V. I. Karpman and V. V. Solov'ev, A perturbational approach to the two-solitons system, *Physica* 3D: 487–502, 1981.

Experimental results are presented in

F. M. Mitschke and L. F. Mollenauer, Experimental observation of interaction forces between solitons in optical fibers, *Optics Letters*, 12: 355–357, 1987.

C. De Angelis and S. Wabnitz, Interactions of orthogonally polarized solitons in optical fibers, *Optics Communications*, 125: 186–196, 1996.

A theoretical analysis of soliton interaction in amplified transmission links is also reported in Chapter 5.

11. Simulation of soliton systems is described in several papers. Some examples of performance evaluation of transmission systems are found in

J. V. Wright and S. F. Carter, Constraints on the design of long-haul soliton systems, *Proceedings of Nonlinear Guide-Wave Phenomena*, pp. 6–9, Cambridge, England, September 4–6, 1991.

V. Veluppillai, J.-B. Thomine, and F. Pirio, Numerical simulations for power margins in soliton transmission systems over transoceanic distances, *Optics Letters*, 19: 1618–1620, 1994.

F. Matera and M. Settembre, Comparison of the performance of optically amplified transmission systems, *IEEE-Journal of Lightwave Technology*, 14: 1– 12, 1996.

F. M. Knox, W. Forysiak, and N. Doran, 10-Gb/s soliton communication systems over standard fiber at 1.55 mm and the use of dispersion management, *IEEE-Journal of Lightwave Technology*, 13: 1955–1962, 1995.

12. A simple description of the soliton limitations in optical communications can be found in

D. Wood, Constraints of the bit rate in direct detection optical communication systems using linear or soliton pulses, *IEEE-Journal of Lightwave Technology*, 8: 1097–1106, 1990.

J. V. Wright and S. F. Carter, Constraints on the design of long-haul soliton systems, *Proceedings of Nonlinear Guide-Wave Phenomena*, pp. 6–9, Cambridge, England, September 4–6, 1991.

J. P. Gordon and L. F. Mollenauer, Effects of fiber nonlinearities and amplifier spacing on ultra-long distance transmission, *IEEE-Journal of Lightwave Technology*, 9: 170–173, 1991.

Details on the effect of the soliton instability can be found in

L. F. Mollenauer, S. G. Evangelides, and J. P. Gordon, Long-distance soliton propagation using lumped amplifiers and dispersion shifted fiber, *IEEE-Journal of Lightwave Technology*, 9: 170–173, 1991.

13. J. V. Wright and S. F. Carter, Constraints on the design of long-haul soliton systems, *Proceedings of Nonlinear Guide-Wave Phenomena*, pp. 6–9, Cambridge, England, September 4–6, 1991.

14. Details on the use of in-line filters in soliton systems can be found in

A. Mecozzi, J. D. Moores, H. A. Haus, and Y. Lai, Solitons transmission control, *Optics Letters*, 16: 1841–1843, 1991.

Y. Kodama and A. Hasegawa, Generation of asymptotically stable optical solitons and suppression of the Gordon-Haus effect, *Optics Letters*, 17: 31– 33, 1992.

L. F. Mollenauer, J. P. Gordon, and S. G. Evangelides, The sliding-frequency guiding filter: An improvement form of soliton jitter control, *Optics Letters*, 17: 1575–1577, 1992.

Y. Kodama and S. Wabnitz, Analysis of soliton stability and interactions with sliding filters, *Optics Letters*, 19: 162–164, 1994.

M. Romagnoli and S. Wabnitz, Bandwidth limits of soliton transmission with sliding filters, *Optics Communications*, 104: 293–297, 1994.

15. F. Matera and M. Settembre, Comparison of the performance of optically amplified transmission systems, *IEEE-Journal of Lightwave Technology*, 14: 1–12, 1996.

16. D. Marcuse, Single-channel operation in very long nonlinear fibers with optical amplifiers at zero dispersion, *IEEE-Journal of Lightwave Technology*, 9: 356–361, 1991.

17. The theoretical performance of optical systems under dispersion management is studied in

E. Lichtman and S. G. Evangelides, Reduction of the nonlinear impairment in ultralong lightwave systems by tailoring the fiber dispersion, *Electronics Letters*, 30: 348–346, 1994.

An experimental investigation can be found in

H. Taga, S. Yamamoto, N. Edagawa, Y. Yoshida, S. Akiba, and H. Wakabayashi, Performance evaluation of different types of fiber-chromatic dispersion equalization for IM-DD ultralong-distance optical communication systems with Er-doped fiber amplifiers, *IEEE-Journal of Lightwave Technology*, 12: 1616–1621, 1994.

18. Studies on the soliton systems in links with dispersion management can be found in Gabitov, E. G. Shapiro, and S. K. Turitsyn, Optical pulse dynamics in fiber links with dispersion compensation, *Optics Communications*, 134: 317–329, 1997.

S. Wabnitz, I. Uzunov, and F. Lederer, Soliton transmission with periodic dispersion compensation: Effects of radiation, *Photonics Technology Letters*, forthcoming.

M. Nagazawa and H. Kubota, Optical soliton communication in a positively and negatively dispersion-allocated optical fiber transmission line, *Electronics Letters*, 31: 216–217, 1995.

P. L. Chu, A Malomed, and G. D. Peng, Soliton amplification and reshaping in optical fibers with variable dispersion, *Journal of the Optical Society of America B*, 13: 1784–1802, 1996.

F. M. Knox, W. Forysiak, and N. Doran, 10- Gbit/s soliton communication systems over standard fiber at 1.55 mm and the use of dispersion management, *IEEE-Journal of Lightwave Technology*, 13: 1995–1962, 1995.

N. J. Smith, F. M. Knox, N. Doran, K. J. Blow, and I. Bennion, Enhanced power solitons in optical fibers with periodic dispersion management, *Electronics Letters*, 32: 54–55, 1996.

N. J. Smith, W. Forysiak, and N. Doran, Reduced Gordon-Haus jitter due to enhanced power solitons in strongly dispersion managed systems, *Electronics Letters*, 32: 54–55, 1996.

19. K. Tajima, Completion of soliton broadening in nonlinear optical fibers with loss, *Optics Letters*, 12: 54–56, 1987.

20. P. V. Mamyshev and L. F. Mollenauer, Pseudo-phase-matched four-wave mixing in soliton wavelength-division multiplexing transmission, *Optics Letters*, 21: 396–398, 1996.

21. D. Cuomo, G. Ferri, and G. Galasso, A dispersion compensating fiber for the improvement of standard optical networks, *Proceedings of Eleventh EFOC&N'93*, pp. 245–248, The Hague, June 30–July 2, 1993.

22. E. Iannone, F. Matera, A. Mecozzi, and M. Settembre, Performance evaluation of very long span direct detection intensity and polarization modulated systems, *IEEE-Journal of Lightwave Technology*, 14: 261, 1996.

23. E. Lichtman and S. G. Evangelides, Reduction of the nonlinear impairment in ultralong lightwave systems by tailoring the fibre dispersion, *Electronics Letters*, 30: 346–348, 1994.

24. F. M. Knox, W. Forysiak, and N. Doran, 10-Gbit/s soliton communication systems over standard fiber at 1.55 mm and the use of dispersion management, *IEEE-Journal of Lightwave Technology*, 13: 1955–1962, 1995.

25. Details on the effect of the PMD can be found, besides in Chapter 2.1.3, in the following references:

C. D. Poole and R. E. Wagner, Phenomenological approach to polarization dispersion in long single-mode fibers, *Electronics Letters*, 22: 1029–1030, 1986.

F. Curti, B. Daino, G. De Marchis, and F. Matera, Statistical treatment of the polarization mode dispersion in long single-mode fibers, *IEEE-Journal of Lightwave Technology*, 8: 1162–1166, 1990.

F. Matera and C. G. Someda, Random birefringence and polarization dispersion in long single-mode fibers, in *Anisotropic and Nonlinear Optical Waveguides*, C. G. Someda and G. Stegeman (eds.), Elsevier Science, Amsterdam, 1992.

C. De Angelis, A. Galtarossa, G. Gianello, F. Matera, and M. Schiano, Time evolution of polarization mode dispersion in long terrestrial links, *IEEE-Journal of Lightwave Technology*, 10: 552–555, 1992.

S. G. Evangelides, L. F. Mollenauer, J. P. Gordon, and N. S. Bergano, Polarization multiplexing with solitons, *IEEE-Journal of Lightwave Technology*, 10: 28–35, 1992.

F. Matera and M. Settembre, Compensation of the polarization mode dispersion by means of the Kerr effect for non return to zero signals, *Optics Letters*, 20: 28–30, 1995.

26. Detailed works on the optical polarization modulated systems can be found in
S. Betti, F. Curti, G. De Marchis, and E. Iannone, *Coherent Optical Communication Systems*, Wiley Series in Microwave and Optical Engineering, Wiley, New York, 1994.
S. Betti, F. Curti, G. De Marchis, and E. Iannone, Polarization modulated coherent optical communication systems, *Fiber and Integrated Optics*, 10: 291–307, 1992.
The first demonstration of a soliton system with polarization multiplexing is reported in
S. G. Evangelides, L. F. Mollenauer, J. P. Gordon, and N. S. Bergano, Polarization multiplexing with solitons, *IEEE-Journal of Lightwave Technology*, 10: 28–35, 1992.

27. E. Iannone, F. Matera, A. Mecozzi, and M. Settembre, Performance evaluation of very long span direct detection intensity and polarization modulated systems, *IEEE-Journal of Lightwave Technology*, 14: 261–272, 1996.

28. Details on the polarization effects in fiber links can be found in
E. Lichtman, Performance limitations imposed on all-optical ultralong lightwave systems at zero-dispersion wavelength, *IEEE-Journal of Lightwave Technology*, 13: 898–905, 1995.
E. Lichtman, Limitations imposed by polarization-dependent gain and loss on all-optical ultralong communication systems, *IEEE-Journal of Lightwave Technology*, 13: 906–913, 1995.
The effect of the polarization hole burning is thoroughly considered in
N. S. Bergano, V. J. Mazurczyk, and C. R. Davidson, Polarization hole-burning in erbium-doped fiber amplifier transmission systems, *Proceedings of ECOC'94*, pp. 621–628, Florence September 25–29, 1994.

29. An analysis of the SOA gain dependence on the signal polarization and a good reference list can be found in
T. Saito and T. Mukai, Travelling-wave semiconductor laser amplifiers, in *Coherence, Amplification, and Quantum Effects in Semiconductor Lasers*, Y. Yamamoto (ed.), Wiley, New York, 1991.
Practical independence of signal polarization can be obtained in state-of-the-art devices; see, for example,
P. Doussiere, et al., 1550 nm polarization independent DBR gain clamped SOA

with high dynamic input power range, *Proceedings of ECOC'96*, pp. 3.169–3.172, Oslo, Norway, September 15–19, 1996.

30. N. S. Bergano, V. J. Mazurczyk, and C. R. Davidson, Polarization hole-burning in erbium-doped fiber amplifier transmission systems, *Proceedings of ECOC'94*, pp. 621–628, Florence, September 25–29, 1994.

31. Work on soliton propagation in the presence of polarization effects, and in particular, the condition for the peak power of the fundamental soliton, can be found in
L. F. Mollenauer, K. Smith, J. P. Gordon, and C. R. Menyuk, Resistence of solitons to the effects of polarization dispersion in optical fibers, *Optics Letters*, 9: 1218–1221, 1989.
P. K. Wai, C. R. Menyuk, and H. H. Chen, Stability of solitons in randomly varying birefringent fibers, *Optics Letters*, 16: 1231–1233, 1991.
P. K. Wai, C. R. Menyuk, and H. H. Chen, Effects of randomly varying birefringence on soliton interactions in optical fibers, *Optics Letters*, 16: 1735–1737, 1991.
S. G. Evangelides, L. F. Mollenauer, J. P. Gordon, and N. S. Bergano, Polarization multiplexing with solitons, *IEEE-Journal of Lightwave Technology*, 10: 28–35, 1992.
P. K. Wai and C. R. Menyuk, Polarization decorrelation in optical fibers with random varying birefringence, *Optics Letters*, 19: 1517–1519, 1994.

32. Polarization control based on optical tracking of the receiver SOP has been mainly studied with coherent detection. A review of the methods for polarization tracking, in which the main properties required by a practical device are discussed along with all the conventional polarization control techniques, in
T. Okoshi, Polarization-state control schemes for heterodyne or homodyne optical fiber communications, *IEEE-Journal of Lightwave Technology*, LT-3: 1232–1236, 1985.
A more recent review reporting the results obtained from the operation of the considered devices in the framework of a practical transmission system is
N. G. Walker and G. R. Walker, Polarization control for coherent communications, *IEEE-Journal of Lightwave Technology*, LT-8: 438–458, 1990.

33. The concept of electronic compensation of fiber-induced polarization fluctuations was introduced in conjunction with polarization modulation and coherent detection in
S. Betti, F. Curti, B. Daino, G. De Marchis, and E. Iannone, State of polarization and phase noise independent coherent optical transmission system based on Stokes parameters detection, *Electronics Letters*, 24: 1460- -1461, 1988.
The method has been also applied to other modulation formats where coherent detection is used; see, for example,
S. Betti, F. Curti, G. De Marchis, and E. Iannone, Phase noise and polarization state insensitive coherent optical systems, *IEEE-Journal of Lightwave Technology*, LT-5: 561–572, 1987.
S. Betti, F. Curti, G. De Marchis, and E. Iannone, Double FSK coherent optical system exploiting the orthogonal polarization modes of a single-mode fiber, *Microwave and Optical Technology Letters*, 2: 325–327, 1989.
More complex electronic systems are needed to track the received field polarization if multilevel modulation is used. Examples of this systems are reported in the first three papers cited in ref. [34].

34. Different papers have been published dealing with multilevel polarization modulation. See, for example,

S. Betti, G. De Marchis, F. Curti, and E. Iannone, Multilevel coherent optical system based on Stokes parameters modulation, *IEEE-Journal of Lightwave Technology*, LT-9: 1702–1716, 1991.

S. Benedetto and P. Poggiolini, Performance evaluation of multilevel polarization shift keying modulation schemes, *Electronics Letters*, 26: 244–246, 1990.

S. Betti, G. De Marchis, and E. Iannone, Polarization modulated direct detection optical transmission systems, *IEEE-Journal of Lightwave Technology*, 10: 1985–1997, 1992.

M. Midrio, P. Franco, F. Matera, and M. Romagnoli, Polarization-multilevel soliton transmission, *Electronics Letters*, 31: 1473–1474, 1995.

Besides polarization modulation, even quadrature modulation allows very efficient multilevel optical systems to be designed. See, for example,

S. Betti, F. Curti, G. De Marchis, and E. Iannone, Exploiting fiber optics transmission capacity: 4-quadratures multilevel signalling, *Electronics Letters*, 26: 992–993, 1990.

R. Cusani, E. Iannone, A. M. Salonico, and M. Todaro, An efficient multilevel coherent optical system: M-4Q-QAM, *IEEE-Journal of Lightwave Technology*, LT-10: 777–786, 1992.

35. Some interesting papers on soliton systems with alternate polarization are

F. Favre and D. Le Guen, 20 Gbit/s soliton transmission over 19 Mm using sliding frequency guiding filters, *Electronics Letters*, 31: 991–992, 1995.

F. Favre, De Le Guen, and M. L. Moulinard, Robustness of 20 Gbit/s 63 km-span, and Mn sliding filter controlled recirculating loop soliton transmission, *Electronics Letters*, 31: 1600–1601, 1995.

F. Matera, M. Romagnoli, and B. Daino, Alternate polarization soliton transmission in standard dispersion fiber links with no in-line controls, *Electronics Letters*, 31: 1172–1174, 1995.

36. L. F. Mollenauer, K. Smith, J. P. Gordon, and C. R. Menyuk, Resistence of solitons to the effects of polarization dispersion in optical fibers, *Optics Letters*, 9: 1218–1221, 1989.

37. E. Iannone, F. Matera, A. Galtarossa, and M. Schiano, Effect of polarization dispersion on the performance of IM-DD communication systems, *IEEE Photonics Technology Letters*, 1247–1249, 1994.

38. F. Matera and M. Settembre, Compensation of the polarization mode dispersion by means of the Kerr effect for nonreturn-to-zero signals, *Optics Letters*, 20: 28–30, 1995.

39. M. Born and E. Wolf, *Principle of Optics*, Pergamon Press, Oxford, 1980.

40. F. Matera, A. Mecozzi, and M. Settembre, Light depolarization due to amplified spontaneous emission and Kerr nonlinearity in long-haul fiber links close to zero dispersion, *Optics Letters*, 20: 1465, 1995.

41. F. Matera, A. Mecozzi, and M. Settembre, Light depolarization in long fiber links, *Electronics Letters*, 31: 473– 475, 1995.

42. S. Betti, F. Curti, B. Daino, G. De Marchis, E. Iannone, and F. Matera, Evolution of the bandwidth of the principal states of polarization in single mode fibers, *Optics Letters*, 16: 467–469, 1991.

43. S. G. Evangelides, L. F. Mollenauer, J. P. Gordon, and N. S. Bergano, Polarization multiplexing with solitons, *IEEE-Journal of Lightwave Technology*, 10: 28–35, 1992.

44. Several details on SOAs can be found in
G. P. Agrawal and N. A. Olsson, Self-phase modulation and spectral broadening of optical pulses in semiconductor laser amplifiers, *IEEE-Journal of Quantum Electron*, 25: 2297, 1989.

45. C. T. H. F. Liendenbaum, J. J. Reid, L. F. Tiemeijer, A. J. Boot, P. I. Kuindersma, I. Gabitov, and A. Mattheus, Experimental long-haul 1300 mm soliton transmission on stenoberol simple-mode fibers using quantum-well laser amplifiers, *Proceedings of ECOC'94*, p. 233, Florence, 1994.

46. Studies on optical systems with SOAs are in the following papers:
F. Matera and M. Settembre, Study of the performance of 1.3 μm transmission systems on standard step-index fibers with semiconductor optical amplifiers, *Optics Communication*, 133: 463, 1997.
F. Matera and M. Settembre, Analysis of performance of 1.3 μm single-channel transmission systems on standard step-index fibers with semiconductor optical amplifiers, *Microwave and Optical Technology Letters*, 13: 1012, 1996.

47. See, for example,
W. Rümelin, Numerical treatment of stochastic differential equations, *SIAM Journal of Numerical Analysis*, 19: 604–613, 1992.

48. The results of the comparison between the two methods are in deliverable ACTS UPGRADE D2111
F. Matera, et al., Cascaded optical communications with in-line semiconductor optical amplifiers, *Journal of Lightwave Technology*, 15: 962, 1997.

CHAPTER EIGHT

WDM Optically Amplified Systems

Over very long distances, TDM systems despite their high transmission capacity do not exploit efficiently the wide bandwidth of single-mode optical fibers. If TDM multiplexing is performed by electronic multiplexers, the electronic bandwidth limits the achievable bit-rate. On the other hand, although optical TDM multiplexing allows bit-rates as high as 100 Gbit/s to be achieved, such technology is not yet mature for application in communication networks.

The state of the art for efficiently exploiting the huge potential capacity of fiber-optic links resides in optical frequency division multiplexing, also called wavelength division multiplexing (WDM). WDM technology is much more mature than optical TDM technology. The first WDM point-to-point systems are already on the market, and all optical switching fabrics based on WDM have been realized up to a prototype level in international research projects [1].

Experiments based on WDM show that by this technique high transmission capacity can be achieved [2–16]. A simple example is useful. If interactions among different channels are neglected, the total capacity of a WDM system is simply given by the channel bit-rate multiplied by the channel number. To limit the interference among channels, the channel spacing $\Delta\nu$ must be much higher than the signal bit-rate, as shown in Section 4.2.3. In practice, the minimum frequency spacing has to be higher than 10 times the signal bit-rate [17–18]. Assuming a bit-rate of 10 Gbit/s, an available bandwidth of 2000 GHz (16.4 nm in the third window), and a channel spacing of 150 GHz, 13 channels can be transmitted for an overall capacity of 130 Gbit/s.

In a short period point-to-point WDM systems will probably become widely used in transport and access areas: Both new systems will be installed and WDM technology will be used to upgrade already existing TDM systems. This last option is particularly interesting because it maximizes the revenue for the investment made in constructing the communication infrastructure.

However, some problems arise when a TDM link has to be upgraded using WDM technology; they are mainly related to the optical amplifiers gain curve and to the dispersion characteristics of the fiber link. If, for example, in-line optical amplifiers have to be changed, the upgrading cost is higher.

As far as the distribution area is concerned, the introduction of WDM does not appear feasible in the immediate future because of the wide diffusion of SCM systems [19].

Probably at some not-so-distant time, all-optical WDM–based networks will be functioning in access, transport, and highways areas. These networks can be expected to allow the overall communication capacity to be greatly increased, reducing the overall network cost. However, they will also present completely new problems to be solved.

This chapter is mainly focused on WDM point-to-point systems operating over long distances, while WDM transmission over short distances was dealt with in Chapter 6. An analysis of the particular transmission problems arising in WDM networks is presented in Chapter 9.

In the preceding chapter numerical simulation was used to evaluate the performances of different transmission systems. In the case of WDM systems, a simulation-based analysis has many limitations. As detailed in Appendix A1, the simulated optical bandwidth is proportional to the sampling rate. Thus, if a wide bandwidth WDM system is to be simulated, the sampling time must be very short; a very large number of points are needed to represent the signal in the time domain, and the computation time becomes very large. As an example, consider a system with 8 channels, a bit-rate of 2.5 Gbit/s, an optical bandwidth of 640 GHz, and a link length of 1000 km. The simulation time on a personal computer with a 133 MHz PENTIUM processor is as long as 5 minutes for each cycle with 32 bits involving 200 steps every 40 km. If a similar system with a bandwidth of 1280 GHz is to be simulated (e.g., a system with 16 channels and a channel spacing of 80 GHz), the computation time rises to more than 18 minutes. Thus, say, 10 cycles would take 3 hours.

Typically any numerical simulation of WDM systems treats only a limited optical bandwidth. In this chapter, to overcome this difficulty, the analysis is integrated with a review of some important experimental results in order to give an idea of the possibilities of the WDM technology.

The chapter is organized as follows: A general discussion of the main phenomena limiting the capacity of a WDM transmission system is presented in Section 8.1, systems operating in a constant dispersion regime are considered in Section 8.2 where both NRZ and soliton transmission are analyzed, while links with fluctuating dispersion are considered in Section 8.3. The chapter ends with a discussion of the main experimental results in Section 8.4.

In all the simulations reported in this chapter, the fiber parameters are those reported in Table 2.1, while the system parameters are those reported in Table 4.1, which is the same as in Chapter 7.

8.1 MAIN PHENOMENA LIMITING THE CAPACITY OF A WDM SYSTEM

Besides the constraints on the capacity of TDM systems, other constraints arise in WDM systems. The capacity of a WDM systems is determined by the total number of channels that can be packed into the available optical bandwidth. Thus two important parameters have to be fixed: the optical bandwidth and the channel spacing.

The optical bandwidth is mainly limited by the in-line optical devices: optical amplifiers, filters, dispersion compensators, and so on [2, 5]. Conventional erbium optical amplifiers present a bandwidth of the order of 30–40 nm, which is much narrower with respect to the bandwidth of optical fibers [2]. Moreover, when a large number of optical amplifiers are cascaded, the link bandwidth is further reduced. For example, if the amplifiers transfer function is parabolic, the overall bandwidth of a cascade of N amplifiers is $B_o = B_a/\sqrt{N}$, where B_a is the bandwidth of one amplifier.

The bandwidth of EDFAs is not parabolic, and the bandwidth reduction can be limited by flattening the amplifier transfer function. This flattening can be obtained by equalizing the amplifier gain [5, 21] or by suitably designing the doped fiber used in the amplifier. In particular, a flouride-doped EDFA [11, 21] will exhibit a wide optical bandwidth and quite a flat frequency response. In exploiting such techniques, optical amplifiers with a bandwidth of the order of 20 THz (about 160 nm at 1.55 μm) could be realized, permitting the design of optical links with lengths of a few thousands kilometer and optical bandwidths of a few THz.

Once the optical bandwidth is fixed, the WDM system capacity depends on the channel spacing. The channel spacing is determined by the condition that channel interactions should be sufficiently reduced to allow a correct transmission.

A simple type of channel interaction is the linear crosstalk analyzed in Section 4.2.3. There it was shown that linear crosstalk can be reduced both by increasing the channel spacing and by adopting a selective optical filter for channel demultiplexing at the receiver. With a narrow demultiplexing filter the channel spacing can be reduced up to a few times the bit-rate if the only channel interaction is linear crosstalk.

In long WDM links, however, another form of channel interaction arises due to nonlinear phenomena occurring during fiber propagation. In particular, the Kerr effect causes both FWM [22] and XPM [23], generating nonlinear crosstalk among the transmitted channels. Moreover, if the optical bandwidth is sufficiently large, even the Raman effect gives rise to channel interaction [24]. As shown in Chapter 2, all these nonlinear effects introduce a crosstalk component located in the same bandwidth of the channel to be selected; thus their effect cannot be limited by increasing the selectivity of the demultiplexing filter.

The effect of fiber FWM can be limited by increasing the frequency spacing among the channels, operating in fibers with high GVD or allocating the channels with a nonuniform frequency spacing [17–18]. The technique of unequal frequency spacing is simple and effective. As shown in Chapter 2, due to FWM, spurious frequencies are generated in particular positions of the signal spectrum. The method of unequal frequency spacing consists in placing the transmitted channels in frequency positions in which no spurious frequency generated by FWM is present. In this case no nonlinear crosstalk is induced by FWM, which has only the effect of reducing the signal power due to the creation of the spurious frequencies out of the signal bandwidth. The resulting penalty is much smaller than that caused by nonlinear crosstalk; thus the impact of FWM is strongly reduced.

A suitable algorithm for the allocation of the channels is proposed in ref. [18]. The disadvantage of this method is that it requires an optical bandwidth much wider than that required by the same WDM signal adopting a uniform channel spacing. Once the minimum adopted spacing and the channel bit-rate are fixed, the occupied optical bandwidth increases rapidly as the number of channels increases, so the method is essentially limited to WDM systems with a small number of channels (as a rule of thumb, of the order of 10 or 12).

As far as XPM is concerned, its effect can be reduced by increasing the channel spacing and the link dispersion. Both of these methods reduce the superposition time for pulses traveling at different frequencies, thus limiting the interaction strength.

Raman-induced crosstalk increases with increasing the channel spacing differently from FWM and XPM; in the absence of other limitations, it sets an upper limit to the usable optical bandwidth. The nonlinear crosstalk due to the Raman effect can be limited by reducing the average optical power along the link, as shown in Section 2.2.4. In particular, equation (2.52) gives the optical power per channel above which a Raman-induced eye penalty of 0.5 dB is suffered by the system. If the transmitted optical power is a few dBs below this value, the Raman-induced crosstalk can be neglected.

A last important observation is that the performances of different channels in the same WDM system are not the same. A first factor causing such behavior is the unavoidable difference among the gains experienced by the different channels while traveling through the amplifiers. This gain imbalance can seriously degrade the performance. Two equalization strategies are possible: The amplifier gain can be equalized so that the different channels have the same power or the same optical SNR at the link output. Since, at least in the linear case, the error probability depends only on the optical SNR (see Section 4.3.2) the second strategy gives generally better results.

Different techniques have been attempted to equalize the amplifier gain. The most natural approach is to design the doped fiber with a flat gain in the considered bandwidth. As discussed before, this is an important research area to which much effort is currently directed. Another possibility is to use an equalizing filter inside the amplifier so as to shape the overall amplifer response. Different filters can be used for this task; for example, grating filters seem to be most promising for this application.

However, even if the amplifiers gain is perfectly equalized in the system bandwidth, the transmission performance of different channels may not be uniform. This occurs as the impact of nonlinear crosstalk fluctuates due to FWM, XPM, and Raman effects. Nonuniform performance is more evident in systems in which the FWM efficiency is high; thus it will be further analyzed later in this chapter with relation to the performance of WDM systems in a regime of low chromatic dispersion.

8.2   WDM SYSTEMS WITH CONSTANT DISPERSION

For transmission systems with a constant GVD, two propagation regimes were defined in Chapter 7: the low dispersion regime and the high anomalous dispersion regime. The terms *low dispersion and high dispersion* were used in different ways with respect to their usual meanings. Dispersion is considered low when the spectral broadening due to the nonlinear interaction between the transmitted signal and the ASE cannot be neglected. When the dephasing introduced by the GVD strongly limits the spectral broadening, the GVD is considered high.

Unlike the TDM case, in WDM systems, the GVD cannot be lowered too much. The lowering of GVD not only causes the signal spectrum to widen, thus increasing linear crosstalk, but also the FWM efficiency increases. This last effect is more dangerous than the spectrum widening, since it generates a nonlinear crosstalk whose power level increases with the cube of the transmitted power.

We assume IM-DD systems in which the demultiplexing process in front of the receiver is performed by a narrowband Fabry-Perot filter to limit the linear crosstalk among channels. The bandwidth of this filter is set to $2R$, where R is the bit-rate.

8.2.1   WDM Systems Adopting NRZ Signaling

WDM systems operating in the regime of low GVD have not been considered since the efficiency of FWM is high, and even XPM cannot be neglected as a source of nonlinear crosstalk. A first consequence is that even in the presence of perfect SNR equalization at the link output, the performance of one channel is different from that of the others. This effect is evident in Fig. 8.1, where a WDM system composed of six channels at a bit-rate of 5 Gbit/s is considered. The channel spacing is 125 GHz, corresponding to an overall optical bandwidth of 630 GHz (about 5 nm at 1.55 μm), and the amplifiers gain is assumed to be constant in all of the optical bandwidth. The link length is 4000 km, and the fiber dispersion is $-1.28\,\text{ps}^2/\text{km}$.

The different performances of the first, third, and sixth channels versus the input peak power per channel are evaluated in terms of the Q factor. The third channel, at the center of the WDM spectrum, exhibits the worst performance

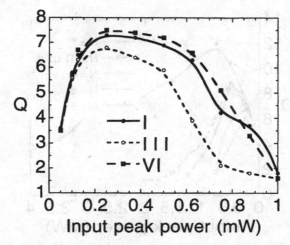

FIGURE 8.1 Q factor versus input average power per channel for each channel of a 6×5 Gbit/s NRZ IM–DD system transmitting over a DS link 4000 km long with amplifiers spaced 40 km apart. NRZ signals are considered, and the channel spacing is 125 GHz. The roman numerals indicate the channel on which the Q factor has been evaluated.

with respect to the channels at the spectrum borders. The performance penalty is manifest as the transmitted power exceeds 0.2 mW at which point the FWM and the XPM effects start to become significant.

Different 5 Gbit/s WDM systems are compared in Fig. 8.2 over a link length of 4000 km. The solid lines were obtained at a fixed channel spacing of 125 GHz increasing channel numbers. The dotted line represents the case of 10 channels with unequal channel spacing. In this case the channels are located according to the method proposed in [17–18], and the frequency spacing are 60, 120, 72, 96, 108, 132, 84, 144, and 156 GHz.

The degradation induced by the FWM is clearly revealed when the number of channels is increased, except for the case of unequal spacing which is quite immune from FWM. The main limitation in the case of unequal spacing is due to the XPM.

Figure 8.2 shows that the achievable capacity of the considered WDM systems is of the order of 30 Gbit/s (not considering the unequal frequency spacing), which is much higher than the maximum capacity attainable under the same conditions by TDM, as shown in Section 7.1.3. Moreover, if the available optical bandwidth is as large as 2 THz (value reached with gain equalization techniques [2, 5, 20–21]) the channel spacing can be increased up to 200 GHz so that 10 channels can be transmitted by uniform channel spacing.

The fact that WDM systems have a higher potential capacity with respect to TDM systems under the same conditions is not limited to the case analyzed in Fig. 8.2, but it is common to a large variety of situations, described in the literature [25].

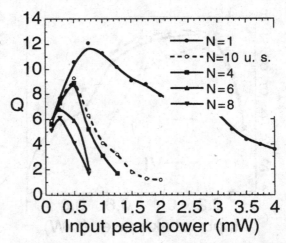

FIGURE 8.2 Q factor versus input average power for $N \times 5$ Gbit/s NRZ systems propagating over a DS link 4000 km long with amplifiers spaced 40 km apart. The number of channels is reported in the figure. Solid lines were obtained considering a uniform frequency channel separation, Δf, equal to 125 GHz. The dotted line represents the case of 10 channels with a nonuniform frequency spacing; in this case the frequency spacing are 60, 120, 72, 96, 108, 132, 84, 144, 156 GHz.

Assuming as a target the system capacity of 10 or 40 Gbit/s, in Tables 8.1–8.4 the system performances in terms of the Q factor are reported for several WDM systems considering both the cases of DS and step-index fibers. Shaded cells in the tables indicate a tolerance power not exceeding 20% with respect to the optimal value required to obtain the specified Q factor.

At 10 Gbit/s different *granularity* of the system can be achieved on both standard and DS fibers. At 40 Gbit/s the *granularity* is too small and it is not recommended.

TABLE 8.1 Maximum Q factor for 40-Gbit/s WDM NRZ systems operating in DS fibers.

	DS fiber				
40 Gbit/s	500 km	1000 km	2000 km	3000 km	4000 km
256 × 155 Mbit/s					
64 × 622 Mbit/s					
16 × 2.4 Gbit/s	41.2	19.6	14.1	8.9	6.2
4 × 10 Gbit/s	16.1	6.8			

Note: Black cells indicate the impossibility of reaching a Q higher than 6, while shaded cells represents critical operation conditions.

TABLE 8.2 Maximum Q factor for 40-Gbit/s WDM NRZ systems operating in step-index fibers.

40 Gbit/s	Step-index fiber				
	500 km	1000 km	2000 km	3000 km	4000 km
256×155 Mbit/s					
64×622 Mbit/s	35.4	17.4	12.2	8.9	6.3
16×2.4 Gbit/s	12.6				
4×10 Gbit/s					

Note: Black cells indicate the impossibility of reaching a Q higher than 6, while shaded cells represents critical operation conditions.

TABLE 8.3 Maximum Q factor for 10-Gbit/s WDM NRZ systems operating in DS fibers.

10 Gbit/s	DS fiber				
	500 km	1000 km	2000 km	3000 km	4000 km
64×155 Mbit/s	>50	11.5	7.2		
16×622 Mbit/s	>50	18.7	14.1	10.6	8.7
4×2.4 Gbit/s	>50	30.1	23.2	13.5	9.5

Note: Black cells indicate the impossibility of reaching a Q higher than 6, while shaded cells represents critical operation conditions.

TABLE 8.4 Maximum Q factor for 10-Gbit/s WDM NRZ systems operating in step-index fibers.

10 Gbit/s	Step-index fiber				
	500 km	1000 km	2000 km	3000 km	4000 km
64×155 Mbit/s	>50	16.7	12.9	9.6	8.3
16×622 Mbit/s	>50	21.5	14.7	11.9	9.7
4×2.4 Gbit/s	15.1				

Note: Black cells indicate the impossibility of reaching a Q higher than 6, while shaded cells represents critical operation conditions.

These results can be simply explained. Two main factors limit the performances of the analyzed WDM systems: nonlinear crosstalk due to FWM and chromatic dispersion. Once the optical bandwidth has been fixed, the minimum channel spacing is determined by the condition of a low FWM efficiency. This is a stronger constraint for the number of transmissible channels on a DS fiber than on a standard fiber.

On the other hand, the effect of GVD mainly introduces an upper limit for the single channel bit-rate according to equation (6.1), which is written as $R < \sqrt{\dfrac{c}{2\lambda^2 DL}}$. Moreover it should be noted that nonlinear compensation of the GVD is not so effective in this case. This is due to the fact that interchannel FWM strongly influences the nonlinear phase shift.

8.2.2 WDM Systems Adopting Soliton Signaling

At least, in principle, both GVD and FWM can be avoided by adopting soliton systems. In a lossless fiber, solitons at different frequencies, not only compensate GVD by the nonlinear frequency shift but also show very weak interaction (see Section 5.4.4). If two solitons traveling at different speeds collide, the frequency separation between them increases during the first part of the collision and decreases during the second part so that no net frequency shift remains when the collision is over [27]. The only effect of the collision is a small time displacement, given by equation (5.387) in soliton units. Taking into account all the collisions a soliton experiences along the link, this displacement reveals itself at the receiver as another form of time jitter. As shown in Section 4.2.1, the total time jitter must be much shorter than the bit-time; thus in practice soliton interaction limits the frequency separation between adjacent soliton channels. In a first approximation, the jitter standard deviation can be assumed equal to the single collision displacement; under this assumption, using the criteria adopted in Chapter 7 for soliton systems in the presence of time jitter, the single collision displacement must be smaller than $0.06\,T$, where T is equal to $1/R$.

Extending equation (5.387) and disregarding the channel spacing range in which the linear crosstalk is too high, the following condition is derived:

$$\Delta\nu > \Delta\nu_{\mathrm{ml}} = \frac{L}{\pi T_s z_0}\left(1 + \sqrt{1 - 0.27\frac{T z_0^2}{T_s L^2}}\right) \tag{8.1}$$

where z_0 is the soliton period, defined in equation (2.67).

In the presence of lumped amplification, as shown in [27] and in Section 5.4.5, the collision between solitons at different frequencies results in a net frequency shift. In general, the process during a collision is quite complex, since sideband instability effects are present due to FWM [28]. The frequency shift can be explained by noticing that the collision takes place while solitons propagate, changing their energy so that the symmetry leading to the zero net frequency shift is broken. The presence of asymmetric soliton collisions also leads to the generation of radiative waves through the creation of sidebands during the collision [28].

The magnitude of the impact of asymmetric collisions on the system performances can be measured, as shown in Section 5.4.5, in terms of the ratio between the collision length and the amplifier spacing. The collision length, defined in equation (5.447) in soliton units, can be rewritten as

$$L_{\text{coll}} = 0.561 \frac{T_s}{\beta_2 \Delta \nu} \tag{8.2}$$

Soliton collisions are particularly detrimental when L_{coll} is of the order of the amplifier spacing L_A. Thus, to ensure a correct system operation, L_{coll} has to be either much longer or much shorter than L_A [27].

From the condition on L_{coll}, a condition on the channel spacing $\Delta \nu$ arises. Assuming a channel spacing that is *quite smaller than* can be quantified as ten times smaller, it results in

$$\Delta \nu < \Delta \nu_M = 0.28 \frac{T_s}{\beta_2 L_A} \quad \text{or} \quad \Delta \nu > \Delta \nu_m = 5.6 \frac{T_s}{\beta_2 L_A} \tag{8.3}$$

where the first condition derives from $L_{\text{coll}} > 2L_A$ and the second from $L_{\text{coll}} < L_A/10$ (compare Fig. 5.16). In a practical WDM system, the condition $\Delta \nu_{m1} < \Delta \nu_M < \Delta \nu_{m2}$ always holds, so the channel spacing can be chosen in two intervals: $(\Delta \nu_{m1}, \Delta \nu_M)$ and $(\Delta \nu_{m2}, \infty)$.

The dependence of the system performances on the channel spacing is shown in Fig. 8.3, where WDM systems transmitting two 5-Gbit/s channels over 9000 km of DS fiber ($\beta_2 = -1.28 \, \text{ps}^2/\text{km}$) are considered. In the figure the Q factor is plotted against the channel spacing in the following cases:

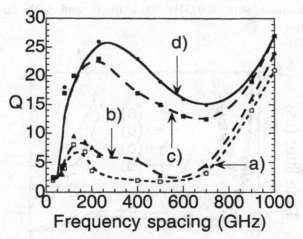

FIGURE 8.3 Q factor versus frequency spacing in a two-channel 5-Gbit/s IM–DD soliton system operating a DS link 9000 km long with amplifiers spaced 40 km apart. Curve (a) is evaluated in the absence of in-line optical filters, (b) with in-line fixed filters and $B_o = 150 \, \text{GHz}$, (c) with in-line fixed filters and $B_o = 120 \, \text{GHz}$, and (d) with in-line sliding filters and $B_o = 100 \, \text{GHz}$ and $\delta f = 0.034 \, \text{GHz}$.

   a in the absence of in-line optical filters ($\Delta T = 6.5$).

   b including fixed in-line filters with $B_o = 150\,\text{GHz}$ ($\Delta T = 6$).

   c including fixed in-line filters with $B_o = 120\,\text{GHz}$ ($\Delta T = 6$).

   d including sliding filters with $B_o = 100\,\text{GHz}$ and $\delta f = 0.034\,\text{GHz}$ ($\Delta T = 5.5$).

In-line filters are constituted by a single Fabry-Perot filter with two resonance frequencies coinciding with the WDM carriers in the case of fixed filters and with the desired sliding frequencies in case *d*.

   The two intervals in which the parameter $\Delta \nu$ can be chosen to obtain good system performances are clear in the figure. Optical in-line filters broaden such intervals, especially when the filter bandwidth is narrow, as shown by curves *c* and *d* [29–30].

   In Fig. 8.4 the time jitter standard deviation is also evaluated under the conditions of the previous figure. With our assumptions, the maximum time jitter in this case is 12 ps (6% of the bit time) [31]. It is noteworthy that in WDM systems the main advantage of sliding versus fixed filters is in the possibility of using a narrower bandwidth better than in the jitter reduction [30].

   Three different WDM soliton systems operating at 5 Gbit/s over 9000 km of DS fiber ($\beta_2 = -1.28\,\text{ps}^2/\text{km}$) are compared in Fig. 8.5 in terms of the Q factor versus the input peak power. The different curves refer to the following cases:

   a for 2 channels without in-line filters and with $\Delta \nu = 80\,\text{GHz}$ ($\Delta T = 6.5$).

   b for 8 channels with 120 GHz fixed filters and with $\Delta \nu = 125\,\text{GHz}$ ($\Delta T = 6$).

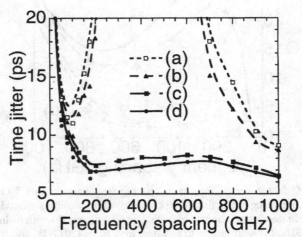

FIGURE 8.4   Standard deviation of the time jitter under the same conditions of Fig. 8.3.

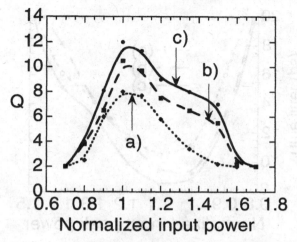

FIGURE 8.5 Q factor versus the normalized input power for WDM IM–DD soliton systems operating in a DS link 9000 km long with amplifiers spaced 40 km apart. Curve (a) refers to a two-channel system without filters with a frequency spacing, $\Delta\nu$, equal to 80 GHz, curve (b) to 8 channels with fixed filters and $\Delta\nu = 125$ GHz, and curve (c) to 10 channels with sliding filters and $\Delta\nu = 100$ GHz. In case (b) the filter bandwidth is 120 GHz, while in case (c) the filter bandwidth is 100 GHz and the frequency shift, δf, is 0.034 GHz.

c for 10 channels with sliding filters with a bandwidth 100 GHz and a frequency shift equal to 0.034 GHz and with $\Delta\nu = 100$ GHz ($\Delta T = 5.5$).

The time jitter standard deviation is also evaluated for these systems, and the results are reported in Fig. 8.6.

8.2.3 Comparison among Different WDM Systems in the Constant GVD Regime

Assuming a DS fiber link with a GVD of $-1.28 \, \text{ps}^2/\text{km}$ and fixing the single channel bit rate to 5 Gbit/s, the maximum capacity of NRZ and soliton WDM systems has been evaluated as reported in terms of propagation distance in Fig. 8.7. The maximum capacity was obtained as follows: The optical bandwidth was assumed to be 1 THz, and for each number of channels, the Q factor and the time jitter were evaluated versus the transmitted power. As usual, a maximum value Q_{max} of the Q factor was obtained for a certain optimum value of the transmitted power. Then the maximum capacity was calculated by multiplying the channel bit-rate (5 Gbit/s) by the maximum number of channels which ensures a value of Q_{max} higher than six and a time jitter to the bit-time ratio lower than 0.06.

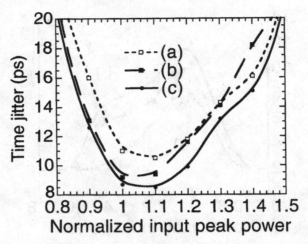

FIGURE 8.6 Standard deviation of the time jitter in the same conditions of Fig. 8.5.

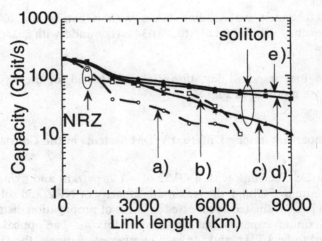

FIGURE 8.7 Maximum capacity versus propagation distance for WDM optical systems with a channel bit-rate of 5 Gbit/s operating in a DS fiber link with amplifiers spaced 40 km apart. Curve (a) refers to a NRZ WDM system with equally spaced frequency channels, curve (b) to a NRZ WDM system with unequally spaced frequency channels, curve (c) to an equally spaced WDM soliton system in the absence of in-line filtering, (d) to an equally spaced WDM soliton system with fixed in-line filtering, and (e) to an equally spaced WDM soliton system with sliding filtering.

The different curves refer to the following cases:

a refers to an NRZ system with equally spaced channels.
b refers to an NRZ system with unequally spaced channels.
c refers to a soliton system in the absence of in-line filtering.
d refers to a soliton system with fixed in-line filtering.
e refers to a soliton system with sliding filtering.

The channel spacing is assumed to be uniform in all cases but *b*. The channel spacing can be simply obtained by dividing the optical bandwidth by the channel number which can be deduced from Fig. 8.7. Exceptions are cases *b* and *c*. In the case *b* the algorithm described in [17, 18] has been applied with a minimum channel spacing of 25 GHz. In curve *c*, where the frequency interval $(\Delta\nu_{m1}, \Delta\nu_M)$ is larger than the available optical bandwidth, the channel spacing is evaluated as 1 THz/N; in the other cases, where $(\Delta\nu_{m1}, \Delta\nu_M)$ is equal or narrower than the available bandwidth, the channels are located inside a bandwidth B_D which is one-half of the interval $(\Delta\nu_{m1}, \Delta\nu_M)$, and the channel spacing is given by B_D/N. For soliton systems ΔT and filter parameters were chosen according to the results of Fig. 8.5.

In this case only soliton systems can transmit up to 7000 km, since NRZ systems are disrupted by nonlinear interaction among channels even if unequal channel spacing is used.

For distances between 6000 and 7000 km, the highest capacity is attained by WDM soliton systems with in-line filtering, while for distances between 2000 and 4000 km, WDM NRZ systems with unequal channel spacing exhibit a capacity similar to that of WDM soliton systems with in-line filtering. The main limitation of the WDM systems with unequal frequency spacing is given by the required optical bandwidth; considering an available bandwidth of 1 THz, a minimum frequency spacing of 25 GHz and a slot with a frequency interval of 5 GHz [18], the number of channels at 5 Gbit/s is limited to 18. This is the reason why cuive *b* is reported for a capacity up to 90 Gbit/s.

For a distance lower than 1000 km, FWM results are less detrimental and NRZ systems with equally spaced channels are more efficient than those adopting unequal channel spacing, since they occupy a narrower optical bandwidth.

A similar curve is reported in Fig. 8.8 in the case of a step-index fiber. The channel bit-rate is 1 Gbit/s. The different curves refer to these cases:

a refers to a NRZ system with uniform channel spacing.
b refers to a soliton system in the absence of in-line filtering.
c refers to a soliton system with fixed in-line filtering.
d refers to a soliton system adopting sliding filtering.

The in-line filter parameters are the same as those of Fig. 8.7.

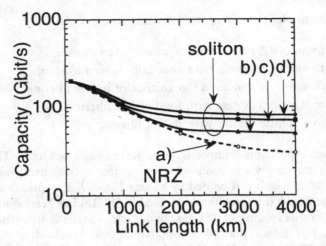

FIGURE 8.8 Maximum capacity versus fiber link for WDM systems in the case of conventional step-index fiber links with amplifiers spaced 40 km apart. The channel bit-rate is 1 Gbit/s. Curve (a) refers to a NRZ WDM system with equally spaced frequency channels, curve (b) to an equally spaced NRZ WDM soliton system in the absence of in-line filtering, (c) to an equally spaced WDM soliton system with fixed in-line filtering, and (d) to an equally spaced WDM soliton system with sliding filtering. The normalized in-line filter parameters are those of Fig. 8.7.

Unlike DS fibers, step-index fibers can support a much larger number of channels even where uniform channel spacing is adopted. This is due to the fact that the FWM efficiency is reduced by the phase mismatch that is naturally provided by the conventional step-index fibers. For this reason unequally channel spacing results are not reported. The higher value of GVD limits the number of pulse collisions, so the action of the filter is not as effective as in the case of DS fiber. Thus in conventional step-index fibers, very high-capacity systems can be implemented by using WDM systems without optical in-line filters that require sophisticated technology.

8.3 WDM SYSTEMS WITH FLUCTUATING CHROMATIC DISPERSION

In Section 7.1.4 it was shown that chromatic dispersion management can largely enhance system performance. The FWM efficiency can be kept low by a high local GVD, thus limiting the spectrum widening, while the GVD detrimental effect is minimized by the low (or zero) average value.

Since FWM and GVD are the main disturbing phenomena present also in WDM system, WDM systems adopting dispersion management can be expected to show much improved performances. In particular, the bit-rate of each channel can be considerably increased in contrast to the case of constant

GVD. Among all the possible dispersion maps that can be implemented, we will analyze only the sawtooth dispersion distribution, which is defined in Section 7.1.4.

Both NRZ and soliton signals can be used in systems adopting dispersion management. We do not report a comparison among these systems, which can be quite complex due to the large number of parameters; we consider rather some representative cases showing the potentialities of this technique.

8.3.1   WDM Systems Adopting NRZ Signaling and Sawtooth GVD Distribution

The Q factor of an NRZ WDM system with a bit-rate of 10 Gbit/s is shown in Fig. 8.9 in terms of the transmitted average power of a single channel. Three channels are transmitted over a distance of 9000 km with a channel spacing of $\Delta\nu = 100$ GHz. The link is constituted by DS fibers with a GVD equal to $+1.28\,\mathrm{ps}^2/\mathrm{km}$ and a step-index fiber with a GVD equal to $-20\,\mathrm{ps}^2/\mathrm{km}$; the period of the GVD distribution is 80 km, and the average GVD is zero for the channel in the middle of the WDM spectrum. The average GVD of the other channels is determined by the third-order dispersion and the channel spacing. In this case, for the first and the third channels it is equal to $-0.068\,\mathrm{ps}^2/\mathrm{km}$ and $0.068\,\mathrm{ps}^2/\mathrm{km}$, respectively. The amplifier spacing is 40 km.

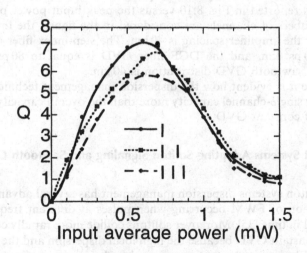

FIGURE 8.9 Q factor versus the transmitted peak power for each channel of a 3×10 Gbit/s WDM NRZ system, operating in a link with dispersion management 9000 km long. The channel spacing is 100 GHz. The link is constituted by DS fibers with a GVD equal to $1.28\,\mathrm{ps}^2/\mathrm{km}$ and a step-index fiber with a GVD equal to $-20\,\mathrm{ps}^2/\mathrm{km}$; the period of the GVD distribution is 80 km, and the average GVD is zero for the channel in the middle of the WDM spectrum. The amplifier spacing is 40 km.

Due to the different value of the average GVD, the three channels have different performances. The Q factor of the third channel never rises to six due to the propagation in the normal region. In this case the phase shifts due to GVD and the Kerr effect accumulate, producing unacceptable intersymbol interference at the receiver. On the contrary, in the case of the first channel, propagation in the anomalous region allows partial nonlinear compensation of the GVD to be achieved.

Even if only channels one and two achieve good performances, Fig. 8.9 gives a first proof of the effectiveness of dispersion management in WDM systems: A capacity of 20 Gbit/s can be transmitted over 9000 km.

It is to be noted that the main factor limiting the capacity of the system considered in Fig. 8.9 is the third-order dispersion instead of the optical bandwidth, as in the case of WDM systems with constant GVD. This is true for almost all the WDM systems adopting a sawtooth GVD distribution; thus to obtain a higher capacity a compensation of the third-order dispersion has to be performed. Third-order dispersion compensation can be achieved by using particular fibers with a negative dispersion slope [33] or suitably designed optical filters [34].

The use of dispersion management also allows WDM systems with a bit-rate of 10 Gbit/s to be realized by adopting step-index fibers cables. In this case dispersion compensation has to be performed by means of DCFs, as discussed in Section 7.1.4.

The Q factor of a WDM system adopting a bit-rate of 10 Gbit/s and step-index fibers is reported in Fig. 8.10 versus the peak input power per channel. Different numbers of channels are considered in the figure, the link length is 4000 km, and the amplifier spacing is 40 km. The step-index fiber has a GVD equal to $-20 \, \text{ps}^2/\text{km}$, and the DCF fiber GVD is equal to $80 \, \text{ps}^2/\text{km}$. The period of the sawtooth GVD distribution is 40 km.

In this case it is evident how the dispersion management technique is effective in raising single-channel capacity more than the overall capacity compared to the case of constant GVD.

8.3.2 WDM Systems Adopting Soliton Signaling and Sawtooth GVD Distribution

In WDM soliton systems dispersion management has several advantages, such as the reduction of FWM occurring when pulses at different frequencies are superimposed in time [35]. Moreover solitons collide more rapidly compared to the case of constant GVD because the high local dispersion and the asymmetry effects caused by power fluctuations are reduced.

Figure 8.11 shows the Q factor for a $2 \times 20 \, \text{Gbit/s}$ system. The fiber link is 10,000 km long, and it is realized by alternating 50-km-long DS fibers ($\beta_2 = -1.28 \, \text{ps}^2/\text{km}$) and 2.6-km-long DCFs ($\beta_2 = 20 \, \text{ps}^2/\text{km}$, $\alpha = 0.6$ dB/km); the average GVD is $-0.26 \, \text{ps}^2/\text{km}$. In Fig. 8.11 both the case where a compensating fiber is placed after the optical amplifier (a) and where it is

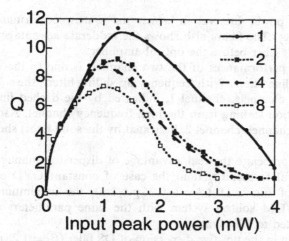

FIGURE 8.10 Q factor versus the input peak power of a WDM system adopting a bit-rate of 10 Gbit/s and step-index fibers for each channel of a 8 × 10 Gbit/s WDM NRZ system. The link length is 4000 km, and the amplifier spacing is 40 km. The step-index fiber has a GVD equal to $-20\,\mathrm{ps}^2/\mathrm{km}$, and the DCF fiber GVD is equal to $80\,\mathrm{ps}^2/\mathrm{km}$. The period of the sawtooth GVD distribution is 40 km.

before the optical amplifier (*b*) are considered. The solitons pulse width is $T_F = 11\,\mathrm{ps}$, the frequency spacing among the channels is 256 GHz, the in-line Fabry-Perot filters have a bandwidth of 146 GHz, and the sliding frequency is 16 GHz/Mm. In the figure the performances of both channels are reported. The best system performance is attained for the soliton peak

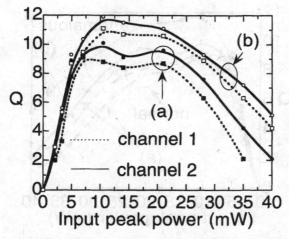

FIGURE 8.11 Dependence of the Q factor on the input peak power in the two channels for 2 × 20 Gbit/s soliton transmissions and DCFs placed (a) after or (b) before the amplifiers, respectively.

power, but the power interval in which the system performances remain acceptable is large. The figure also shows the moderate advantage of locating the compensating fiber before the optical amplifier.

The different performances of the two channels is due to the direction of sliding of the in-line filters. With frequency up-sliding filters, the solitons in the lower-frequency channels (channel 1, indicated by the dashed lines) are disturbed by radiation leaking from the high-frequency channel. As a result the high-frequency channel (channel 2, indicated by the solid lines) shows the best performances.

In order to appreciate the real advantage of dispersion management technique, it is useful to recall that in the case of constant GVD equal to the average value of Fig. 8.11 ($\beta_2 = -0.26 \, \text{ps}^2/\text{km}$), the maximum span of a 2×20 Gbit/s WDM soliton system with the same parameters as those of Fig. 8.11 is limited to 5000 km.

If propagation in the positive dispersion of DS fiber ($\beta_2 = 1.28 \, \text{ps}^2/\text{km}$) and in the negative dispersion of compensating fiber ($\beta_2 = -20 \, \text{ps}^2/\text{km}$) is considered, somewhat worse performance is obtained, as is evident in Fig. 8.12. It is mainly due to the nonelasticity of the rapid soliton-soliton collisions that occur in the DS section with normal GVD.

The dependence of the system performances on the channel spacing is reported in Fig. 8.13 where channel 1 is considered under the operating condition of curve b of Fig. 8.11. It is assumed that exactly the same soliton characteristic power is transmitted.

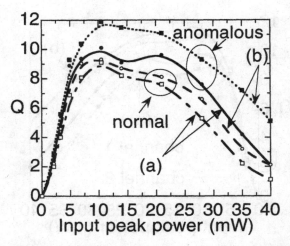

FIGURE 8.12 Dependence of the Q factor on the input peak power in channel 2, for 2×20 Gbit/s transmissions, DCFs placed (a) after or (b) before the amplifiers, and DSF with normal or anomalous GVD.

For $\Delta\nu < 100$ GHz the system performances are limited by linear crosstalk. Since only two channels are transmitted, the crosstalk power is almost one-half with respect to those considered in the analysis of Section 4.2.3, and the linear crosstalk penalty is small for $\Delta\nu > 100$ GHz. Above this separation the system performances improve, increasing $\Delta\nu$ until reaching the condition in which the collision length is comparable with the amplifier spacing. However, also under the worst condition, $\Delta\nu = 256$ GHz, the system shows very good performances. For $\Delta\nu > 256$ GHz the Q factor increases again with the channel spacing due to the increasing collision length. By adding two more channels in the worst-case frequency spacing condition ($\Delta\nu = 256$ GHz), propagation can be attained, but the tolerance range of input powers is more restricted compared to the two-channel case. The comparison between the 2×20 Gbit/s and a 4×20 Gbit/s systems is reported in Fig. 8.14 in terms of the Q factor for the lowest-frequency channel (the most degraded channel) versus the transmitted power under the same conditions as those of Fig. 8.11. The frequency spacing among the channels is 256 Gbit/s. The figure shows that four channels can also propagate in resonant conditions, even though the tolerance in terms of input power is worse with respect to the case of two channels.

The compensation ratio—that is, the relative value of the chromatic dispersion in the DS fiber and in the DCFs—is another key design factor in WDM systems. Tolerance of GVD fluctuations is a relevant question for the stability of dispersion-compensated soliton WDM systems.

These points were investigated in the 2×20 Gbit/s soliton system. The amplifier spacing was fixed at 52.6 km, the values of the length and of the GVD of the DCFs were varied, keeping constant the average GVD value ($\beta_2 = -0.26$ ps^2/km). The dependence of the Q factor, evaluated on the

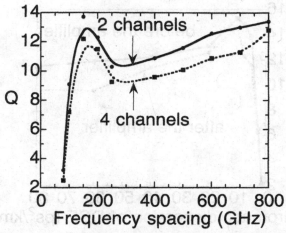

FIGURE 8.13 Dependence of the Q factor on the channel frequency spacing, DCFs before the amplifiers, and 2×20 or 4×20 Gbit/s transmissions.

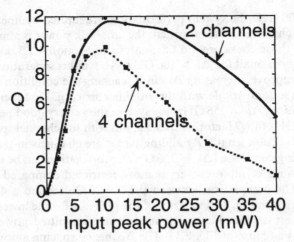

FIGURE 8.14 Dependence of the Q factor on the input peak power in channel 2, DCFs before the amplifiers, and 2×20 or 4×20 Gbit/s transmissions.

worst-performance channel (the lowest-frequency channel), on the GVD of DCFs is illustrated in Fig. 8.15. The cases of a DCFs placed in front and after the optical amplifiers are both included in the figure.

Figure 8.15 reveals that the optimal value of the DCF GVD is $10\,\mathrm{ps^2/km}$. For larger values the Q factor slowly decreases but remains higher than 6 ($Q > 10$); for lower GVD values the performances remain acceptable down to $5\,\mathrm{ps^2/km}$ and then rapidly degrade. In particular, for $\beta_2 < 3\,\mathrm{ps^2/km}$ the Q

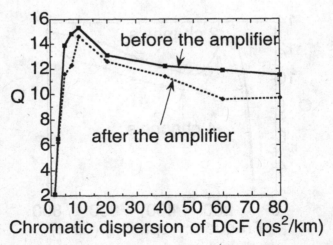

FIGURE 8.15 Dependence of the Q factor on the GVD of the DCF, for 2×20 Gbit/s transmissions and a fixed amplifier spacing.

factor is lower than 6. This system degradation is due to the large loss and nonlinearity of the long DCFs, which require an input peak power substantially higher than the value used in the simulations. Moreover the results of [35] indicate that whenever the length of the two concatenated fibers with alternating signs of GVD are comparable, the input peak power must be raised above the average soliton power to obtain good transmission.

8.4 WDM SYSTEMS EXPERIMENTS

Several system experiments have shown that an overall capacity of the order of 100 Gbit/s can be transmitted over more than 1000 km of fiber cable by the WDM technique [3–5]. The method to obtain the highest capacity depends on the characteristics of the link and in particular on its GVD. For example, in links encompassing step-index fibers, it is preferable to use systems with many channels at a low-bit rate (622 Mbit/s or 2.5 Gbit/s), whereas in DS fibers the best results are obtained with a lower number of channels but with higher bit-rate (2.5, 5, and 10 Gbit/s) [25]. In both cases dispersion management permits a large increase of capacity [2][5].

Also the choice of signal format depends on several links parameters. As shown by simulations, soliton WDM systems allow us to better exploit fiber capacity, but also NRZ signals offer very high potentiality. A promising solution seems the one using different signal formats, both soliton and NRZ, within the same system, depending on the value of the GVD in which the channel propagates. This has been succesfully demonstrated by Bergano and coworkers in experiments in which 32 channels at 5 Gbit/s were transmitted over 9300 km [3] using NRZ signals in the normal dispersion region and solitons in the anomalous region.

As far as the channel spacing is concerned, the WDM experiments can be divided into two classes: low dense WDM systems and high dense WDM systems. In the first case the channel spacing is of the order of few nanometers, at least one order of magnitude greater than the bit-rate. In the second case the channel spacing is only a few times the bit-rate (e.g., $\Delta \nu = 5R$, or 50 GHz if 10 Gbit/s channels are used).

In the case of low dense systems, the bandwidth of the optical selection filter in front of the receiver is much larger than the bit-rate. As a result the penalty due to the greater ASE power impinging the photodiode is quite limited (see Section 4.3.2), and a very accurate control of the transmitted wavelengths is not required.

On the other hand, the system capacity can theoretically be increased by adopting high dense WDM. However, two major problems arise: crosstalk and stabilization of the optical sources. Linear crosstalk can be limited by increasing the receiver selectivity. The most efficient way to obtain high-frequency selective receivers is by coherent detection: If nonlinear crosstalk is negligible, the channel spacing can be reduced up to few times the bit-rate by adopting a high-selectivity electrical IF filter (see Section 4.3.2).

As far as FWM–induced crosstalk is considered, it can be reduced by increasing the local GVD of the fiber: For example, the frequency spacing among the channels could be reduced below 40 GHz using links encompassing step-index fibers. Then a WDM system composed of many low-speed channels could be designed, as shown in Tables 8.2 and 8.4.

Once the channel spacing has been fixed, the capacity of the WDM system is mainly fixed by the available optical bandwidth. A single doped fiber amplifier has a somewhat flat bandwidth between 1530 and 1560 nm, but when a large number of optical amplifiers are cascaded the overall gain curve can be quite different. An example is the 3-dB bandwidth of a cascade of 140 amplifier which measures only about 3.5 nm [5].

Since the fluctuation of the overall link gain over the bandwidth of the WDM comb has to be limited, an important device for high-capacity WDM systems is *gain equalization*; the principle of this device is quite simple: It must compensate the unideal gain curve of the optical amplifiers.

Optical equalizers are optical filters whose transfer function should be approximately the inverse of the gain curve of one or more optical amplifiers. This devices can be located at each amplifier position or at every N amplifers, where N is of the order of 10 [20].

Another way to increase the available optical bandwidth is to replace erbium-doped amplifiers with amplifiers that have a wider bandwidth, such as fluoride-based erbium doped amplifiers [11, 21].

Finally, in experimental low-dense WDM systems, dispersion compensation is quite difficult. In these systems the compensation of the GVD is apparently not enough to limit dispersion-induced penalty because of the importance of third-order dispersion. Only one channel can operate at the zero average dispersion wavelength, while the other channels present an average nonzero GVD. Therefore for wide bandwidth WDM systems some sort of third-order chromatic dispersion compensation is needed. For short links, in which the interplay among dispersive and nonlinear effects is not important, such compensation can be directly performed at the receiver. In the case of long links such compensation must be periodically carried out by in-line devices. At the state of the art, third-order dispersion compensation is obtained by either using fibers [33] or gratings [34].

In Table 8.5 we report some essential characteristics of WDM experiments. The experimental results show that the capacity obtained by NRZ and by soliton systems is equivalent, and in both cases the capacity of 10^{18} m bit/s (e.g., 50 Gbit/s over 20000 km or 1 Tbit/s over 1000 km) is not far away. Moreover systems with an overall capacity of $10^{17} \times$ m bit/s (e.g., 100 Gbit/s over 1000 km) seems ready for engineering and industrial production within a few years.

A closing note is that from a theoretical standpoint soliton systems adopting dispersion management and sliding filters are able to offer substantially higher

TABLE 8.5   Main characteristics of some important experiments on high-capacity WDM systems.

Capacity (Gbit*Mm)	N	Bit-rate (Gbit/s)	L and L_A (km)	Signal	Notes	Lab.	Ref.
1488	32	5	9300, 45	NRZ + soliton	DS fiber	AT&T	[3]
1000	5	20	10,000, 50	soliton	DS fiber plus synchronous modulation		[4]
900	20	5	9000, 45	NRZ	Dispersion management plus gain equalization	AT&T	[5]
400	4	10	10,000, 50	soliton	Unequally spaced channels, filtering, and in-line synchronous modulation	NTT	[6]
600	3	20	10,000, 50	soliton	Unequally spaced channels, filtering, and in-line synchronous modulation	NTT	[7]
200	8	2.5	10,000, 45	soliton	DS fiber	AT&T	[8]
120	8	2.5	6000, 75	NRZ	DS fiber	Alcatel	[9]
80	8	10	1000, 66	NRZ	DS fiber	KDD	[10]
60	4	10	1500, 100	NRZ	DS fiber	KDD	[10]
60	10	10	600	NRZ	DS fiber, unequally spaced channels, and fluoride-based erbium-doped fibers	NTT	[11]
48	8	20	300, 150	NRZ	Dispersion reversal	AT&T	[12]
38.2	128	0.622	480, 40	FSK, NRZ	DS fiber	NTT	[13]
37	4	2.5	3711, 90	NRZ	Unequally spaced channels, GVD compensation at the output, and DS fiber	Alcatel	[14]

Note: The link length is indicated by L, the amplifier spacing by L_A, and the number of channels by N.

capacity than NRZ systems. This difference between soliton and NRZ transmission is not apparent in the experimental results, however, which demonstrates that, at the state of the art, the more complex soliton systems present nonnegligible realization problems.

REFERENCES

1. See, for example,
   R. E. Wagner, R. C. Alferness, A. A. M. Saleh, and M. S. Goodman, MONET: Multiwavelength optical networking, *IEEE-Journal of Lightwave Technology* 14: 1349–1355, 1996.
   G. R. Hill, A transport network layer based on optical network elements, *IEEE-Journal of Lightwave Technology* 11: 667–679, 1993.
   S. Johansson, Transport network involving a reconfigurable WDM network layer—a European demonstration, *IEEE-Journal of Lightwave Technology* 14: 1341–1348, 1996.
   G.-K. Chang, G. Ellinas, J. K. Gamelin, M. Z. Iqbal, and C. A. Brackett, Multiwavelength reconfigurable WDM/ATM/SONET network testbed, *IEEE-Journal of Lightwave Technology* 14: 1320–1340, 1996.
   M. Journal O'Mahony, D. Simeonidou, A. Yu, and J. Zhou, The design of a European optical network, *IEEE-Journal of Lightwave Technology*, 13: 817–828, 1995.
   M. W. Chbat, A. Leclert, E. Jones, B. Mikkelsen, H. Fevrier, K. Wünstel, N. Flaarønning, J. Verbeke, M. Puleo, H. Melchior, P. Demeester, J. Mørk, and T. Olsen, WDM for transmission and routing in wide-scale networks: The optical pan-European network approach (ACTS project OPEN), *Proceedings of NOC'96*, pp. 182–186, Broadband Superhighway Book, Heidelberg, June 25–28, 1996.

2. H. Taga, N. Edagawa, Y. Yoshida, S. Yamamoto, and H. Wakabayashi, IM–DD long distance multichannel WDM experiments using Er-doped fiber amplifiers, *IEEE-Journal Lightwave Technology*, 12: 1448–1453, 1994.

3. N. S. Bergano, et al., Long-haul WDM transmission using optimum channel modulation: A 160 Gb/s (32 × 5 Gb/s) 9,300 km demonstration, *Proceedings of OFC'97*, Dallas, February 16–21, 1997, PD21.

4. M. Nagazawa, K. Suzuki, H. Kubota, A. Sahara, and E. Yamada, 100 Gbit/s WDM (20 Gbis/s × 5 channels) soliton transmission over 10000 km using in-line synchronous modulation and optical filtering, *Proceedings of OFC'97*, Dallas, February 16–21, 1997, PD21.

5. N. S. Bergano and C. R. Davidson, Wavelength division multiplexing in long-haul transmission systems, *IEEE-Journal Lightwave Technology* 14: 1299–1308, 1996.

6. M. Nagazawa, S. Suzuki, H. Kubota, Y. Kimura, E. Yamada, K. Tamura, T. Komukai, and T. Imai, 40 Gbit/s WDM (10 Gbit/s × 4 unequally spaced channels) soliton transmission over 10000 km using synchronous modulation and narrow band optical filtering, *Electronics Letters* 32: 828–830, 1996.

7. M. Nagazawa, S. Suzuki, H. Kubota, and E. Yamada, 60 Gbit/s (20 Gbit/s × 3 unequally spaced channels) soliton transmission over 10000 km using in-line synchronous modulation and optical filtering, *Electronics Letters* 32: 1686–1688, 1996.

8. M. N. Bruce, Soliton WDM transmission of 8 × 2.5 Gb/s, error free over 10 Mm, *Proceedings of OFC'95*, San Diego, February 26–March 3, 1995, PD21.

9. O. Gautheron, G. Bassier, V. Letellier, G. Grandpierre, and P. Bollaert, 8 × 2.5 Gbit/s WDM transmission over 6000 km with wavelength add/drop multiplexing, *Electronics Letters* 32: 1019–1020, 1996.

10. H. Taga, Long distance transmission experiments using the WDM technique, *IEEE-Journal of Lightwave Technology* 14: 1287–1298, 1996.

11. S. Yoshida, S. Kuwano, Y. Yamada, T. Kanamori, N. Takachio, and K. Iwashita, 10 Gbit/s × 10 channel transmission experiment over 600 km with 100 km amplifier spacing employing cascaded fluoride-based erbium droped fibre amplifiers, *Electronics Letters* 31: 1678–1679, 1995.

12. A. H. Gnauk, A. R. Chraplyvy, R. W. Tkach, and R. M. Derosier, 160 Gbit/s (8 × 20 Gbit/s WDM) 300 km transmission with 50 km amplifier spacing and span-by-span dispersion reversal, *Electronics Letters* 30: 1241–1243, 1994.

13. K. Oda, M. Fukutoku, H. Toba, and T. Kominato, 128 channel, 480 km FSK-DD transmission experiments using 0.98 mm pumped erbium-doped fibre amplifiers and a tunable gain equalizer, *Electronics Letters* 30: 982–983, 1994.

14. M. S. Chaudhry, D. Simeonidou, K. P. Jones, N. H. Taylor, and P. R. Morkel, 3711, 4 × 2.5 Gbit/s WDM transmission over installed RIOJA submarine cable system, *Electronics Letters* 31: 1588–1589, 1995.

15. S. Yoshida, S. Kuwano, M. Yamada, T. Kanamori, N. Takachio, and K. Iwashita, 10 Gbit/s × 10 channel transmission experiments over 600 km with 100 km repeater spacing employing cascaded fluoride-based erbium droped fibre amplifiers, *Electronics Letters* 31: 1678–1679, 1995.

16. J. C. Fegger et al., 10 Gbit/s WDM transmission measurements on an installed optical amplifier undersea cable system, *Electronics Letters* 31: 1676–1678, 1995.

17. F. Forghieri, R. W. Tkach, A. R. Chraplyvy, and D. Marcuse, Reduction of four-wave mixing cross talk in WDM systems using unequally spaced channels, *Proceedings of OFC'93*, pp. 252–253, San Jose, February 21–26, 1993.

18. F. Forghieri, R. W. Tkach, and A. R. Chraplyvy, WDM systems with unequally spaced channels, *IEEE-Journal of Lightwave Technology* 13: 889–897, 1995.

19. Applications of the SCM optical techniques to broadband and video distribution are reviewed in the papers
   R. Olshansky, V. A. Lanzisera, and P. M. Hill, Subcarrier multiplexed lightwave systems for broad-band distribution, *IEEE-Journal of Lightwave Technology* 7: 1329–1341, 1989.
   T. E. Darcie, Subcarrier multiplexing for lightwave networks and video distribution systems, *IEEE-Journal on Selected Areas in Communications* 8: 1240–1248, 1990.
   R. Olshansky, R. Gross, and M. Schmidt, Subcarrier multiplexed coherent light-wave systems for video distribution, *IEEE-Journal on Selected Areas in Communication* 8: 1268–1275, 1990.
   Shigeki Watanabe et al., Optical coherent broad-band transmission for long-haul and distribution systems using subcarrier multiplexing, *IEEE-Journal of Lightwave Technology* 11: 116–127, 1993.

20. A. R. Chraplyvy, R. W. Tkach, A. R. Reichmann, P. D. Magill, and J. A. Nagel,

End-to-end equalization experiments in amplified WDM lightwave systems, *IEEE Photonics Technology Letters* 6: 754–756, 1994.

21. B. Clesca, D. Baylat, C. Coeurjolly, L. Berthelon, L. Hamon, and J. L. Beylat, Over 25 nm, 16 wavelength-multiplexed signal transmission through four fluoride-based fiber amplifier cascade and 440 km, *Proceedings of OFC'94*, paper PD20, 1994.

22. The impact of the FWM effect in optical systems is shown in
    G. P. Agrawal, *Nonlinear Fiber Optics*, Academic Press, San Diego, 1989.
    N. Shibata, R. P. Braun, and R. G. Waarts, Phase-mismatch dependence of efficiency of wave generation through four-wave mixing in a single-mode optical fiber, *IEEE-Journal of Quantum Electronics* 23: 1205–1210, 1987.
    A. R. Chraplyvy, Limitations on lightwave communications imposed by optical-fiber nonlinearities, *IEEE-Journal of Lightwave Technology* 8: 1548–1557, 1990.
    A. Chraplyvy, Systems impact of fiber nonlinearities, *Short Course Notes of OFC'94*, February 21, 1994.
    R. W. Tkach, A. R. Chraplyvy, F. Forghieri, A. H. Gnauck, and R. M. Derosier, Four-photon mixing and high-speed WDM systems, *IEEE-Journal of Lightwave Technology* 13: 841–849, 1995.

23. The XPM effect in optical systems is analyzed in
    D. Marcuse, A. R. Chraplyvy, and R. W. Tkach, Dependence of the cross-phase modulation on channel number in fiber WDM systems, *IEEE-Journal of Lightwave Technology* 12: 885–890, 1994.

24. Details on the limitations of optical system in presence of Raman effect can be found in
    A. R. Chraplyvy, Limitations on lightwave communications imposed by optical-fiber nonlinearities, *IEEE-Journal of Lightwave Technology* 8: 1548–1557, 1990.
    A. Chraplyvy, Systems impact of fiber nonlinearities, *Short Course Notes of OFC'94*, February 21, 1994.

25. Results of numerical simulation of WDM systems can be found in
    G. De Marchis, F. Matera, and M. Settembre, Comparison of performance of long-haul high-capacity optical systems, *Electronics Letters* 29: 1777–1778, 1993.
    F. Matera and M. Settembre, Comparison of the performance of optically amplified transmission systems, *IEEE-Journal of Lightwave Technology* 14: 1–12, 1996.
    F. Matera, M. Settembre, B. Daino, and G. De Marchis, Investigation on the implementation of high capacity optically amplified systems in links up to 4000 km long, *Optics Communications* 119: 289–295, 1995.

26. F. Matera and M. Settembre, Nonlinear compensation of chromatic dispersion for phase- and intensity-modulated signals in the presence of amplified spontaneous emission noise, *Optics Letter* 19: 1198–1200, 1994.

27. L. F. Mollenauer, S. G. Evangelides, and J. P. Gordon, Wavelength division multiplexing with solitons in ultra-long distance transmission using lumped amplifiers, *IEEE-Journal of Lightwave Technology* 9: 362–367, 1991.

28. A. Hasegawa, Y. Kodama, *Solitons in Optical Communications*, Clarendon Press, Oxford, 1995, Chapter 10.

29. A. Mecozzi and H. A. Haus, Effect of filters on soliton interactions in wavelength-division-multiplexing systems, *Optics Letters* 17: 988–990, 1992.

30. S. Wabnitz and Y. Kodama, Effect of filtering on the dynamics of multisoliton

collisions in a periodically amplified wavelength division multiplexed soliton system, *Optics Communications* 113: 395–400, 1995.

31. J. V. Wright and S. F. Carter, Constraints on the design of long-haul soliton systems, *Proceedings of Nonlinear Guide-Wave Phenomena*, pp. 6–9, Cambridge, England, September 4–6, 1991.

32. E. Lichtman and S. G. Evangelides, Reductions of the nonlinear impairment in ultralong lightwave systems by tailoring the fibre dispersion, *Electronics Letters* 30: 348–346, 1994.

33. For a review of all-fibers dispersion-compensating devices based on distributed resonant coupling, see
F. Oullette, J. F. Cliche, and S. Gagnon, All fiber devices for chromatic dispersion compensation based on chirped distributed resonant coupling, *IEEE-Journal of Lightwave Technology* 12: 1728–1738, 1994.

34. M. Stern, J. P. Heritage, and E. W. Chase, Grating compensation of third-order fiber dispersion, *IEEE-Journal of Quantum Electronics* 28: 2742–2747, 1992.
S. Wabnitz, Stabilization of sliding-filtered soliton wavelength division multiplexing transmissions by dispersion-compensating fibers, *Optics Letters* 21: 638–640, 1996.

35. F. Matera and S. Wabnitz, Periodic dispersion compensation of soliton wavelength division multiplexed transmissions with sliding filters, *Optical Fiber Technology* 3: 7–20, 1997.

CHAPTER NINE

Transmission in All-Optical Networks

In Chapters 2 through 8 the structure and performances of high-capacity optical systems were analyzed by assuming point-to-point transmission. This is the case when the optical transmission system is embedded in a network adopting electronic switching. Signal regeneration then occurs in the network nodes, and point-to-point links connect different nodes.

In Chapter 1 it was shown that in increasing the capacity of transmission systems, the only suitable way to achieve high-capacity networks at an affordable cost is by the introduction of some optics in the systems performing routing and switching. The exploitation of optics to accomplish network functions different from pure transmission would be carried out by different techniques in the various network areas. The debate on the most suitable solution for each area is open: Many possibilities have been explored, but it is not completely clear which will be useful in practice.

Since the purpose of this book is to analyze the impact of nonlinear transmission in optical networks, we will take an abstract transmission model and trace the route of a signal through an all-optical network using suitable values for the model parameters.

In this model we assume that all-optical signal regenerators are not available. At the state of the art, the commercial development of the all-optical regenerator seems to be far off in the future, even though a large amount of research has been devoted to this device [1]. When a reliable and feasible all-optical regenerator becomes available, point-to-point transmission between regenerating network nodes will be reproduced.

We do not consider networks adopting subcarrier multiplexing, which is a hybrid electrooptic transmission technique. Subcarrier multiplexing is important in the distribution area, and the main transmission impairments are related to the imperfect behavior of the transmitter (which causes nonlinear distortions, chirping, and clipping [2, 3, 4]) and to linear propagation effects, namely fiber attenuation and dispersion.

Our abstract model is suitable in analyzing transmission through a large variety of all-optical networks adopting TDM or WDM, and this practically includes all networks areas.

After the introduction of the model in Section 9.1, the model will be set up in Section 9.2 to analyze optical networks in the local area, while in Sections 9.3 and 9.4 it will be exploited to analyze transmission through optical networks in the access and in the transport areas.

9.1 MODELING TRANSMISSION THROUGH AN OPTICAL NETWORK

One of the main characteristics of the considered all-optical networks is that the signal is processed only by analog devices. Thus the propagation route of a signal through the network can be viewed as a transmitter, a chain of analog all-optical devices separated by fiber links, and a receiver. The scheme of such a chain is shown in Fig. 9.1, where d_h indicates the hth in-line device, L_h the length of the hth fiber link, and K the overall number of fiber links. In the example of Fig. 9.1, no fiber is placed between the transmitter and d_0 and between d_K and the receiver. This is the common situation; however, a fiber link can be simply introduced at these two points.

Both linear and nonlinear processing can be performed inside the device d_h. At the state of the art, seven functions seem to be enough to represent the main network systems:

- *Loss* is present, when splitting the signal, when combining different signals to form a WDM comb, when switching by a space switch matrix, and so on.
- *Gain* is needed to compensate both processing and propagation losses; gain is provided by optical amplifiers.
- *Noise generation* is unavoidable when optical gain is present.
- *Filtering* is needed both to reduce the noise and to separate different channels if WDM is adopted.
- *Spectral inversion* is a useful technique for dispersion and nonlinear effects compensation (as shown in Chapter 6); it is produced by some kinds of frequency converters.

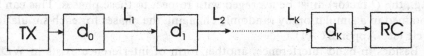

FIGURE 9.1 Transmission model of the signal route through an optical network. TX is a transmitter, RC a receiver.

- *WDM multiplexing* is needed in WDM systems to reconstruct the multiplexed signals after demultiplexing and separate processing of the channels.
- *Interference generation* is always present when the route of a signal crosses that of another signal.

The first five functions are easy to represent with ideal devices, which appear as building blocks of d_h.

In certain interesting cases the model of d_h is rather simple, a sort of equivalent amplifier providing a maximum gain G_h and a frequency transfer function $H_h(\nu)$. Noise is taken into account by a noise figure F_h. The noise figure F_h can be determined from the noise spectral density S_n at the device output with the formula $F_h = S_n/[|G_h - 1|h\nu]$, where the internal gain and losses (G_{int} and α_{int}) are related to the device net gain G_h by the equation $G_h = G_{int}\alpha_{int}$. If the internal losses are high, F_h can be much greater than 3 dB, and G_h can be even less than one provided that the internal gain is not high enough to compensate the losses. Then F_h is zero if no internal gain is present and otherwise, nonzero.

In a WDM network, multiplexing may be needed at the output of the d_h. WDM multiplexing is simply operated by adding channels at different wavelengths, and it can be represented as a generator of the WDM channels plus an adder.

Unlike the other effects, interference is not easy to model in an abstract way: It depends on the nature of the analyzed network. Thus, if an abstract model is wanted, some assumptions have to be made.

In general, we have two different kinds of interference. In-band interference is present where the considered channel crosses a network element together with one or more optical channels at the same wavelength. Due to the imperfect separation between the different channels, a certain percentage of the power of the other channels is transferred on the considered channel, constituting in-band interference.

We will assume that the device in which in-band interference occurs is perfectly linear; this is generally the case, since this device is usually a multiplexer. Under this assumption the in-band interference term added to the signal can be simulated by another signal at the same frequency which is filtered by a suitable transfer function. The optical phase and the phase of the modulating function of the interference contribution is random with respect to the parameters of the considered signal; thus the performance evaluation parameters (e.g., the Q factor) must be averaged with respect to these phases. This can be obtained in a simulation by randomly changing the phases for each simulation cycle (see Appendix A1).

Besides in-band interference, another form of interference exists in WDM systems: out-band interference. Out-band interference is due to the fact that when filtering the considered channel to obtain demultiplexing, a part of the power of the adjacent channels enters the useful bandwidth. This is due both to the ideally infinite tails of the signal spectrum and to the tails of the optical filters. These kinds of interference, which are generally less dangerous than in-band interference, are accurately reproduced by propagating the WDM comb along the chain shown in Fig. 9.1.

The resulting abstract model of d_h can be obtained by cascading one or more of these ideal devices:

- Filters
- Amplifiers
- Attenuators
- In-band interference generators
- White gaussian noise generators
- Spectrum inverters
- WDM comb generators

Once the abstract transmission model is assembled, the resulting fiber link can be analyzed with the techniques adopted for point-to-point links, as shown in Chapter 4. In particular, simulation will be adopted in this chapter based on the split-step method described in detail in Appendix A1. This technique will be adopted for all the considered systems, both in the cases of NRZ and soliton transmission in order to obtain comparable results.

9.2 TRANSMISSION IN LOCAL AREA NETWORKS

First example we want to carry out is transmission in a local area network. However, since we are interested in the impact of nonlinear effects during propagation, we do not consider networks in very limited areas (e.g., a building) nor adopt low-speed transmission (e.g., 622 Mbit/s). We concentrate our attention on networks in a large local area, managing high-speed signals by means of packet switching.

We assume that all digital processing needed to route the packets through the network is performed ideally. Thus we do not state if it is performed electronically, by detecting the packet header, or by some of the numerous all-optical techniques that has been proposed in the literature [5].

As far as electronic processing is concerned, it is to be noted that networks with electronic processing of the packet header can be competitive with networks with all-optical processing due to the low cost of digital electronics. If needed, to allow simpler processing, the packet header can be transmitted to a lower bit-rate with respect to the information field.

It is assumed that optical packet storage is operated inside the nodes. A large amount of research has been dedicated to study optical memories: At the state of the art it appears that fiber memories are almost ripe for application, so we will consider fiber memories [6]. A fiber memory is essentially a fiber ring of sufficient length to delay the signal for the wanted time interval; in our model fiber memory is represented by a fiber piece plus the loss of the input/output directional coupler.

Relying on these assumptions, we will analyze transmission through different kinds of local networks.

9.2.1 Passive Star Network

The simplest possible example of a local area network is the passive star network. In this configuration all the network interfaces units (NIUs) are connected to a passive optical star and associated to a different optical frequency. Each NIU has a tunable transmitter and a fixed frequency receiver. At the output of the star center, a fixed filter selects on each branch the frequency of the receiver connected with that branch; an EDFA is present in front of each receiver. The scheme of the resulting network is shown in Fig. 9.2.

When the ith user wants to transmit a packet to the jth user, she tunes the transmitter on the frequency ν_j associated with the jth user. Obviously a protocol is needed to detect or to avoid contentions of different packets directed toward the same receiver [7].

The main advantage of the passive star network is its simplicity: The huge bandwidth of the optical fiber is exploited directly to obtain a simple network operation. On the other hand, this network architecture has several disadvantages. First of all, the useful bandwidth is limited by the transmitter's tunability range, and not by the fiber. For example, if the tunability range is 7 nm (about 854 GHz at 1.55 μm), the channel speed is 10 Gbit/s, and the channel spacing is 200 GHz, the maximum number of users is 4, while it becomes 16 at a bit-rate of 2.5 Gbit/s with a channel spacing of 50 GHz. Moreover high bit-rate trans-

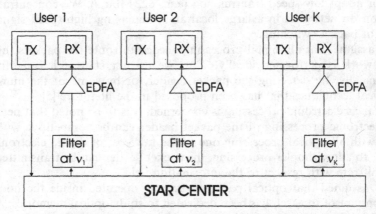

FIGURE 9.2 Scheme of a passive star optical network.

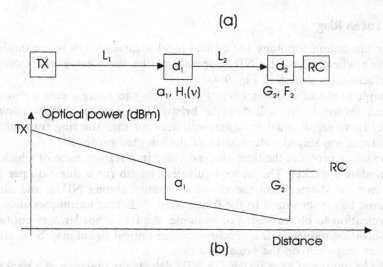

FIGURE 9.3   Abstract transmission model of the passive star network (a) and optical power level along a signal route (b).

mitters that can be tuned rapidly (on a packet basis) over a wide tuning interval are quite expensive, while the user terminal should be as cheap as possible.

The abstract model for the transmission route through the considered network is shown in Fig. 9.3a. In this figure two in-line devices are shown: d_1 and d_2. The first is characterized by a loss a_1 equal to the loss at the star center ($a_1 = 1/K_u$ if there are K_u users), and a transfer function $H_1(\nu)$ that represents the filter at the star's center. The second device is characterized by a gain and a noise factor: It represents the EDFA amplifier in front of the receiver. In Fig. 9.3b the optical power level along the link is also shown.

As long as the distance between each NIU and the star center is of the order of 10 km and the bit-rate is limited below 40 Gbit/s, no relevant nonlinear effect arises during transmission in the passive star LAN. The overall number of NIUs that can be connected coincides with the maximum number of optical frequencies that the network can support. Since the optical bandwidth is limited by the tuning range of the transmitters, increasing the number of frequencies means decreasing the channel spacing. This can be done up to the linear crosstalk limit (see Section 4.2.3); thus high-frequency selectivity is important in these networks. The frequency selectivity can be attained by using highly selective optical filters (e.g., three-stage Fabry-Perot filters) or adopting coherent detection. In particular, coherent detection allows the channel spacing to be reduced down to a few times the bit-rate (e.g., $4R$), thus reaching the higher network capacity.

At high bit-rates fiber dispersion becomes important. Here any of the dispersion-compensating methods analyzed in Chapter 6 can be applied if needed.

9.2.2   Token Ring

Another interesting topology for optical local area networks is the unidirectional ring where all of the NIUs operate at the same wavelength. Such a network scheme is shown in Fig. 9.4.

A simple protocol allowing correct data exchange along a ring network is the token protocol. We will describe briefly this protocol in the following example. However, it must be taken into account that the ring transmission performances are almost independent of the adopted protocol.

In the token protocol the time axis is divided into frames, each of which can accommodate a packet. The frame equivalent length (time duration per light speed) must be shorter than the shortest distance among NIUs, and all the NIUs must be synchronized to the frame train. Different techniques allow the synchronization to be realized: For example, the frame header may contain a synchronization sequence or a synchronization optical signal may be available at another frequency on the same fiber ring.

At the beginning of each frame, the NIU detects the presence of a packet. If the packet is present, it is stored in an optical memory and the header is processed. If the packet is directed to the NIU the information field is detected; otherwise, the packet has to be forwarded to the next NIU along the ring.

In order to transmit a packet without causing conflicts, the token technique is used. A particular packet, called token, signals that a certain node can transmit. When the node receives the token, it transmits the packet to be forwarded and those locally generated; then the NIU retransmits the token

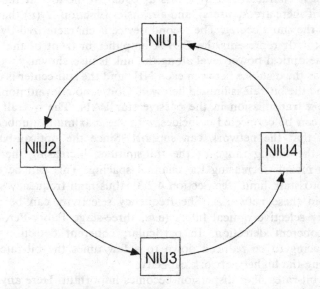

FIGURE 9.4   Ring topology; NIU indicates a network interface unit.

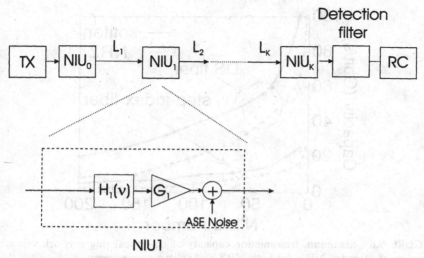

FIGURE 9.5 Abstract transmission model of an optical ring network.

to the next node. Since only one node at a time has the token, conflict between packets is impossible.

The abstract model of the longest signal route through the network is shown in Fig. 9.5, where it is assumed that optical amplification is performed at the output of each NIU. Note that the equivalent length of the optical memory inside the NIUs depends not only on the packet length but also on the number of loops the packet does. We assume that the maximum number of 10 loops is fixed. Then the dispersion and the nonlinear behavior of the fiber memory can be neglected and the NIU considered simply as an attenuating device.

To provide the transmission performances of the token ring, we assume that the NIUs are equally spaced along the ring, with a NIU spacing of 10 km, which is significant for a network in a limited metropolitan area. The fiber parameters are reported in Table 2.1, the amplifier inversion factor is assumed to be equal to 3 dBs.

The transfer function $H(\nu)$ of a single NIU is assumed to have a Fabry-Perot shape with a 3-dB bandwidth of 500 GHz (about 4 nm at $\lambda = 1.55 \, \mu m$), and a Fabry-Perot filter is placed in front of a receiver having a 3-dB bandwidth of 40 GHz. The maximum transmission capacity of the network (maximum bit-rate at which the error probability of 10^{-9} can be attained) is reported in Fig. 9.6 with relation to the number of cascaded NIUs both for NRZ and for soliton transmission. Both DS and step-index fibers are considered.

In Fig. 9.6 it is evident that the potential capacity of the network is large. For example, a capacity of 60 Gbit/s can be shared among 10 NIUs using NRZ signaling, and the NIUs number can be increased up to 40 if solitons are used.

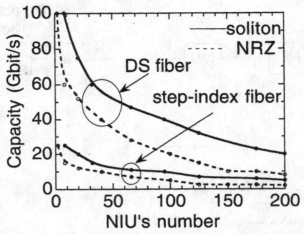

FIGURE 9.6 Maximum transmission capacity of an optical ring network versus the number of cascaded NIUs for both NRZ and soliton transmission.

The real bottleneck of the network is the processing inside the NIUs. Such high capacity cannot be completely exploited until all-optical packet processing becomes available.

9.2.3 Shuffle Multihop Network

A multihop network is a network in which a packet, before arriving to the destination node, has to be processed by other nodes different from the transmitting one [8]. In this general sense, even the token ring considered in Section 9.2.2 is a multihop network. If particular logical topologies are adopted, the routing algorithm of a multihop network can be considerably simplified. It is not within the scope of this book to analyze in detail all the logical topologies that can be adopted for multihop networks. We rather limit the discussion to the most common topology, which is the perfect shuffle. A multihop network with a perfect shuffle topology is called a shuffle multihop network (SMN).

In an SMN each NIU has p inputs and p outputs. The logical topology of a network with $K_u = kp^k$ NIUs is constituted by a matrix of k columns of p^k NIUs each. The NIUs belonging to adjacent columns are connected by a perfect shuffle permutation [9] so that a path composed at most by $2k$ NIUs ($2k - 1$ hops) exists to connect each couple of NIUs. An example of a logical SMN topology with $p = 2$ and $k = 3$ is given in Fig. 9.7.

The general scheme of an NIU for an SMN is reported in Fig. 9.8 in the case of $p = 2$. Each NIU is subdivided into p equal parts. Each part, which presents an input and the possibility of transmitting on p different outputs, performs the operations of reception, address control, routing, and transmission by means of the devices R, AC, RT, and T, respectively.

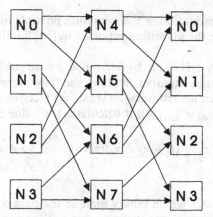

FIGURE 9.7   Logical topology of a shuffle multihop network with $p = 2$ and $k = 3$. N indicates the NIUs.

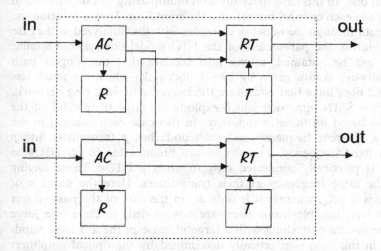

FIGURE 9.8   Functional scheme of the NIU of a SMN. AC indicates the address-processing device, RT the routing and transmission device, R the receiver, and T the transmitter.

Each NIU reads the header of an incoming packet and checks, by means of the AC device, if it belongs to the node. If its does, the packet is detected by the receiver R; if not, it is retransmitted by transmitters T.

If the output ports of each NIU are numbered from 0 to p and the address of the packet is coded in a base p number, the routing process can be based on the observation of only one digit of the address; this process is performed by RT. There exist several possible routing algorithms that deliver packets from a source to the final destination in the minimum possible number of hops. As an example we consider the algorithm reported in [10] limiting our description to $p = 2$.

In an SMN each NIU can be identified by an address (x, y), where x represents the column number and y the row number ($x = 0 \ldots k - 1$, $y = 0 \ldots k - 1$). When an arbitrary NIU (x_r, y_r) receives a packet with a destination address $(x_d, y_d) \neq (x_r, y_r)$, it calculates the value ξ given by

$$\xi = \begin{cases} (k + x_d - x_r)\mathrm{mod}(k) & \text{if } x_d \neq x_r \\ k & \text{if } x_d = x_r \end{cases} \tag{9.2}$$

The value of ξ indicates which binary digit of the packet address indicates the correct output port to retransmit the packet.

The logical SMN topology can be realized by different physical topologies. The most direct solution is to adopt a physical topology completely coincident with the logical one. In this case space division multiplexing can be adopted in the network, since each channel travels on a different fiber. In this solution the transmission path through the network coincides with that analyzed in the case of a ring. Obviously the parameters of the NIU model change, but similar performances can be obtained, taking into account that the longest path through the network is now given by $2k - 1$ fiber links—which are much less than the $K_u - 2$ fiber links that constitute the longest path in a ring network.

An alternative SMN approach which exploits the large bandwidth of the optical fiber, is based on the star topology. In this case each channel in the network has a different frequency and each node has p transmitter and p receivers at a fixed frequency. The signals are broadcasted to all NIUs so that an NIU is physically connected only to other p NIUs: Those having receivers at the same frequency as their transmitters. Here the number of used frequencies is pK_u, whereas it is only K_u in the case of the passive star network. However, tunable transmitters are not needed, so there is a large decrease in the system cost along with a large increase in the available bandwidth, which in this case is practically determined by the optical amplifiers being used.

As far as the processing inside the NIU is concerned, it is to be noticed that the signal is received and retransmitted at two different wavelengths. This is easy if electronic processing is present inside the NIU, while all-optical wavelength conversion is needed to implement all-optical NIUs. At the state of the art, several methods to obtain all-optical wavelength conversion have been proposed [11]; some are almost ready for commercial application if a wide market for them should emerge.

Based on these assumptions is the transmission model shown in Fig. 9.9 for longest signal route through an SMN with a star topology. It is to be noted that two different devices alternate along the link: the NIU (e.g., d_0 in the figure) and the star center (e.g., d_1 in the figure). To simulate $2k - 1$ hops by

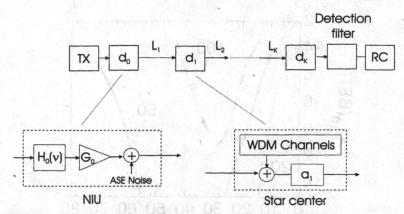

FIGURE 9.9 Abstract model of the transmission through an SMN realized by a star physical topology. The star center loss is indicated with a_1.

the model of Fig. 9.9, $2k$ NIUs and $2k - 1$ star centers must be cascaded. In the NIU the incoming WDM comb is demodulated by a fixed filter, the packet is stored in an optical memory, and then it is amplified and forwarded to the next node. At the star center there is a loss corresponding to the signal splitting, and the WDM comb is formed again.

It should be noted that the scheme of the NIUs is similar to that adopted in the token ring. Of course the model parameters are completely different. The overall transfer function of the NIUs is essentially determined by the input filter that selects the opportune frequency in the incoming wavelength comb, and the overall gain and noise factor have to take into account all the optical devices.

In order to set up the transmission performances of the SMN, we assume that the NIUs are 10 km apart from a 64×64 star coupler with a loss of 18 dB. The fiber is assumed to be either a DS fiber or a step-index fiber whose parameters are reported in Table 2.1, and the amplifier inversion factor is assumed to be equal to 3 dB.

The transfer function $H(\nu)$ of a single NIU is assumed to have a Fabry-Perot shape with a 3-dB bandwidth of $20R$, while a Fabry-Perot filter is placed in front of the receiver having a 3 dB bandwidth of 40 GHz.

As a first example, we report the network performance considering TDM transmission in the case of a DS fiber ($\beta_2 = -1.28 \, \text{ps}^2/\text{km}$). The error probability, evaluated starting from the Q factor via gaussian approximation (see Section 4.2), is reported in Fig. 9.10 against input peak power for a network adopting soliton transmission at 10 Gbit/s ($1/T_F = 5R$) for different numbers of NIUs, N. A minimum error probability occurs for a peak power of the order of soliton power P_k; above P_k the strong power fluctuations during signal propagation induce instability effects, and the error probability increases with the increasing transmitted power.

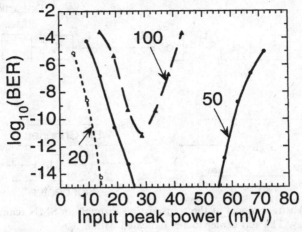

FIGURE 9.10 Error probability versus input peak power for a network using a soliton transmission at 10 Gbit/s in a link encompassing a DS fiber. Reported is the number of NIUs crossed by the signal.

For $N > 100$ the nonlinear interaction between ASE and the signal is the main transmission impairment. In fact, in the absence of ASE for $N = 100$ and $P = P_k$, the signal undergoes a negligible degradation (the eye-opening penalty is EOP $= 0.9$ dB). Figure 9.10 shows also that for $N = 50$ the error probability is lower than 10^{-12} in a wide interval of the transmitted power. This is important in designing practical systems, robust with respect to fluctuations in the component characteristics and to aging.

The case of multiwavelength soliton transmission at 10 Gbit/s ($1/T_F = 5R$) is analyzed in Fig. 9.11 for $N = 50$. The number N_W of different wavelengths is also reported in the figure. To limit the FWM effects, the channels are allocated covering all the available optical bandwidth; as a result the frequency spacing is $1000/N_W$ (GHz). If the number of WDM channels is increased, the system performances deteriorates, mainly due to FWM. The maximum capacity that can be reached in this kind of link is 40 Gbit/s.

In Fig. 9.12 the maximum capacity of a network adopting DS fibers is reported versus the number of nodes for different bit-rates. The maximum number of wavelengths in WDM systems depends on the available optical bandwidth and on the minimum frequency spacing, $\Delta \nu$, needed to avoid cross-talk at the receiver. In IM–DD systems $\Delta \nu$ can be assumed equal to $5R$. Thus, for a limited number of NIUs (short propagation distance) and an optical bandwidth of 1000 GHz, the maximum capacity is 200 Gbit/s, independently of the signal bit-rate.

If the signal propagates through a long fiber link, the maximum number of WDM channels is also influenced by FWM. This phenomenon is taken into

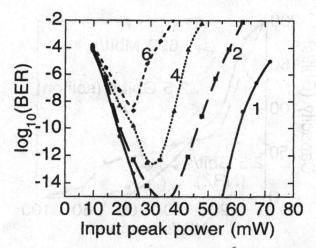

FIGURE 9.11 Error probability versus input peak power for a network adopting WDM soliton systems transmitting in a link encompassing a DS fiber. The signal bit-rate of each channel (wavelength) is 10 Gbit/s.

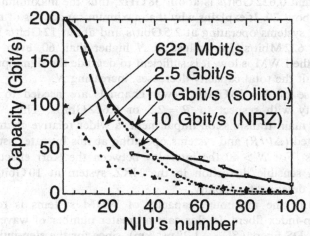

FIGURE 9.12 Maximum capacity considering different channel bit-rates versus NIUs number for a network adopting WDM systems operating in links encompassing a DS fiber. In the cases of 2.5 Gbit/s and 622 Mbit/s, the performance of NRZ and soliton signal are very similar, and therefore we have not distinguished between the soliton and NRZ signals.

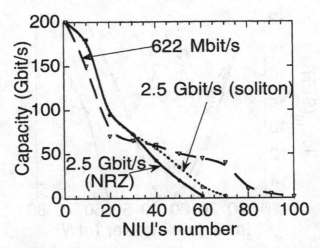

FIGURE 9.13 Maximum capacity considering different channel bit-rates versus the NIU number for a network adopting WDM systems operating in links encompassing a step-index fiber. In the case of 622 Mbit/s, the performance of NRZ and soliton signal are very similar, and therefore we have not distinguished between the soliton and NRZ signals.

account in the figure, since the channel spacing is optimized to obtained the higher capacity attainable with the assigned overall bandwidth of 1000 GHz.

In a large interval of N (about $30 < N < 60$) the optimum channel spacing for $R = 2.5$ and 0.622 Gbit/s is about 38 GHz; thus the maximum number of channels is about 20. It explains why the maximum capacity is of the order of 50 Gbit/s for systems operating at 2.5 Gbit/s and about 12 Gbit/s for systems operating at 622 Mbit/s. For values of N higher than 60, even though the efficiency of the FWM is low, it is sufficient to degrade the signal propagation, and as a result the total capacity decreases, increasing N.

In the case of $R = 10$ Gbit/s, fewer channels are needed to obtain the same capacity with respect to $R = 2.5$ or 0.622 Gbit/s. Thus, as long as FWM is the main transmission impairment, a wider relative channel spacing can be adopted ($\Delta \nu / R$) and systems operating at this bit-rate assure the best performances. For $N > 40$ the interplay between the Kerr effect and GVD introduces a sensible distortion in the NRZ system at 10 Gbit/s; thus its performance deteriorates rapidly.

In Fig. 9.13 the maximum capacity of WDM systems is reported for standard step-index fibers. In this case a greater number of wavelengths can be used with DS fibers ($\beta_2 = -1.28$ ps^2/km), since for the step-index fiber the FWM bandwidth (equation 2.45), $\Delta \nu_{3\,dB}$, is about 8.5 GHz. On the other hand, as described in the preceding section on TDM systems, scaling down a factor four in the bit-rate, the behaviour of the signals operating in links with step-index fibers becomes similar to that of systems operating with DS fibers.

9.3   TRANSMISSION IN THE ACCESS AREA

Networks in the access area are quite different from local area network both in the functions and in the physical structure. We will assume that slow reconfiguration is needed in the access and in the transport areas and that WDM multiplexing is adopted.

We will analyze only one topology for the access area network: the ring containing optical add/drop multiplexers (OADM; see Section 1.2.3). This is an important topology, since it derives from the concept of the SDH ring, which is the main high-capacity electrical network in the access area. It is to be noted that optical rings adopting OADM based on optical TDM have been proposed in the literature, but the enabling technologies are not yet competitive with the WDM solution.

In this section we will analyze the transmission performances of an optical ring in the access area, after a brief review of some possible OADM structures. The problem of the optical protection against faults will not be considered. This is an important problem, as in the case of all very high-capacity public networks, and the interested reader can find a wide literature in the references [12].

9.3.1   Possible OADM Structures

The function of an OADM is essentially to drop from a WDM comb a set of selected channels and replace them with locally generated channels. The channels that are not involved in the add/drop operation have to be transmitted without distortion through the node. The simpler structure of the OADM is shown in Fig. 9.14. At the input the WDM comb is demultiplexed, for example, by a grating or by an interferometric demultiplexer. Each WDM channel feeds a 2×2 switch connected by a local receiver and a local transmitter on the frequency of that channel. If the switch is in the bar state, the channel crosses the OADM transparently, but, if the switch is in the cross state, the input channel is dropped and a local channel is added to the WDM comb. At the OADM output the power of all the output channels is equalized by a set of optical attenuators; the WDM comb is reconstructed by a multiplexer, and the WDM signal is amplified to compensate for the losses.

In-band interference can be characterized by an interference parameter C/S, which can be defined as the ratio between the interfering power and the signal power at the OADM output, after the reconstruction of the WDM comb. In-band interference is generated by the finite insulation of the 2×2 switch in the cross state and by the imperfect channel selection operated by the input demultiplexer. On the ground of the physical characteristics of these devices, C/S can be easily evaluated.

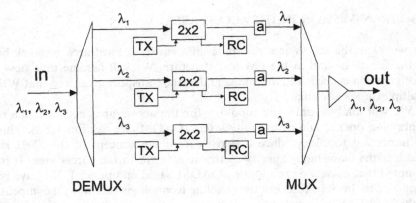

FIGURE 9.14 Structure of an OADM based on demultiplexing of the input WDM comb and separate processing of all the channels. The transmitters are indicated with TX, the receivers with RC, and the switches with 2 × 2. The optical attenuators for the power equalization are indicated by a.

Different technologies allow an all-optical 2 × 2 switch to be realized: Optomechanical components [13], thermooptic components [14], and switches based on LiNbO$_3$ technology [15] are already on the market, while devices based on InP technology [16] will become available soon.

A particular realization of the structure of Fig. 9.14 is also shown in Fig. 9.15 [17]. The OADM is based on a form of demultiplexer realized by interferential filters. A third OADM structure, based on the properties of acoustooptic filters [18], is shown in Fig. 9.16 [19]. In this case WDM demultiplexing is not needed, and the acoustooptic filter works as a 2 × 2 WDM switch. The acoustooptic filter realized to accomplish this function is often called a λ-switch. In a λ-switch, a radio frequency f_j is associated with each frequency of the input WDM comb; if the switch is driven with f_j, it is in the bar state for the optical channel at the frequency ν_j and in the cross state for the other input channels. Moreover, since the switch behaves linearly, if a set of radio frequencies drives the switch, the corresponding set of optical frequencies sees the switch in the bar state, while for the remaining input channels, the switch is in the cross state. The OADM of Fig. 9.16 is the most direct way to exploit the characteristics of the λ-switch.

It is beyond the scope of this section to compare different OADM architectures. Other architectures have been proposed that are not described in this section, and if a comparison were carried out, system- and cost-related aspects would have to be considered besides the transmission performances. The examples presented may be helpful in understanding the model used for the evaluation of transmission performances.

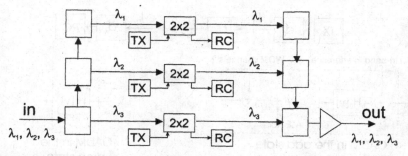

FIGURE 9.15 Structure of an OADM based on multiplexers and demultiplexers realized by interference filters. The attenuators for power equalization are not shown in the figure for simplicity.

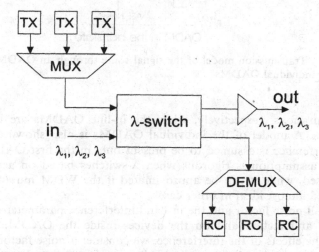

FIGURE 9.16 Structure of an OADM based on an acoustooptic switch.

9.3.2 Transmission Performance of an OADM Ring

To evaluate the transmission performances of the OADM ring, the ring length and the OADM position must be fixed. We will assume equally spaced OADM, with spacing $L = 100$ km.

Two other important parameters are the overall link capacity (number of WDM channels times the bit-rate) and the granularity of the WDM multiplexing (number of WDM channels). We will consider an interesting case: a ring capacity of 20 Gbit/s obtained by eight channels at a bit-rate of 2.5 Gbit/s. The channel spacing is assumed to be equal to 0.5 nm.

A diagram showing the longest path along the OADM ring is provided in Fig. 9.17. In this model the first and the last OADMs are in the add and in the

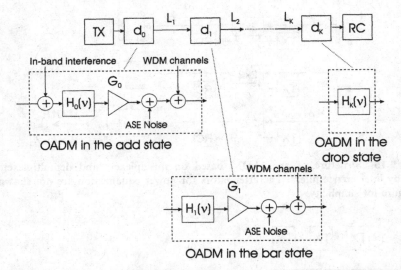

FIGURE 9.17 Transmission model of the signal route through an OADMs ring and models of the individual OADMs.

drop configurations, respectively, while the in-line OADMs are in the bar configuration. A model of the individual OADMs is also shown. Note that in-band interference is assumed to be present only in the first OADM of the chain. This assumption is rigorous when λ-switches based on acoustooptic filters are used, while it can be approximated if the WDM mux/demux can be considered almost ideal in other cases.

The overall noise factor and the in-band interference parameter depend on the OADM architecture and on the devices inside the OADM. To better investigate the effects of the interference, we consider a noise factor of 3 dB.

In Fig. 9.18 we plot the error probability against the peak power at the OADM output for a network with four OADM nodes. The WDM comb is composed by 8 NRZ channels with a bit-rate of 2.5 Gbit/s; the transmission is carried out over a DS fiber cable with $\beta_2 = 1.28\,\text{ps}^2/\text{km}$. The amplifier spacing is 50 km. Different values of the interference parameter (interfering power normalized to the signal power) are considered:

a   for the absence of interference

b   for $-30\,\text{dB}$

c   for $-20\,\text{dB}$

d   for $-15\,\text{dB}$

The importance of in-band interference is evident from the fact that curve *c* shows a minimum error probability far below 10^{-15}, while curve *d* does not approach 10^{-9}. The value of the interference parameter for which the minimum error probability is almost 10^{-9} is about $-17\,\text{dB}$. The presence of a minimum

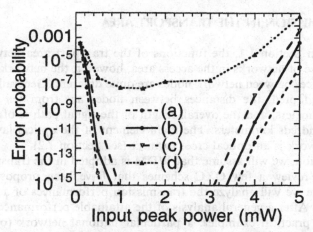

FIGURE 9.18 Error probability versus the peak power at the OADM output for a network with four OADM nodes. DS fibers are adopted, and different values of the interfering power normalized to the signal power are considered: (a) absence of interference, (b) −30 dB, (c) −20 dB, (d) −15 dB.

error probability is due to the FWM: When the FWM becomes the main transmission impairment, the error probability is an increasing function of the transmitted power.

As expected, the minimum error probability does not exist in Fig. 9.19, which is analogous to Fig. 9.18 but for the use of a step fiber with $\beta_2 = -20 \, \text{ps}^2/\text{km}$. Rather the impact of in-band crosstalk is almost the same.

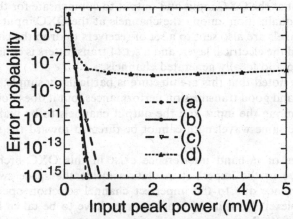

FIGURE 9.19 Error probability versus the peak power at the OADM output for a network with four OADM nodes. Step-index fibers are adopted, and different values of the interfering power normalized to the signal power are considered: (a) absence of interference, (b) −30 dB, (c) −20 dB, (d) −15 dB.

9.4 TRANSMISSION IN THE TRANSPORT AREA

As specified in Section 1.1, the functions of the transport area networks are similar to those of networks in the access area; however, the network topology and the distances between network nodes are quite different. Generally, a mesh topology is adopted, the distances between nodes are from ten to a few hundreds kilometers, and the overall length of the signal path is of the order of several hundreds kilometers. The main element of the optical layer of the transport network is an optical cross connect (see Section 1.2).

In this section, we will assume that WDM is adopted in the transport area. First we will review a few OXC schemes that have been proposed in the literature; then we will analyze the transmission performances of a transport area network. After a general analysis of the attainable performances, we will consider two practical examples: a particular national network (one of the possible schemes of the high-speed Italian optical network) and a particular network covering a continental area (the Pan-European optical network proposed in the frame of the research projects COST 239 and ACTS OPEN promoted by the European Economic Community).

9.4.1 Optical Cross-Connect Architecture

A simple OXC architecture allowing the reconfiguration of the input/output routing is presented in Fig. 9.20 [21]. At the input of the OXC K_f fibers are present; on each fiber a WDM comb of K_c channels is carried. The incoming channels are demultiplexed, and all the channels at the same wavelength are sent to the same space switch matrix. After the space switching the wavelength channels are WDM multiplexed on the different output fibers. Optical amplifiers are present at the OXC input and output to compensate for the losses and attain power equalization among the channels at the OXC input and output. The input channels are also sent to a set of receivers to drop the channels to be directed toward the electrical layer, and a set of transmitters is connected to the output fibers to add locally generated channels.

It should be noted that this architecture is particularly simple, offering low internal losses and good transmission performances, but it does not allow every permutation among the input and the output channels to be realized. In fact channels at the same wavelength cannot be directed toward the same output fiber.

Two sources of in-band interference exist in this OXC architecture: an interference due to the imperfect insulation inside the space switch matrix, and an interference due to the imperfect channel selection operated by the input demultiplexer. Both interference sources have to be taken into account to evaluate the interference parameter of this structure.

Different technological alternatives are available to realize the space switch matrix, the key element of this OXC [13, 14, 15, 16], and 4×4 or 8×8 optical matrixes are already on the market.

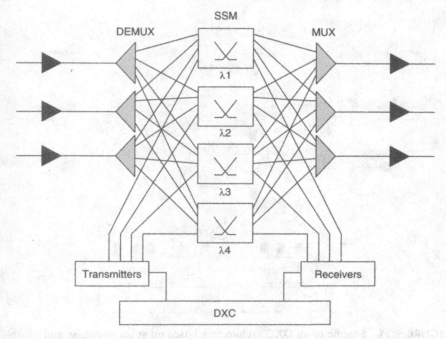

FIGURE 9.20   Scheme of an OXC architecture based on space switching.

A second possible OXC scheme is presented in Fig. 9.21 [22]. Each of the K_f fibers at the OXC input carries a comb of K_c WDM channels, and K_c channels are added dropped to the electrical level through local transmitters and receivers. The incoming channels are discriminated through the optical splitters (SC) and the tunable optical filters (TF). Then they are routed, by optical space switched matrixes, either to the proper output or to the set of optical receivers. Locally generated channels directly enter the optical switch to be routed to the proper output. The combiners blend the wavelength channels into the output fibers. At the node input and output optical amplifiers allow the losses to be compensated and power equalization to be attained.

It is worth noting that contentions could arise when channels entering the OXC at the same frequency from different fibers are sent to the same output. Contentions can be avoided by assigning a fixed wavelength to each optical path throughout the network. In this case channels with the same wavelength are always routed to different outputs. Otherwise, contentions may be solved by detecting one of the channels by a local receiver and retransmitting it at a different wavelength. However, this way the OXC transparency is lost, and some node resources are turned away from the add/drop function. The problem can be overcome by introducing wavelength conversion inside the OXC. The block diagram of this second architecture, in which one wavelength converter is placed after any output of the switch matrixes, is shown in Fig. 9.22

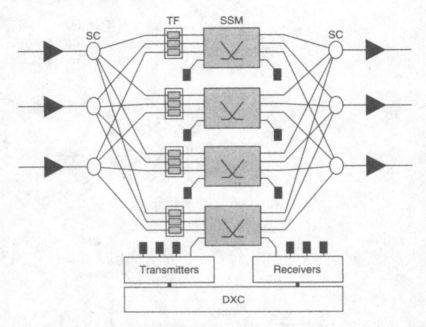

FIGURE 9.21 Scheme of an OXC architecture based on space switching and tunable filters. Star couplers are indicated by SC, tunable filters by TF, and space-switching matrixes by SSM.

[23]. In this case contentions are avoided by changing the wavelength of one of the channels.

The transmission model of the architecture of Fig. 9.22 is shown in Fig. 9.23 assuming that spectral inversion occurs due to wavelength conversion. In the absence of spectral inversion, the corresponding block is deleted. It should be noted that, in the scheme of Fig. 9.23, ASE noise is inserted in two different points, where the converter ASE, if present, is taken into account at the OXC output.

The scheme of Fig. 9.23 can also be adopted to model the OXC architectures introduced in [24] and incorporating the concept of the delivery and coupling switch [25]. This concept is not described here for sake of brevity, but it allows a class of OXC architectures to be introduced having good switching and transmission performances.

A last OXC architecture, based on frequency switching, is shown in Fig. 9.24 [26]. At the OXC input, the WDM combs carried by the input fibers are wavelength translated so as to occupy contiguous parts of the optical spectrum. This operation can be carried out by K_f wavelength converters, since there exist optical wavelength converters that are able to translate an entire WDM comb [27]. After translation, the signals are amplified by a set of amplifiers and feed the inputs of an $(K_f + 1) \times (K_f + 1)K_c$ star coupler. At the coupler center all

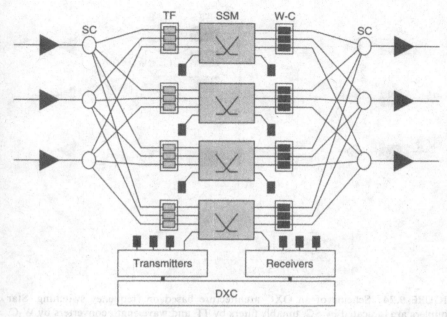

FIGURE 9.22 Scheme of an OXC architecture based on space switching and tunable filters that adopt wavelength conversion. Star couplers are indicated by SC, tunable filters by TF, space-switching matrixes by SSM, and wavelength converters by W-C.

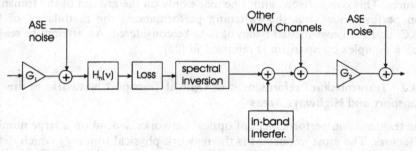

FIGURE 9.23 Transmission model of an OXC; the spectral inversion block can be absent.

the incoming optical channels are wavelength multiplexed onto a single comb. At the coupler output, any tunable filter selects one channel, and the succeeding wavelength converter sets its wavelength to a suitable value to allow multiplexing onto the selected output fiber. Multiplexing occurs by optical couplers, so at the OXC output the signal is amplified by a set of optical amplifiers and power equalization is also attained. The abstract model for the OXC of Fig. 9.24 is similar to that of Fig. 9.23 but eliminates spectral inversion, if this is the case, and adds a loss at the OXC input.

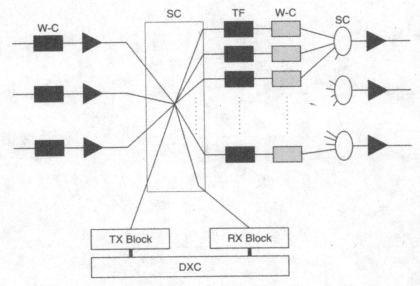

FIGURE 9.24 Scheme of an OXC architecture based on frequency switching. Star couplers are indicated by SC, tunable filters by TF and wavelength converters by W-C.

We do not discuss the complex problem of comparing different OXC architectures. This comparison cannot be made only on the ground of the transmission performances, but the switching performances, the modularity of the OXC, and even cost-related issues have to be considered. An attempt to realize such a complex comparison is reported in [28].

9.4.2 Transmission Performances of Optical Transport Networks in the Transport and Highways Areas

The transmission performances of optical networks depend on a large number of factors. The most important is the network physical topology which determines the number of OXC input/output ports. Also the OXC spacing along a signal path is determined by the network topology.

Besides the network topology, traffic requirements influence the transmission performances, since they determine the required capacity per fiber link. Then besides the network characteristics, the OXC architecture, the type of devices adopted inside the OXC, and the adopted transmission format influence the transmission performances.

It is beyond the scope of this section to derive general results that can be applied to every possible network topology within a fixed geographic area. To give an idea of the performances that can be attained by all-optical transport networks and to underline the major transmission impairments, we will examine two particular cases: a possible topology for the Italian high-speed transport network [29] and the so-called Pan-European network proposed in the frame of the COST 239 and ACTS OPEN EU research projects [22]. In the first case we consider a typical network on a national scale, with a maximum signal path of the order of 1600 km; the second example concerns a continental network with a maximum path length of more than 4000 km. Thus the Pan-European network can be considered a network in the communications highways area.

As far as the OXC architecture is concerned, we will limit our analysis to an architecture that can be represented by the scheme of Fig. 9.23 without the spectral inversion device. The parameters of the OXC are selected considering, in particular, the architecture of Fig. 9.21. This architecture has been chosen because it has been practically realized at the prototype level in the frame of the RACE MWTN research project, and it has been proposed in the frame of ACTS OPEN for the Pan-European network [22]. In particular, the filter inside the OXC is supposed to be a double-stage Fabry-Perot filter with a bilateral bandwidth equal to $20R$, being R the channel bit-rate, and the gain of the EDFAs inside the OXC are assumed to be equal. A double-stage Fabry-Perot filter with a bilateral bandwidth equal to $4R$ is considered in front of each receiver. The space switching matrixes are assumed to be passive devices with a loss equal to 10 dB for a 5×5 matrix.

An Example of Transmission through a Transport Network in a National Area

The network topology considered in this section is shown in Fig. 9.25, where the distances among the OXCs are also reported. The number of input/output ports if the OXCs in the network are all equal must be five: four for the fiber links and one for the local channels (see Fig. 9.20).

To evaluate the network transmission performances, we have selected the longest path along the network which, as is indicated in the figure, is a path from Torino to Catania that results 1785 km long. Transmission occurs in the third window, and the fiber dispersion is $\beta_2 = -1.3 \, ps^2/km$ for the bit-rates of 2.5 and 10 Gbit/s and $\beta_2 = -20 \, ps^2/km$ for the bit-rate of 622 Mbit/s. The other fiber parameters are reported in Table 2.1. The amplifier inversion factor is 3 dB. According to the scheme of Fig. 9.20, we assume a filter bandwidth of 1 nm. The in-band interference power depends on the filter bandwidth, the signal bandwidth, and on the channel spacing, and it can be evaluated from the knowledge of these three parameters. In our simulations we neglect the contribution of the in-band interference noise due to switches presented in the model of Fig. 9.20.

The maximum value of the Q factor (obtained for the optimum transmitted power) is reported in Fig. 9.26 versus the number of channels per fiber in the case $R = 2.5 \, Gbit/s$. Different channel separations are considered, ranging from 100 to 400 GHz.

As the number of channels is increased, the system performance gets worse due both to the increasing of the OXC internal losses and to nonlinear

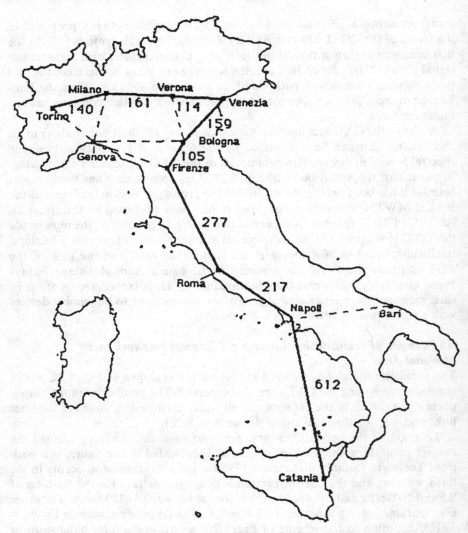

FIGURE 9.25 Possible physical topology of the high-speed Italian transport network; the lengths of the different fiber links are reported in kilometers and the path considered in the simulations is indicated by the solid line.

crosstalk. Regarding the OXC loses, the dimension of the space switch matrixes adopted in the selected OXC architecture depends only on the number of input/output ports; thus the switching fabric loss does not increase. On the other hand, the losses of the splitters and the combiners are directly proportional to the number of frequencies per fiber. As far as the nonlinear crosstalk is concerned, the power of the FWM components increases with the number of channels, causing a greater crosstalk penalty.

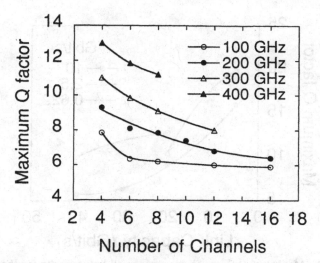

FIGURE 9.26 Maximum Q factor versus the number of channels per fiber for the Italian network. Different channel separations are considered, ranging from 100 to 400 GHz, and $R = 2.5$ Gbit/s.

The effect of the different crosstalk mechanisms is evident in Fig. 9.26 where the maximum Q factor is shown to increase with the channel spacing. The main crosstalk component is the linear in-band crosstalk; the nonlinear crosstalk due to FWM is nonnegligible, especially since the channel number is high. However, the linear out-band crosstalk is negligible in all the considered cases, which is due to the large channel spacing that is necessary to limit the in-band crosstalk penalty.

The maximum Q factor is reported versus the overall link capacity in Fig. 9.27 for different channel bit-rates. The overall optical bandwidth is fixed to 2500 GHz (about 29.5 nm at 1.55 μm), and the channel spacing is selected so to occupy all the available bandwidth. It is worth noting that in the situation analyzed in Fig. 9.27, the 10-Gbit/s signals can be transmitted over 1785 km due to the nonlinear dispersion compensation occurring in the anomalous propagation regime. Such compensation allows transmission to exceed the dispersion limit (about 1000 km).

Figure 9.27 shows that for a required capacity lower than about 20 Gbit/s, the best performances are attained by adopting a bit-rate of 622 Mbit/s and a step-index fiber. The low bit-rate channels are more robust to the effect of optical noise and SPM, while the step-index fiber prevents a high FWM–induced penalty. These advantages overcompensate the higher OXC internal loss due to the high number of channels needed when $R = 622$ Mbit/s.

When the critical capacity of about 20 Gbit/s is exceeded, the best transmission performances are attained by increasing the bit-rate and adopting DS fibers. In fact a high bit-rate allows the OXC losses to be reduced and the

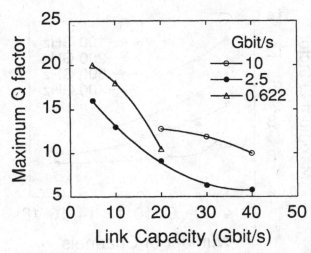

FIGURE 9.27   Maximum Q factor versus the overall link capacity for the Italian network. The optical bandwidth is 2500 GHz.

channel spacing to be increased, so reducing the impact of linear crosstalk. However, increasing the bit-rate and decreasing the fiber dispersion causes the Kerr effect to become more prominent, introducing both nonlinear distortion and nonlinear crosstalk.

From this discussion an optimum bit-rate can be expected to exist for each assigned link capacity. For example, in the case of Fig. 9.27, the maximum Q factor of a single TDM channel at 30 Gbit/s is lower than 4 while three channels at a bit-rate of 10 Gbit/s can be transmitted with a Q factor equal to about 14. However, the exact knowledge of the optimum bit-rate is not critical for a mesh network, since it depends on the analyzed signal path. Once determined what bit-rates ensure good transmission performances in all the networks, the bit-rate could be chosen according to flexibililty and management considerations.

An Example of Transmission through a Transport Network in the Highways Area

Another network topology that we consider in this section is that of the Pan-European network proposed in the frame of the COST 239 and ACTS OPEN EU research projects [22]. Its network topology is shown in Fig. 9.28. To analyze the transmission performances of this network, we select two signal paths, which are indicated in the figure by bold lines. The cities in which the nodes are placed along these paths and the relative distances are reported in Table 9.1.

Since the geographic area covered by the network is quite wide, dispersion management is used throughout the network. The design of dispersion

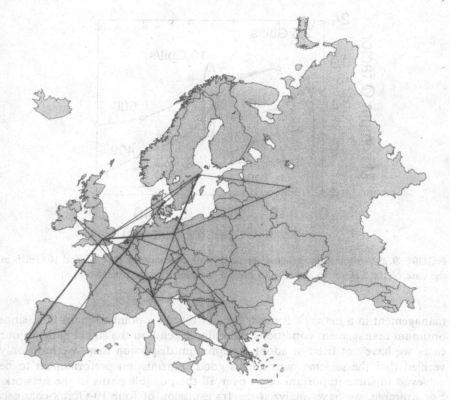

FIGURE 9.28 Physical topology of the pan-European optical network proposed in the framework of the ACTS OPEN program. The two sample paths are indicated by the bold lines.

TABLE 9.1 Detailed description of the routes followed by the two signal paths in the trans-European network considered in this section.

Signal path	Link	Link length (km)
Rome–Amsterdam	Rome–Milan	630
	Milan–Zurich	310
	Zurich–Paris	580
	Paris–Brussels	290
	Brussels–Amsterdam	230
Lisbon–Stockholm	Lisbon–London	2240
	London–Berlin	1030
	Berlin–Stockholm	1090

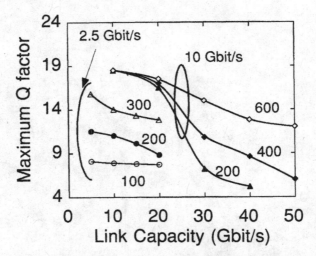

FIGURE 9.29 Maximum Q factor versus the link capacity for $R = 2.5$ and 10 Gbit/s in the case of the Lisbon to Stockholm link of the pan-European network.

management in a network is more difficult that in a point-to-point link, since optimum management conditions generally depend on the signal path. In our cases we have not tried to adopt an optimum dispersion map, we have only verified that the selected map allows good transmission performances to be achieved in some important cases over all the possible paths in the network. For example, we have analyzed the transmission of four 10-Gbit/s channels

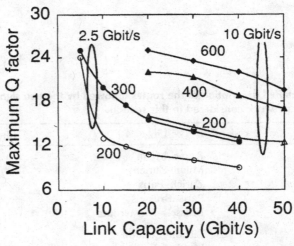

FIGURE 9.30 Maximum Q factor versus the link capacity for $R = 2.5$ and 10 Gbit/s in the case of the Rome to Amsterdam link of the pan-European network.

with a spacing of 600 GHz and of eight 2.5-Gbit/s channels with a spacing of 200 GHz. The transmission is operated by a DS fiber with a dispersion coefficient $\beta_2 = 1.28 \, \text{ps}^2/\text{km}$ at $1.55 \, \mu\text{m}$, while dispersion compensation is attained by step-index fibers with $\beta_2 = -20 \, \text{ps}^2/\text{km}$. In all the network links the amplifier spacing is 40 km, and the period of the compensation is 120 km, which is the step-index placed at the input of the corresponding amplifier (see Section 7.1.4). The OXC parameters are the same as those adopted in Section 9.4.3.

The longest transmission performance between the chosen paths (from Liston to Stockholm) is analyzed in Fig. 9.29. In particular, the maximum value of the Q factor (attained for the optimum transmitted power) is reported with relation to the overall link capacity for the bit-rates of 2.5 and 10 Gbit/s and for different channel spacing. Fig. 9.30, for the path from Rome to Amsterdam is analogous to Fig. 9.29.

It is evident from these figures that the best transmission performances are attained adopting, whenever possible, a bit-rate of 10 Gbit/s. Especially in the first case the ASE power must be as low as possible, since this reduces the signal spectrum widening (see Section 7.1.1). The ASE power is lower if the losses inside the OXC are kept low, that is, if the bit-rate is high so to transmit the wanted capacity by a smaller number of channels. Moreover decreasing the number of channels reduces even the FWM–induced crosstalk.

It is also to be noticed that by increasing the bit-rate above 10 Gbit/s, the system performance does not increase further. This is due to the fact that the Kerr induced distortion and the ASE-signal interaction gets strong as the bit-rate increases.

REFERENCES

1. Different schemes have been proposed for all-optical regeneration. Most of these schemes are based on the principle of the nonlinear optical loop mirror. Examples of such schemes capable of performing amplification, retiming and reshaping of high-speed signals are reported in
   J. K. Lucek and K. Smith, All-optical signal regenerator, *Optics Letters* 18: 1226–1228, 1993.
   Another interesting system that is able to compensate for fiber dispersion and the Kerr effect during fiber propagation and allows ASE noise rejection to be obtained is reported in
   K. Mori, T. Morioka, and M. Saruwatari, Optical parametric loop mirror, *Optics Letters* 20: 1424–1426, 1995.
2. Theoretical and experimental results concerning this phenomenon are reported in
   P. P. Iannone and T. E. Darcie, Multichannel intermodulation distortion in high-speed GaINAsP lasers, *Electronics Letters* 23: 1361–1362, 1987.
   J. A. Chiddix, H. Laor, D. M. Pangrac, L. D. Williamson, and R. W. Wolfe, AM video on fiber in CATV systems: Need and implementation, *IEEE-Journal on Selected Areas in Communications* 8: 1229–1239, 1990.

3. M. R. Phillips, T. E. Darcie, D. Marcuse, G. E. Bodeep, and N. J. Frigo, Nonlinear distortion generated by dispersive transmission of chirped intensity-modulated signals, *IEEE-Photonics Technology Letters* 3: 481–483, 1991.

4. Experimental results on the impact of clipping on SCM systems are reported in
K. Fujito, M. Fuse, and K. Maeda, AM/digital multi-channel optical video distribution, *Proceedings of ECOC'95*, pp. 367–371, Brussels, September 17–21, 1995.
An accurate theoretical analysis of the impact of clipping is given in
G. Aureli, S. Betti, E. Bravi, V. C. Di Biase, and M. Giaconi, Effect of clipping impulsive noise in a hybrid optical-fiber/coaxial network for transmission of sub-carrier-multiplexing (SCM) analog and digital channels, *Microwave and Optical Technology Letters* 11: 313–316, 1996.

5. For a recent review of all-optical packet switching, see the paper
S. P. Monacos, J. M. Morookian, L. Davis, L. A. Bergman, S. Forouhar, and J. R. Sauer, All-optical WDM packet networks, *IEEE-Journal of Lightwave Technology* 14: 1356–1370, 1996.
For an experiment showing the operation of all-optical packet switching, see
D. J. Blumenthal, R. J. Feuerstein, and J. R. Sauer, First demonstration of multi-hope all-optical packet switching, *IEEE-Photonics Technology Letters* 6: 457–460, 1994.
For examples of all-optical processing in packet networks adopting solitons, see
C. E. Soccolich, M. N. Islam, B. J. Hong, M. Chbat, and J. R. Sauer, Application of ultrafast gates to a soliton ring network, *Proceedings of Nonlinear Guided Wave Phenomena*, pp. 366–369, Cambridge, England, September 2–4, 1991.
M. Settembre and F. Matera, All-optical implementations for high-capacity TDMA networks, *Fiber and Integrated Optics* 12: 173–186, 1993.
F. Matera, E. Ripani, and M. Settembre, Proposal for an all optical soliton shuffle multihop network, *Electronics Letters* 28: 1570–1571, 1992.
A large class of techniques for all-optical processing of nonsoliton signals rely on logic gates realized by semiconductor amplifiers. See, for example,
A. D'Ottavi, E. Iannone, and S. Scotti, Address recognition in all-optical packet switching by FWM in semiconductor amplifiers, *Microwave and Optical Technology Letters* 10: 228–230, 1995.

6. In principle, if some small attenuation is accepted, a simple fiber loop connected with the transmission line via a 3-dB coupler can be regarded as optical memory. This principle is exploited, for example, in
N. A. Whitaker, Jr., M. C. Gabriel, H. Avramopoulos, and A. Huang, All-optical, all-fiber circulating shift register with an inverter, *Optics Letters* 16: 1999–2001, 1991.
H. Avramopoulos and N. A. Whitaker, Jr., Addressable fiber loop memory, *Optics Letters* 18: 22–24, 1993.
If long-term storage is needed, the fiber losses have to be compensated as far as the signal degradation due to fiber dispersion and nonlinear effects. An example of very long-term memory for soliton signals is reported in
H. A. Haus and A. Mecozzi, Long-term storage of a bit stream of solitons, *Optics Letters* 17: 1500–1502, 1992.
Different methods have been deviced for nonsoliton signals, see, for example,
M. Calzavara, P. Gambini, M. Puleo, B. Bostica, P. Cinato, and E. Vezzoni, Optical fiber-loop memory for multiwavelength packet buffering in ATM

switching applications, *Proceedings of OFC'93*, pp. 19–20, San Jose, February 21–26, 1993.

K. Nonaka, H. Tsuda, K. Hirabayashi, and T. Kurokawa, Digitally regenerating optical-fiber-loop memory, and T. Kurokawa, Digitally regenerating optical-fiber-loop memory with a side-injection ight-controlled bistable laser diode, *Proceedings of ECOC'93*, pp. 285–288, Montreaux, Switzerland, September 12–16, 1993.

7. Different protocols suitable for passive star networks are introduced and analyzed in
I. M. I. Habbab, M. Kaverhad, and C. E. W. Sandberg, Protocols for very high speed optical fiber local area networks using a passive star topology, *IEEE-Journal of Lightwave Technology*, 5: 1782–1794, 1987.
N. Mehravari, Performance and protocol improvements for very high speed optical fiber local area networks using a passive star topology, *IEEE-Journal of Lightwave Technology* 8: 520–530, 1990.

8. A large variety of different multihop networks have been introduced in different network areas. See, for example,
C. A. Brackett, A. S. Acampora, J. Sweitzer, G. Tangonam, M. T. Smith, W. Lennon, K. C. Wang, and R. H. Hobbs, A scalable multiwavelength multihop optical network: A proposal for research on all-optical networks, *IEEE-Journal of Lightwave Technology* 11: 736–753, 1993.
A. S. Acampora, A multihop local lightwave network, *Proceedings of GLOBECOM'87*, pp. 1459–1467, Tokyo, Japan, November 1987.

9. The Shuffle Multihop network is described, for example, in
A. S. Acampora, M. J. Karol, "An overview of Lightwave packet networks", IEEE-Network, pp. 29–41, January 1989;
A. S. Acampora, M. J. Karol, G. Hluchyi, "Terabit Lightwave Networks: the multihop approach", AT&T Technical Journal, vol. 66, no 6, pp. 21–34, 1987.

10. See the first reference of [9].

11. A recent review on the different methods proposed to obtain wavelength conversion in multiwavelength optical networks is
S. J. B. Yoo, Wavelength conversion technologies for WDM network applications, *IEEE-Journal of Lightwave Technology* 14: 955–966, 1996.
In particular, considering wavelength converters based on semiconductor amplifiers, the structure and possible performances of wavelength converters based on cross-gain and cross-phase modulation are reported in
B. Glance et al., High performance optical wavelength shifters, *Electronics Letters* 28: 1714–1715, 1992
M. Schilling et al. Monolithic Mach-Zehnder interferometer based optical wavelength converter operated at 2.5 Gb/s with extinction ratio improvement and low penalty, *Proceedings of ECOC'95*, pp. 647–650, 1995.
T. Durhuus, B. Mikkelsen, C. Joergensen, S. L. Danielsen, and K. E. Stubkjaer, All-optical wavelength conversion by semiconductor optical amplifiers, *IEEE-Journal of Lightwave Technology* 14: 942–954, 1996.
For an analysis of wavelength converters based on four-wave mixing, see, for example,
A. D'Ottavi, E. Iannone, A. Mecozzi, S. Scotti, P. Spano, R. Dall'Ara, J. Eckner, and G. Guekos, Efficiency and noise performances of wavelength converters based

on FWM in semiconductor optical amplifiers, *IEEE-Photonics Technology Letters* 7: 357–359, 1995.

Another interesting technique for obtaining wavelength conversion is difference frequency generation in passive waveguides. This technique is described, for example, in

S. J. B. Yoo, C. Caneau, R. Bhat, M. A. Koza, A. Rajhel, and N. Antoniades, All-optical wavelength conversion by quasi-phasematched difference frequency generation in AlGaAs waveguides, *Applied Physics Letters* 66: 2609–2611, 1966.

12. Problems related to integrity and fault restoration in high capacity access and transport networks are discussed in

L. Nederlof, K. Struyve, C. O'Shea, H. Misser, Du Yonggang, and B. Tamayo, End-to-end survivable broadband networks, *IEEE-Communication Magazine* 33: 63–70, 1995.

T.-H. Wu, Emerging technologies for fiber network survivability, *IEEE-Communication Magazine* 33: 58–74, 1995.

Some particular aspects related to networks with an optical path layer and integrated strategies to manage different transmission formats are discussed in the excellent paper

K. Sato, S. Okamoto, and H. Hadama, Network performance and integrity enhancement with optical path layer technologies, *IEEE-Journal on Selected Areas in Communications* 12: 159–170, 1994.

Other interesting related issues are discussed in

P. Dumortier, Survivability in optical backbone networks: Complementary or inter-layer competition? *Proceedings of NOC'96*, Broadband Superhighway Book, pp. 274–280, Heidelberg, June 25–28, 1996.

Particular survivability solutions for all-optical ring networks in the access and transport areas are discussed in

S. Johansson, Transport network involving a reconfigurable WDM network layer—a European demonstration, *IEEE-Journal of Lightwave Technology* 14: 1341–1348, 1996.

L. Blain, A. Hamel, T. Jakab, and A. Sutter, Comparison of classical and WDM based ring architecture, *Proceedings of NOC'96*, Broadband Superhighway Book, pp. 261–268, Heidelberg, June 25–28, 1996.

13. For an example of commercial device of this class, see

Fiber optic components, test instruments and OSP products, Catalog 1995/1996 from JDS FITEL.

14. Different technologies can be adopted to obtain thermooptic switch matrixes. For example, polymer technology is described in

N. Keil, H. H. Yao, C. Zawadzki, and B. Strebel, Rearrangeable nonblocking polymer waveguide thermo-optic 4×4 switching matrix with low power consumption at $1.55\,\mu\text{m}$, *Electronics Letters* 31: 403–404, 1995.

Another technology is that described in

M. Okuno, K. Kato, Y. Ohmori, M. Kawachi, and T. Matsunaga, Improved 8×8 integrated optical matrix switch using silica-based planar lightwave circhits, *IEEE-Journal of Lightwave Technology* 12: 1597–1606, 1994.

The same technology has been used to realize the delivery and coupling switches reported in

M. Koga, Y. Hamazumi, A. Watanabe, S. Okamoto, H. Obara, K. Sato, M. Okuno, and S. Suzuki, Design and performance of an optical-path cross-connect system based on wavelength path concept, *IEEE-Journal of Lightwave Technology* 14: 1106–1119, 1996.

15. A good example of LiNbO3 technology is described in
    P. Granestrand, B. Lagenström, H. Olofsson, J.-E. Falk, and B. Stolz, Pigtailed tree-structured 8×8 LiNbO$_3$ switch matrix with 112 digital optical switches, *IEEE Photonics Technology Letters* 6: 71–73, 1994.

16. The basic technology for this kind of switches is reported in
    M. Gustavsson, B. Lagenström, L. Thylén, M. Janson, L. Lundgren, A.-C. Mörner, M. Rask, and B. Stolz, Monolithically integrated 4×4 InGaAsP/InP laser amplifier gate switch arrays, *Electronics Letters* 28: 2223–2225, 1992.
    Improved performance devices are reported in
    W. H. van Berlo, M. Janson, L. Lundgren, A.-C. Mörner, J. Terlecki, M. Gustavsson, P. Granestrand, and P. Svensson, Polarization insensitive, monolithic 4×4 InGaAsP/InP laser, *IEEE-Photonics Technology Letters* 7: 1291–1293, 1995.
    E. Almström, C. P. Larsen, L. Gillner, W. H. van Berlo, M. Gustavsson, and E. Berglind, Experimental and analytical evaluation of packaged 4×4 InGaAsP/InP semiconductor optical amplifier gate switch matrices for optical networks, *IEEE-Journal of Lightwave Technology* 14: 996–1004, 1996.

17. This type of OADM and the ONTC program are described, for example, in
    G.-K. Chang, G. Ellinas, J. K. Gamelin, M. Z. Iqbal, and C. A. Brackett, Multiwavelength reconfigurable WDM/ATM/SONET network testbed, *IEEE-Journal of Lightwave Technology* 14: 1320–1340, 1996.
    The same OADM structure is used in the Prometeo project, the first Italian demonstrator of all-optical ring network in the access area.

18. A recent and complete review of the possible structure and performances of acoustooptic filters designed for λ-selective switch is presented in the paper
    D. A. Smith, R. S. Chakravarthy, Z. Bao, E. Baran, J. L. Jackel, A. d'Alessandro, D. J. Fritz, S. H. Huang, X. Y. Zou, S.-M. Hwang, A. E. Willner, and K. D. Li, Evolution of the acusto-optic wavelength routing switch, *IEEE-Journal of Lightwave Technology* 14: 1005–1019, 1996.

19. The OADM structure discussed in this section is inspired by the system proposed in the context of the RACE MWTN project of the European Community. The same OADM structure is being used in the Prometeo project, in the first Italian demonstrator of all-optical ring network in the access area. See, for example,
    S. Johansson, Transport network involving a reconfigurable WDM network layer— a European demonstration, *IEEE-Journal of Lightwave Technology* 14: 1341–1348, 1996.

20. Classical studies comparing an accurate model on the interference and the gaussian approximation in point-to-point systems are
    L. G. Kazowsky and J. L. Gimlett, Sensitivity penalty in multichannel coherent optical communications, *IEEE-Journal of Lightwave Technology* 6: 1353–1365, 1988.
    G. Jacobsen, Multichannel system design using optical preamplifiers and accounting for the effects of phase noise, amplifier noise and receiver noise, *IEEE-Journal of Lightwave Technology* 10: 367–376, 1992.
    A paper in which in-band interference in the optical network is considered by comparing its accurate modeling with a kind of gaussian approximation is

E. L. Goldstein, L. Eskildsen, and A. F. Elrefaie, Performance implications of component crosstalk in transparent lightwave networks, *IEEE-Photonics Technology Letters* 6: 657–660, 1994.

21. This seems to be the simplest architecture for a reconfigurable OXC; it is presented, for example, in
    R. Ramaswami and K. N. Sivarajan, Routing and wavelength assignment in all-optical networks, *IEEE-Transaction on Networking* 3: 489–500, 1995.
    The OXC architecture is also one of the architectures considered in the MONET project. See, for example,
    R. E. Wagner, R. C. Alferness, A. A. M. Saleh, and M. S. Goodman, MONET: Multiwavelength Optical Networking, *IEEE-Journal of Lightwave Technology* 14: 1349–1355, 1996.

22. This OXC scheme has been proposed in the context of the RACE MWTN EU project. See, for example,
    G. R. Hill, A transport network layer based on optical network elements, *IEEE-Journal of Lightwave Technology* 11: 667–679, 1993.
    S. Johansson, Transport network involving a reconfigurable WDM network layer—a European demonstration, *IEEE-Journal of Lightwave Technology* 14: 1341–1348, 1996.
    The same OXC scheme has been proposed, first in the context of COST 139, then in the context of the ACTS OPEN (both projects sponsored by the European Community) for the Pan-European optical network. See, for example,
    M. J. O'Mahony, D. Simeonidou, and A. Yu, J. Zhou, The design of a European optical network, *IEEE-Journal of Lightwave Technology* 13: p. 817–828, 1995.
    M. W. Chbat, A. Leclert, E. Jones, B. Mikkelsen, H. Fevrier, K. Wünstel, N. Flaarønning, J. Verbeke, M. Puleo, H. Melchior, P. Demeester, J. Mørk, and T. Olsen, WDM for transmission and routing in wide-scale networks: The optical pan-European network approach (ACTS project OPEN), *Proceedings of NOC'96*, Broadband Superhighway Book, pp. 182–186, Heidelberg, June 25–28, 1996.

23. For example, for this OXC architecture, see the last paper of [22] and
    R. Sabella, E. Iannone, and E. Pagano, Optical transport networks employing all-optical wavelength conversion: Limits and features, *IEEE-Journal of Selected Areas in Communications* 14: 1996.
    E. Iannone, R. Sabella, L. de Stefano, and F. Valeri, All-optical wavelength conversion in optical multicarrier networks, *IEEE Transactions on Communications* 44: 716–724, 1996.
    A generalization of this architecture is presented in
    R. Ramaswami and K. N. Sivarajan, Routing and wavelength assignment in all-optical networks. *IEEE-Transaction on Networking* 3: 489–500, 1995.
    The experimental results on such an OXC are reported in
    J. Graf. O. Jahreis, O. Möller, U. Pauluhn, D. Popescu, B. Stilling, and F. Derr, OCC'95-optical cross-connecting with supervision and node management, *Proceedings of NOC'96*, Broadband Superhighway Book, pp. 166–173, Heidelberg, June 25–28, 1996.
    A similar OXC is also considered in the MONET project; references on this project are reported in [21].

24. OXC architectures based on the concept of the delivery and coupling switch were introduced in

A. Watanabe, S. Okamoto, and K. Sato, Optical path cross-connect node architecture with high modularity for photonic transport networks, *IEICE Transaction on Communications* E77-B: 1220–1229, 1994.
Experimental results on such an OXC are reported in
M. Koga, Y. Hamazumi, A. Watanabe, S. Okamoto, H. Obara, K. Sato, M. Okuno, and S. Suzuki, Design and performance of an optical-path cross-connect system based on wavelength path concept, *IEEE-Journal of Lightwave Technology* 14: 1106–1119, 1996.

25. Experimental delivery and coupling switches have been reported in
M. Koga, A. Watanabe, S. Okamoto, K. Sato, and M. Okuno, 8 × 16 delivery-and-coupling-type optical switches for a 320-Gb/s throughput optical path cross-connect system, *Proceedings of OFC'96*, pp. 259–260, San Jose, February 25–March 1, 1996.

26. This OXC architecture is reported in
R. Sabella and E. Iannone, A new modular optical path cross-connect, *Electronics Letters* 32: pp. 125–126, 1996.

27. Wavelength converters based on coherent nonlinear effects, such as four-wave mixing or second harmonic generation, have this potential. Experiments in which an entire WDM comb is simultaneously translated are reported in
K. Inoue, T. Hasegawa, K. Oda, and H. Toba, Multichannel frequency conversion experiment using fibre four-wave mixing, *Electronics Letters* 29: 1708–1710, 1993.
R. Schnabel, U. Hilbk, T. Hermes, P. Meibner, C. Helmolt, K. Magari, F. Raub, W. Pieper, F. J. Westphal, R. Ludwig, L. Kuller, and H. G. Weber, Polarisation insensitive frequency conversion of a 10 channels OFDM signal using four-wave mixing in a semiconductor laser amplifier, *IEEE-Photonics Technology Letters* 6: 56–58, 1994.

28. Two papers providing a comprehensive comparison among different OXC architectures are
S. Okamoto, A. Watanabe, and K.-I. Sato, Optical path cross-connect node architectures for photonic transport network, *IEEE-Journal of Lightwave Technology* 14: 1410–1422, 1996.
E. Iannone and R. Sabella, Optical path technologies: A comparison among different cross-connect architectures, *IEEE-Journal of Lightwave Technology* 14: 2184–2196, 1996.

29. See, for example, the first paper of ref. [23].

Simulation of High-Capacity Optical Transmission Systems

In this appendix we describe the numerical method we have used throughout this book to study the propagation of optical signals in fiber links and to evaluate the performances of optical communication systems [1].

To refer to a concrete case, we will assume an amplified optical link such as those described in Section 4.3. The link can be viewed as a periodic alternation of fiber pieces of length L_A and optical repeaters; the optical repeaters are composed by an optical in-line amplifier and, in this case, by optical filters, dispersion compensation devices, and so on. Optical amplifiers and the other optical devices are assumed to have a zero length. The only device whose length has been taken into account is the dispersion-compensating fiber.

Optical amplifiers provide optical gain and add optical noise. In the case of EDFAs the optical gain can be assumed to be constant, depending only on the signal average power and not on the particular transmitted message. In contrast, in the case of semiconductor amplifiers, the instantaneous gain depends on the particular transmitted message.

The signal propagation is simulated by alternating the numerical solution of the equations describing fiber propagation with a numerical model of the optical repeater. This model produces the signal at the repeater output, including the ASE noise and starts from the input signal.

At the link output the optical signal is transformed into an electrical current by a PIN photodiode and is electrically filtered. The signal timing is recovered by a baseband PLL, and a threshold decision device decides the received bit [2]. Before the photodiode, an optical Fabry-Perot filter is generically located to limit the presence of the ASE noise. Its bandwidth depends on the kind of system; as an example, for single-channel soliton systems we consider the bandwidth, B_F, to be equal to $4R$, while for NRZ signals the bandwidth is chosen to equal the FWHM bandwidth of the signal spectrum at the link output. For NRZ systems operating in links either with high GVD or with dispersion management, B_F ranges between $4R$ and $10R$.

In the program the PIN photodiode and the electrical front end are simulated by a square law device and a gaussian noise source that introduces the receiver noise. The electrical filter at the receiver is a second-order Butterworth filter whose bandwidth is $0.8R$. No optimization of the electrical filter shape or of its bandwidth is carried out. To carry out synchronization, the delay between the output signal and the transmitted one is estimated by finding the maximum of their correlation. Once the overall delay is known, the sampling instants are easily determined.

After synchronization, the eye opening penalty of the received optical signal is evaluated and compared with the eye opening of the transmitted signal to obtain the eye penalty, as defined by equation (4.15). The Q factor is also evaluated to take into account the noise effects. Since the jitter can be an important phenomenon for some optical systems, the average jitter and its variance can be evaluated too. In particular, the jitter variance is evaluated by detecting the shift of the pulse peak at the fiber output [3].

In the following sections the most important steps of the simulation algorithm will be described in detail.

A1.1   TRANSMITTED SIGNAL

The transmitter is not simulated by a physical model: It is assumed to be ideal but for a finite bandwidth, reflecting in a finite rising time of the transmitted pulses. Practically, we assume the use of lasers externally modulated so that the chirp contribution can be ignored. In the case of the NRZ transmission, the signal amplitude is modulated by a train of risen cosine pulses with a rising time $T_r = T/4$, where T is the inverse of the bit-rate R. The temporal and the spectral profile of a NRZ signal at 5 Gbit/s are reported in Fig. A1.1.

In the soliton case, the transmitted signal is assumed to be a train of ideal solitons. The temporal width of the soliton pulses has been optimized to obtain the best system performance. In Fig. A1.2 the temporal and the spectral profiles of a soliton signal at 5 Gbit/s with a 3-dB width (FWHM width) are equal to 40 ps.

Besides soliton signals, another pulse shape used in this book for the RZ format is the supergaussian shape defined by $\exp(-[T/T_0]^{2m}/2)$. At $m = 1$ we have the conventional gaussian pulse.

In the case of single-channel systems the transmitted bit-stream consists of several repetitions of the following sequence of 32 bit: 01011000101111011010100000101110. Such a sequence is sampled to obtain 2048 samples.

In the case of WDM signals, the transmitted bit-stream for the generic channel is obtained by randomly translating the sequence adopted in the single-channel case. The overall optical bandwidth of the simulated signal

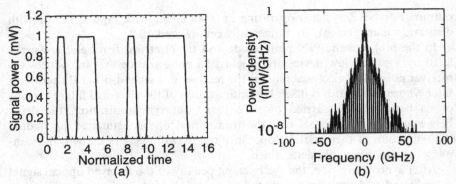

FIGURE A1.1 Temporal and the spectral profile of an NRZ signal at 5 Gbit/s.

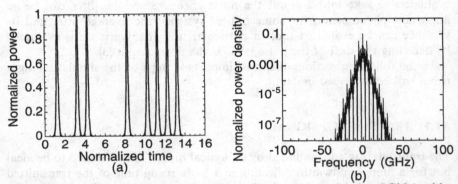

FIGURE A1.2 Temporal and the spectral profile of a soliton signal at 5 Gbit/s with a FWHM time equal to 40 ps.

depends on the sampling time. For example, to simulate a total bandwidth of about 1000 GHz the bit number, N_b, and the number of time samples, N_P, for $R = 10$ Gbit/s is $N_b = 32$ and $N_P = 4096$ (corresponding to a bandwidth of 1280 GHz). If the bit-rate is 1 Gbit/s instead of 10 Gbit/s, to avoid the use of a large quantity of computer memory, we have chosen $N_b = 16$ and $N_P = 16,384$.

A1.2 SIMULATION OF SIGNAL PROPAGATION

The simulation of the fiber propagation is based on the solution of the non-linear Schrödinger equation by the split-step method [4]. In order to accurately simulate signal propagation, the polarization effects must be taken into account; as a consequence the solution of the coupled nonlinear Schrödinger equations (2.77) is required. Unfortunately, this solution is computationally complex, and the program implementing it requires long machine time to be executed. This is also caused by the presence of random mode coupling, which requires adopting a smaller step in the propagation direction.

However, if some propagation conditions are verified, the signal propagation can be simulated by the scalar Schrödinger equation. These conditions are often verified in conventional optical systems, and therefore the performance of a large variety of optical systems can be evaluated numerically solving equation (2.57) instead of equation (2.77).

A1.2.1 Scalar Propagation Problem

Several methods can be used to numerically solve equation (2.57). Most can be classified in two categories: as finite difference methods or as pseudospectral methods. Among the pseudospectral methods, one of the most important is the split-step Fourier method (SSFM), whose results are very interesting under the point of view of the calculation time, thanks to the use of the fast-Fourier transform (FFT).

The SSFM is based on the division of the fiber into small steps in the propagation length. The fiber propagation is simulated by approximately solving the propagation equation in each step, starting from the field at the output of the previous step.

The approximate solution of equation (2.57) in a single step can be expressed by rewriting the equation as

$$\left[\frac{\partial}{\partial \zeta} + i\hat{D}_{\beta 2} + \hat{D}_a + i\hat{D}_\gamma + \hat{D}_{\beta 3}\right] U(\zeta, \tau) = 0 \tag{A1.1}$$

where the *real* operators appearing in equation (A1.1) are defined as

$$\hat{D}_{\beta 2} = \frac{1}{2}\,\text{sign}(\beta_2)\,\frac{\partial^2}{\partial \tau^2}, \quad \hat{D}_a = \frac{L_D}{2L_a}$$

$$\hat{D}_\gamma = -\frac{L_D}{L_{NL}}\,|U|^2, \quad \hat{D}_{\beta 3} = -\frac{1}{6}\frac{L_D}{L_D'}\frac{\partial^3}{\partial \tau^3} \tag{A1.2}$$

The formal solution of equation (A1.2) can be written as

$$U(\zeta, \tau) = \exp[i\hat{D}_\gamma L_s]\exp[[i\hat{D}_{\beta 2} + \hat{D}_a + \hat{D}_{\beta 3}]L_s] U(0, \tau) \tag{A1.3}$$

where L_s is the length of the step. The second member of equation (A1.3) shows that the field at the step output can be determined by applying two operators to the field at the step input: first a linear operator taking into account dispersion and attenuation, then a nonlinear operator accounting for the Kerr effect. It is to be noted that equation (A1.3) does not constitute a real solution of (A1.1), since the nonlinear operator depends on the local value of the field and not only on the field at the step input. However, if the step is sufficiently small, the nonlinear operator can be approximately evaluated starting from the field at the step input, and the dispersive and the nonlinear contribution in the step can be evaluated separately.

In particular, in the dispersive step the signal is Fourier transformed by the FFT and hence multiplied by the transfer function of the fiber piece. In agreement with equation (2.18), such transfer function can be written as

$$F(\omega) = \exp[-\alpha L_s]\exp[i(\beta_2\omega^2 L_s/2 + \beta_3\omega^3 L_s/6)] \qquad (A1.4)$$

The resulting signal is Fourier anti-transformed and is multiplied by the operator representing the Kerr effect, which in agreement with equation (A1.3), is

$$e^{i\hat{D}_\gamma L_s} = \exp\left[i\gamma\,\frac{e^{-\alpha L_s} - 1}{\alpha L_s}\,P_o(0,\tau)L_s\right] \qquad (A1.5)$$

where $P_o(0,\tau)$ is the optical power at the step input. It is to be noted that if the nonlinear operator were evaluated using the power $P_o(0,\tau)$, the nonlinear effect would be overestimated, since the optical power attenuates during propagation along the step. The term $[\exp(-\alpha L_s) - 1]/(\alpha L_s)$ is inserted to evaluate the nonlinear operator using the average optical power in L_s.

The accuracy of the method depends on the length L_s. A good rule is that L_s has to be much shorter than any characteristic lengths of the fiber $(L_D, L_{NL}, L_D', \ldots)$. In case of single-channel systems, we have verified that in all the cases analyzed in this book a step length equal to $L_D/20$ ensures good accuracy. In the case of WDM systems, a larger number of steps is required to take into account the effects of pulse collisions [5]. In this case we have considered $L_s = 0.2\,\text{km}$ in order to obtain a maximum phase shift per step of the order of $0.05\,\text{rad}$.

A1.2.2 Vector Propagation Problem

The solution of the coupled nonlinear Schrödinger equations, which take into account the polarization effects, is much more complex than that of the scalar propagation equation. The SSFM can be applied only in particular propagation regimes, when suitable approximations can be carried out, as detailed in Section 2.4.

The first case is the so-called low-birefringence condition when $\Delta\beta T_0^2 \ll \beta_2$ (the symbols in this section are the same as those of Section 2.4). In this case the effect of the birefringence is low in a dispersion length, and the term $\exp(-2i\Delta\beta z)$ can be set to one in equations (2.77) [6].

The second case is the so-called high-birefringence condition when $\Delta\beta T_0^2 \gg \beta_2$. In this case the exponential term $\exp(-2i\Delta\beta z)$ fluctuates rapidly in the dispersion length, and its average effect tends to vanish. In the case of high birefringence, we neglect the exponential term in equations (2.77) and write the propagation equations, in absence of random mode coupling, as

$$\left(\frac{\partial A_x}{\partial z} + \frac{\beta'_x}{2}\frac{\partial A}{\partial t}\right) + \frac{i}{2}\beta_2\frac{\partial^2 A_x}{\partial t^2} + \frac{\alpha_x}{2}A_x = i\gamma(|A_x|^2 + \tfrac{2}{3}|A_y|^2)A_x$$

$$\left(\frac{\partial A_y}{\partial z} + \frac{\beta'_y}{2}\frac{\partial A_y}{\partial t}\right) + \frac{i}{2}\beta_2\frac{\partial^2 A_y}{\partial t^2} + \frac{\alpha_y}{2}A_y = i\gamma(|A_y|^2 + \tfrac{2}{3}|A_x|^2)A_y \tag{A1.6}$$

The high-birefringence approximation is generally satisfied by optical fibers for telecommunications; thus equations (A1.6) can be used in a simulation program if the effect of random mode coupling is reproduced in some way.

Equations (A1.6) can be written in the form

$$\left[\frac{\partial}{\partial\zeta} + i\hat{D}_{\text{Lin}} + i\hat{D}_{\text{NL}}\right]\underline{U}(\zeta,\tau) = 0 \tag{A1.7}$$

where $\underline{U}(\zeta,\tau)$ is a vector whose components are the field components $U(\zeta,\tau)$ and $V(\zeta,\tau)$ normalized to the square root of the peak power, $\hat{D}_{\text{Lin}}$ is the linear operator, and $\hat{D}_{\text{NL}}$ is the nonlinear operator. The elements of the linear operator frequency response **F** are

$$F_{11}(\omega) = \exp(-\frac{\alpha_x}{2}L_s)\exp[i(\beta'_x\omega + \tfrac{1}{2}\beta_2\omega^2 + \tfrac{1}{6}\beta_3\omega^2)L_s]$$

$$F_{22}(\omega) = \exp(-\frac{\alpha_y}{2}L_s)\exp[(\beta'_y\omega + \tfrac{1}{2}\beta_2\omega^2 + \tfrac{1}{6}\beta_2\omega^3)L_s] \tag{A1.8}$$

$$F_{21}(\omega) = F_{12}(\omega) = 0$$

while the nonlinear operator can be represented by the matrix

$$e^{i\hat{D}_{\text{NL}}L_s} = \begin{bmatrix} \exp[i\gamma\kappa P_x(0,\tau)L_s & + & \tfrac{2}{3}i\gamma\kappa P_y(0,\tau)L_s] & 0 \\ 0 & & \exp[\tfrac{2}{3}i\gamma\kappa P_x(0,\tau)L_s & + & i\gamma\kappa P_y(0,\tau)L_s] \end{bmatrix} \tag{A1.9}$$

where P_x and P_y are the optical powers along the two linear polarization components and the constant κ, taking into account the effect of the fiber attenuation on the nonlinear operator, given by $\kappa = [\exp(-\alpha L_s) - 1]/(\alpha L_s)$ as in the scalar case.

The random mode coupling can be simulated if the integration step of the SSFM is chosen approximately equal to the characteristic length L_h of the random coupling defined in Section 2.1.3. Under this condition the coupling between the polarization modes can be neglected when solving the propagation within a single step. At the output of each step, before evaluating the propagation in the next step, the birefringence axes are randomly rotated and a random phase shift is added [7].

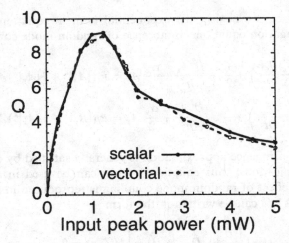

FIGURE A1.3   Comparison between the Q factors evaluated with the scalar and the vector method in the case considered in Fig. 7.28 for $L_c = 80$ km and when the fiber operating in normal dispersion is located at the amplifier output.

Since the numerical solution of the vector propagation problem is considerably more complex than the solution of the scalar problem, in this book the scalar approximation is adopted in all cases where polarization evolution does not significantly affect system performance. This is the case of an IM-DD system where the transmitted pulses all have the same polarization state and the PMD can be neglected. The first condition is almost always verified in practical systems. The second holds with a very good approximation where the DGD at the link output is smaller that 10% of the bit-interval. Under these conditions the values of the Q factor, of the EOP, and of the jitter variance obtained from the scalar and vector simulations practically coincide. For example, in Fig. A1.3 we report a comparison between the Q factors evaluated by the two methods considered in Fig. 7.28 for $L_c = 80$ km, where the fiber operating in normal dispersion is located at the amplifier output.

A1.3 AMPLIFIER SIMULATION

Two types of optical amplifiers have been considered in this book: EDFAs and SOAs. The behavior of each amplifier type is different: Gain saturation is a slow process in EDFAs, while it is a fast process in SOAs. As a consequence a static model can be adopted in the first case, while amplifier dynamics must be taken into account in the second.

A1.3.1 Simulation of Erbium-Doped Fiber Amplifiers

In the case of EDFAs, the amplification process is simply obtained by multiplying the electrical field for the total gain G and by adding an independent noise term to each spectral component of the signal. The real and imaginary parts of the noise spectral components are independent gaussian variables with variance [8] $\sigma^2 = Fh\nu(G - 1)\Delta\nu/2$, where F accounts for incomplete population inversion (see Section 3.1), h is the Plank constant, ν is the frequency, and $\Delta\nu$ is the bandwidth occupied by each Fourier component of the discrete Fourier spectrum. The total gain G of booster amplifiers and receiver preamplifiers is set depending on the system characteristics. In the case of in-line amplifiers, the total gain is $G = W^2G_0$, where $G_0 = e^{\alpha L_A}$ is the gain that exactly compensates for the fiber loss and W is the excess gain necessary to compensate for the losses of other optical devices present at the amplifier location [9]. The EDFA gain has been assumed to be flat in all of the considered optical bandwidths.

A1.3.2 Simulation of Semiconductor Amplifiers

In the simulation of optical links adopting SOAs, some problems arise that do not exist in using fiber amplifiers. To take into account the fast saturation mechanism, the amplifier must be simulated by solving equations (3.18) and (3.20), taking into account the relation (3.29) between the local gain and the local refraction index. Moreover noise terms have to be added to the equation describing the field propagation and the carrier dynamics. As a consequence equations (3.18) and (3.20) become stochastic equations, and the standard Runge-Kutta method can no longer be applied to solve them. Numerical solution of stochastic equations is possible by using the rules of Ito differential calculus [10], but a simulation program adopting such an algorithm is more time-consuming than a simulation of systems adopting fiber amplifiers.

The ASE noise can also be included in the amplifier simulation in a more heuristic way. The amplifier can be simulated neglecting the presence of ASE. At the SOA output the noise is added to the signal in the time domain: A complex noise sample is added to each signal sample, whose real and imaginary parts are independent gaussian variables. The variance of these variables is evaluated by equation (3.10) using the instantaneous value of the amplifier gain and of the carrier density.

This second simulation method is much less complex than the first, since the SOA equations can be solved by a standard Runge-Kutta method, but its accuracy has to be verified. To do this, we have analyzed a large number of different fiber links adopting SOAs [11–12]: In all the practical cases the results coming out from the two simulation models are in very good agreement. Therefore the heuristic model has been used throughout this book.

Another interesting numerical problem arises when using the Runge-Kutta method to solve the SOA equations in a simulation program adopting the SSFM for fiber propagation. Since the SSFM exploits the FFT, it implicitly assumes that the signal is periodic outside the simulated time window. On the other hand, since the Runge-Kutta algorithm performs a numerical solution in the time domain, it implicitly assumes that the signal is zero outside the simulated time window. Thus it is not possible to simply put one after the other simulation blocks adopting the two algorithms.

In our simulation program this problem is solved by imposing a periodic input to the SOA. To do this, the input signal is presented at the SOA several times as if it was the period of a periodic function. This procedure ends when the SOA gain is equal, within a given accuracy, at the beginning and at the end of the period. Generally, the period has to be repeated three or four times so that the simulation complexity is not increased.

A1.4 OPTICAL FILTERS

Different types of optical filters can be used in optical communication systems, for example, Fabry-Perot filters, grating filters, interference filters, and acousto-optic filters. In our simulation program no optimization of the shape of optical filters has been tried: Optimizing the optical filter shape is not generally a crucial issue.

In all the simulations optical Fabry-Perot filters are adopted. The transfer function of a Fabry-Perot filter can be written as [13]

$$H_{FP}(\nu) = \frac{1 - \Re}{1 - \Re \exp[i2\pi(\nu - \nu_0)/\Delta\nu]} \tag{A1.10}$$

where $\Re$ is the reflectivity of the interferometer mirrors and $\Delta\nu$ is the spacing between adjacent resonant frequencies. To obtain a good quality filter, the interferometer loss has to be small; this implies that $1 - \Re \ll 1$. In this approximation the 3-dB bandwidth B_0 of the Fabry-Perot filter is given by

$$B_0 = \frac{1 - \Re}{\pi\sqrt{\Re}} \Delta\nu \tag{A1.11}$$

Generally, the factor $\mathscr{F} = (\pi\sqrt{\Re})/(1 - \Re)$ which depends only on the mirror's reflectivity is called *filter finesse*, so that the filter is characterized by $\Delta\nu$ and the *finesse* $\mathscr{F}$.

Around a given resonance peak the transfer function of a Fabry-Perot filter can be approximated by a Lorentzian function, writing

$$H_{FP}(\nu) \approx \frac{1}{1 - [i4\pi(\nu - \nu_0)/B_0]} \tag{A1.12}$$

In Fig. A1.4 the square module of the Fabry-Perot transfer function and of the Lorentzian approximation are compared. Unless there is a strong linear interference between adjacent frequency multiplexed signals, requiring one to correctly model the shape of the filters tails, the Lorentzian approximation is quite good, and it is often used in simulations. In the optical receiver the electrical filter is a second-order Butterworth filter, whose bandwidth was fixed at $0.8R$.

A1.5 STATISTICAL ESTIMATION OF THE PERFORMANCE EVALUATION PARAMETERS

Three main performance evaluation parameters are estimated by the simulation programs exploited for this book: the Q factor, the eye-opening penalty (EOP) and the jitter variance. The definitions of these parameters are reported in Chapter 4.

The time jitter can be estimated in different ways depending on how the pulse positions is evaluated. In our simulation program the jitter is calculated by detecting the maximum of the received pulse. This technique has been proved to be more accurate than that based on the pulse center of mass, especially when using in-line soliton control [3].

As detailed in Section A1.1, a simulation cycle is performed by transmitting a sequence of 32 bits, so it is important to determine how many cycles are needed to obtain a small statistical error. This analysis has been carried out by concentrating the attention on the estimation of the Q factor, since it is the most critical parameter under this point of view.

In Fig. A1.5 the Q factor is shown versus the number of simulation cycles in three typical cases from Chapter 7:

a in the case of Fig. 7.28, for $L_c = 80$ km and $P_o = 1$ mW.
b in the case of Fig. 7.15a for $R = 2.5$ Gbit/s with a pulse duration of 25 ps.
c in the case of Fig. 7.21 for $R = 15$ Gbit/s and $P_n = P_o/P_k = 1$.

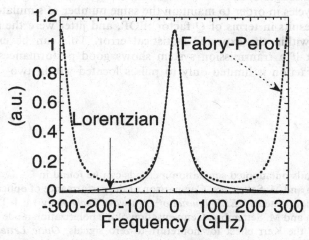

FIGURE A1.4 Square module of the Fabry-Perot transfer function and of the Lorentzian approximation.

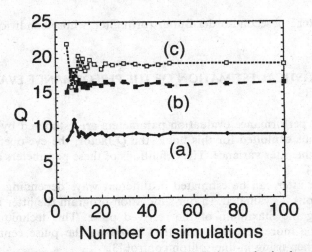

FIGURE A1.5 Q factor versus the number of simulation cycles in three typical cases from the simulations of Chapter 7: (a) is the case of Fig. 7.28 for $L_c = 80$ km and $P_o = 1$ mW, (b) is the case of Fig. 7.15a for $R = 2.5$ Gbit/s with a pulse duration of 20 ps, and (c) is the case of Fig. 7.21 for $R = 15$ Gbit/s and $P_n = P_o/P_k = 1$.

From Fig. A1.5 it can be noted that the Q factor stabilizes to its asymptotic value before 25 cycles in all cases; thus the use of 25 cycles allows a good estimation of the Q factor to be obtained.

A last important parameter to take into account is the length of the bit sequence transmitted in a simulation cycle. In Section A1.1 it was stated that we have used a 32-bit sequence. In some critical cases we repeated the simulation by using longer sequences (64 and 128 bits) but reducing the number of simulation cycles in order to maintain the same number of simulated bits. The simulation results in terms of Q factor, EOP, and jitter were the same in the three cases within the expected statistical error. This can be explained by the fact that if a transmission system shows good performances, the inter-symbol interaction is limited only to pulses located within two- or three-bit intervals.

REFERENCES

1. More details on adopted simulation methods can be found in
   F. Matera and M. Settembre, Comparison of the performance of optically amplified transmission systems, *IEEE-Journal of Lightwave Technology* 14: 1–12, 1996.
   F. Matera and M. Settembre, Compensation of the polarization mode dispersion by means of the Kerr effect for nonreturn-to-zero signals, *Optic Letters* 20: 28–30, 1995.
   F. Matera and M. Settembre, Nonlinear evolution of amplitude and phase

modulated signals and performance evaluation of single-channel systems in long haul optical fiber links, *Journal of Optical Communications* 17: 1–20, 1996.

E. Iannone, F. Matera, A. Mecozzi, and M. Settembre, Performance evaluation of very long span direct detection intensity and polarization modulated systems, *IEEE-Journal of Lightwave Technology* 14: 261, 1996.

The simulation programs used for this book have been tested in the context of the EU ACTS projects ESTHER and UPGRADE. Within these projects, a first testing phase was devoted to a comparison of simulation programs from different research institutes in a large number of situations. In particular, besides Fondazione Ugo Bordoni with which the authors are affiliated, CNET laboratories (France, Lannion), the University of Kaserlautern, and Lucent Technology (Germany, Numberg) participated in this phase. The results are reported in the annual report of ACTS ESTHER for 1995 and in the document ACTS UPGRADE D211.

In a second phase a successful comparison was made between the simulation and experimental results. The transmission experiments where performed at CNET and at the University of Eindoven; the results of this second phase are reported in the documents ESTHER WP04/401 and UPGRADE D2121.

2. A large number of books and papers report on studies of different IM-DD receiver models. The model used in the simulation programs is based on these references:

G. P. Agawal, *Fiber-Optic Communication Systems*, Wiley, New York, 1995, chs. 4 and 5.

S. Betti, G. De Marchis, and E. Iannone, *Coherent Optical Communications Systems*, Wiley, New York, 1995, ch. 5.

More complete analyses can be found, for example, in

M. J. Howes and D. V. Morgan, eds., *Optical Fiber Communications*, Wiley, New York, 1980.

T. Van Muoi, Receiver design for high-speed optical-fibre systems, *IEEE-Journal of Lightwave Technology* 2: 243–266, 1984.

More recent implications about the impact of new technologies on IM-DD systems can be found in

Technical Staff of CSELT, *Fiber Optics Communication Handbook*, 2d ed., F. Tosco (ed.), TAB Books, McGraw-Hill, New York, 1990, part IV.

L. Kazovsky, S. Benedetto, and A. Willner, *Optical Fiber Communication Systems*, Artech House, Boston, 1996, ch. 3.

3. In the EU ACTS Project ESTHER different methods were considered in evaluating time jitter, and the simulation results were compared with experimental ones. In particular, two different algorithms for the evaluation of the jitter were compared: an algorithm based on the detection of the shift in the pulse envelope peak and an algorithm based on the evaluation of the position of the pulse center of mass. The two algorithms were applied to the jitter variance in the experimental and simulation results, obtaining different values of jitter variance. These values were then used to fit the experimental and simulation results with a theoretical evaluation of error probability which, as shown in Section 4.2.1, does not take into account pulse distortion. The results obtained using the algorithm based on the pulse envelope peak translation reproduce the experimental error probability much better than those obtained using the other algorithm. Therefore in this book the jitter variance is estimated from the simulation by measuring the position of the peaks of the pulses constituting the signal envelope.

4. There is a wide literature regarding the split-step algorithms and in general the quasi-spectral methods of solution of nonlinear propagation equations. The algorithm we have implemented was inspired by the method presented in G. P. Agrawal, *Nonlinear Fibre Optics*, Academic Press, New York, 1989.

5. F. Matera and M. Settembre, Comparison of the performance of optically amplified transmission systems, *IEEE-Journal of Lightwave Technology* 14: 1–12, 1996.

6. F. Matera and M. Settembre, Compensation of the polarization mode dispersion by means of the Kerr effect for nonreturn-to-zero signals, *Optics Letters* 20: 28–30, 1995.

7. C. De Angelis, F. Matera, and S. Wabnitz, Soliton instabilities from resonant random mode coupling in birefringent optical fibers, *Optics Letters* 17: 850–852, 1992.

8. D. Marcuse, Single-channel operation in very long nonlinear fibers with optical amplifiers at zero dispersion, *IEEE-Journal of Lightwave Technology* 9: 356–361, 1991.

9. See, for example,
   L. F. Mollenauer, J. P. Gordon, and S. G. Evangelides, The sliding-frequency guiding filter: An improvement form of soliton jitter control, *Optics Letters* 17: 1575–1577, 1992.
   Y. Kodama and S. Wabnitz, Analysis of soliton stability and interactions with sliding filters, *Optics Letters* 19: 162–164, 1994.
   M. Romagnoli and S. Wabnitz, Bandwidth limits of soliton transmission and sliding filters, *Optics Communications* 104: 293–297, 1994.

10. See, for example,
    C. W. Gardiner, *Handbook of Stochastic Methods*, Springer, Berlin, 1983.
    N. G. Van Kampen, *Stochastic Processes in Physics and Chemistry*, North Holland, Amsterdam, 1983.

11. Document D2121, ACTS UPGRADE project.

12. M. Settembre et al., Cascaded optical communication systems with in-line semiconductor optical amplifiers, *IEEE-Journal of Lightwave Technology* 15: 962–967, 1997.

13. M. Born and E. Wolf, *Principle of Optics*, Pergamon Press, 6th ed., 1980, pp. 324–329.

Useful Topics for the Evaluation of Error Probability

The scope of this appendix is to review some mathematical topics that are frequently useful when evaluating the performances of optical transmission systems. We do not try to give a rigorous presentation of the mathematical material, which can be found in the references we mention. We give some definitions and equations in order to provide a quick reference for the reader. In the first section the Marcum Q function is introduced. It is useful in analyzing the performances of transmission systems adopting quadratic law detectors.

In the second section an algorithm allowing the characteristic function of a quadratic form of gaussian variables to be evaluated is detailed. This algorithm is used, for example, in Chapter 4, for optical systems adopting in-line amplifiers.

Finally in the third section the procedure is considered for the evaluation of the error probability starting from the characteristic function of the decision variable. First the Cauchy equation is introduced, and then the main algorithms used to find approximate solutions are analyzed.

A2.1 MARCUM Q FUNCTION

The Marcum Q function was firstly defined by J. I. Marcum in a paper related to radar theory [1], and it is very useful when analyzing receiver schemes based on asynchronous demodulation. A review about the use of the Marcum function for the error probability evaluation for this kind of demodulators can be found in the classic book of M. Schwartz, W. R. Bennett, and S. Stein [2]. The short review of the definition and the main properties of the Q function presented in this appendix is inspired by this book, and more complete and detailed analysis can be found in the paper of S. Stein [3]. For the tables of the Q function and a detailed analysis of algorithms for their numerical calculation, see, for example, ref. [4].

The Q function is a real function of two variables, defined as

$$Q(a, b) = \int_b^\infty e^{-(a^2+x^2)/2} I_0(ax) x \, dx \qquad \text{(A2.1)}$$

or equivalently as

$$Q(\sqrt{2a}, \sqrt{2b}) = \int_b^\infty e^{-(a+x)/2} I_0(2\sqrt{ax}) dx \qquad \text{(A2.2)}$$

where $I_0(x)$ represents the zeroth order modified Bessel function.

From the definition of the Q function, the following properties can be directly demonstrated

$$Q(0, b) = e^{-b^2/2} \qquad \text{(A2.3)}$$

$$Q(a, 0) = 1 \qquad \text{(A2.4)}$$

Exploiting the known representation of $I_0(x)$ as inverse Laplace transform,

$$I_0(x) = \frac{1}{2\pi i} \int_{c-i\infty}^{c+i\infty} \frac{1}{s} \exp\left(\frac{x}{2} \frac{s^2+1}{s}\right) ds \qquad (c > 0) \qquad \text{(A2.5)}$$

where s is the Laplace variable conjugate to x and c a real positive number. A representation of the Q function as an inverse Laplace transform can be derived by two alternative forms:

$$Q(a, b) = -\frac{1}{2\pi i} \exp\left(-\frac{a^2+b^2}{2}\right) \int_{c-i\infty}^{c+i\infty} \frac{1}{s(s-1)} \exp\left(\frac{sb^2}{2} + \frac{a^2}{2s}\right) ds \qquad \text{(A2.6)}$$

where $0 < c < 1$ and

$$Q(a, b) = -\frac{1}{2\pi i} e^{-(a+b)} \int_{c-i\infty}^{c+i\infty} \frac{1}{(s-1)} \exp\left(as + \frac{b}{s}\right) ds \qquad (c > 1) \qquad \text{(A2.7)}$$

The equations above are useful when using the characteristic function to derive the error probability for a communciation system in the form of Cauchy integral. Indeed a class of Cauchy integrals can be reduced to the above integral and solved by the Q function. Moreover, in exploiting the representation of the Q function as an inverse Laplace transform, the following symmetry and antisymmetry formulas can be derived:

$$Q(a,b) + Q(b,a) = 1 + e^{-(a^2+b^2)/2}I_0(ab) \tag{A2.8}$$

$$Q(a,a) = \tfrac{1}{2} + \tfrac{1}{2}e^{-a^2}I_0(a^2) \tag{A2.9}$$

$$1 + Q(a,b) - Q(b,a) = \frac{b^2 - a^2}{b^2 + a^2} \int_{b^2+a^2}^{\infty} e^{-x}I_0\left(\frac{2abx}{b^2 + a^2}\right)dx \tag{A2.10}$$

Since the Marcum function is often used in evaluating the error probability of communication systems, it is useful to derive an asymptotic approximation for large values of its arguments. Such a formula can be directly derived from the definition (A2.1) of the Q function by the asymptotic approximation of the zeroth order modified Bessel function:

$$I_0(x) \approx \frac{e^x}{\sqrt{2\pi x}}\left[1 + o\left(\frac{1}{x}\right)\right] \tag{A2.11}$$

Substituting equation (A2.11) into equation (A2.1) and assuming that $b \gg 1$ and $b \gg b - a$, the following asymptotic expression can be derived:

$$Q(a,b) \approx \tfrac{1}{2}\,\text{erfc}\left(\frac{b - a}{\sqrt{2}}\right) \tag{A2.12}$$

where erfc( ) is the complementary error function.

An asymptotic expression can be also derived for the combination of Q functions in equation (A2.10), which often appears in the expression for the error probability. Assuming that $b \gg 1$, $a \gg 1$, and $b \gg b - a$, the following expression can be derived:

$$1 + Q(a,b) - Q(b,a) \approx \text{erfc}\left(\frac{b - a}{\sqrt{2}}\right) \tag{A2.13}$$

A2.2 DISTRIBUTION OF HERMITIAN QUADRATIC FORMS OF GAUSSIAN RANDOM VARIABLES

In the performances of communication systems with asynchronous demodulation, the decision variable frequently turns out to be a hermitian quadratic form of gaussian variables. Thus the evaluation of the probability distribution of such random variables is analyzed in details.

Starting from the general case where the random variables z_k $(k = 1,\ldots,n)$ of the quadratic form are complex, indicating by $\underline{Z}$ the line vector whose elements are the variables z_k and by $[\mathbf{F}] = [f_{k,j}]$ an $n \times n$ complex matrix, the quadratic form ξ of the random variables z_k can be defined as

$$\xi = \underline{Z}[\mathbf{F}]\underline{Z}^T = \sum_{h=1}^{n}\sum_{k=1}^{n} f_{h,k}z_h z_k^* \tag{A2.14}$$

where * indicates the conjugate of a complex number and T the transpose conjugate of a complex matrix. The matrix $[\mathbf{F}]$ is named the characteristic matrix of the quadratic form. It is to be noted that if $f_{h,k} = f_{k,h}^*$ the value of the quadratic form is real whatever the complex value of the random variables z_k and the elements of the matrix $[\mathbf{F}]$. If such a condition is satisfied, $[\mathbf{F}]$ results in a hermitian matrix and the quadratic form is named hermitian. It is to be noted that if real random variables and matrixes are considered, a hermitian matrix reduces to a symmetric matrix.

If $\xi \geq 0$ whatever the value of $\underline{Z}$, the matrix $[\mathbf{F}]$ and the quadratic form ξ are said to be semidefinite positive, while if ξ is strictly greater than zero, they are said to be definite positive. The same definitions hold also for semidefinite and definite negative with $\leq$ and $<$ instead of $\geq$ and $>$, respectively.

Exploiting the properties of hermitian matrixes [2], it is possible to obtain an analytical expression for the characteristic function of a hermitian quadratic form of gaussian variables, which is useful in the evaluation of transmission system performance. The characteristic function of a random variable can be defined as a Fourier transform of its probability density function, or as its bilateral Laplace transform, or as its unilateral Laplace transform if the considered variable assumes only positive values. In this book the bilateral Laplace transform is used to define the characteristic function. Therefore, indicating by s the Laplace variable conjugated to ξ and by $p(\xi)$ the probability density function of ξ, the characteristic function $G_\xi(s)$ results can be defined as

$$G_\xi(s) = \int_{-\infty}^{\infty} p(\xi)e^{-\xi s}\, d\xi \tag{A2.15}$$

Once the characteristic function has been determined, the probability density of ξ can be obtained by the Cauchy-Riemann inversion formula expressed by

$$p(\xi) = \frac{1}{2\pi i} \int_{c-i\infty}^{*c+i\infty} G_\xi(s)\, e^{s\xi}\, ds \tag{A2.16}$$

where c is an arbitrary abscissa value in the complex s plane internal to the absolute convergence strip of the Laplace integral (A2.15) and the star above the integral sign stands for principal Cauchy value, since it is not ensured that it converges as a Lebesgue integral. In the case where the inversion formula converges as a Lebesgue integral, such a specification can be omitted.

Starting from the above definitions, the characteristic function of a hermitian quadratic form of gaussian variables can be written according to the following two equivalent forms:

$$G_\xi(s) = \frac{\exp\{-\frac{1}{2}\langle \underline{Z}\rangle^T([\mathbf{R}]^T)^{-1}[[\mathbf{I}] - ([\mathbf{I}] + 2s[\mathbf{R}]^T[\mathbf{F}])^{-1}]\langle \underline{Z}\rangle\}}{[\det([\mathbf{I}] + 2s[\mathbf{R}]^T[\mathbf{F}])]^\rho} \tag{A2.17}$$

$$G_\xi(s) = \frac{\exp\{-s\langle \underline{Z}\rangle^T([\mathbf{F}]^{-1} + 2s[\mathbf{R}]^T)^{-1}\langle \underline{Z}\rangle\}}{[\det([\mathbf{I}] + 2s[\mathbf{R}]^T[\mathbf{F}])]^\rho} \tag{A2.18}$$

where $\langle \underline{Z}\rangle$ indicates the average value of the vector $\underline{Z}$ and $\rho = 1$ in case where $\underline{Z}$ is a vector of complex random variables, while $\rho = \frac{1}{2}$ if the components of $\underline{Z}$ are real. The matrix $[\mathbf{R}]$ is the correlation matrix of the random variables z_k, and in the case of complex variables, it is defined as

$$[\mathbf{R}] = \frac{1}{2}\langle [\underline{Z} - \langle \underline{Z}\rangle]^T[\underline{Z} - \langle \underline{Z}\rangle]\rangle \tag{A2.19}$$

It is worth noting that the expression of the characteristic function reduces to a simpler form if the average values of all the variables z_k are equal to zero. In this case equations (A2.17) and (A2.18) reduce to

$$G_\xi(s) = \frac{1}{[\det([\mathbf{I}] + 2s[\mathbf{R}]^T[\mathbf{F}])]^\rho} \tag{A2.20}$$

A2.3   CAUCHY EQUATION AND ITS APPROXIMATIONS

In this section a method is reviewed in order to evaluate the error probability of a communication system starting from the knowledge of the characteristic function of the decision variable. In Section A2.2 it was shown that if the decision variable can be written as a hermitian quadratic form of gaussian variables, the characteristic function can be expressed in closed form. However, that is not the only case where the method of the characteristic function is useful.

It is supposed that the error probability P_e is expressed according to the relationship

$$P_e = \text{Pr}\{\xi < \xi_{\text{th}}\} = \int_{-\infty}^{\xi_{\text{th}}} p(\xi)\,d\xi \tag{A2.21}$$

where ξ is a random variable whose characteristic function $G_\xi(s)$ is known, $p(\xi)$ the probability density function of ξ, and ξ_{th} the decision threshold. Moreover it is assumed that $p(\xi)$ is continuous and derivable over all the real axis and that its tails decrease at least exponentially so that the convergence strip of $G_\xi(s)$ can be written as $-h_1 < \text{Re}(s) < h_2$, where $h_1 \geq 0$ and $h_2 > 0$.

It is to be noted that these hypotheses are verified if the decision variable is a hermitian quadratic form of gaussian variables. They imply that the inverse Laplace transform, expressed by equation (A2.16) converges as Lebesgue integral so that the asterisk above the integral can be neglected.

Substituting equation (A2.15) in equation (A2.21) and exploiting the theorem of order inversion for the integral obtains

$$P_e = \frac{1}{2\pi i} \int_{c-i\infty}^{c+i\infty} G_\xi(\xi) \int_{-\infty}^{\xi_{th}} e^{s\xi} \, d\xi \, ds \qquad (A2.22)$$

where c is a positive real number such that $0 < c < h_2$.

The inner integral can be readily calculated so as to obtain the so-called Cauchy formula providing the error probability P_e as function of the characteristic function of the decision variable,

$$P_e = \frac{1}{2\pi i} \int_{c-i\infty}^{c+i\infty} \frac{G_\xi(\xi)}{s} e^{s\xi_{th}} \, ds \qquad (A2.23)$$

The integral (A2.23) can be exactly solved in several interesting cases by the residues theorem [6]; in the other cases accurate asymptotic approximations can be found for its solution.

The method of the characteristic function for the evaluation of the error probability is useful not only because it leads to the Cauchy integral which can be exactly solved in several important cases but also because it allows good asymptotical approximations when the Cauchy integral cannot be exactly solved. Since optical transmission systems operate at low error probability values (of the order of 10^{-9}), asymptotical approximations for the error probability are quite accurate, and they are often used.

The simplest approximation is given by the so-called Chernov bound [7]. Starting from the expression (A2.21) for the error probability, it can be noted that for each positive real number x, the following inequality holds:

$$P_e = \text{Pr}\{\xi < \xi_{th}\} = \int_{-\infty}^{\xi_{th}} p(\xi) \, d\xi \leq \int_{-\infty}^{\infty} e^{-x(\xi-\xi_{th})} p(\xi) \, d\xi \qquad (A2.24)$$

The integral involving the exponential function can be expressed in terms of the characteristic function of the random variable x, evaluated along the real axis so as to obtain

$$P_e \leq e^{x\xi_{th}} G_\xi(x) \qquad (x > 0) \qquad (A2.25)$$

To obtain the tightest upperbound expression, (A2.25) must be minimized with respect to x. This is equivalent to minimizing its natural logarithm, obtaining the following equation:

$$\frac{d}{dx}\{\log[G_\xi(x)] - x\xi_{th}\} = \frac{1}{G_\xi(x)}\frac{dG_\xi(x)}{dx} + \xi_{th} = 0 \tag{A2.26}$$

which permits one to find a positive value of x ensuring an optimum estimate for the error probability.

The Chernov bound is sufficiently accurate to be useful in several applications but a quite tighter upperbound can be obtained that requires a little additional computational effort with respect to Chernov bound. This method is the saddle-point approximation of the Cauchy integral (A2.23), also named the *steepest descendent method* [8].

If the equation

$$\frac{d}{ds}\log\left[\frac{G_\xi(x)e^{s\xi_{th}}}{s}\right] = \frac{1}{G_\xi(x)}\frac{dG_\xi(x)}{dx} + \xi_{th} - \frac{1}{s} = 0 \tag{A2.27}$$

has a real root in the strip $0 < \text{Re}\{s\} < h_2$ (which is the holomorphic field of the argument of the Cauchy integral, since the denominator has a pole in $s = 0$ and $G_\xi(s)$ is holomorphic in $-h_1 < \text{Re}\{s\} < h_2$), it is a saddle point. The first step to approximately integrate the Cauchy integral using the saddle-point method is to consider the smallest positive real radix x_0 of equation (A2.27) and then to choose an integration path crossing the real axis at this point. It can be shown that the main contribution to the integral is that around the saddle point, so a good approximation can be obtained by evaluating the integral near the saddle point.

In the vicinity of the saddle point, the integrand can be approximated as a gaussian function. Therefore, assuming that

$$\Phi(s) = \log[G_\xi(s)] + \xi_{th}s - \log(s) \tag{A2.28}$$

near the saddle point, it turns out that

$$\frac{G_\xi(s)}{s}e^{s\xi_{th}} - \exp[\Phi(x_0) - \tfrac{1}{2}y^2\Phi''(x_0)] \tag{A2.29}$$

where $y = \text{Im}\{s\}$ and $\Phi''(s)$ is the second derivative of $\Phi(s)$ with respect to s. Substituting this approximation into the Cauchy integral, the approximate value for the error probability is obtained:

$$P_e = -\frac{1}{2\pi i}\int_{x_0-i\infty}^{x_0+i\infty}\exp[\Phi(x_0) - \tfrac{1}{2}y^2\Phi''(x_0)]d(x_0 + iy) = \frac{e^{\Phi(x_0)}}{\sqrt{2\pi\Phi''(x_0)}} \tag{A2.30}$$

Expression (A2.30) represents the saddle-point approximation of the Cauchy integral. Although the starting points of the Chernov bound and of the saddle-point approximations seem different, it is meaningful that equations (A2.26) and (A2.27) for the determination of the real value x_0 in the two cases are rather similar. As a matter of fact, if the Chernov-bound approximation is detailed, it can be viewed as a further approximation of the saddle-point method, which therefore is expected to be more accurate. Such an analysis, which is beyond the purpose of this appendix, can be found in the paper of Schumacher and O'Reilly [9] and in the references that they indicate.

A simple example of the use of the Chernov and saddle-point bounds allows us to make an immediate comparison between these methods. Assuming ξ as a random variable whose probability density function is given by

$$p(\xi) = \begin{cases} e^\xi, & \xi < 0 \\ 0, & \xi > 0 \end{cases} \tag{A2.31}$$

the error probability P_e is given by

$$P_e = \Pr\{\xi < -\delta\} = e^{-\delta} \tag{A2.32}$$

where $\delta \gg 2$.

To evaluate P_e by the Chernov bound, a positive real number x_0 must be determined, solving equation (A2.26), whose result is expressed in this case as $\delta - \delta x_0 = 1$. Once x_0 has been determined, from equation (A2.25) it turns out that $P_e \approx (e\delta)e^{-\delta}$, which is obviously an asymptotic approximation for P_e. The ratio between this approximate value and that given by (A2.32) is equal to $(e\delta)$, so it increases with increasing δ. For example, for $\delta = 12$ the correct value is 6.14×10^{-6}, while that derived by the Chernov bound is equal to 2×10^{-4}.

To evaluate P_e using the saddle-point approximation, a positive real number x_0 must be determined in solving equation (A2.27), which in this case is expressed as $x_0^2 \delta + (2 - \delta)x_0 - 1 = 0$. The only positive solution is given by

$$x_0 = \frac{\delta - 2}{2\delta} + \sqrt{\frac{4 + \delta^2}{4\delta^2}} \approx \frac{\delta - 1}{\delta} \tag{A2.33}$$

where the approximation takes into account the hypothesis that $\delta \gg 2$. It comes out that the value of x_0 is about the same as derived by the Chernov bound. Using the approximate value for x_0, from equation (A2.30) the following relationship can be derived:

$$P_e \approx \frac{e}{\sqrt{2\pi}} \frac{\delta}{\sqrt{\delta^2 - 2\delta}} e^{-\delta} \tag{A2.34}$$

This is a better result with respect that derived by the Chernov bound. In particular, the percentage difference between the exact and approximate values for P_e tends to $e/\sqrt{2\pi} \approx 1.0844$ when δ tends to infinity.

REFERENCES

1. J. I. Marcum, A statistical theory of target detection by pulsed radar, *IRE Transaction on Information Theory* 6: 56–267, 1960.

2. M. Schwartz, W. R. Bennett, and S. Stein, *Communication Systems and Techniques*, McGraw-Hill, New York, 1966, ch. 8 and app. A.

3. S. Stein, The Q-function and related integrals, Research report 467, Applied Research Laboratory, Sylvania Electronic Systems, July 1965.

4. A tabulation of the Marcum Q function can be found in
   J. I. Marcum, Tables of the Q function, Rand Corporation Res. Memo RM 399, January 1950;
   This report can be directly obtained from the Rand corporation.
   A numerical algorithm for computing the Q functions is reported in
   A. R. Di Donato and M. P. Jarnagin, A method for computing the circular coverage function, *Mathematics of Computation* 16: 347–355, 1962.

5. L. C. Maximon, On the representation of indefinite integrals containing Bessel functions by simple Neumann series, *Proceedings of American Mathematics Society* 7: 1054–1062, 1956.

6. The exact solution of the Cauchy integral in the case of envelope detection is reported, for example, in ref. [2]. Several examples of exact solutionss of the Cauchy equations related to problems of optical communications are reported in the book (e.g., see the section of Chapter 6 on polarization modulation)
   S. Betti, G. De Marchis and E. Iannone, *Coherent Optical Communication Systems*, Wiley, New York, 1995.
   Another interesitng case of exact solution of the Cauchy integral is reported in the Appendix B of the paper
   J. Salz, Coherent lightwave communcations, *AT&T Technical Journal* 64: 2153–2209, 1985.
   In this paper the author reports that documentation for this calculation was provided by B. F. Logan, Jr. from AT&T Research Laboratories.

7. The Chernov bound was introduced in
   H. Chernov, A measure of asymptotic efficiency of tests of a hypothesis based on the sum of observations, *Mathematical Statistics Annals* 23: 493–507, 1952.
   The simple approach reported in the appendix is derived from
   J. G. Proakis, *Digital Communications*, 2d ed., McGraw-Hill, New York, 1989, ch. 1.

8. The steepest descendent approximation, also called the saddle-point approximation, is a classic method of mathematical physics essential due to P. Debye in a paper appearing in the *Mathematical Annals* in 1909. A complete and rigorous review of this method can be found in
   P. M. Morse and H. Feshbach, *Methods of Theoretical Physics*, Part I, McGraw-Hill, New York, 1935.

An important application of this method for the analysis of error probability in communication systems can be found in the paper

J. E. Mazo and J. Salz, Probability of error for quadratic detectors, *Bell System Technical Journal* 44: 2165–2187, 1965.

In this reference the saddle-point approximation is exploited in conjunction to the expression of the characteristic function of a hermitian quadratic form of gaussian variables to analyze a general class of quadratic detectors. Examples of application of the saddle point approximation to the analysis of optical IM-DD and coherent systems are

C. W. Helstrom, Performance analysis of optical receivers by the saddlepoint approximation, *IEEE-Transaction on Communications* 27: 186–191, 1979.

S. Betti, G. De Marchis, E. Iannone, and F. Matera, Dichroism effect on polarisation modulated optical systems using Stokes parameters coherent detection, *IEEE-Journal of Lightwave Technology* 8: 1762–1768, 1990.

9. K. Shumacher and J. J. O'Reilly, Relationship between the saddlepoint approximation and the modified Chernov bound, *IEEE-Transaction on Communications* 38: 270–272, 1990.

List of Acronyms

AM	Amplitude modulation
APD	Avalanche photodiodes
ASE	Amplified spontaneous emission
ASN	Average amplitude signal-to-noise ratio
ATM	Asynchronous transfer mode
DCF	Dispersion compensating fiber
DFB	Distributed feedback laser
DGD	Differential group delay
DS	Dispersion shifted fiber
DST	Dispersion supported transmission
DXC	Digital cross-connect
EDFA	Erbium-doped fiber amplifier
FEC	Forward error correction
FTTC	Fiber to the curb
FTTH	Fiber to the home
FWHM	Full width at half maximum
FWM	Four-wave mixing
GVD	Group velocity dispersion
IM-DD	Intensity modulation-direct detection
LAN	Local area network
MPSI	Mid-point spectral inversion
NA	Numerical aperture of an optical fiber
NIU	Network user interface
NRZ	Nonreturn to zero
NSE	Nonlinear Shrödinger equation
OADM	Optical add/drop multiplexer
OPC	Optical phase conjugation
OTDM	Optical time division multiplexing
OXC	Optical cross-connect
PDH	Plesiochronous digital hierarchy

PIN	Photodiodes with a p-i-n band structure
PLL	Phase lock loop
PM-DD	Polarization modulation-direct detection
PMD	Polarization mode dispersion
PSK	Phase shift keying
PSP	Principal state of polarization
RZ	Return to zero
SBS	Stimulated Brillouin scattering
SDH	Synchronous digital hierarchy
SMN	Shuffle multihope network
SOA	Semiconductor optical amplifier
SOP	State of polarization
SPM	Self-phase modulation
SRS	Stimulated Raman scattering
SSFM	Split step Fourier method
TDM	Time division multiplexing
WDM	Wavelength division multiplexing
XPM	Cross-phase modulation

Index

WILEY SERIES IN MICROWAVE AND OPTICAL ENGINEERING

KAI CHANG, Editor
Texas A&M University

ANALYSIS OF MULTICONDUCTOR TRANSMISSION LINES • *Clayton R. Paul*

INTRODUCTION TO ELECTROMAGNETIC COMPATIBILITY • *Clayton R. Paul*

INTRODUCTION TO HIGH-SPEED ELECTRONICS AND OPTOELECTRONICS • *Leonard M. Riaziat*

NEW FRONTIERS IN MEDICAL DEVICE TECHNOLOGY • *Arye Rosen and Harel Rosen (eds.)*

NONLINEAR OPTICS • *E. G. Sauter*

FREQUENCY SELECTIVE SURFACE AND GRID ARRAY • *T. K. Wu (ed.)*

ACTIVE AND QUASI-OPTICAL ARRAYS FOR SOLID-STATE POWER COMBINING • *Robert A. York and Zoya B. Popović (eds.)*

OPTICAL SIGNAL PROCESSING, COMPUTING AND NEURAL NETWORKS • *Francis T. S. Yu and Suganda Jutamulia*

Printed in the United States
By Bookmasters